COLLOQUIA MATHEMATICA
SOCIETATIS JÁNOS BOLYAI, 26.

MATHEMATICAL LOGIC IN COMPUTER SCIENCE

Edited by:

B. DÖMÖLKI
and
T. GERGELY

NORTH-HOLLAND PUBLISHING COMPANY
AMSTERDAM – OXFORD – NEW YORK

Budapest, Hungary, 1981

ISBN North-Holland: 0444 85440 1
ISBN Bolyai: 963 8021 46 2
ISSN Bolyai: 0139 3383

Joint edition published by

JÁNOS BOLYAI MATHEMATICAL SOCIETY
and
NORTH-HOLLAND PUBLISHING COMPANY
Amsterdam – Oxford – New York

In the U.S.A. and Canada:
NORTH-HOLLAND PUBLISHING COMPANY
52 Vanderbilt Avenue
New York, N.Y. 10017

Printed in Hungary
ÁFÉSZ, VÁC
Sokszorosító üzeme

PREFACE

Since the very early days of computing, mathematical logic and its related areas have always served as one of the most important scientific disciplines to support an exact mathematical foundation of computer science. Governed by this idea Bolyai János Mathematical Society organized a Colloquium on Mathematical Logic in Computer Science, in Salgótarján between 10-15 September, 1978. The Organizing Committee was on the opinion to keep the number of participants on a reasonably low level so as to ensure intensive discussions and to avoid parallel sections.

Out of the papers prepared for the Colloquium 31 were accepted for publication in this volume. Most of the papers were updated before publishing.

The volume includes some papers the author of which could not attend the Colloquium in person.

It would be a difficult and rather fruitless task to try to evaluate or even survey all the papers. A short classification, however, may help the orientation of the reader in the rather wide field of interests covered by the volume.

The main bulk of the papers is dealing with applications of the notions, methods and results of mathematical logic to specific problems relevant to theory of programming, connected mainly to the description of language semantics. The papers by *De Bakker-Zucker*, *Engeler*, and *Obtulowicz-Wiweger* fall into this category, together with *Grabowski* and *Mirkowska*, dealing with algorithmic logic and *Chkhenkeli*, giving logical foundations to important application area. Also in this class, a set of papers are using model-theoretical methods for the mathematical foundation of certain computer science concepts, such as in the papers of *Bertoni-Mauri-Miglioli*, *Gergely-Úry*, *Márkusz-Szőts* and *Tiuryn*. The paper by *Pliuskevicius* deals with the proof theory for programming logic.

Algebraic logic and different algebraic approaches to the theory of programming are represented by papers of *Andréka-Sain*, *Arbib-Manes*, *Courcelle*, *Hoehnke*, *Meseguer* and *Németi*.

One of the most direct and straightforward applications of logic in programming can be found in a rather new direction of very high level programming languages called "Logic Programming". Papers of *Balogh*, *Brown*, *Bruynooghe* and *Pereira-Monteiro* describe some theoretical and practical aspects of logic programming languages. Papers by *Champeaux*, *Gergely-Vershinin* and *Štěpánková* deal with different aspects of the logic based problem solving.

Papers concerning the design and application of programming languages and methodologies form the last group of papers in our classification. Here we have papers about the logical foundation of a programming language *(Dávid-Sárközy)*, of a specification method *(Aszalós)*, parallel computations *(Radziszowski* and *Stapp)*, abstract data types *(Banachowski* and *Bergstra-van der Weide)* and tree transformations *(Říha)*.

We hope that present volume achieves our aim as to support further communication between mathematical logicians and computer scientists.

We are indebted to all those whose work helped the successful realization of the Colloquium and we also thank those who contributed to the publication of present volume.

The Editors

CONTENTS

MONDAY (September 11)

10.30 *Opening*

10.45-11.45

J. De Bakker (The Netherlands): Current issues in programming theory.

R. Pliuskevicius (SU): On the Genzen-type theory for program analysis and synthesis.

Break

13.30-15.00

I. Németi (Hungary): Algebraic observations in connection with least fixed point semantics.

J. Meseguer (USA): Factorizations, completions and colimits for ω-posets.

G. Gati (Switzerland): Investigation of graph theoretical problems by generalized Galois theory.

Break

15.30-17.00

B. Courcelle (France): Equational theories and equivalences of programs.

J.A. Bergstra (The Netherlands): On the operational comparison of algebraic datatypes.

B. Wojdylo (Poland): Universal algebraic approach to programming languages.

20.00 *Welcome party*

TUESDAY (September 12)

8.30-10.00

A. Obtulowicz (Poland): Algebraic theories of lambda calculi.

J. Aszalós (Hungary): Axiomatic description of multi-level program specifications.

Z. Márkusz - M. Szőts (Hungary): On semantics of programming defined by universal algebraic tools.

Break

10.30-12.00

D. de Champeaux (The Netherlands): Other directions for automatic theorem proving.

L. Farinas Del Cerro (France): Mechanization in non-classical logics.

P. Gburzynski (Poland): Computer experiments with resolution principle.

13.30-15.30

P. Ecsedi-Tóth (Hungary): On logical definition of programming semantics.

G. Dávid (Hungary): Structure logic.

T. Chkhenkeli (SU): About many-sorted logical approach to the information retrieval languages.

K.P. Vershinin (SU): A theorem proving strategy for formulae with restricted quantifiers.

16.00 Excursion

WEDNESDAY (September 13)

8.30-10.00

H. Andréka (Hungary): The model theory of any logic of programming can be proved to be non-classical.

T. Gergely (Hungary): May the theory of programming be first order?

L. Úry (Hungary): Logical base for correct programming.

Break

10.30-12.30

A. Salwicki (Poland): Algorithmic theories of data structures representing sets.

A. Kreczmar (Poland): Algorithmic and dynamic logic.

G. Mirkowska (Poland): Algorithmic properties of non-deterministic programs.

J. Jaromczyk (Poland): Some applications of algorithmic logic.

F.M. Brown (UK): A semantic theory for logic programming

Break

13.30-15.30

P. Szeredi (Hungary): Loops in PROLOG.

L.M. Pereira (Portugal): The declarative and operational semantics of parallelism and co-routining in logic programs.

K. Balogh (Hungary): An interactive program verifier.

M. Bruynooghe (Belgium): Intelligent backtracking for an interpreter of Horn clause logic programs.

Break

15.30-17.30

W. Ratajczak (Poland): LOGLAN = a high level programming language.

D. Szczepánska (Poland): Simultation type in Loglan-77.

H. Oktaba (Poland): The semantics of parallel in Loglan-77.

A. Řiha (Czechoslovakia): A certain type of dependency tree transformations.

20.00 *Panel discussion*
Applicability of logical approach in programming practice.

THURSDAY (September 14)
8.30-10.00

P.A. Miglioli - M. Ornaghi (Italy): A logically justified model for non-deterministic and parallel computations.

S. Radziszowski (Poland): Complexity of parallel computations.

L. Stapp (Poland): Some remarks on correctness proving for parallel programs.

Break
10.30-12.00

A.I. Litwinuk (Poland): CPL-systems programming language

I. Futó - F. Darvas - J. Szeredi - Gy. Mátrai (Hungary): Knowledge-based programs for drug design problem solving.

M. Grabowski (Poland): Full weak second order logic versus algorithmic logic.

13.30 *Excursion*

FRIDAY September 15

8.30-10.00

A. Bertoni - G.C. Mauri - P.A. Miglioli (Italy): Model theoretic aspects of abstract data specification.

L. Banachowski (Poland): On implementations of programs with abstract data types.

V.V. Alexandrov - A.O. Poliakov (SU): Data representation based on recursive algorithms.

Break

10.30-12.00

I. Ratkó (Hungary): On optimization problems of logical expressions in programming language.

O. Štěpánková (Czechoslovakia): Normal form of proof of certain formulas of situation calculus.

I. Sain (Hungary): Injective classes of Arbib-Manes automata.

LIST OF PARTICIPANTS

V. Alexandrov
Leningrad, 199164 Mendeleyevscaya linija 1, Leningrad Research Computer Centre of the USSR Academy of Sciences
USSR

H. Andréka
MTA MKI, Budapest, Reáltanoda u. 13-15., 1053
HUNGARY

J. Aszalós
SZÁMKI, Budapest, Csalogány u. 30-32., 1015
HUNGARY

I. Bach
MTA SZTAKI, Budapest, Kende u. 13-17., 1111
HUNGARY

A. Bagyinszki
ELTE TTK, Numerikus és Gép.Mat. Tanszék, Budapest, Muzeum krt. 6-8., 1088
HUNGARY

L. Banachowski
Wydzial Matematyki Instytut Informatyki, 00-901 Warszawa, PKiN pok. 850
POLAND

K. Balogh
Budapest, Család u. 19., 1039
HUNGARY

I. Bartalos
JATE Kibernetikai Laboratórium, Szeged, Árpád tér 2., 6720
HUNGARY

G. Belovári
Budapest, Alkotás u. 25., II., VII.53., 1123
HUNGARY

B. Bódy
OTSzK, Budapest, Angol u. 27., 1149
HUNGARY

M. Bruynooghe
Katholike Universiteit Leuven, Afd. Toegepaste Wiskunde en Programmatie, Celestijnenlaan 200 B, 3030 Heverlee
BELGIUM

B. Courcelle
IRIA, Domaine de Voluceau - Rocquencourt,
B.P. 105 - 78150 Le Chesnay
FRANCE

J. Cser
MTA KFKI, Budapest, Pf. 49, 1525
HUNGARY

G. Dávid
MTA SZTAKI, Budapest, Kende u. 13-17., 1111
HUNGARY

B. Dömölki
SZKI, Budapest, Akadémia u. 17., 1054
HUNGARY

P. E.-Tóth
MTA SZAB Székház, Szeged, Somogyi u. 7., 6720
HUNGARY

I. Fábián
Budapest, Szakasits Á. u. 40/a, 1115
HUNGARY

L. Farinas Del Cerro
Centre National de la Recherche Scientifique,
Laboratoire d'Informatique pour les Sciences de l'Homme,
31, chemin Joseph Aiguier 13274 Marseille Cedex 2
FRANCE

E. Farkas
MTA SZTAKI, Budapest, Kende u. 13-17., 1111
HUNGARY

Zs. Farkas
SZKI, Budapest, Akadémia u. 17., 1054
HUNGARY

I. Fekete
ELTE TTK, Numerikus Tanszék, Budapest, Muzeum krt. 6-8., 1088
HUNGARY

I. Futó
SZKI, Budapest, Akadémia u. 17. 1054
HUNGARY

G. Gati
Institut für Informatik, Eidgenössische Technische Hochschule, 8092 Zürich
SWITZERLAND

P. Gburzynski
Institute of Computer Science, Warsaw University, 00-901 PKiN, VIII.p. Warsaw
POLAND

T. Gergely
SZÁMKI, Budapest, Pf. 227, 1536
HUNGARY

E. Gesztelyi
KLTE Számitástudományi Tanszék, Debrecen Pf. 12, 4010
HUNGARY

M. Grabowski
00-905 Warszawa OS. Przyjaźń 27
POLAND

S. Horváth
ELTE TTK Numerikus és Gépi Mat. Tanszék, Budapest, Muzeum krt. 6-8., 1088
HUNGARY

P. Hunya
JATE Kibernetikai Laboratórium, Szeged, Árpád tér 2., 6720
HUNGARY

Á. K. Petróczki
Budapest, Koszta J.u. 16., 1124
HUNGARY

E. Knuth
MTA SZTAKI, Budapest, Kende u. 13-17., 1111
HUNGARY

L. Kozma
SZÁMKI, Budapest, Pf. 227, 1536
HUNGARY

P. Köves
SZKI, Budapest, Akadémia u. 17., 1054
HUNGARY

M. Kővári
MTA KFKI, Budapest, Pf. 49, 1525
HUNGARY

A. Kreczmar
Wydzial Matematyki Uniwesytetu Warszawskiego, Warszawa, Palac Kultury i Nauki 8.p.
POLAND

T. Langer
SZÁMKI, Budapest, Pf. 227, 1536
HUNGARY

P. Lapis
MTA SZTAKI, Budapest, Kende u. 13-17., 1111
HUNGARY

I. Láng
MTA KFKI, Budapest, Pf. 49, 1525
HUNGARY

A. Litwiniuk
Warsaw University, PKiN VIII p., 00-901 Warszawa
POLAND

I. Losonczi
MTA SZTAKI, Budapest, Kende u. 13-17., 1111
HUNGARY

P. Lukács
Budapest, Váci u. 65., 1134
HUNGARY

G. Mauri
Universitá di Milano, Istituto di Cibernetica, Via Viotti 5, 20133 Milano
ITALY

Zs. Márkusz
MTA SZTAKI, Budapest, Kende u. 13-17., 1111
HUNGARY

E. Merényi
MTA KFKI, Budapest, Pf. 49, 1525
HUNGARY

P. Miglioli
Universitá di Milano, Istituto di Cimernetica, Via
Viotti 5, 20133 Milano
ITALY

G. Mirkowska
Warszawa, Kochanowskiego 28/9
POLAND

R. Nagy
BHG Fejl. Int. Budapest, Petczvál J. u. 31., 1115
HUNGARY

M. Naszódi
MTA SZTAKI, Budapest, Kende u. 13-17., 1111
HUNGARY

M. Náray
SZKI, Budapest, Akadémia u. 17., 1054
HUNGARY

I. Németi
MTA MKI, Budapest, Reáltanoda u. 13-15., 1053
HUNGARY

A. Obtulowicz
Instytut Matematyczny P.A.N., ul. Sniadeckich 8.
Skrytka Pocztowa 137, 00950 Warszawa
POLAND

H. Oktaba
Wadzial Matmatyki,Uniwersytet Warszawski, Warszawa,
PKiN VIII p.
POLAND

M. Ornaghi
Universitá di Milano, Istituto di Cibernetica, Via
Viotti 5, 20133 Milano
ITALY

L.M. Pereira
Ministério do Equipamento Social e do Ambiente,
Laboratório Nacional de Engenharia Civil, Informatica,
Av Brasil, Lisboa -5.
PORTUGAL

R. Pliuskevicius
Institute of Mathematics and Cybernetics of the Acad. of Sci. of the Lithuanian SSR, 232600 Vilnius, 54 Poželos str.
USSR

V. Ponomarev
Leningrad Research Computer Centre of the USSR Acad. of Sci. Leningrad, 199164, Mendeleyevscaya linija 1.
USSR

F. Potári
TÁKI, Budapest, Gábor Á. u. 65., 1026
HUNGARY

I. Ratkó
MTA SZTAKI, Budapest, Victor Hugo u. 18-20., 1132
HUNGARY

S. Radziszowski
Uniwersytet Warszawski, Instytut Informatyki
00-950 Warszawa PKiN,
POLAND

W. Ratajczak
Wydzial Matematyki i Mechaniki University of Warsaw,
POLAND

Z. Renc
MFF UK Malostranské Nám. 25, 11800 Praha 1
CZECHOSLOVAKIA

A. Riha
Balbinova 5, 12000 Praha 2
CZECHOSLOVAKIA

A. Salwicki
Mathematical Institute of Polish Acad. of Sci., Sniadeckich 8, 00-950 Warsaw
POLAND

E. Sánta-Tóth
SZKI, Budapest, Akadémia u. 17., 1054
HUNGARY

A. Sárközy
MTA SZTAKI, Budapest, Kende u. 13-17., 1111
HUNGARY

E. Simon
JATE, Somogyi B.u.7., Szeged, 6720
HUNGARY

L. Stapp
Marszalkowska 28 m. 126, 00576 Warszava
POLAND

P. Štěpánek
Dept. of Cybernetics and Op. Res., Charles Univ.,
Malostranské Nám. 25, 11800 Praha 1
CZECHOSLOVAKIA

O. Štěpánková
Institute of Computational Techniques of CVUT, Horská 3,
12800 Praha 2
CZECHOSLOVAKIA

M. Sunovecz
SZÁMKI, Budapest, Csalogány u. 32., 1015
HUNGARY

D. Szczepańska
Wydz. Matematyki, Uniwersytet Warszawski, Warszawa,
PKiN VIII p.,
POLAND

K. Szenes
Budapest, Kerékgyártó u. 7/b, 1147
HUNGARY

P. Szeredi
SZKI, Budapest, Akadémia u. 17, 1054
HUNGARY

M. Szijártó
ELTE TTK Gépi és Num. Mat. Tanszék, Budapest,
Muzeum krt. 6-8., 1088.
HUNGARY

P. Szőke
SZÁMKI, Budapest, Pf. 227, 1536
HUNGARY

M. Szőts
SZÁMKI, Budapest, Pf. 227, 1536
HUNGARY

D. Szubbocsev
KLTE Számoló Központ, Debrecen, Pf. 58, 4010
HUNGARY

R. Treer
ÁSZSZ, Budapest, Pf. 232, 1536
HUNGARY

A. Udvari
ÁSZSZ, Budapest, Pf. 232, 1536
HUNGARY

L. Úry
SZÁMKI, Budapest, Pf. 227, 1536
HUNGARY

D. Varga
Budapest, Angol u. 27., 1149
HUNGARY

E. Vass
MTA KFKI, Budapest, Pf. 49, 1525
HUNGARY

T.P. van der Weide
Opaalstraat 96, 2332 TL Leiden,
THE NETHERLANDS

J. Zajdon
Budapest, Gárdonyi G. u. 34., 1026
HUNGARY

Z. Zsombok
ÁSZSZ, Budapest, Pf. 232, 1536
HUNGARY

COLLOQUIA MATHEMATICA SOCIETATIS JÁNOS BOLYAI
26. Mathematical Logic in Computer Science
Salgótarján (HUNGARY), 1978.

CONNECTIONS BETWEEN ALGEBRAIC LOGIC AND INITIAL ALGEBRA SEMANTICS OF CF *LANGUAGES*

H. Andréka and I. Sain

Introduction

It appears that many people have been doing the same or almost the same thing independently under quite different titles. This thing is: to analyse *semantic* and related aspects of languages *in general*, and to do this "systematically" by applying some well organized branch of mathematics which is flexible enough like *universal algebra* (or category theory). Some of these works are: Montague [30], Andréka-Gergely-Németi [5],[6] §.II, Rasiowa [39]p.168, ADJ [1] especially §3.1. p.75, Andréka [3], Kaphengst-Reichel [23], Andréka-Németi [7],[9], Goguen [19], and many other works related to ADJ team's initial algebra approach. Among the trade offs of this approach are an understanding of the relationship between different languages and theories, and an ability to work with the category of theories (or languages) and theory morphisms, see e.g. Burstall-Goguen [13]. (This understanding is needed in many fields of computer science, e.g. in structured programming and specifications, representation of knowledge, AI.)

Some of the quoted papers appeared in the field of "Algebraic Logic", so one may ask: What do they have to do with "Language Theory"? Well, there is a part of algebraic logic sometimes called *"Universal Algebraic Logic"* with the aim of producing a unified theory of a hopefully large variety of languages, their semantics, their model theories etc..

Here we try to sum up what is common in all the above quoted works and try to make a "bridge" between them and to make an introduction to them. One of our theses will be that the Lawvere style Algebraic Theories approach, used by the ADJ-followers, and Algebraic Logic (e.g. Cylindric Algebras) are practically the same things, more precisely, they are complementary: there is a strong connection between them which can be utilized to the benefit of both sides. Our attitude is not completely new: This complementarity has already been pointed out and investigated in Daigneault [16].

The present paper could be divided to 3 main parts: *§1-§2:* Preparations, *§3-6:* Methodology of Algebraic semantics. Applications and illustrations of §3-6 are in §4,7, 9. §8 deals with the CS case.

In §1 the general concept of a language with semantics is recalled and discussed. §2 contains the *algebraic* concept of a language with semantics. §3 explains the *process of algebraization* of a language with semantics in detail. §4 shows how the fundamental notions of Algebraic Logic were obtained by executing the "algorithm" described in §3. In §3,5,6 the methodology of doing and applying algebraic semantics is expounded. Dynamic Cylindric Algebras (á la Pratt) are in §7. The Context Sensitive case is treated in §8. §9 deals with algebras corresponding to not very classical logics.

The second part of this paper is Németi [32] in this volume. [32] contains a detailed example for the general methodology outlined in the present paper. Results and constructions that were claimed or promised in §4-6,9 are elaborated in [32]. Trains of thought that are vague in §4-6 are filled with concrete content in [32]. Basic results on the algebraization of First order Theories by generalized *Cylindric set Algebras* is one of the main themes of [32]. Not only models but also arbitrary sets of models are represented by generalized Cylindric set Algebras in [32]. Theories and theory morphisms (interpretations) in the sense of Burstall-Goguen [13] are shown to profit from Cylindric Algebra theory.

The backbone of the present paper is §3. There a process of algebraization is elaborated, i.e. an "algorithm" of doing algebraic semantics is described; in the rest of the paper this algorithm is executed in more and more detail. During this repeated execution of the algorithm of §3 several seemingly different branches turn out to be rather strongly connected to each other, e.g. Cylindric Algebras, the theory of their Regularity, Theory-morphisms, "Burstall-Goguen's Putting Theories together", Pratt's Dynamic Algebras. Throughout references are made to §3 and Example 1. The first section where Initial Algebra Semantics and Cylindric Algebra theory meet is §4.

Acknowledgement: Thanks are due to Peter Burmeister for his improving this paper enormously.

NOTATIONS

Throughout this list of notations A and B are arbitrary classes.

"Sets" is the class of all sets. We let Sets $\stackrel{d}{=}$ "Sets" .

Ord is the class of all ordinals.

ω denotes the set of *natural numbers (finite ordinals)*.

$n = \{0,1,\ldots,n-1\}$ if $n\in\omega$. In general, an ordinal is the set of smaller ordinals.

$\mathrm{Sb}\, A \overset{d}{=} \{X\in \text{Sets} : X\subseteq A\}$.

$\cup A \overset{d}{=} \{x : (\exists a\in A)\, x\in a\}$.

$A{\sim}B \overset{d}{=} \{x\in A : x\notin B\}$.

$A\times B \overset{d}{=} \{\langle a,b\rangle : a\in A,\ b\in B\}$.

${}^{A}B$ denotes the class of all functions from A into B.

$f : A \to B$ means $f\in {}^{A}B$.

Let $f : A \to B$ and $g : B \to C$. Then

$g\circ f$ denotes the *composition* of the functions g and f i.e. $g\circ f : A \to C$ and $(g\circ f)(x) = g(f(x))$ for every $x\in A$.

A function is considered to be a class of pairs, i.e.
$f = \{\langle x,f(x)\rangle : x\in A\}$, if $f : A \to$ "Sets".

$\mathrm{Rng}\, f$ is the *range* of f, i.e. $\mathrm{Rng}\, f \overset{d}{=} \{f(x) : x\in A\}$, if $f : A \to$ "Sets".

$\langle f(a) : a\in A\rangle$ and

$\langle f(a)\rangle_{a\in A}$ denote functions defined as: $\langle f(a)\rangle_{a\in A} \overset{d}{=}$
$\overset{d}{=} \langle f(a) : a\in A\rangle \overset{d}{=} \{\langle a,f(a)\rangle : a\in A\}$.

Let $f : A \to B$ and $H \subseteq A$. Then

$H{\restriction}f$ denotes the *restriction* of f to domain H, i.e. $H{\restriction}f : H \to B$ and $H{\restriction}f \overset{d}{=} \langle f(x) : x\in H\rangle$.

A^* denotes the set of all *finite strings* of elements of A, if A is a set. I.e. A^* is the free monoid generated by A.

t denotes a *similarity type*. A similarity type is a function $t\in {}^{\Sigma}\omega$. If $f\in\Sigma$ then $t(f)$ is said to be the *arity* (or rank or type) of the symbol f, see [20]. Many authors use the inverse $\langle \Sigma_n : n\in\omega\rangle$ of t and they call it an *operator domain* or *signature*, e.g. Goguen, Mal'cev, Cohn. Here

$\Sigma_n = \{f \in \Sigma : t(f) = n\}$ for every $n \in \omega$. The letter Σ stands for "signature" or "similarity type".

Many-sorted similarity types are more complicated:

g denotes a many-sorted or heterogeneous similarity type, i.e. $g \in {}^{\Sigma}(N^*)$ where N is an arbitrary set called the set of *sorts* of g. If $g(f) = \langle n_1, \ldots, n_k, n_{k+1} \rangle$ then the arguments of the function symbol f are of sort $n_1, \ldots, n_k$ respectively and its result is of sort n_{k+1}. See e.g. [1],[25],[28] Def.1.19, or "Summing up §3.A" here.

The German capitals

$\mathfrak{A}, \mathfrak{B}, \mathfrak{N}$ denote *models* in the general sense of Def.1 in §1. I.e. a German capital denotes an element of some class M such that M is the class of models of some language $L = \langle S, M, k \rangle$ in the sense of Def. 1 in §1. If $L = L_1$ then the *universe* of $\mathfrak{A} \in M_1$ is denoted by A.

$L_1 = \langle S_1, M_1, k_1 \rangle$ is the *classical first order language* defined in Example 1 in §1.

M_1 is the class of *classical first order models.*

S_1 is the set of all *first order formulas.*

k_1 is the meaning function of L_1, see Def.1 in §1.

$\models$ denotes the usual *relation of satisfaction*, see Example 1 in §1. I.e.: $\models \subseteq M_1 \times S_1 \times$ "Sets" is a ternary relation and for every $\mathfrak{A} \in M_1$, $\varphi \in S_1$, $q \in {}^{\omega}A$ $\mathfrak{A} \models \varphi[q]$ means that the valuation q of the free variables satisfies the formula φ in $\mathfrak{A}$.

Capitals underlined by tildes like

$\underset{\sim}{A}$ denote *algebras*. Their universes are denoted by the corresponding capitals without underlining, e.g. the *universe* of the algebra $\underset{\sim}{A}$ is A.

$\underset{\sim}{\mathfrak{A}}$ denotes algebras too.

I,**H**,**S** and **P** denote operators on classes of algebras. Let K be a class of algebras. Then **I** K, **H** K, **S** K and **P** K are the classes of all *isomorphic* copies, all *homomorphic* images, all *sub*algebras and all algebras isomorphic to *products* of elements of K respectively. See e.g. [20].

Th K is the set of all first order formulas valid in K if K is a class of algebras, see [20] or [29] as $\Theta\rho$ K .

Mod T is the class of all algebras in which T is valid, if T is a set of formulas, see [29] p.393.

Crs_α is the class of all *cylindric-relativized set algebras*, see Def.4 in §9 and [21].

Gs_α is the class of all *generalized cylindric set algebras*, see Def.5 in §9 and [21].

Lrg is the class of all *locally finite regular generalized cylindric set algebras*. Lrg is defined in Németi [32] Def.1 in this volume.

Lr is the class of all *locally finite regular cylindric set algebras*. Lr is defined in Németi [32] Def.1 in this volume.

§1. The concept of a language with semantics

The following definition originates from Abstract Model Theory, cf. [26],[24] p.133 ("Abstract Logic"),[29] p.417("General Logic"),[40],[15](Chap.1 "Basic notions of Abstract Model Theory")Def.1.1. A basic problem of Abstract Model Theory formulated as Problem 9 of Makowsky [26]p. 147 reads as: "Extend the framework of Abstract Model Theory such thattopological model theory and even other model theories, e.g. Kripke models, fit nicely into it." A detailed exposition of Abstract Model Theory is Sain [40].

DEFINITION 1 (The notion of a language)

L is defined to be a *language* (with semantics) iff L is a triple $L = \langle S,M,k \rangle$ such that S is a nonempty *set*, M is a nonempty *class*, and k is a *function* with domain $S \times M$, i.e. there exists a class R such that $k : S \times M \to R$. □

Conventions, remarks about Definition 1

Note that by Def.1 every triple consisting of a set, a class, and a function defined on their Cartesian product is a language.

If $L = \langle S,M,k \rangle$ is a language then we use the following names for its parts S,M and k :

S is called the *syntactic language*, in short the *syntax* of L . (Its members are often called expressions, sentences or formulas.)

M is called the class of *models* (or possible worlds or possible interpretations) of L .

k is called the *meaning* function, $k : S \times M \to$ "Sets", and for every $\varphi \in S$ and $\mathfrak{A} \in M$ we say that $k(\varphi,\mathfrak{A})$ is *the meaning* of φ in the model $\mathfrak{A}$. I.e. $k(\varphi,\mathfrak{A})$ is the meaning of the syntactic expression (or sentence or whatever) in the world (or interpretation or model) $\mathfrak{A}$. Other authors often use the word *denotation* instead of meaning for $k(\varphi,\mathfrak{A})$.

Examples for languages

First we give examples that are *not* intuitive, they only illustrate what the definition says word by word. Ord denotes the class of all ordinals.

1. $\langle \mathrm{Ord},\mathrm{Ord},+ \rangle$ is *not* a language.

   It is true that $+: \mathrm{Ord} \times \mathrm{Ord} \to \mathrm{Ord}$ is a function since the addition is defined on ordinal numbers but the first member of the triple is *not* a set.

2. $\langle \omega, Ord, + \rangle$ is a language.
ω is the *set* of all finite ordinals, Ord is a class and $+: \omega \times Ord \to Ord$ is a function. So it does satisfy our Def.1.

3. $\langle \omega, \omega, + \rangle$ is a language since sets are also classes.

4. Let $\prec$ be a well ordering of the reals R. Let for every $r, m \in R$

$$k(r,m) \stackrel{d}{=} \begin{cases} 1 & \text{if } r \prec m \\ 0 & \text{otherwise} \end{cases} .$$

Now $\langle R, R, k \rangle$ is a language.

5. Let t be the similarity type of arithmetic. Let Eq be the set of all equations of type t over the variables $\{x_i : i \in \omega\}$. Let Alg be the class of all algebras of type t. E.g. $\langle \omega, +, \cdot, 0, 1 \rangle \in Alg$. If $\underset{\sim}{A}$ is an algebra and $e \in Eq$ is an equation then let $k(e, \underset{\sim}{A})$ be the set of all solutions of the equation e in the algebra $\underset{\sim}{A}$. Now clearly $\langle Eq, Alg, k \rangle$ is a language in the sense of Def.1. Note that $k(e, \underset{\sim}{A}) = \{q \in {}^{\omega}A : \underset{\sim}{A} \models e[q]\}$, see the notations.

6. Let t be as above. Let Terms be the set of all terms of type t without variable symbols. I.e. $(1+1+1) \in Terms$, $(0+1+0) \in Terms$ but $(x+1) \notin Terms$. For any term $\tau \in Terms$ and algebra $\underset{\sim}{A}$ let $q(\tau, \underset{\sim}{A})$ be the element of $\underset{\sim}{A}$ denoted by τ. E.g. $q((1+1), \langle \omega, +, \cdot, 0, 1 \rangle) = 2 \in \omega$. (Recall that if there are no variable symbols in a term τ then it denotes an element of the universe A of the algebra $\underset{\sim}{A}$.) Now $\langle Terms, Alg, q \rangle$ is a language.

End of Examples.

Examples 1-4 above were rather naive and ad-hoc. We only wanted to illustrate what *is* said and what *is not* said in Def.1. In Def.2 below we introduce restrictions on the *presentation* of a language. (I.e. we restrict the ways one can "give" or "define" a fixed language L .) Examples 5 and 6 above do satisfy the conditions of Def. 2 i.e. they are well presented languages. As a contrast, Example 4 above is not well presented.

DEFINITION 2 (The notion of a well presented language)

A language $L = \langle S,M,k \rangle$ is *well presented* if (1)-(3) below hold.

(1) S is defined by a *generative grammar* or Chomsky grammar (see e.g. [18] or [22] p.11). I.e.:
An *alphabet* X is fixed. We denote the set of finite strings of elements of X by X^* . A *grammar* G is given such that the syntactic language defined by G is $S \subseteq X^*$. I.e. S consists of those strings of symbols (from X^*) which can be derived by G .

(2) M is a *class definable in set theory*. I.e. there is a set theoretical formula $\mu(x)$ (a formula of the language say ZFC) such that M is the collection of all such sets $\mathfrak{A}$ for which $\mu(\mathfrak{A})$ is true:
$$M = \{\mathfrak{A}\in\text{"Sets"} : \mu(\mathfrak{A})\}.$$

(3) k is a function (class of pairs) *definable in set theory*. I.e. there is a set theoretical formula $\varkappa(x,y,z)$ such that k is the class of those triples $\langle \varphi,\mathfrak{A},m \rangle$ of sets for which $\varkappa(\varphi,\mathfrak{A},m)$ is true, $\varphi\in S$, $\mathfrak{A}\in M$ and m is an arbitrary set. I.e. $k = \{\langle \varphi,\mathfrak{A},m \rangle : : \varkappa(\varphi,\mathfrak{A},m), \varphi\in S, \mathfrak{A}\in M, m\in\text{"Sets"}\}$. □

Remark

If $\varkappa$ and μ are absolute in the set theoretical sense then L has certain nice properties, see [40]. Typical examples when this is *not* the case are:

- Higher order logic with classical model theory,

- Program schemes with classical models as defined e.g. in Manna [27] Chap.4. (The latter is called "program schemes with standard time" in [10],[11].)

About absoluteness and "stableness" of languages see [40], especially Motivations 2.1, Def.2.6, Thm.2.1 there.

In the following examples "Models" denotes the class of all classical first order models of a fixed similarity type (see e.g. [14] or [20]) and "First o. formulas" is the set of all first order formulas of the same similarity type (see the above references).

Examples for well presented languages

1. ⟨First o. formulas, Models, $\models$⟩ .
2. ⟨Modal formulas, Kripke models, $\models$⟩.
3. ⟨Dynamic logic formulas, Dynamic Models, $\models$⟩.
4. ⟨Program schemes, Models, "traces"⟩.
5. ⟨Program schemes, Models, "computed functions"⟩ .
6. ⟨Program schemes, Continuous algebras, "least fixed points"⟩ .

Example 1 above will be explained below in detail. For the rest of the examples the reader is referred to Ex.2: [15,37], Ex.3:[10,11,37,38], Ex.4:[27,10,11,40], Ex.5: [27], Ex.6:[1].

Example 1

Consider the language

$$L_1 \stackrel{d}{=} \langle \text{First o. formulas, Models}, \models \rangle .$$

In (i) and (ii) below we shall define L_1 more precisely:

(i) Let t be a fixed similarity type. From now on

S_1 denotes the set of all first order formulas of type t ,

M_1 denotes the class of all classical first

order models of type t , and

k_1 denotes the meaning function

$k_1 : S_1 \times M_1 \to$ "Sets", which will be defined in (ii) below.

Throughout the paper the language L_1 is defined as:

$$L_1 \stackrel{d}{=} \langle S_1, M_1, k_1 \rangle .$$

By a little abuse of notation we shall sometimes write

$L_1 = \langle S_1, M_1, \models \rangle$ where $\models$ is the usual relation of satisfaction:

For any $\varphi \in S_1$, $\mathfrak{A} \in M_1$ with universe A , and for any valuation $q : \omega \to A$ of the free variables, $\mathfrak{A} \models \varphi[q]$ is defined to hold iff the *valuation* q of the free variables satisfies the formula φ in $\mathfrak{A}$, or, in other words, iff φ is true under the valuation q in $\mathfrak{A}$. See e.g. Henkin-Monk-Tarski[20] p.44.

By now at least the $\langle S_1, M_1, \models \rangle$ version of L_1 is defined precisely. To this we add that S_1 is defined by a CF grammar in §4. Next we turn to define the meaning function k_1 .

(ii) *The meaning function* k_1 .

For $Sb\ C$, ${}^{\omega}A$, and $Sb\,{}^{\omega}A$ see the list of Notations. Recall that the universe of $\mathfrak{A} \in M_1$ is always denoted by A .

In defining k_1 , the idea is to observe that $\models$ is a ternary relation, and that any ternary relation $R \subseteq A \times B \times C$ is actually a binary function $R_1 : A \times B \to Sb\ C$ defined as:

$R_1(a,b) \stackrel{d}{=} \{c \in C : \langle a,b,c \rangle \in R\}$ for any $a \in A$, $b \in B$.

The function $k_1 : S_1 \times M_1 \to$ "Sets" is defined as follows: Let $\varphi \in S_1$ and $\mathfrak{A} \in M_1$ be arbitrary. We define:

$$k_1(\varphi, \mathfrak{A}) \stackrel{d}{=} \{q \in {}^{\omega}A : \mathfrak{A} \models \varphi[q]\} .$$

I.e. $k_1(\varphi,\mathfrak{A})$ is the set of those valuations q which satisfy φ in $\mathfrak{A}$. About the meanings observe that for fixed φ and $\mathfrak{A}$, $k_1(\varphi,\mathfrak{A}) \in Sb\,{}^{\omega}A$.
Further, let $\mathfrak{A}\in M_1$ be fixed. Then we define

$$k_1(-,\mathfrak{A}) : S_1 \to Sb\,{}^{\omega}A$$

to be the unary function obtained from k_1 by fixing its second argument. I.e.:

$$k_1(-,\mathfrak{A})(\varphi) \overset{d}{=} k_1(\varphi,\mathfrak{A}) \quad \text{for every } \varphi\in S_1 .$$

Then $k_1(-,\mathfrak{A}) : S_1 \to Sb\,{}^{\omega}A$ shows that the meanings are subsets of ${}^{\omega}A$ if $\mathfrak{A}$ is fixed.
We often interchange k_1 and $\models$ because

$$q \in k_1(\varphi,\mathfrak{A}) \quad \text{iff} \quad \mathfrak{A}\models \varphi[q] .$$

I.e. k_1 and $\models$ contain the same information.

If $k_1(\varphi,\mathfrak{A}) = {}^{\omega}A$ then we say that φ is *valid* in $\mathfrak{A}$. (See e.g. [29]Def.11.5 where $k_1(\varphi,\mathfrak{A})$ is denoted by "$\varphi^{\mathfrak{A}}$".) Consulting the 5th example following Def.1 might help to understand the definitions of

$$k_1 : S_1\times M_1 \to \text{Sets} \quad \text{and} \quad \models\subseteq M_1\times S_1\times \text{Sets}$$

above. Note that in addition to $k_1 : S_1\times M_1 \to \text{Sets}$ we also know that $k_1 : S_1\times M_1 \to Sb({}^{\omega}\text{Sets})$.
The fact that L_1 is well presented is proved and illustrated in detail in [40].

(iii) *Illustration* of L_1 and $k_1 : S_1\times M_1 \to \text{Sets}$.
Let $\mathfrak{N} = \langle \omega,\le \rangle \in M_1$ be the model of type $t=\{\langle \le,2\rangle\}$ consisting of the natural numbers ω together with their usual ordering $\le$. To be more precise, we should write $\mathfrak{N}=\langle N,\mathfrak{N}(\le)\rangle$ where $N \overset{d}{=} \omega$ and $\mathfrak{N}(\le) \overset{d}{=}$ $\overset{d}{=} \{\langle n,m \rangle\in{}^{2}\omega : n\le m\}$. Then $(x_0\le x_1) \in S_1$ is a formula of $\mathfrak{N}$, also $(\neg\exists x_1\neg(x_0\le x_1)) \in S_1$ etc.

Recall that if $q\in{}^{\omega}N$ then $q=\langle q_0,q_1,q_2,\dots,q_i,\dots\rangle_{i<\omega}=$ $=\langle q_i : i<\omega \rangle$. That is we often write q_i instead of $q(i)$. The meaning of the formula $(x_0\le x_1)$ in the model $\mathfrak{N}$ is

$$k_1(\ (x_0 \leq x_1)\ ,\ \mathfrak{N}\ )\ =\ \{q \in {}^{\omega}N\ :\ q_0 \leq q_1\}\ .$$

A more complex meaning is $k_1(\neg\exists x_1\neg(x_0 \leq x_1)\ ,\ \mathfrak{N}\ ) =$ $= \{q \in {}^{\omega}N : q_0 = 0\}$. Similarly $k_1(\ (x_3 = x_7)\ ,\ \mathfrak{N}\ ) =$ $= \{q \in {}^{\omega}N : q_3 = q_7\}$. Clearly $k_1(\ (x_0 \leq x_1)\ ,\ \mathfrak{N}\ ) \subseteq {}^{\omega}N$. $k_1(\ \neg(x_0 \leq x_0)\ ,\ \mathfrak{N}\ ) = 0$ and $k_1(\ (x_0 \leq x_0)\ ,\ \mathfrak{N}\ ) = {}^{\omega}N$ are the smallest and the greatest elements of $Sb\ {}^{\omega}N$ $= \{Y : Y \subseteq {}^{\omega}N\}$. Clearly all meanings $k_1(\varphi, \mathfrak{N})$ of formulas φ in $\mathfrak{N}$ are elements of $Sb\ {}^{\omega}N$.

See §4 below for more information on Example 1.

End of Example 1

Throughout this paper we shall frequently refer to the above Example 1, especially to part (ii) of it.

§2. The algebraic concept of a language with semantics

In the universal algebraic approach i.e. in *initial algebra semantics* the concept of a language shows a pattern somewhat different from the above one, see [30],[19] [1] and other ADJ works.

There a language with semantics is a triple

$\langle \underset{\sim}{Fr}, M, K \rangle$, where

First of all a similarity type g for (possibly heterogeneous) algebras is fixed; and then

$\underset{\sim}{Fr}$ is a *free algebra* (or equivalently an initial algebra) of type g ,

M is a *class of algebras* of type g called meaning algebras,

K is a class of *homomorphisms* from the algebra $\underset{\sim}{Fr}$ into elements of M .

The important difference is that in the algebraic approach everything is *inside* of a single similarity class of universal algebras (sometimes everything is inside of a variety or even quasivariety).

The connection between the two notions lies in the following process:

1 First one has a language L in the sense of Definitions 1 and 2.

2 Then one *transforms* L *into* the frame of *algebra* and obtains $L' = \langle \underset{\sim}{F}r, M, K \rangle$ (all inside of the same similarity class of algebras). The reason for this is that in algebra we have tools and methods. (The whole thing is analogous to the Laplace transform or Z-transform in Linear System Theory.)

3 Then one investigates L′ inside of algebra and gets results.

4 Finally one *translates* the results *back* to L.

The crucial points are 2 and 4 transforming into algebra and back as we shall see.

Step 2 is called *algebraization* of the language L.

In 1 -4 above we only sketched a process which will be elaborated in sections 3, 5, 6.

Detailed exposition of the "Algebraic Concept of a Language with Semantics" $\langle \underset{\sim}{F}r, M, K \rangle$, quoted above, can be found e.g. in Andréka-Németi [7] §1 p.8, Andréka-Gergely-Németi [5], [6] Def.2.1 and ADJ [1] §1.

In the following we shall restrict ourselves to the case when the syntactic language S is given by a CF grammar. The case CS is treated at the end of this paper. The general case was treated by Kaphengst-Reichel [23], they found that if the grammar is not CF then partial algebras are really useful. For the theory of the latter cf. e.g. Andréka-Németi [8], Németi-Sain [35], Pasztor [36] Burmeister [12], [4], [31].

The contents of §3.A below is explained more briskly but precisely in §.3.1 of ADJ[1] p.75. Intuitive explanations of §3.A below can be found in Goguen [19] §2.1-2.3 , §3, §3.1.

§3. *The process of algebraization*

Suppose that a well presented language $L = \langle S, M, k \rangle$ is given, cf. Definition 2. What is the general method of turning L into algebras $\langle \underset{\sim}{Fr}, M, K \rangle$?

§3.A *Turning* S *into an algebra* $\underset{\sim}{Fr}$

Before turning anything into algebra, what should the "all encompassing similarity type" g be? (Certainly not that of M, if M happens to have such a thing at all. In L_1 the class M_1 did have a similarity type t.)

Recall from Definition 2 that $S \subseteq X^*$ where X is an alphabet and the syntactic language S is defined by a grammar G. There are *several grammars* defining the same language S. Therefore we: *choose* a grammar G defining S. The *choice* of G is important and we shall return to this point later (cf. e.g. First Criterion of Adequateness below).

We shall use the following terminologies about grammars: If $u \in X^*$ is such that there is a non-terminal symbol A of the grammar G such that u can be derived from A by the grammar G, then we say that "u is of syntactic category A". (Of course, u can be of several syntactic categories.) "u is of syntactic category A" will be denoted by $A \vdash_G^* u$ (see e.g. Kain

[22]p. 11 where $\vdash_G^*$ is denoted by $\overset{*}{\underset{G}{\Rightarrow}}$.)

Recall that we decided to restrict ourselves to CF grammars. So G, chosen above, is a CF grammar. CF abbreviates "context free". G consists of rules $\{R_i : i \in I\}$ of the form:

$(R_i) : A_o \vdash u_o A_1 u_1 \ldots A_n u_n$, where for every $j \leq n$:
$u_j \in X^*$ and
$A_j \notin X^*$ further A_j is said to be a nonterminal symbol of G i.e. A_j is a so called *syntactic category* of G.

Let N denote the set of all nonterminal symbols of G (i.e. N is the set of all syntactic categories of G, e.g.: $A_o, \ldots, A_n \in N$ in the above rule (R_i)). (Of course $N \cap X^* = 0$.)

Now we *translate the grammar* G *into an* N-*sorted similarity type* g of heterogenous universal algebras. For heterogenous universal algebras see e.g. Matthiessen [28],ADJ[1], Goguen[19],[25].

Every syntactic category i.e. nonterminal symbol A_j of G will be a *sort* of g, i.e. N will be the set of sorts of the similarity type g.

The reader is asked to consult Example 2 and the "Summing up §3.A" below.

To every rule $(R_i):A_o \vdash u_o A_1 u_1 \ldots A_n u_n$ of the above kind we correlate an operation symbol f_i of arity:

$$\langle A_1, \ldots, A_n, A_o\rangle \text{ i.e.: } g(f_i) \overset{d}{=} \langle A_1, \ldots, A_n, A_o\rangle.$$

This means that in every algebra $\underset{\sim}{U}$ of similarity type g the function symbol f_i denotes a function

$$F_i : U_{A_1} \times \ldots \times U_{A_n} \to U_{A_o} \quad \text{where } U_A \text{ is the}$$

universe of sort A in the algebra $\underset{\sim}{U}$ for every syntactic category A. By this we have a similarity type g corresponding to the grammar G, cf. "Summing up §3.A" below.

Now we translate $S \subseteq X^*$ into an algebra $\underset{\sim}{S}$ of similarity type g.

Let $A \in N$ be a syntactic category of G. Then the universe S_A of sort A in $\underset{\sim}{S}$ is defined to be the set of all strings $u \in X^*$ of syntactic category A. I.e.:

$$S_A \stackrel{d}{=} \{u \in X^* : A \vdash_G^* u\}.$$

In short: "the syntactic categories are the universes" of the algebra $\underset{\sim}{S}$.

Let the i-th rule of G be

$(R_i) : A_o \vdash u_o A_1 u_1 \ldots A_n u_n$ where $u_j \in X^*$ and $A_j \in N$ for $i \leq n$.

Recall that there is an operation symbol f_i corresponding to this rule (R_i) and $g(f_i) = \langle A_1, \ldots, A_n, A_o \rangle$.

Now we define the operation

$$F_i : S_{A_1} \times \ldots \times S_{A_n} \to S_{A_o}$$

of $\underset{\sim}{S}$ corresponding to the operation symbol f_i as:

$F_i(a_1, \ldots, a_n) \stackrel{d}{=} u_o a_1 u_1 \ldots a_n u_n$ for every $a_1 \in S_{A_1}, \ldots, a_n \in S_{A_n}$.

Clearly: $F_i(a_1, \ldots, a_n) \in S_{A_o}$ as required.

By this we have completed the definition of the algebra $\underset{\sim}{S}$ of similarity type g.: If $G = \langle N, X, \langle R_i \rangle_{i \in I} \rangle$ then $\underset{\sim}{S} = \langle \langle S_A \rangle_{A \in N}, \langle F_i \rangle_{i \in I} \rangle$ as defined above.

EXAMPLE 2.

Consider the grammar consisting of the rules $(R_\wedge)$, $(R_\neg)$, (R_i), $(i \in \omega)$ defined below:

$(R_\wedge)$: Formula $\vdash$ (Formula $\wedge$ Formula)

$(R_\neg)$: Formula $\vdash$ $\neg$Formula

(R_i) : Formula $\vdash p_i$, for every $i\in\omega$.
Here $\{),(,\wedge,\neg,p_i : i\in\omega\}$ are the terminal symbols and the only nonterminal i.e. the only syntactic category is "Formula".

The rule $(R_\wedge)$ is of form $A_o \vdash u_o A_1 u_1 A_2 u_2$ where $A_o=A_1=A_2=$"Formula", $u_o=$"(", $u_1=$"$\wedge$", and $u_2=$")". This is the grammar of propositional language with propositional variables $\{p_i : i\in\omega\}$. The corresponding algebra is

$\underset{\sim}{S} = \langle S, F_\wedge, F_\neg, F_i\rangle_{i\in\omega}$ where S is the set of all propositional formulas with propositional variables $\{p_i : i\in\omega\}$ and for every $\varphi,\Psi \in S$, the operations are:

$$F_\wedge(\varphi,\Psi) \overset{d}{=} (\varphi\wedge\Psi),$$
$$F_\neg(\varphi) \overset{d}{=} \neg\varphi,$$

and for every $i\in\omega$, $F_i \overset{d}{=} p_i$ is a constant.

A disadvantage of the above example is that there is only one syntactic category involved there namely "Formula". Hence the algebra $\underset{\sim}{S}$ has only one sort. If we take first order logic *with terms* i.e. with function symbols then we have two syntactic categories "Formula" and "Term" and then $\underset{\sim}{S}$ has two sorts:

"Formula" and "Term".

Then the derivation rule $(R_=)$: Formula $\vdash$ (Term=Term) gives rise to an operation symbol $f_=$ with arity <Term, Term, Formula>. For more detail see the proof of Claim 3 in Example 3 in §3.B.
End of Example 2.

The grammar in the above Example 2 is extended in sec.4 to a grammar G_1 defining the syntactic language S_1 as defined in Example 1.
Note that: If we add parse trees then $\underset{\sim}{S}$ is *absolutely free**/ (in the terminology of ADJ[1]:

*/ In the standard universal algebraic sense, cf Henkin-Monk-Tarski[20].

initial in the similarity class of all algebras of similarity type g), if not then $\underset{\sim}{S}$ is relatively free. Adding parse trees means to extend X by the symbols $\{f_i : i\in I\} \cup \{),(\}$ and to define the operations as: $F_i(a_1,\ldots,a_n) \overset{d}{=} f_i(u_o a_1 \ldots a_n u_n)$. For more on this see §3.1 of ADJ[1]. In this "Parse Trees Added approach" the so called *ambiguous semantics* can be treated nicely, too, see Montague[30].

Summing up §3.A.

1. Let $G = \langle N,X,\langle R_i : i\in I\rangle\rangle$ be a CF grammar. Here N and X are the sets of nonterminals and terminals respectively, and R_i $(i\in I)$ are the derivation rules.

Then we associate an N-sorted similarity type g to the grammar G. The associated similarity type g is a function $g : I \to N^*$. I.e. the operation symbols of g are the elements of I, and the sorts of g are the elements of N.

If $i\in I$ and $g(i) = \langle A_1,\ldots,A_{n+1}\rangle$ then we say that the arguments of the function symbol "i" are of sorts $A_1,\ldots,A_n$ respectively and its result is of sort A_{n+1}.

An algebra $\underset{\sim}{U}$ of similarity type g is a pair: $\underset{\sim}{U} = \langle\langle U_A\rangle_{A\in N}, \langle F_i\rangle_{i\in I}\rangle$ subject to the condition that, for every $i\in I$:

$$F_i : U_{A_1} \times \ldots \times U_{A_n} \to U_{A_{n+1}} \quad \text{where } g(i)=\langle A_1,\ldots,A_{n+1}\rangle.$$

2. The similarity type $g : I \to N^*$ is defined as: For every $i\in I$, if the rule R_i is of form

$$(R_i) : A_{n+1} \vdash u_o A_1 u_1 \ldots A_n u_n \quad \text{where } A_o,\ldots,A_{n+1} \in N \text{ and } u_o,\ldots,u_n \in X^*$$

then $g(i) \overset{d}{=} \langle A_1,\ldots,A_{n+1}\rangle$.

A distinguished algebra of similarity type g is $\underset{\sim}{S}$ defined as:

$\underset{\sim}{S} = \langle\langle S_A \rangle_{A \in N}, \langle F_i \rangle_{i \in I}\rangle$ where for every $A \in N$ and $i \in I$: $S_A \overset{d}{=} \{u \in X^* : A \vdash_G^* u\}$ and $F_i : S_{A_1} \times \ldots \times S_{A_n} \to S_{A_{n+1}}$ is defined as:

$$F_i(a_1,\ldots,a_n) \overset{d}{=} u_o a_1 \ldots a_n u_n \text{ for every } a_1 \in S_{A_1},\ldots a_n \in S_{A_n},$$

where the rule R_i is $A_{n+1} \vdash u_o A_1 u_1 \ldots A_n u_n$. Sometimes we denote the function symbol "i" by "f_i".

Remark (Preparations for §3.B)

Let $G = \langle N, X, \langle R_i \rangle_{i \in I}\rangle$ be a grammar, let g be the associated N-sorted similarity type, and $\underset{\sim}{S}$ be the associated "syntactic" algebra of type g. Let $\underset{\sim}{U}$ be an *arbitrary* algebra of type g and let $h : \underset{\sim}{S} \to \underset{\sim}{U}$ be a homomorphism. Then we can think of $\underset{\sim}{U}$ as a meaning algebra for $\underset{\sim}{S}$. We can consider the elements m of $\underset{\sim}{U}$ to be the *meanings* of elements of $\underset{\sim}{S}$. I.e. for any syntactic expression $w \in S$ we may say that the value $h(w)$ is the meaning of the expression w in the "world" $\underset{\sim}{U}$. Then $h : \underset{\sim}{S} \to \underset{\sim}{U}$ will satisfy Frege's Principle of Compositionality for semantics, see van Emde Boas-Janssen [17].

In short: any algebra $\underset{\sim}{U}$ of type g can be considered to be a *meaning algebra* for $\underset{\sim}{S}$. Let Alg_g denote the similarity class of all algebras of type g. Then $\langle \underset{\sim}{S}, Alg_g, \ldots\rangle$ *is* a language in the "Algebraic Semantics" sense of §2. Recall, however, that in our *original situation* a language $L = \langle S,M,k\rangle$ *with semantics* was given.

Below we shall see how to select *those* meaning algebras from Alg_g which really do correspond to some possible world or model $\mathfrak{A} \in M$ of our original language $L = \langle S,M,k \rangle$. These algebras will be called *possible meaning algebras*. The class of all possible meaning algebras corresponding to M will be denoted by $\mathcal{M}$. $\mathcal{M} \subseteq Alg_g$. Those algebras of type g which are not in $\mathcal{M}$ will be called impossible meaning algebras.

Then we shall have languages $\langle \underset{\sim}{S}, \mathcal{M}, \ldots \rangle$ in the sense of Algebraic Semantics (§2) associated to languages $\langle S,M,k \rangle$ in the sense of Abstract Model Theory.

§3.B. *Turning* M *and* k *into algebras and homomorphisms of type* g

Now we briefly return to Example 1. In (ii) of Example 1 $k_1(-,\mathfrak{A})$ (for a fixed classical first order model $\mathfrak{A}$) was defined to be a function $k_1(-,\mathfrak{A}) : S_1 \to Sb\ {}^{\omega}A$ where $Sb\ {}^{\omega}A$ is defined in the list of Notations. As far as only the Boolean connectives $\wedge, \neg$ are concerned, see Example 2, $k_1(-,\mathfrak{A})$ is a homomorphism from $\underset{\sim}{S}_1$ into the Boolean set algebra $\langle Sb\ {}^{\omega}A, \cap, - \rangle$ where $-X \overset{d}{=} {}^{\omega}A \sim X$, for ${}^{\omega}A \sim X$ see the list of Notations.

In sec.4 a grammar G_1 is given for S_1. The Boolean connectives correspond to the first 2 rules of G_1. Of course, we have to deal with the other rules of G_1 too, i.e. with the other connectives of S_1. Instead of going into this here, we return to the general case.

The general case:

First we recall some concepts of Universal Algebra:

Let $\underset{\sim}{B}$ be an algebra. Then B is its universe. A function h : B → is a homomorphism on $\underset{\sim}{B}$ iff there exists an algebra $\underset{\sim}{C}$ such that $h : \underset{\sim}{B} \to \underset{\sim}{C}$ is a homomorphism. The smallest such $\underset{\sim}{C}$ is called the image of $\underset{\sim}{B}$ by h. In other words, h is a homomorphism on $\underset{\sim}{B}$ iff its kernel Ker(h) is a congruence of $\underset{\sim}{B}$. Then the image of $\underset{\sim}{B}$ by h is isomorphic to the factor-algebra or quotient algebra $\underset{\sim}{B}$/Ker(h). See e.g. Henkin-Monk-Tarski [20].

Recall from the begining of §3 that we are in the process of algebraization of some language L = <S,M,k> in the sense of Definitions 1, 2. S was turned into an algebra $\underset{\sim}{S}$ of similarity type g in §3.A. The algebraization of M and k goes as follows:

Recall from Def.1 that $k : S \times M \to$ "Sets". For every $\mathfrak{A} \in M$ define the function

$k(-,\mathfrak{A}) : S \to$ "Sets" by

$k(-,\mathfrak{A})(\varphi) \overset{d}{=} k(\varphi,\mathfrak{A})$ for every $\varphi \in S$.

In λ-notation: $k(-,\mathfrak{A})$ is the function $\lambda\varphi k(\varphi,\mathfrak{A})$.

So $k(-,\mathfrak{A})$ is a mapping on the universe of $\underset{\sim}{S}$. There are two possibilities: either $k(-,\mathfrak{A})$ is a *homomorphisms* on $\underset{\sim}{S}$ or it is *not*.

If there exists an $\mathfrak{A} \in M$ such that $k(-,\mathfrak{A})$ is not a homomorphism on $\underset{\sim}{S}$ then we say that G is *not adequate* to the semantics <M,k> of L, and then we go back to the beginning and *choose another grammar* G′ defining the same syntactic language S, and then we begin the whole process with G′.

DEFINITION 3 (First Criterion of Adequateness)

Let $L = \langle S, M, k \rangle$ be a language in the sense of Definitions 1,2. Now, a grammar G defining S is said to be *adequate* to the semantics of L if the function $k(-,\mathfrak{A})$ is a homomorphism on the algebra $\underset{\sim}{S}$ for every $\mathfrak{A} \in M$. □

Suppose that the First Criterion of Adequateness is satisfied. The *image* of the algebra $\underset{\sim}{S}$ by the homomorphism $k(-,\mathfrak{A})$ will be denoted by $\underset{\sim}{\mathfrak{A}}$. Then $k(-,\mathfrak{A}): \underset{\sim}{S} \rightarrow \underset{\sim}{\mathfrak{A}}$ is a surjective homomorphism between two algebras of type g. Thus for every $\mathfrak{A} \in M$ the associated $\underset{\sim}{\mathfrak{A}}$ is *an algebra of type* g. We call $\underset{\sim}{\mathfrak{A}}$ the *meaning algebra* associated to the model $\mathfrak{A}$. Note that if $\mathfrak{A} \in M$ is fixed then the universe of the meaning algebra $\underset{\sim}{\mathfrak{A}}$ is $\{k(\varphi,\mathfrak{A}) : \varphi \in S\}$ by the definition of $\underset{\sim}{\mathfrak{A}}$.

The case of Example 1.

In Example 1. the language $L_1 = \langle S_1, M_1, k_1 \rangle$ was defined. If $\mathfrak{A} \in M_1$ is a classical model with universe A then the universe of the meaning algebra $\underset{\sim}{\mathfrak{A}}$ is a subset of $Sb\ {}^{\omega}A$. This shows that the universe of the meaning algebra $\underset{\sim}{\mathfrak{A}}$ is different from the universe of the original model $\mathfrak{A}$.

The general case:

Now $\mathcal{M}$ is defined to be the class of all meaning algebras associated to the models in M:

$$\mathcal{M} \stackrel{d}{=} \{\underset{\sim}{\mathfrak{A}} : \mathfrak{A} \in M\}.$$

K is defined to be the class of all homomorphisms from the algebra $\underset{\sim}{S}$ into elements of $\mathcal{M}$:

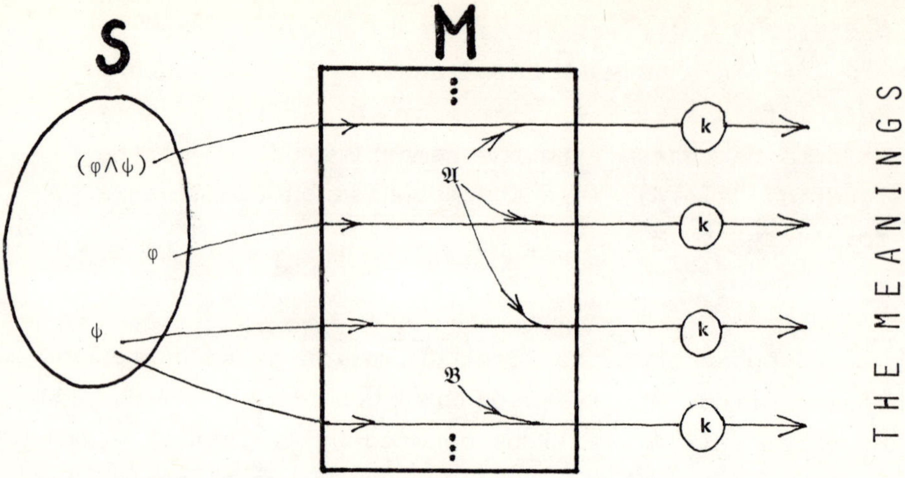

The original language ⟨ S,M,k ⟩.

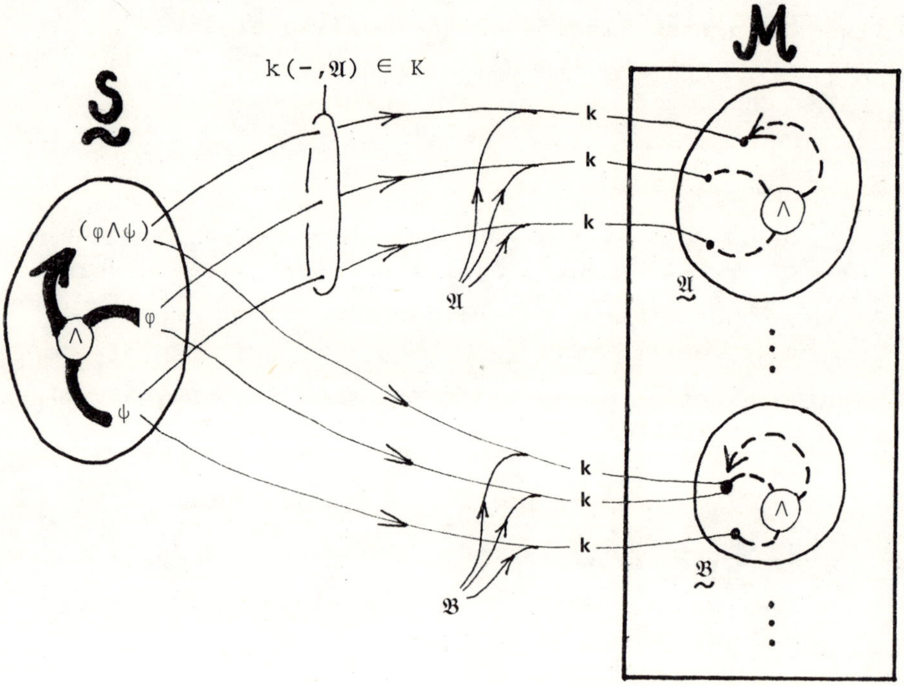

Algebraization of S (and G) into $\underset{\sim}{S}$ and ⟨ M,k ⟩ into $\mathcal{M}$.
Illustration to §3.B.

$K \stackrel{d}{=} \{h : \underset{\sim}{S} \stackrel{h}{\to} \underset{\sim}{\mathfrak{A}} \in \mathcal{M}\}$.

Note that $K = \{k(-,\mathfrak{A}) : \mathfrak{A} \in M\}$. Hence for any "meaning homomorphism" $h \in K$ there is $\mathfrak{A} \in M$ such that $h = k(-,\mathfrak{A})$. I.e.:
For syntactic expressions $\varphi \in S$ the new meanings $h(\varphi)$ in $\langle \underset{\sim}{S}, \mathcal{M}, K\rangle$ coincide with the old meanings $k(\varphi,\mathfrak{A})$ in $L = \langle S,M,k\rangle$.

The case of Example 1.

For L_1 denote the above class of homomorphisms by K_1. Now if $h \in K_1$ then there is a set A such that $h : S_1 \to Sb^{\omega}A$. The meaning algebras $\mathcal{M}_1$ corresponding to M_1 are Boolean algebras enriched with some additional operations. See §4. These enriched Boolean algebras will be defined in §9. Def.4,[32]Def.1(iv).

By this the process of algebraization is finished: We have transformed the language $L = \langle S,M,k\rangle$ into $\langle \underset{\sim}{S}, \mathcal{M}, K\rangle$ which is a language in the algebraic sense of §2.

The First Criterion of Adequateness (of G to L) formulated in Def.3 above is an exact, mathematical version of Frege's famous Principle of Compositionality.

Recall that if t is a similarity type then Alg_t denotes the similarity class of all algebras of type t.

EXAMPLE 3 (Grammars which are NOT ADEQUATE)

(i)

We define the language $L \stackrel{d}{=} \langle S,M,k\rangle$ as follows:
Let the alphabet X be $X \stackrel{d}{=} \{f,q\}$ and let

$$S \stackrel{d}{=} \{f,q\}^* \sim \{\lambda\} \quad \text{where}$$

λ denotes the empty sequence (i.e. λ is the unit element of the free monoid $\{f,q\}^*$).

Let

$M \overset{d}{=} \{\langle A,f',q',0'\rangle : A \in \text{"Sets"} \text{ and } f',q' \in {}^{A}A \text{ and } 0' \in A\}$.

More precisely, let $M \overset{d}{=} Alg_t$ for the similarity type

$$t = \{\langle f,1\rangle,\langle q,1\rangle,\langle 0,0\rangle\}.$$

The meaning function $k : S \times M \to \text{"Sets"}$ is defined by induction: Let $\mathfrak{A} = \langle A,f',q',0'\rangle \in M$ be arbitrary. Then

$$k(f,\mathfrak{A}) \overset{d}{=} f'(0') \quad \text{and}$$
$$k(q,\mathfrak{A}) \overset{d}{=} q'(0').$$

Let $\tau \in S$. Then:

$$k(f\tau,\mathfrak{A}) \overset{d}{=} f'(k(\tau,\mathfrak{A})) \quad \text{and}$$
$$k(q\tau,\mathfrak{A}) \overset{d}{=} q'(k(\tau,\mathfrak{A})).$$

E.g. $k(qfqf,\mathfrak{A}) = q'(f'(q'(f'(0'))))$.

By this the meaning function $k : S \times M \to \text{"Sets"}$ is defined. Hence $L = \langle S,M,k\rangle$ is defined.

Consider the two grammars G_2 and G_3 defined as:

$$G_2 \overset{d}{=} \langle\{T\},X,\langle R_c,R_d,R_f,R_q\rangle\rangle \quad \text{and}$$
$$G_3 \overset{d}{=} \langle\{T\},X,\langle R_c,R_d,R_F,R_Q\rangle\rangle \quad \text{where}$$

$(R_c) : T \vdash f$

$(R_d) : T \vdash q$

$(R_f) : T \vdash Tf$

$(R_q) : T \vdash Tq$

$(R_F) : T \vdash fT$

$(R_Q) : T \vdash qT$.

Clearly G_2 and G_3 generate the same syntactic language S. The question is: which of the two grammars G_2 and G_3 is adequate to the language L?

The corresponding similarity types g_2 and g_3 are:

$$g_2 = \{\langle c,0\rangle, \langle d,0\rangle, \langle f,1\rangle, \langle q,1\rangle\} \quad \text{and}$$
$$g_3 = \{\langle c,0\rangle, \langle d,0\rangle, \langle F,1\rangle, \langle Q,1\rangle\}.$$

The syntax algebras $\underset{\sim}{S}_2$ and $\underset{\sim}{S}_3$ corresponding to G_2 and G_3 are the following:

$\underset{\sim}{S}_2 = \langle S, c_2, d_2, f_2, q_2 \rangle$ where

$c_2 = f \in S$,

$d_2 = q \in S$,

$f_2 : S \to S$ and $q_2 : S \to S$ are defined as:

$f_2(\tau) \overset{d}{\equiv} \tau f$ and $q_2(\tau) \overset{d}{\equiv} \tau q$ for every $\tau \in S$.

$\underset{\sim}{S}_3 = \langle S, c_3, d_3, F_3, Q_3 \rangle$ where

$c_3 = f \in S$,

$d_3 = q \in S$,

$F_3 : S \to S$ and $Q_3 : S \to S$ are defined as:

$F_3(\tau) \overset{d}{\equiv} f\tau$ and $Q_3(\tau) \overset{d}{\equiv} q\tau$ for every $\tau \in S$.

See the Figure!

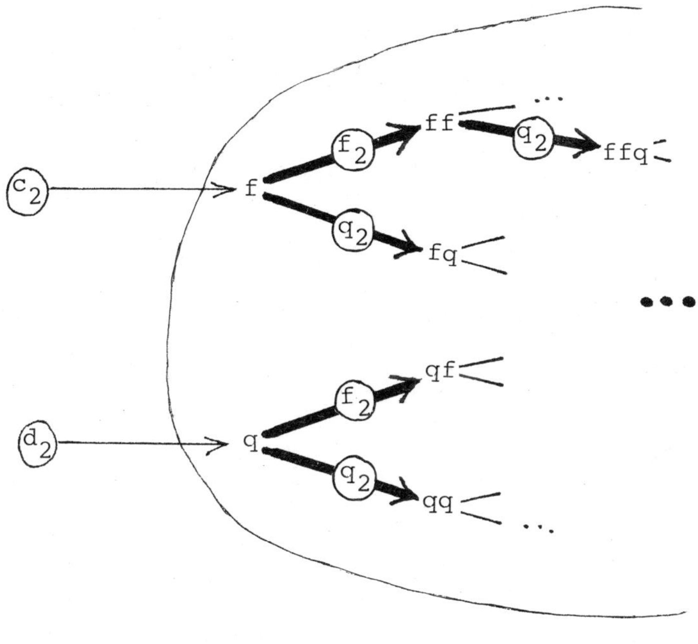

The algebra $\underset{\sim}{S}_2$

CLAIM 1.

The grammar G_2 *is not adequate to the language* **L**.

PROOF:

Let $\mathfrak{A} = \langle\{0,1,2,3\}, f', q', 0'\rangle \in Alg_t$ be such that

$f'(0') = 0$

$f'(1) = 2$

$f'(2) = 3$

$q'(0') = 1$

$0' = 0$

Let $h \overset{d}{=} k(-,\mathfrak{A})$.

We shall show that $h(ff) = h(f)$ while $h(q_2(ff)) \neq$
$\neq h(q_2(f))$.

See the Figure!

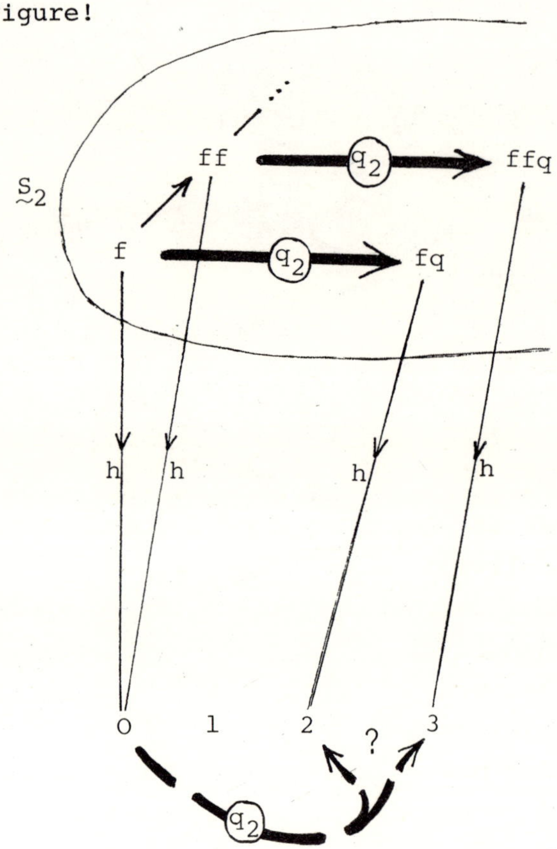

We shall show that $h(ff) = h(f)$ while $h(q_2(ff)) \neq h(q_2(f))$ which implies that h is not a homomorphism on $\underset{\sim}{S}_2$ namely it is not compatible with the operation q_2 of $\underset{\sim}{S}_2$:

$$h(f) = k(f,\mathfrak{A}) = f'(0') = 0.$$

$$h(ff) = f'(h(f)) = f'(f'(0')) = f'(0') = 0 \quad .$$

$$h(q_2(ff)) = h(ffq) = f'(h(fq)) = f'(f'(h(q))) = f'(f'(q'(0'))) = f'(f'(1)) = 3 \; ,$$

$$h(q_2(ff)) = h(fq) = f'(h(q)) = f'(q'(0')) = f'(1) = 2.$$

Thus really $h(ff) = h(f)$ and $h(q_2(ff)) \neq h(q_2(f))$. This proves that G_2 is not adequate to L since $k(-,\mathfrak{A})$ is *not* a homomorphism on $\underset{\sim}{S}_2$. □

The reasons for G_2 being inadequate are the followings:
There are two strings $f \in S$ and $ff \in S$ in our language S. They have the *same* meaning in $\mathfrak{A}$ namely 0. There is an operation $q_2 : S \to S$ in our syntax algebra $\underset{\sim}{S}_2$ originating from the rule $(R_q) : T \vdash Tq$. But then $q_2(f)$ and $q_2(ff)$ have *different* meanings.

The rule $(R_q) : T \vdash Tq$ says that from the expression ff we can build up ffq; but the meaning $k(ffq,\mathfrak{A})$ of ffq is not a "function of" the meaning $k(ff, \mathfrak{A})$ of ff. Hence Frege's Principle of Compositionality is violated by (R_q).

Roughly: The semantics of L interprets the expressions $\tau \in L$ from "the right to the left" while G_2 analyses them from "the left to the right". As a result, G_2 is not adequate to L.

Consider now G_3. There the syntax algebra is

$$\underset{\sim}{S}_3 = \langle S, c_3, d_3, F_3, Q_3 \rangle \text{ where}$$

$$c_3 = f \; ; \quad d_3 = q \; ; \quad F_3 : S \to S \text{ and } Q_3 : S \to S$$

are such that

$F_3(\tau) = f\tau$ and $Q_3(\tau) = q\tau$ for every $\tau \in S$.

CLAIM 2.

The grammar G_3 *is adequate to* L.

PROOF:

See the Figure!

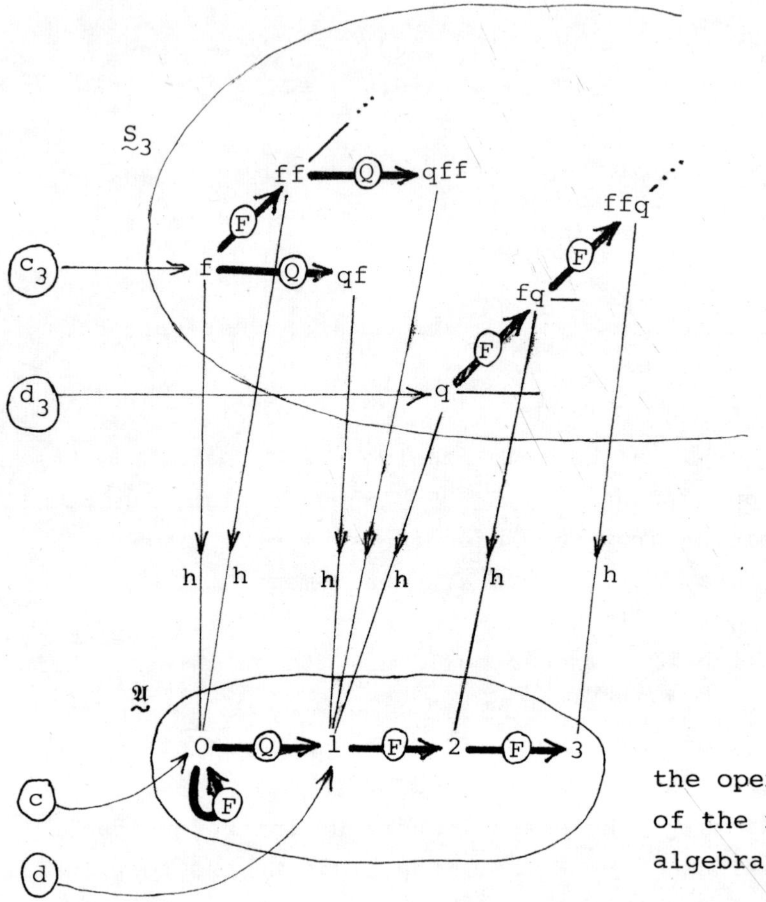

the operations of the new algebra $\underset{\sim}{\mathfrak{A}}$

Illustration to Claim 2:

The function $h: S \to A$ defined in the proof of Claim 1 is a homomorphism on $\underset{\sim}{S}_3$.

Let $\mathfrak{A} = \langle A, f', q', O' \rangle \in M$ be fixed. Let $h \stackrel{d}{=} k(-,\mathfrak{A})$. Let $\tau \in S$. Clearly $h(\tau) \in A$.

$F_3(\tau) = f\tau$ and $Q_3(\tau) = q\tau$. Hence

$h(F_3(\tau)) = h(f\tau) = f'(h(\tau))$ by the definition of k. Similarly $h(Q_3(\tau)) = q'(h(\tau))$.

This proves that h is a homomorphism

$h : \langle S, c_3, d_3, F_3, Q_3 \rangle \to \langle A, f'(O'), q'(O'), f', q' \rangle$.

Hence $k(-,\mathfrak{A})$ is a homomorphism on $\underset{\sim}{S}_3$ for every $\mathfrak{A} \in M$.

□

During the proof of Claim 2 we have executed §3.B to our language $\langle S, M, k \rangle$ and G_3 turning each model $\mathfrak{A} \in M$ into a meaning algebra $\underset{\sim}{\mathfrak{A}} \subseteq \langle A, f'(O'), q'(O'), f', q' \rangle \in Alg_{g_3}$ of type g_3. By letting $\mathcal{M} = \{\underset{\sim}{\mathfrak{A}} : \mathfrak{A} \in M\}$ we obtained an algebraized language $\langle \underset{\sim}{S}_3, \mathcal{M}, K \rangle$ where K is the class of all homomorphisms from $\underset{\sim}{S}_3 \in Alg_{g_3}$ into elements of $\mathcal{M} \subseteq Alg_{g_3}$.

(ii)

As a second example we define the language

$L_\ell = \langle S_\ell, M_\ell, k \rangle$ of Lattice Theory:

Let $\ell : \{\wedge, \vee\} \to \omega$ be the similarity type of lattices. I.e. $\ell = \{\langle \wedge, 2 \rangle, \langle \vee, 2 \rangle\}$.

Hence all lattices are in Alg_ℓ. Now

$M_\ell \stackrel{d}{=} Alg_\ell$.

We define the syntactic language S_ℓ to be the language generated by the grammar G_4 given below.

$G_4 \stackrel{d}{=} \langle N, X, \langle R_j \rangle_{j \in I} \rangle$ where

$N \stackrel{d}{=} \{\text{Formula}, \text{Term}, \text{Op}\}$,

$X \stackrel{d}{=} \{\wedge, \vee,), (, =, x_i : i \in \omega\}$,

$I \stackrel{d}{=} \{\wedge, \vee, =, \text{op}, \text{inf}, \text{sup}, i : i \in \omega\}$,

and the set of derivation rules is:

$(R_\wedge)$: Formula ⊢ (Formula ∧ Formula)

$(R_\vee)$: Formula $\vdash$ (Formula $\vee$ Formula)

$(R_=)$: Formula $\vdash$ (Term = Term)

(R_{op}) : Term $\vdash$ (Term Op Term)

(R_{inf}) : Op $\vdash$ $\wedge$

(R_{sup}) : Op $\vdash$ $\vee$

(R_i) : Term $\vdash$ x_i , for every $i \in \omega$.

Thus we have three syntactic categories in the grammar G_4: Formula, Term, and Op. We denote

$$S_F = \{w : \text{Formula} \vdash\!\frac{*}{G_4}\, w\}\ ,$$

$$S_T \overset{d}{=} \{w : \text{Term} \vdash\!\frac{*}{G_4}\, w\}\ , \quad \text{and}$$

$$S_{Op} \overset{d}{=} \{w : \text{Op} \vdash\!\frac{*}{G_4}\, w\}.$$

I.e., the set of all words of syntactic category Formula is denoted by S_F instead of $S_{Formula}$ etc.. Note that S_T is the set of all terms of similarity type ℓ over the set of variables $\{x_i : i \in \omega\}$;

$$S_{Op} = \{\wedge, \vee\}.$$

Note that e.g.

$$((x_o \vee x_1) = x_1) \vee ((x_o \vee x_1) = x_o)$$

is an element of S_F. (The class of totally ordered lattices is defined by the above formula.)

Thus the syntactic language S_ℓ generated by G_4 is:

$$S_\ell = S_F \cup S_T \cup S_{Op}\ .$$

Now we turn to defining the meaning function

$$k : S_\ell \times Alg_\ell \rightarrow \text{"Sets"}.$$

Let $\mathfrak{A} = \langle A,F \rangle \in Alg_\ell$ be arbitrary.

Let $f \in S_{Op}$. Then

$k(f,\mathfrak{A}) : A \times A \rightarrow A$ is defined as

$k(f,\mathfrak{A}) \overset{d}{=} F(f)$.

Note that

$$k(f,\mathfrak{A}) \in {}^{(^2A)}A.$$

Now let $\tau \in S_T$ be a term of type ℓ.
Let $q \in {}^{\omega}A$ be a valuation of the variables. Then the value $\tau[q]$ of the term τ at the valuation q is defined the usual way. Hence $\tau[q] \in A$.

E.g. $(x_2 \vee x_1)[q] = F(\vee)(q_2,q_1)$ and

$x_i[q] = q_i$

where $q = \langle q_i \rangle_{i<\omega}$.

Now, $k(\tau,\mathfrak{A})$ is defined to be a function

$k(\tau,\mathfrak{A}) : {}^{\omega}A \to A$ such that

$k(\tau,\mathfrak{A})(q) \overset{d}{=} \tau[q]$ for every $q \in {}^{\omega}A$.

Note that

$k(\tau,\mathfrak{A}) \in {}^{({}^{\omega}A)}A$.

Using Def.0.4.46 of Henkin-Monk-Tarski[20], $k(\tau,\mathfrak{A})$ is the polynomial in ω variables determined by τ over $\mathfrak{A}$. It is denoted there by $Pd_{\omega}^{(\mathfrak{A})}\tau$.

By this we have defined the restriction of k to domain

$(S_T \cup S_{Op}) \times Alg_{\ell}$.

Now let $\varphi \in S_F$. Then

$k(\varphi,\mathfrak{A}) \overset{d}{=} \{q \in {}^{\omega}A : \mathfrak{A} \models \varphi[q]\}$.

By this we have defined the meaning function

$k : S_{\ell} \times Alg_{\ell} \to$ "Sets".

Note that $k(\varphi,\mathfrak{A}) \in Sb\ {}^{\omega}A$, for $\varphi \in S_F$.

Now the language $L_{\ell} = \langle S_{\ell}, M_{\ell}, k\rangle$ is defined (and L_{ℓ} is well presented).

CLAIM 3.

The grammar G_4 *is adequate to* L_{ℓ}.

PROOF:

Let g_4 be the similarity type corresponding to the grammar G_4. Then $g_4 : I \to N^*$.

Let $\underset{\sim}{S}^4$ be the syntax algebra corresponding to the grammar G_4. Then:

$$\underset{\sim}{S}^4 = \langle\langle S_F, S_T, S_{Op}\rangle, \langle F_i : i \in (\{\wedge, \vee, =, op, inf, sup\} \cup \omega)\rangle\rangle$$

where e.g.

$$F_\wedge : S_F \times S_F \to S_F,$$
$$F_= : S_T \times S_T \to S_F,$$
$$F_{op} : S_T \times S_{Op} \times S_T \to S_T,$$
$$F_{sup} \in S_{Op},$$
$$F_i \in S_T \quad \text{for every } i \in \omega.$$

Let $\mathfrak{A} = \langle A, F\rangle \in Alg_\ell$. We shall define an algebra $\underset{\sim}{B} \in Alg_{g_4}$ and we shall prove that $\underset{\sim}{B} = \underset{\sim}{\mathfrak{A}}$.:

$$\underset{\sim}{B} \overset{d}{=} \langle\langle B_F, B_T, B_{Op}\rangle, P\rangle$$

where

$$B_F \subseteq Sb\ {}^\omega A, \quad B_T \subseteq {}^{({}^\omega A)}A, \quad \text{and} \quad B_{Op} \subseteq {}^{({}^2 A)}A,$$

and they are defined as:

$$B_F \overset{d}{=} \{k(\varphi, \mathfrak{A}) : \varphi \in S_F\} \subseteq Sb\ {}^\omega A;$$
$$B_T \overset{d}{=} \{k(\tau, \mathfrak{A}) : \tau \in S_T\} \subseteq {}^{({}^\omega A)}A,$$ i.e. B_T is the set of all polynomial functions (in ω variables) over $\mathfrak{A}$;

$$B_{Op} \overset{d}{=} \{k(f, \mathfrak{A}) : f \in S_{Op}\} = \{k(\wedge, \mathfrak{A}), k(\vee, \mathfrak{A})\} =$$
$$= \{F(\wedge), F(\vee)\} \subseteq {}^{({}^2 A)}A.$$

The operations of $\underset{\sim}{B}$ are defined as:

$$P(\wedge) \overset{d}{=} \langle X \cap Y : \langle X, Y\rangle \in B_F \times B_F\rangle,$$
$$P(\vee) \overset{d}{=} \langle X \cup Y : \langle X, Y\rangle \in B_F \times B_F\rangle,$$
$$P(=) \overset{d}{=} \langle \{q \in {}^\omega A : \tau[q] = \sigma[q]\} : \langle\tau, \sigma\rangle \in B_T \times B_T\rangle,$$
$$P(op) \overset{d}{=} \langle\langle f(\tau[q], \sigma[q]) : q \in {}^\omega A\rangle : \langle\tau, f, \sigma\rangle \in B_T \times B_{Op} \times B_T\rangle,$$
$$P(inf) \overset{d}{=} F(\wedge) (\in B_{Op}),$$
$$P(sup) \overset{d}{=} F(\vee) (\in B_{Op}),$$
$$P(i) \overset{d}{=} \langle q(i) : q \in {}^\omega A\rangle (\in B_T), \text{ for every } i \in \omega.$$

Clearly $\langle B_F, P(\vee), P(\wedge)\rangle \subseteq \langle Sb\,{}^\omega A, \cup, \cap\rangle$ is a familiar algebra: it is a distributive lattice with unit element ${}^\omega A$. Further $P(op) : B_T \times B_{Op} \times B_T \to B_T$ is such that for any two polynomials $\tau, \sigma \in B_T \subseteq {}^{({}^\omega A)}A$ and operation $f \in B_{Op} \subseteq {}^{({}^2 A)}A$ we have $P(op)(\tau, f, \sigma) \in {}^{({}^\omega A)}A$ is a new polynomial, namely $P(op)(\tau, f, \sigma)(q) = f(\tau[q], \sigma[q])$

for every $q \in {}^{\omega}A$. Now it is easy to check that $k(-,\mathfrak{A}) : \underset{\sim}{S}^4 \to \underset{\sim}{B}$ is an onto homomorphism and hence $\underset{\sim}{B}$ is the meaning algebra $\underset{\sim}{\mathfrak{A}}$ associated to $\mathfrak{A}$.

This proves that G_4 is adequate to L_ℓ. □

Next we shall define a new grammar G_5 generating the same language S_ℓ.:

$$G_5 \overset{d}{=} \langle N, X, \langle R_j \rangle_{j \in I'} \rangle \quad \text{where}$$

N,X were defined above,

$$I' \overset{d}{=} \{f, =, op, inf, sup, i : i \in \omega\},$$

the derivation rules $(R_=)$, (R_{op}), (R_{inf}), (R_{sup}), (R_i) were defined above, and

$$(R_f) : \text{Formula} \vdash (\text{Formula Op Formula}).$$

Thus the list of derivation rules of G_5 is:

(R_f) : Formula ⊢ (Formula Op Formula)

$(R_=)$: Formula ⊢ (Term = Term)

(R_{op}) : Term ⊢ (Term Op Term)

(R_{inf}): Op ⊢ ∧

(R_{sup}): Op ⊢ ∨

The nonterminals and terminals of G_4 and G_5 coincide. Further:

$$F \overset{*}{\underset{G_5}{\vdash}} w \quad \text{iff} \quad F \overset{*}{\underset{G_4}{\vdash}} w\,,$$

$$T \overset{*}{\underset{G_5}{\vdash}} w \quad \text{iff} \quad T \overset{*}{\underset{G_4}{\vdash}} w\,,$$

$$Op \overset{*}{\underset{G_5}{\vdash}} w \quad \text{iff} \quad Op \overset{*}{\underset{G_4}{\vdash}} w\,.$$

Let $\underset{\sim}{S}^4$ and $\underset{\sim}{S}^5$ be the syntax algebras defined by G_4 and G_5 respectively. Clearly:

$$S_F^4 = S_F^5 \;,\quad S_T^4 = S_T^5 \quad \text{and} \quad S_{Op}^4 = S_{Op}^5\,.$$

The operations of $\underset{\sim}{S}^5$ are the same as those of $\underset{\sim}{S}^4$, see the proof of Claim 3, with the only exception that $F(\wedge)$ and $F(\vee)$ are missing and they are replaced with a single new operation

$$f' : S_F \times S_{Op} \times S_F \to S_F$$

defined as

$$f'(\varphi, op, \Psi) \overset{d}{=} (\varphi \text{ op } \Psi) \quad \text{for} \quad \varphi, \Psi \in S_F \text{ and } op \in S_{Op} .$$

The operation f' is obtained from the new rule (R_f).

CLAIM 4.

The grammar G_5 *is not adequate to* L_ℓ .

PROOF:

Let $\mathfrak{A} = \langle A, F \rangle \in Alg_\ell$ be such that: $F(\wedge) = F(\vee)$ and $|A| > 1$. Let $h \overset{d}{=} k(-, \mathfrak{A})$. Now $(x_o = x_1) \in S_F$ and $h((x_o = x_1)) = \{q \in {}^\omega A : q(0) = q(1)\}$. Further $h((x_2 = x_3)) = \{q \in {}^\omega A : q(2) = q(3)\}$. Clearly $h((x_o = x_1)) \neq h((x_2 = x_3))$. Further $h((x_o = x_1) \vee (x_2 = x_3)) = \{q \in {}^\omega A : q(0) = q(1) \text{ or } q(2) = q(3)\}$. I.e. $h((x_o = x_1) \vee (x_2 = x_3)) = h((x_o = x_1)) \cup h((x_2 = x_3))$ and similarly $h((x_o = x_1) \wedge (x_2 = x_3)) = h((x_o = x_1)) \cap h((x_2 = x_3))$. Let $\varphi \overset{d}{=} ((x_o = x_1) \vee (x_2 = x_3))$ and $\Psi \overset{d}{=} ((x_o = x_1) \wedge (x_2 = x_3))$. Now obviously $h(\varphi) \neq h(\Psi)$ since if $q(0) = q(1)$ and $q(2) \neq q(3)$ then $q \in h(\varphi)$ but $q \notin h(\Psi)$. In the algebra $\underset{\sim}{S}^5$ we have an operation $f' : S_F \times S_{Op} \times S_F \rightarrow S_F$ corresponding to the rule (R_f). Clearly $\varphi = f'((x_o = x_1), \vee, (x_2 = x_3))$ and $\Psi = f'((x_o = x_1), \wedge, (x_2 = x_3))$. Recall that $h(\wedge) = F(\wedge)$ and $h(\vee) = F(\vee)$. Since $F(\wedge) = F(\vee)$, by our choice of $\mathfrak{A}$, we have $h(\wedge) = h(\vee)$. Since $h(\varphi) \neq h(\Psi)$ we have $h(f'((x_o = x_1), \wedge, (x_2 = x_3))) \neq h(f'((x_o = x_1), \vee, (x_2 = x_3)))$ which together with $h(\wedge) = h(\vee)$ implies that h is *not* a homomorphism on $\underset{\sim}{S}^5$ namely the function $h : S_F \cup S_T \cup S_{Op} \rightarrow$ "Sets" is not compatible with the operation $f' : S_F \times S_{Op} \times S_F \rightarrow S_F$ of $\underset{\sim}{S}^5$. □

(iii)

We define the *implication language* $L = \langle S, M, k \rangle$ as follows:

Let the alphabet X be $X \overset{d}{=} \{p_i : i \in \omega\} \cup \{\rightarrow\}$ and let the syntactic language S be $S \overset{d}{=} \{p_{i_o} \rightarrow \ldots \rightarrow p_{i_n} : n \in \omega$ and $\{i_o, \ldots, i_n\} \subseteq \omega\}$. We define $M \overset{d}{=} {}^\omega 2$.

The meaning function $k : S \times M \to$ "Sets" is defined by induction:

Let $\mathfrak{A} = \langle \mathfrak{A}(i) : i \in \omega \rangle \in {}^{\omega}2$ be arbitrary. Then:

$k(p_i, \mathfrak{A}) \overset{d}{=} \mathfrak{A}(i)$,

$$k(p_i \to \varphi, \mathfrak{A}) \overset{d}{=} \begin{cases} 0 & \text{if } (\mathfrak{A}(i)=1 \text{ and } k(\varphi, \mathfrak{A})=0) \\ 1 & \text{otherwise} \end{cases}$$

Now the implication language $L = \langle S, M, k \rangle$ is defined.

Consider the grammars G_6 and G_7 defined as:

$G_6 = \langle N, X, \langle R_i \rangle_{i \in I_6} \rangle$,

$G_7 = \langle N, X, \langle R_i \rangle_{i \in I_7}$, where $N \overset{d}{=} \{F\}$,

$I_6 \overset{d}{=} \omega \cup \{\to\}$

$I_7 \overset{d}{=} \omega \cup \{1i : i \in \omega\}$,

and the derivation rules are:

$R_{\to} : F \vdash F \to F$

$R_i : F \vdash p_i$ for every $i \in \omega$,

$R_{1i} : F \vdash p_i \to F$ for every $i \in \omega$.

Clearly both G_6 and G_7 generate the syntactic language S.

CLAIM 5.

The grammar G_6 *is not adequate to the implication language* L.

The grammar G_7 *is adequate to* L.

PROOF:

We prove that G_6 is not adequate to L. $\underset{\sim}{S}^6$ denotes the syntax algebra corresponding to G_6. Then $\underset{\sim}{S}^6 = \langle S, F_{\to}, F_1, F_2, \ldots \rangle$. Let $\mathfrak{A} \in M$ be such that $\mathfrak{A} = \langle 0,1,0,1,1,\ldots \rangle$. Let $h \overset{d}{=} k(-, \mathfrak{A})$ i.e. $h(\varphi) = k(\varphi, \mathfrak{A})$ for every $\varphi \in S$. We shall show that h is not a homomorphism on $\underset{\sim}{S}^6$ by proving that h is not compatible with the operation $F_{\to}$ of $\underset{\sim}{S}^6$.

$h(p_o \to p_1) = h(p_3 \to p_4) = 1$ and $h(p_2) = 0$ by the definition of $\mathfrak{A}$ and h. Thus $F_{\to}(h(p_o \to p_1), h(p_2)) = F_{\to}(h(p_3 \to p_4)$,

$h(p_2))$. But $h(F_{\rightarrow}(p_o \rightarrow p_1, p_2)) \neq h(F_{\rightarrow}(p_3 \rightarrow p_4, p_2))$ since $h(F_{\rightarrow}(p_o \rightarrow p_1, p_2)) = h(p_o \rightarrow p_1 \rightarrow p_2) = 1$ and $h(F_{\rightarrow}(p_3 \rightarrow p_4, p_2)) = h(p_3 \rightarrow p_4 \rightarrow p_2) = 0$. Hence $h = k(-, \mathfrak{A})$ is not a homomorphism on $\langle S, F_{\rightarrow} \rangle$ and hence it is not a homomorphism on $\underset{\sim}{S}^6$. We have seen that G_6 is not adequate to the implication language L. It is easy to prove that G_7 *is* adequate to L. □

Of course, the choice of G might turn out to have been unlucky: we may find our algebras $\underset{\sim}{S}$ and M cumbersome to handle and then we may be forced to turn back to choose another grammar G′.

As an example of a succesful choice of G, the theory of Cylindric Algebras (see e.g. Henkin-Monk-Tarski[20], Andréka-Gergely-Németi[5],[6]) is the result of *an* algebraization of L_1. See Example 1 ! To M_1 there corresponds the class Lr of locally i -finite cylindric algebras* as was proved in Andréka[2], Andréka-Gergely-Németi [5] (III.2, pp.17-18, and III.6), and in Andréka-Gergely-Németi [6] p.30 following Def. 4.1. In Henkin-Monk-Tarski[21] these algebras are called "Locally finite regular cylindric set algebras" and their class Lr is denoted by "$Cs_\omega^{reg} \cap Lf_\omega$". More details about the algebraization of L_1 can be found in §5 and in [32].

In case of the simplest kind of *first order Kripke models* the corresponding class M of meaning algebras is $Gs_\omega^{reg} \cap Lf_\omega$. The latter is the class of those generalized cylindric set algebras, in short Gs (in the sense of Henkin-Monk-Tarski[20]p.171), every element of which is *i -finite* in the sense of Andréka-Gergely-Németi[6] bottom of p.18. Some results on this algebraization of Kripke semantics will be discussed in [32].

---

* In some of the quoted papers Lr is denoted by cursive Lv or by German L.

Warning: The definition of an i -finite element given in Andréka[2] and Andréka-Gergely-Németi[5] is different from that given in Andréka-Gergely-Németi[6]! E.g.: III. 2 of Andréka-Gergely-Németi[5] is true for i -finite Gs if i -finite is understood in the sense of Andréka-Gergely-Németi[6], but not if it is taken in the sense of Andréka-Gergely-Németi[5].

In computer science the above outlined process of algebraization has been executed rather succesfully by V.R. Pratt recently. He obtained the variety of *Dynamic Algebras* by applying the above process to Dynamic Logic, see Pratt[37],[38]. As Pratt[38] explains, Dynamic Algebras are somewhere in between Cylindric Algebras and Boolean Algebras. The present authors feel that Pratt's theory of Dynamic Algebras is a breakthrough in the fields: Logic of Actions, Logics for Programverification, Dynamic Logic, Robotologic etc.. We feel this because the quoted theory gives an unusually clear insight into the very basic structure underlying all the quoted logics. It also has a nice representation theory via Kripke models satisfying all the criteria that will be stated in §6 (Applicability) of the present paper.

§4. A widely used example: Algebraization of L_1.

The grammar G_1 from which the theory of Cylindric Algebras was obtained is the following:

$(R_\neg)$: Formula $\vdash$ $\neg$ Formula

$(R_\wedge)$: Formula $\vdash$ (Formula$\wedge$Formula)

$(R_{\exists_i})$: Formula $\vdash$ $\exists x_i$ Formula, for every $i \in \omega$.

$(R_{=_{ij}})$: Formula $\vdash$ $x_i = x_j$, for every $i,j \in \omega$.

(R_P) : Formula $\vdash$ $P(x_1 \ldots x_n)$, for every predicate symbol P with arity n .

For $L_1 = \langle S_1, M_1, k_1 \rangle$ see Example 1.
This grammar G_1 defines the syntax S_1 of first order logic.

Let us see how the general procedure given in §3 works.: The set N_1 of nonterminals of G_1 is {Formula}. Therefore the associated similarity type g_1 is only one-sorted. The set T_1 of function symbols of g_1 is:

$T_1 = \{\neg, \wedge, \exists_i, =_{ij}, P : i,j \in \omega$ and P is a predicate symbol$\}$.

Then $g_1 : T_1 \to \{\text{Formula}\}^*$ is the following:

$g_1(\wedge) = \langle \text{Formula}, \text{Formula}, \text{Formula} \rangle$,

$g_1(\neg) = g_1(\exists_i) = \langle \text{Formula}, \text{Formula} \rangle$, and

$g_1(=_{ij}) = g_1(P) = \langle \text{Formula} \rangle$.

Now the syntax algebra $\underset{\sim}{S}_1$ is the following:

$\underset{\sim}{S}_1 = \langle S_{\text{Formula}}, \langle F_i \rangle_{i \in T_1} \rangle$ where

$S_{\text{Formula}} = S_1$ and

$F_\wedge : S_1 \times S_1 \to S_1$ i.e. $F_\wedge$ is a binary function on S_1 such that $F_\wedge(\varphi, \Psi) = (\varphi \wedge \Psi)$ for every $\varphi, \Psi \in S_1$,

$F_\neg : S_1 \to S_1$, etc.

For example, the value of the term $\wedge(\exists_1(P), =_{12})$ in the algebra $\underset{\sim}{S}_1$ is: $F_\wedge(F_{\exists_1}(F_P), F_{=12}) = (\exists x_1 P(x_1 \ldots x_n) \wedge x_1 = x_2)$.
By this we have executed §3.A. Next comes the execution of §3.B. There it reads that we can go on only if G_1 is adequate to L_1 in the sense of Definition 3. G_1 is adequate to L_1. This is checked in detail in the proof of Proposition 1 (iii) in Németi[32]. If we continue mechanically the translation process according to §3.B, we shall obtain from M_1 and k_1 the class $Lr = M_1$ of locally i -finite cylindric set algebras (see Def.4, and the references at the end of §3). This continuation can be found in Németi[32] in the proof of Proposition 1 (iii).

Remark

For reasons beyond the scope of this paper, the constants obtained from the rules (R_p) are often dropped in Cylindric Algebra Theory and then the set of operations is: $T = \{\neg, \wedge, \exists_i, =_{ij} : i,j \in \omega\}$ with arities 1,2,1,0 respectively. The constants corresponding to (R_P) then return later in the form of the so called dimension restricting defining relations, cf. Andréka-Gergely-Németi [6]p.18 and Chap.V.Def.5.1 there.

More on Cylindric Algebras and the algebraization of L_1 can be found in Németi [32].

§5. The algebraic investigations

The definitions of **HSP**M and **SP**M for an arbitrary class of algebras can be found in any textbook on Universal Algebra, e.g. Henkin-Monk-Tarski[20]. Recall that **HSP**M is the equationally axiomatizable hull of the class M of algebras and **SP**M is the smallest class of algebras containing M in which all kinds of free algebras exist in the most general sense. **HSP**M is called the *variety generated by* M while **SP**M is called the *epireflective hull of* M.

The main bulk of investigation of $\langle \underset{\sim}{S}, M, K \rangle$ (see sections 2 and 3) is the study of the variety **HSP**M generated by M. Instead of $\underset{\sim}{S}$, the free algebra $\underset{\sim}{S}'$ of **HSP**M is considered which is the reflection (see McLane 71 p.89) of $\underset{\sim}{S}$ in **HSP**M. Denote the reflecting homomorphism by $r : \underset{\sim}{S} \twoheadrightarrow \underset{\sim}{S}' \in$ **HSP**M. The kernel Ker(r) of r consists exactly of the synonym pairs of S with respect to its original semantics in L. Thus $\underset{\sim}{S}'$ is the nonredundant syntax of L. For this and other reasons **HSP**M is important. Cf. Chap.II in Andréka-Gergely-Németi[6].

Often varieties $V \supseteq M$ are searched for such that the minimal elements of V are in **HSP**M. (An algebra is minimal if it is generated by its constants.) V may be different from $\mathbf{HSP}\mathcal{M}$ and on this price might be immensely simpler. This is the case with cylindric algebras: A. Tarski found a variety CA_ω containing $\mathcal{M}_1$=Lr which can be defined by 7 simple schemes of equations. This variety CA_ω satisfies our requirement with respect to minimal elements, cf. Andréka-Gergely-Németi [6] Cor.3.14 (a). The equations defining $\mathbf{HSP}\mathcal{M}_1$ are much more complicated, cf. Andréka-Gergely-Németi [6] Th.3.5. J.D.Monk proved in 1969 that there is no simple equational definition of $\mathbf{HSP}\mathcal{M}_1$. The reason for the weak requirement that *only the minimal elements* of V need to behave well is due to the fact that (by our constructing method) $\underset{\sim}{S}$ is always minimal and therefore its reflection $\underset{\sim}{S}'$ in V is determined by the minimal elements of V .

§6. Applicability

For $\underset{\sim}{S}$, $\underset{\sim}{S}'$, $\mathbf{HSP}\mathcal{M}$, V etc. see sections 3 and 5. To be able to apply the algebraic results about $\underset{\sim}{S}$, $\underset{\sim}{S}'$, $\mathbf{HSP}\mathcal{M}$, V etc. it is important to make and keep a distinction:

The elements of $\mathcal{M}$ are called meaning algebras while the elements of V, $\mathbf{HSP}\mathcal{M}$, and $\mathbf{SP}\mathcal{M}$ are called *pseudo meaning algebras*. The homomorphisms of $\underset{\sim}{S}$ into an algebra $\underset{\sim}{A}\in V$ are called *pseudo meaning homomorphisms* if $\underset{\sim}{A}\notin\mathcal{M}$ while the elements of K are *real meaning homomorphisms*.

This distinction between real interpretations and pseudo interpretations should be preserved throughout the algebraic studies. It often happens both in the theory of program schemes and in algebraic logic that people completely forget $\mathcal{M}$ and consider *the elements of* V *or* $\mathbf{HSP}\mathcal{M}$ as *the* interpretations. (In Cylindric Algebra Theory $\mathbf{HSP}\mathcal{M}$ is the class R of representable cylindric algebras and V is CA_ω .) Then it happens that we arrive at very nice results about the existence of an interpretation $h : \underset{\sim}{S} \to \underset{\sim}{A} \in V$ with beautiful properties. So we are happy *but when we want to translate* the "Grand Result" to the original language $L = \langle S, M, k \rangle$ *then* we find that there is completely *nothing in* **M** that would correspond to the pseudo interpretation $h : \underset{\sim}{S} \to \underset{\sim}{A}$. The reason of this is that $\underset{\sim}{A}$ was not in M and *only* the elements of M are translations of elements of **M** , the original models.

During the study of an algebraized language $\langle \underset{\sim}{S}, M, K \rangle$ it is very tempting to say that we do not need an infinite class of meaning algebras M :"There is a unique algebra $\underset{\sim}{A}$ which is equivalent to M ." Namely, let $\underset{\sim}{A}$ be $\underset{\sim}{S}'$, the reflection of $\underset{\sim}{S}$ in **HSP**M or, take the direct product of all elements of M and let $\underset{\sim}{A}$ be the minimal subalgebra of this product. Then there is a *unique meaning algebra*, perhaps one interpretation only and the assymetry of *one* syntax S but *many* interpretations $\mathfrak{A} \in M$ is tidily replaced by a one-one $(\underset{\sim}{S}, \underset{\sim}{A})$ relationship. (We comitted this folly in Andréka[2].) This is wrong for several reasons the most outstanding of which is the following. $\underset{\sim}{A}$ is not in M (but only in **SP**M) and therefore *there is nothing* in the original semantics **M** that would correspond to $\underset{\sim}{A}$. More formally: $(\forall \mathfrak{B} \in M)\underset{\sim}{\mathfrak{B}} \neq \underset{\sim}{A}$! I.e. $\underset{\sim}{A}$ is only a pseudo interpretation. More details in this distinction can be found e.g.in Andréka-Gergely-Németi[6], Andréka-Németi[7], Henkin-Monk-Tarski[21] (regular and nonregular algebras), Montague[30].

Examples

1) The most difficult parts of Andréka-Gergely-Németi [6] in establishing properties (e.g. completeness) of the "finitary logics of infinitary relations" are Cor.3.23(d), Cor.3.13, and the Appendix. These results are such statements about M_1=Lr which statements would be easy to

prove if Lr would be replaced by $R = \mathbf{HSP}M_1 = \mathbf{HSP}\,Lr$. However, results about R cannot be used to establish properties of the *original* language, the "finitary logics of infinitary relations" (see loc.sit. Th.6.3 and the remark above it, and the last two lines of p.19).

2) In some instances results have been published saying that a certain programming language P_1 is more powerful than another one, P_2. The interpretation used in the proof was an element $\underset{\sim}{A}$ of $\mathbf{HSP}M$ such that $\underset{\sim}{A}\notin M$. So we still do not know what their result means in real programming situations $L = \langle S,M,k \rangle$.

§7. Dynamic Cylindric Algebras

Dynamic Cylindric Algebras will be abbreviated as DCA-s. DCA-s correspond to first order Dynamic Logic just as Dynamic Algebras of Pratt[38] correspond to propositional Dynamic Logic and as Cylindric Algebras correspond to classical first order logic. The latter correspondence has been explained in detail in Example 1, §4, and [32].

We shall use the notations CA_ω, $\underset{\sim}{Bl}$, Zd of Henkin-Monk-Tarski[20]. A DCA is a heterogenous algebra $\underset{\sim}{D} = \langle \underset{\sim}{B},\underset{\sim}{A},\Diamond \rangle$ where $\underset{\sim}{B}\in CA_\omega$ and $\langle \underset{\sim}{Bl}\,\underset{\sim}{B},\underset{\sim}{A},\Diamond \rangle$ is a Dynamic Algebra in the sense of Pratt[38]. I.e.: $\Diamond : A\times B \rightarrow B$ is a binary operation, such that $(\forall a\in A)$ [the function $\langle \Diamond(a,x) : x\in B \rangle$ is an endomorphism of $\langle B,+,0 \rangle$]. Further we postulate that if $\underset{\sim}{D}\in DCA$ then $(\forall p\in A, \forall b\in Zd\,\underset{\sim}{B})\ \Diamond(p,b) = b$. Here intuitively: The elements of A are actions or programs, the elements of B are statements (first order formulas) and $\Diamond(p,\varphi)$ means that the action p may bring about the truth of formula φ. I.e. $\Diamond(p,\varphi)$ means that the program p is correct w.r.t. the output condition φ. The DCA $\underset{\sim}{D}$ is said to be representable iff $\underset{\sim}{D} \cong \langle \underset{\sim}{B},\underset{\sim}{A},\Diamond \rangle$ where $\underset{\sim}{B}\in Lrg$ and $A\subseteq {}^{B}B$ moreover $(\forall p\in A,\forall b\in B)\Diamond(p,b) = p(b)$. Then we say that $\langle \underset{\sim}{B}, \underset{\sim}{A}, \Diamond \rangle$ is a set DCA. Let $\langle \underset{\sim}{B}, \underset{\sim}{A}, \Diamond \rangle$ be a set DCA and let $p \in A$, $Y \in B$. Let U be a subbase of $\underset{\sim}{B}$. Let $X = \Diamond(p,Y) \cap {}^{\omega}U$. Now we say that $Y \cap {}^{\omega}U$ is the set of all outputs the program may result if started with inputs from X. Observe that the elements of X and Y are ω-sequences from U and hence they can be considered as inputs and outputs of programs, since a subbase U of an Lrg is always considered to be the universe of a model.

§8. Generalization to the CS *case*

The reader is asked to consult the section "Summing up §3.A."

Let $G = \langle N, X, \langle R_i : i \in I \rangle \rangle$ be a CS grammar, with nonterminals N etc.. We shall associate to G an N-sorted similarity type $g : I \to N^*$.

Let G' be the grammar obtained from G by dropping the context conditions, i.e. if (R_i) is $\alpha A_o \beta \vdash \alpha \gamma \beta$ then (R_i') is $A_o \vdash \gamma$ and then $G' = \langle N, X, \langle R_i' : i \in I \rangle \rangle$. Now G' is CF. Hence §3.A associates a similarity type g' to G'. Now we define the similarity type g associated to the CS grammar G to be g'.

Recall the parse-tree algebra defined in the Note following Example 2 in §3.A..

Let $\underset{\sim}{W} = \langle \langle W_A \rangle_{A \in N}, \langle F_i \rangle_{i \in I} \rangle$ be the term algebra i.e. word algebra of type g generated by the constants. I.e. $\underset{\sim}{W}$ is the initial algebra of Alg_g (i.e. $\underset{\sim}{W}$ is the algebra freely generated by 0 in Alg_g). Now, $\underset{\sim}{W}$ is the parse-tree algebra of G', i.e. the elements of W are the parse-trees of G'. (It is well known that parse-trees can be defined to be terms. It was shown in ADJ[1] that the parse-tree algebra $\underset{\sim}{D}$ defined in §3.A is isomorphic to $\underset{\sim}{W}$.)

The basic difference between G and G' is that there are parse-trees of G' which are not parse-trees of G (because of the context conditions). Let $W = \cup\{W_A : A\in N\}$. Then W is the set of all parse-trees of G'. Let $T\subseteq W$ be the set of all parse-trees of G. I.e.: $T \stackrel{d}{=} \{u\in W :$ there is $A\in N$ such that u is a parse-tree starting from A of $G\}$.

Let S be the set of all subterms of elements of T. I.e.: $S \stackrel{d}{=} \{u\in W : u$ is a subterm of some term $w\in T\}$.

In Partial Algebra Theory S is said to be the smallest initial segment of $\underset{\sim}{W}$ containing T.

Let $S_A \stackrel{d}{=} W_A \cap S$, for all $A\in N$. Let $P_i \stackrel{d}{=} F_i \cap S_{A_1}\times\dots\times S_{A_{n+1}}$ for all $g(i) = \langle A_1,\dots,A_{n+1}\rangle$.

Then $\underset{\sim}{S} \stackrel{d}{=} \langle\langle S_A\rangle_{A\in N}, \langle P_i\rangle_{i\in I}\rangle$ is a partial algebra. ($\underset{\sim}{S}$ is a so called relative subalgebra of $\underset{\sim}{W}$.) The partial algebra $\underset{\sim}{S}$ is free in the sense of P. Burmeister and J.Schmidt (for references see [12]).

Consider the equations $Eq \stackrel{d}{=} \{p=p : p\in T\}$. Eq defines a variety V of partial algebras (where we use Burmeister's strong validity) see Andréka-Németi[8], Németi[31] or §5 of Németi-Sain[35]. Note that $Alg_g \subseteq V$ and $V = \mathbf{H_w S_s P}\ V$, see the above quotations or Pasztor[36]. The partial algebra $\underset{\sim}{S}$ is *initial* in the variety V.

From now on $G = \langle N, X, \langle R_i : i\in I\rangle, B\rangle$ is a CS grammar where $B\in N$ is the so called start symbol (or sentence symbol, see Kain[22]) and B is such that if B occurs in a rule (R_i) then B occurs on the left side only and (R_i) is a CF-rule. Consider the universe S_B of the Parse-tree algebra $\underset{\sim}{S}$.

We claim that the following statements 1-3 are true.

1) $S_B \subseteq T$.

2) Moreover: $S_B = T\cap W_B$. I.e. S_B is the set of all parse-trees of G starting from the syntactic category B.

3) Further let $A \in N$. Then S_A consists of all subtrees with root A of parse-trees of G . (I.e.: S_A consists of those parse-trees of syntactic category A of G' which are "used" in some parse-tree of G . I.e.: S_A consists of subtrees of elements of S_B starting from A .)

Let $\underset{\sim}{S}'$ be the syntax algebra associated to the CF grammar G' in §3.A.. Clearly $\underset{\sim}{S}' \in Alg_g \subseteq V$. Since $\underset{\sim}{S}$ is initial in V , there is a unique homomorphism $d : \underset{\sim}{S} \to \underset{\sim}{S}'$. Let $\underset{\sim}{S}'' \subseteq \underset{\sim}{S}'$ be the image of $\underset{\sim}{S}$ along d . I.e.: $d : \underset{\sim}{S} \twoheadrightarrow \underset{\sim}{S}''$ is a surjection. Then $S''_B = \{w \in X^* : B \vdash^{*}_{G} w\}$. Further let $A \in N$ be arbitrary. Then S''_A is the set of those expressions u which occur in some expression $w \in S''_B$ as "subexpressions of syntactic category A" . I.e.: $S''_A = \{u \in X^* : (\exists w \in S''_B)$ u occurs in w as a "subexpression of syntactic category A"$\}$.

Now we feel that the above algebras $\underset{\sim}{S}$ and $\underset{\sim}{S}''$ are in harmony with our intuition about the syntactic structure of the language defined by G . E.g.: Let $\underset{\sim}{U} \in V$ be arbitrary. Then there is a unique homomorphism $h : \underset{\sim}{S} \to \underset{\sim}{U}$, and again, if $w \in S''$ with parse-tree $u \in S$ (i.e. $d(u)=w$) then $h(u)$ can be considered to be the meaning of the expression w parsed as u . Here again Frege's Principle of Compositionality is satisfied.

Then we can proceed along the lines of sections 3,5,6, but this time we have to use Partial Algebra Theory. Partial Algebra Theory does exist, see e.g. Burmeister[12] Németi[31], Andréka-Burmeister-Németi [4], the references above and the ones at the end of §2.

A worked example for the CS *case*

Let $G = \langle N, X, R_i : i \in 9 \rangle$ be the following CS grammar. The terminal symbols are $X \stackrel{d}{=} \{a,b,c\}$. The nonterminal symbols are $N \stackrel{d}{=} \{K,M,B,C,H\}$. The derivation rules R_i together with the associated CF-rules R'_i and the associated (many-sorted) similarity type g are:

$R_0 : K \vdash aM$	$R'_0 : K \vdash aM$	$g(0)=\langle M,K \rangle$
$R_1 : K \vdash aKM$	$R'_1 : K \vdash aKM$	$g(1)=\langle K,M,K \rangle$
$R_2 : M \vdash BC$	$R'_2 : M \vdash BC$	$g(2)=\langle B,C,M \rangle$
$R_3 : CB \vdash CH$	$R'_3 : B \vdash H$	$g(3)=\langle H,B \rangle$
$R_4 : CH \vdash BH$	$R'_4 : C \vdash B$	$g(4)=\langle B,C \rangle$
$R_5 : BH \vdash BC$	$R'_5 : H \vdash C$	$g(5)=\langle C,H \rangle$
$R_6 : aB \vdash ab$	$R'_6 : B \vdash b$	$g(6)=\langle B \rangle$
$R_7 : bB \vdash bb$	$R'_7 : B \vdash b$	$g(7)=\langle B \rangle$
$R_8 : C \vdash c$	$R'_8 : C \vdash c$	$g(8)=\langle C \rangle$

I.e. 0 is a unary function symbol from elements of sort M to elements of sort K ; 6,7 are constant symbols of sort B ; 8 is a constant symbol of sort C , etc.. Note that G generates $\{a^n b^n c^n : n \in \omega\} \subseteq X^*$.

Let $\rho \stackrel{d}{=} 2(6,47)$, $\delta \stackrel{d}{=} 2(358,8)$. Then ρ and δ are terms of sort M , i.e. $\{\rho,\delta\} \subseteq W_M$. Consider the term $\tau \stackrel{d}{=} 1(0\rho,\delta) \in W_K$. Then τ corresponds to the parse-tree of aabbcc illustrated on the figure. Since τ is not only a G′ -derivation but also a G -derivation, we have $\tau \in T$. Therefore $\{\rho,\delta\} \subseteq S_M \sim T$ since these are subterms of $\tau \in T$ but themselves are not G - derivations.

Consider the term $0\delta \in W_K$ corresponding to the "pseudo-derivation" $K \stackrel{\rho}{\vdash} aM \stackrel{\delta}{\vdash} acc$. Now $0\delta \notin S$ since there is no member of the language defined by G of which acc would be a subterm. Then $\delta \in S$, $0\delta \notin S$ show that $\underset{\sim}{S}$ is a partial algebra only since the operation F_0 is not defined on $\delta \in S_M$.

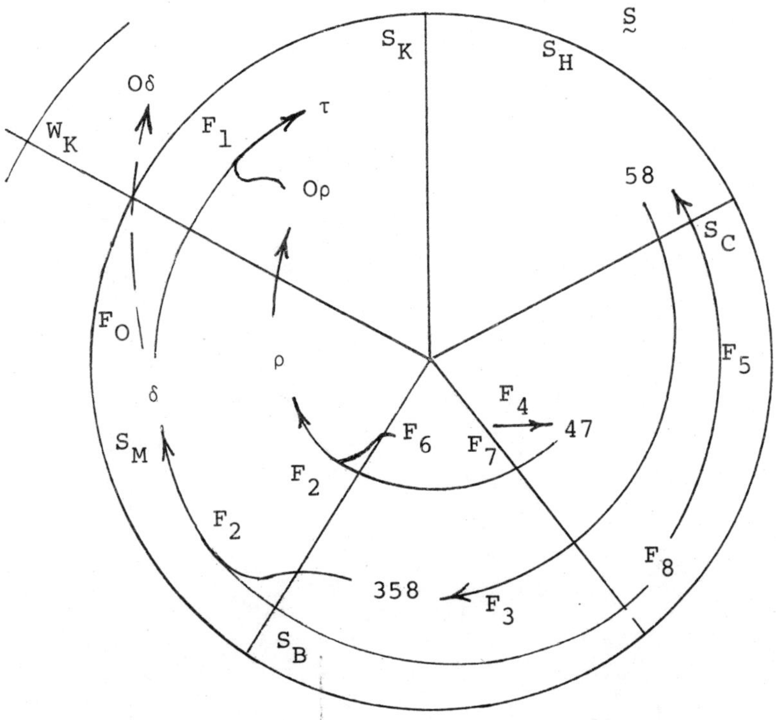

The algebra $\underset{\sim}{S}$ is a partial algebra only.

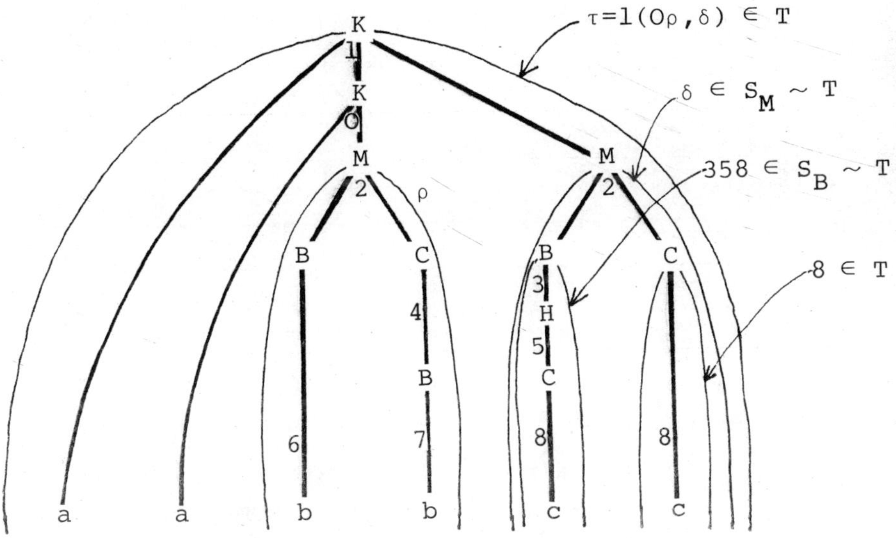

Terms (elements of S) are parse-trees.

§9. Algebras corresponding to not very classical logics are Crs_α *-s.*

Let $g_1 : \{\wedge, \neg, \exists_i, =_{ij} : i,j\in\omega\} \to \omega$ be the similarity type defined in §4. In §4 g_1 was associated to the language L_1 of first order logic with ω variables. Let α be any ordinal. Let L_1^α be the first order language with α variables. Let $g_1^\alpha : \{\wedge, \neg, \exists_i, =_{ij} : i,j\in\alpha\} \to \omega$ be the similarity type associated to L_1^α as in §4. $Alg(g_1^\alpha)$ denotes the similarity class of all algebras of similarity type g_1^α .

Let $\underset{\sim}{A} = \langle A, F \rangle \in Alg(g_1^\alpha)$. Then $\underset{\sim}{A}$ will be denoted as $\underset{\sim}{A} = \langle A, F(\wedge), F(\neg), F(\exists_i), F(=_{ij}) \rangle_{i,j\in\alpha}$ and more concisely we write $\underset{\sim}{A} = \langle A, \cdot, -, c_i, d_{ij} \rangle_{i,j\in\alpha}$. We also write d_{ij}^A and c_i^A for $F(=_{ij})$ and $F(\exists_i)$ resp. Let $\underset{\sim}{A}\in Alg(g_1^\alpha)$. For any $x,y\in A$ we define $x\leq y \iff x\cdot y=x$ in $\underset{\sim}{A}$. Assume $\underset{\sim}{A} \models -(x\cdot -y) = -(y\cdot -y)$. Then we define $1^A \overset{d}{=} -(x\cdot -x)$. We call 1^A the *unit* of $\underset{\sim}{A}$.

The variety $CA_\alpha \subseteq Alg(g_1^\alpha)$ was defined in Henkin-Monk-Tarski [20]. If $\underset{\sim}{A}\in CA_\alpha$ then $\langle A, \cdot, - \rangle$ is a Boolean algebra. CA_0 is the variety of Boolean algebras.

DEFINITION 4 (see the figures at the end of §9)
Let $\underset{\sim}{A}\in Alg(g_1^\alpha)$. We define $\underset{\sim}{A}$ to be a Crs_α iff conditions (i)-(iii) below hold.

(i) $\langle A, \cdot, - \rangle$ is a Boolean set algebra and $\cup A \subseteq {}^\alpha Sets$

(ii) $c_i x = \{q\in\cup A : \alpha\sim\{i\} \upharpoonright q \subseteq p$ for some $p\in x\}$ for all $x\in A$ and $i\in\alpha$.

(iii) $d_{ij} = \{q\in\cup A : q_i=q_j\}$ for all $i,j\in\alpha$.

Crs_α denotes the class of all Crs_α -s.
A Crs_α $\underset{\sim}{A}$ is said to be *full* if $A = Sb(\cup A)$. □

FACT

Let $\underset{\sim}{A} \in Crs_\alpha$. Then $A \subseteq Sb(\cup A)$ and $x \cdot y = x \cap y$ and $-x = \cup A \sim x$ for all $x,y \in A$ in $\underset{\sim}{A}$. 1^A exists and $1^A = \cup A$.

The following result is due to I.Németi.

THEOREM 1

Let $|\alpha|>1$. Then $\mathbf{HSP}\ Crs_\alpha = \mathbf{I}\ Crs_\alpha$.

A direct purely algebraic proof is available from the authors, see also Thm.24. in [32]. □

DEFINITION 5 (see the figures at the end)

$$Gs_\alpha \overset{d}{=} \{\underset{\sim}{A} \in Crs_\alpha : \exists B(\cup A = \cup\{{}^\alpha U : U \in B\} \text{ and } (\forall S \subseteq B)[|S|>1 \Rightarrow \cap S = 0])\} .$$ □

Gs_α is proved to be the algebraic counterpart of L_1^α in Németi [32].

To see that the algebras corresponding to not very classical logics are Crs_α -s, let us take e.g. *two sorted* classical logic. Let $\underset{\sim}{A} \in Crs_{\omega+\omega}$. Then $\underset{\sim}{A}$ corresponds to two sorted logic iff there are sets Y_0, Y_1, U such that $U = Y_0 \cup Y_1$, $Y_0 \cap Y_1 = 0$, and

$$\cup A = \{f \in {}^{\omega+\omega}U : (\forall i<\omega) f_i \in Y_0 \text{ and } (\forall i \geq \omega) f_i \in Y_1\} .$$

Let β be an ordinal and consider β -sorted logic. Let $\alpha \overset{d}{=} \omega \cdot \beta$. (Recall from set theory that $\omega \cdot 1 = \omega$ and $\omega \cdot (\gamma+1) = (\omega \cdot \gamma)+\omega$.) Let $\underset{\sim}{A} \in Crs_\alpha$ with unit V . Then $\underset{\sim}{A}$ corresponds to β -sorted logic iff there is

$\langle Y_i : i<\beta \rangle$ and $U = \cup_{i<\beta} Y_i$ such that $(\forall i<j<\beta)\ Y_i \cap Y_j = 0$ and $V = \{f \in {}^{\alpha}U : (\forall i<\beta)(\forall n \in \omega) f((\omega \cdot i)+n) \in Y_i\}$. Clearly if $\underset{\sim}{A}$ is a Crs_α corresponding to β-sorted logic and $\beta>1$ then $\underset{\sim}{A} \notin \mathbf{I}\, Gs_\alpha$.

Consider next *monadic* α-sorted logic. Monadic means that each sort has exactly one variable. E.g. L_1^1 is monadic 1-sorted logic. The class of meaning algebras corresponding to monadic α-sorted logic is exactly Crs_α^{mon} defined below.

DEFINITION 6 Let $\underset{\sim}{A} \in Crs_\alpha$. We say that $\underset{\sim}{A}$ is *generalized monadic* iff $\cup A = \cup_{i \in I} (P_{j \in \alpha} U_{ij})$ for some system $\langle U_r : r \in I \times \alpha \rangle$ such that $(\forall j \in \alpha)(\forall i,m \in I)[i \neq m \Rightarrow U_{ij} \cap U_{mj} = 0]$. $Crs_\alpha^{mon} \overset{d}{=} \{\underset{\sim}{A} \in Crs_\alpha : \underset{\sim}{A}$ is generalized monadic$\}$. □

$Crs_\alpha \not\subseteq \mathbf{I}\, Crs_\alpha^{mon} \not\subseteq CA_\alpha$ since $Crs_\alpha^{mon} \models c_i c_j c_i x = c_j c_i x$ and $Crs_\alpha^{mon} \not\models c_0 d_{01} = 1$. Note that $Gs_\alpha \subseteq Crs_\alpha^{mon} \subseteq Crs_\alpha$.

THEOREM 2 Let $\alpha>1$. Then $\mathbf{I}\, Crs_\alpha^{mon} = \mathbf{HSP}\ Crs_\alpha^{mon}$.

The proof is roughly the same as that of Thm.24 in [32]. □

These Crs_α-s arise from many-sorted logics by translating (i.e. reducing) many-sorted logic to classical one-sorted logic. Keeping the above examples in mind, one can imagine that other Crs_α-s will arise from reducing other non-classical logics to classical logic. Reducing non-classical logics to classical ones is a standard and very fruitfully applied method in the theory of non-classical logics.

Example for a Crs_2*:*

$U \overset{d}{=} \{0,1,2\}$. $V \subseteq {}^2U$ is defined as $V \overset{d}{=} \{\langle 0,1\rangle,\langle 1,2\rangle\}$.

V =

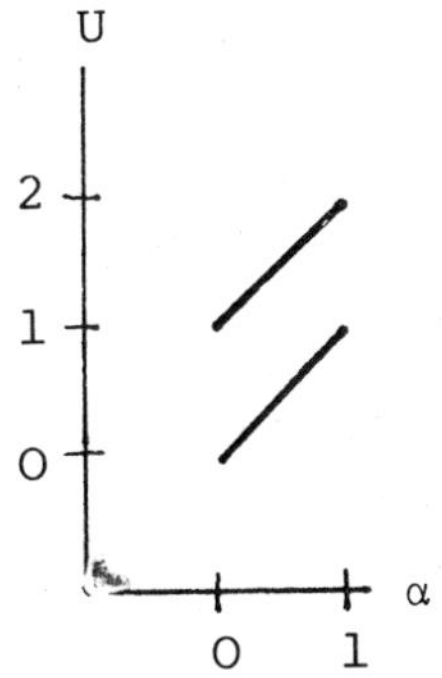

I.e. V =

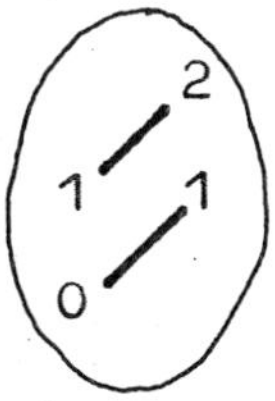

Let $\underset{\sim}{A}$ be the full Crs_2 with unit V . Then $\underset{\sim}{A}$ =

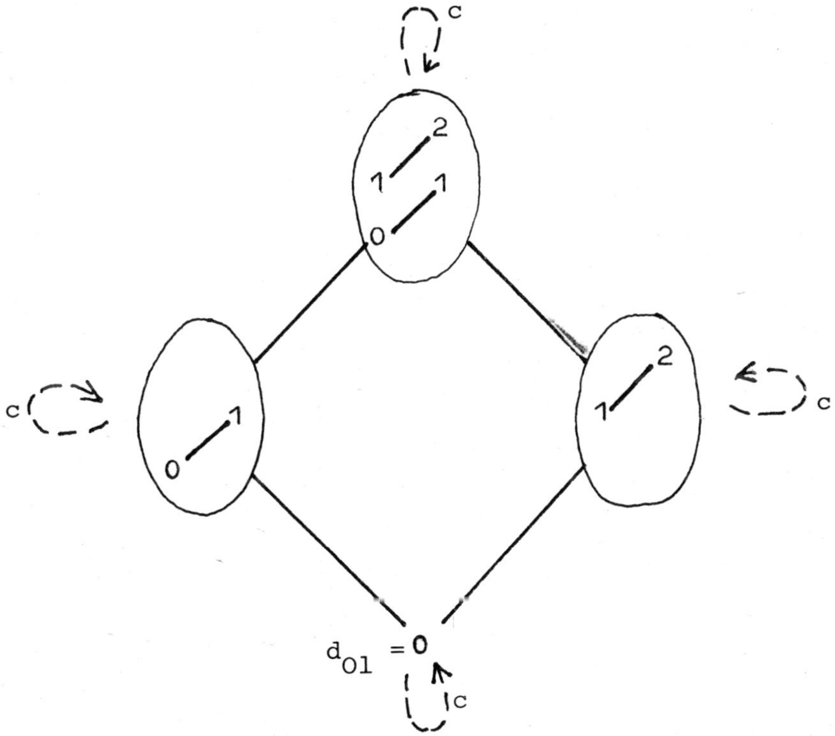

Clearly $\underset{\sim}{A} \models (\forall x)c_1x=x$ and $\underset{\sim}{A} \models d_{01}=0$. Clearly $\underset{\sim}{A} \notin Gs_2$.

As a contrast we show a Gs_2*:*

Let $U = \{0,1,2,3\}$ and $V = {}^2\{0,1\} \cup {}^2\{2,3\}$. Let $\underset{\sim}{A}$ be the minimal Crs_2 with unit V. Then $\underset{\sim}{A} \in Gs_2$.

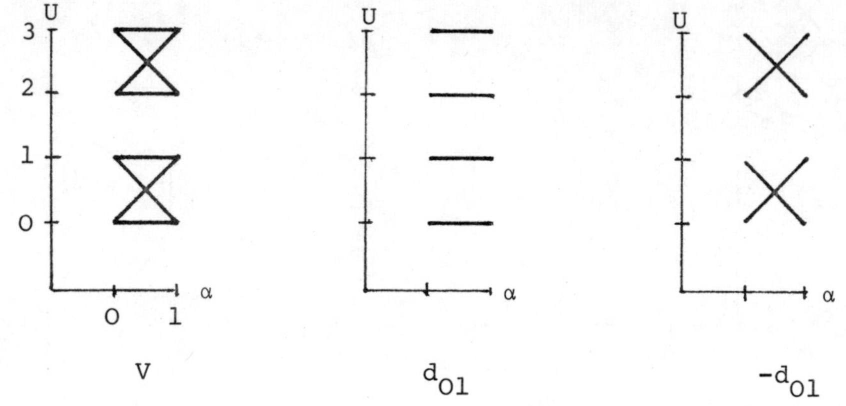

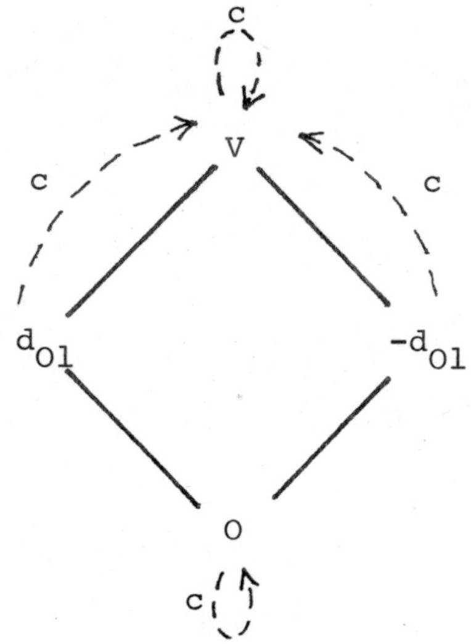

Another Gs_2*:* $U \stackrel{d}{=} \{0,1,2\}$. $V \stackrel{d}{=} {}^2\{0,1\} \cup {}^2\{2\}$. Let $\underset{\sim}{A}$ be the minimal Crs_2 with unit V. Then $\underset{\sim}{A} \in Gs_2$. See the figure on the next page!

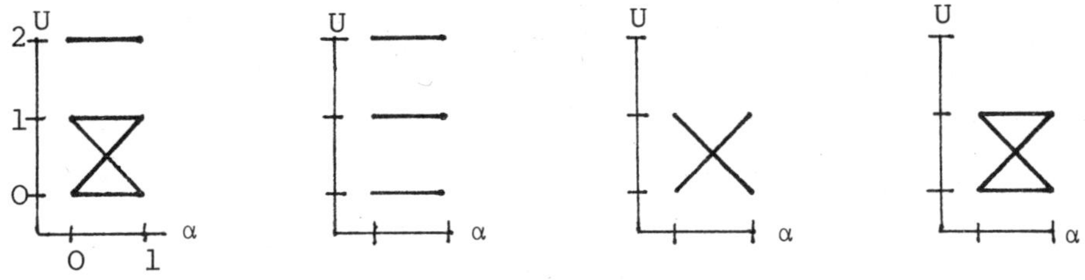

V $\qquad$ d_{01} $\qquad$ $-d_{01}$ $\qquad$ c_0-d_{01}

$\underset{\sim}{A}$ =

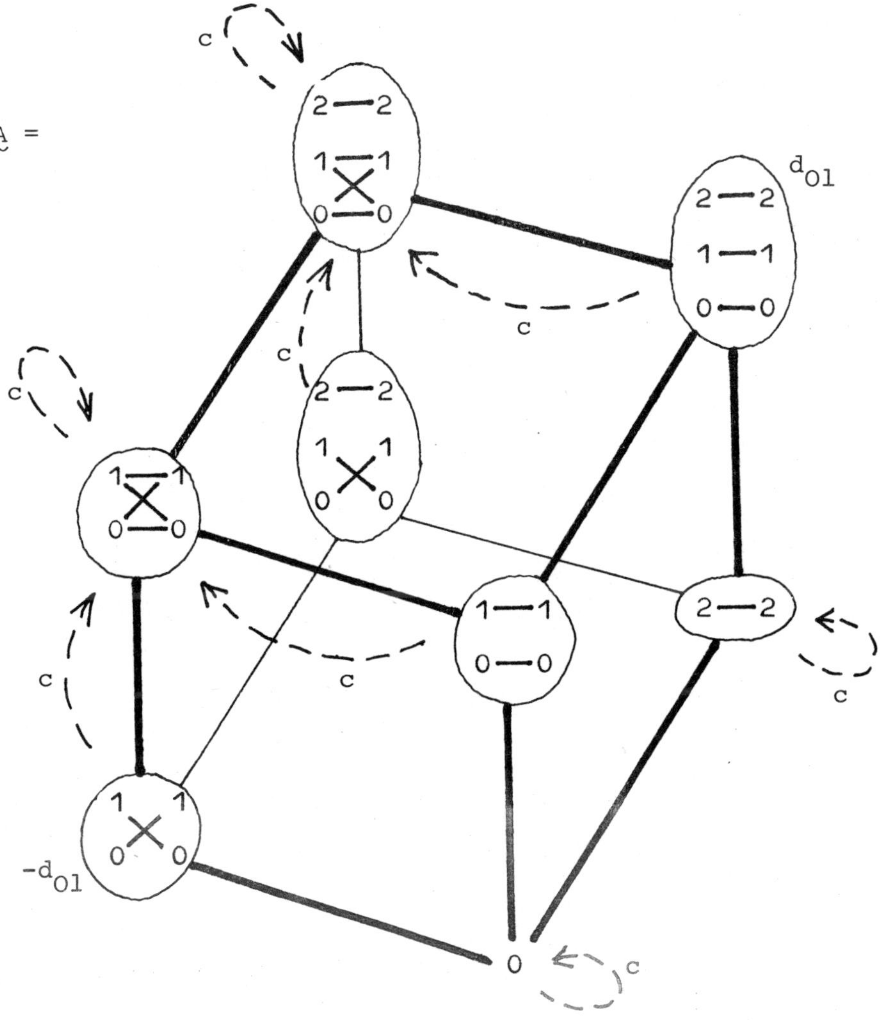

REFERENCES

[1] ADJ(Goguen,J.A. Thatcher,J.W. Wagner,E.G. Wright,J.) : Initial algebra semantics and continuous algebras. *Journal of ACM* vol.24, No.1, 1977, pp.68-95.

[2] Andréka,H.: Algebraic logic. Dissertation,Eötvös L. Univ., Budapest, 1972.

[3] Andréka,H.: Universal algebraic logic. Dissertation, Hung.Acad.Sci. Budapest, 1975.

[4] Andréka,H. Burmeister,P. Németi,I.: On quasivarieties of partial algebras. Technische Hochschule Darmstadt Preprint No.557, 1980.

[5] Andréka,H. Gergely,T. Németi,I.: Purely algebraic construction of logics. Logic Semester'73, Banach Center Warsaw, 1973.

[6] Andréka,H. Gergely,T. Németi,I.: On universal algebraic construction of logics. *Studia Logica* vol.36, No.1-2, 1977, pp.9-47.

[7] Andréka,H. Németi,I.: Simple purely algebraic proof of completeness of logics. *Algebra Universalis* vol. 5, 1975, pp.8-15.

[8] Andréka,H. Németi,I.: Generalization of variety and quasivariety concept to partial algebras through category theory. *Dissertationes Mathematicae*, No.204.

[9] Andréka,H. Németi,I.: Varieties definable by schemes of equations. *Algebra Universalis* vol.11, No.1.

[10] Andréka,H. Németi,I. Sain,I.: Henkin-type semantics for program schemes. In: Budach,L.(ed.) *Fundamentals of Computation Theory FCT'79*, Akademie Verlag, Berlin, 1979, pp.18-24.

[11] Andréka,H. Németi,I. Sain,I.: A complete logic of programs. *Theoretical Computer Science*, to appear.

[12] Burmeister,P.: A survey of partial algebra theory. Technische Hochschule Darmstadt, Preprint,1980.

[13] Burstall,R.M. Goguen,J.A.: The semantics of CLEAR,

a specification language. In: Bjørner,D.(ed.) *Abstract Software Specifications*, Lecture Notes in Comp.Sci. vol.86, Springer Verlag, 1980, pp.292-332.

[14] Chang,C.C. Keisler,H.J.: *Model theory*. North-Holland, 1973.

[15] Dahn,B.I.: Prädikatenkalküle der ersten Stufe für Kripkemodelle. Dissertation(B), Humbold Univ., Sect. Math.,Berlin, 1979.

[16] Daigneault, A.: Lawvere's elementary theories and cylindric algebras.*Fundamenta Mathematica* vol.66, 1970, pp.307-328.

[17] van Emde Boas,P. Janssen,T.M.V.: The impact of Frege's principle for semantics of programming. Univ.of Amsterdam, Dept.Math., Report 79-0 7, 1979.

[18] Ginsburg,S.: *The mathematical theory of context-free languages*. McGraw-Hill, New York, 1962.

[19] Goguen,J.A.: Some ideas in algebraic semantics. In: *Proc.3rd IBM Symp. on Math.Found. of Computer Sci.* Kobe, Japan 1978.

[20] Henkin,L. Monk,J.D. Tarski,A.: *Cylindric algebras, Part I*. North - Holland, 1971.

[21] Henkin,L. Monk,J.D. Tarski,A.: Cylindric setalgebras and related structures. Univ.Colorado, Boulder, Preprint, 1979.

[22] Kain,R.Y.: *Automata theory: machines and languages*. McGraw-Hill, New York, 1972.

[23] Kaphengst,H. Reichel,H.: Initial algebra semantics for CS languages. In: Karpiński,M.(ed.) *Fundamentals of Computation Theory*, Lecture Notes in Comp.Sci. vol.56, Springer Verlag, 1977, pp.120-126.

[24] Lindström,P.: On characterizing elementary logic. In: Stenlund,S.(ed.) *Logical theory and semantic analysis*, D.Reidel Publ.Co., Dordrecht-Holland,1974, pp.129-146.

[25] Lugowski,H.: *Grundzüge der Universellen Algebra.* Teubner-Texte z. Mathematik, Leipzig, 1976.

[26] Makowsky,J.A.: Topological model theory. In: *Model theory and applications*, Proc.C.I.M.E., Edizioni Cremonese, Roma, 1975, pp.122-150.

[27] Manna,Z.: *Mathematical theory of computation.* McGraw-Hill, 1974.

[28] Matthiessen,G.: Theorie der heterogenen Algebren. Univ. Bremen, Mathematik-Arbeitspapiere No.3,1976.

[29] Monk,J.D.: *Mathematical Logic.* Springer Verlag,1976.

[30] Montague,R.: Universal grammar. *Theoria* vol.36,1970. pp.373-398.

[31] Németi,I.: From hereditary classes to varieties in abstract model theory and partial algebra. *Beiträge zur Algebra und Geometrie* vol.7, 1978, pp.69-78.

[32] Németi,I.: Connections between cylindric algebras and initial algebra semantics of CF languages, this volume.

[33] Németi,I.: Every free dynamic algebra is separable and representable. *Theoretical Computer Science*, to appear.

[34] Németi,I.: Some constructions of cylindric algebra theory applied to dynamic algebras of programs. *CL&CL*, Budapest. vol.14.

[35] Németi,I. Sain,I.: Cone injective subcategories and Birkhoff-type theorems. In: Csákány,B. Fried,E. Schmidt,E.T.(eds.) *Universal Algebra.* North-Holland, to appear.

[36] Pasztor,A.: Faktorisierungssysteme in der Kategorie der partiellen Algebren. HochschulSammlung Naturwissenschaft Math.Band 1, HochschulVerlag,Freiburg,1979.

[37] Pratt,V.R.: Models of program logics. In: *20th Ann. Symp.IEEE Found. of Computer Sci.*, San Juan, 1979.

[38] Pratt,V.R.: Dynamic algebras. In: *12th ACM Symp. on Theory of Computing*, Los Angeles, 1980.

[39] Rasiowa,H.: *An algebraic approach to nonclassical logics*. North-Holland, 1974.

[40] Sain,I.: There are general rules for specifying semantics: considerations on abstract model theory. *CL&CL*, Budapest, vol.13, 1979, pp.195-250.

Andréka,H., Sain,I.
Mathematical Institute of the
Hungarian Academy of Sciences
Budapest, Reáltanoda u.13-15.
H-1053 Hungary

COLLOQUIA MATHEMATICA SOCIETATIS JÁNOS BOLYAI
26. Mathematical Logic in Computer Science
Salgótarján (HUNGARY), 1978

ABSTRACT SYNTAX AND THE CANONICAL FIXPOINT OF RECURSIVE CALLS[1]

M.A.Arbib, E.G.Manes

1. *Sum-of-Paths Semantics*

Pfn(D,D) - the partial functions $D \to D$ - has a *partially-additive* structure in that, while we cannot "sum" arbitrary families of such maps, we can for *disjoint* $(f_i : i \in I)$ define the partial function $(\Sigma f_i) : D \to D$ by

$$(\Sigma f_i)(d) = \begin{cases} f_j(D) & \text{if } f \in \mathrm{dom}(f_j) \\ \text{undefined} & \text{if no such } j \text{ exists.} \end{cases}$$

With this partial sum, we can express the semantics of familiar flow diagrams by a "sum-of-paths" - decomposing the diagram into paths which can be interpreted by a disjoint family of partial functions, and then taking the sum of these partial functions.

---

[1] / The present paper is an extended abstract of a paper "Abstract Theory of Recursive Calls" which is to appear in the journal literature. A preliminary version of the paper is available as COINS Technical Report 78-18. The full paper contains full proofs, further examples, and extended citation of the literature. The research was supported in part by the National Science Foundation under grant no. MCS 76-84477.

Example

Given $p:D\to\{true,false\}$, we may let $p_t:D\to D$ send d to d with domain $\{d:p(d)=true\}$, and similarly for p_f. Then the *conditional* "*if* p *then* f *else* $g=f\cdot p_t+g\cdot p_f$" a sum-of-paths as shown in 1

1

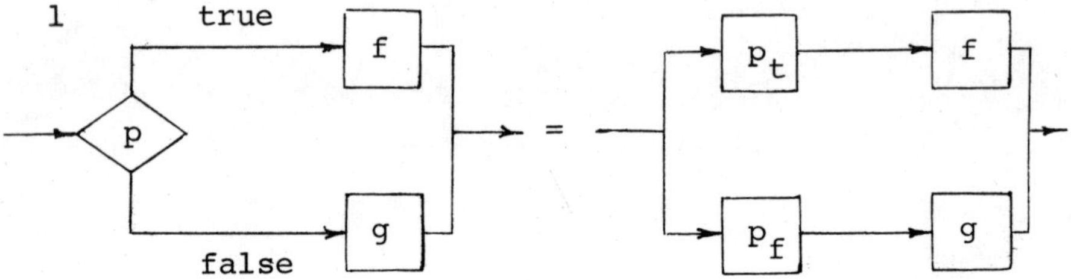

Example

Given $f:D\to D+D$, where $D+D=D\times\{1\}\cup D\times\{2\}$ is the joint union of two copies of D, set

$$f_j:D\to D,\quad d\mapsto\begin{cases} d & \text{if } f(d)=(d,j)\\ \text{undefined} & \text{if not.}\end{cases}$$

Then the iterate f^+ can be expressed as the denumerable sum-of-paths $f^+ = \sum_{n\geq 0} f_2\cdot f_1^n$, with one term for each distinct number $n \geq 0$ of loop traversals in 2.

2

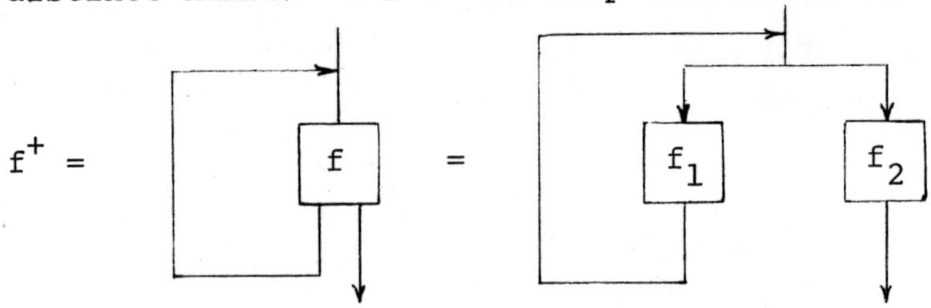

2. Abstract Recursion Schemes

We may regard f^+ as providing the semantics of

the recursive call of 3.

3

$h(a) =$ 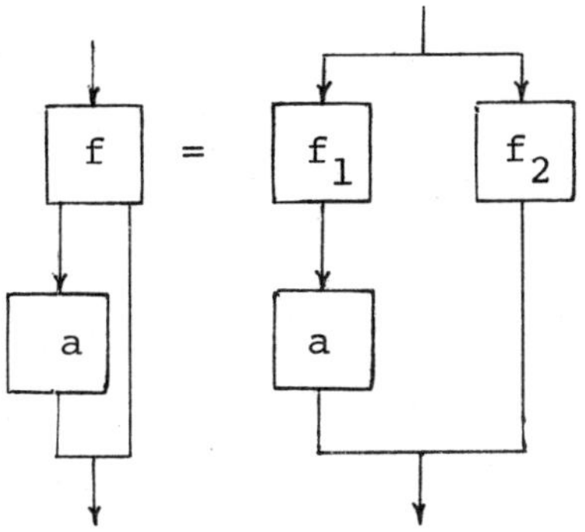

$= f_2 + a \cdot f_1$

If we call a path an n-*substitution path* if it contains n occurences of a variable, then

$H_o = f_2$ is the value of the 0-substitution paths; while

$H_1(a) = a \cdot f_1$ is the value of the 1-substitution paths when the single variable is assigned the value a.

This motivates the general concept of an *abstract recursion scheme* (A,Σ,H). In such a scheme:

(i) (A,Σ) is a partially-additive monoid (we give the formal definition in Section 5 - here it will suffice to think of A as $Pfn(D,D)^n$ for some $n \geq 1$) with Σ as before but on each coordinate.

(ii) H is a collection of maps $H_n : A^n \to A$ $(n \geq 0)$, where each $n \geq 0$ is n-*additive* in the sense that if $\alpha_j = \Sigma(a_i : i \in I_j)$ is defined for each j, $1 \leq j \leq n$, then $\Sigma(H_n(a_{i_1},\ldots,a_{i_n}) : i_j \in I_j,$ $1 \leq j \leq n)$ is defined, and equals $H_n(\alpha_1,\ldots,\alpha_n)$.

(iii) $h(a) = \Sigma_n H_n(a,\ldots,a)$ is defined for each a.

We say an element a of A is a *fixpoint* of (A,Σ,H) if $h(a) = a$.

Example

The recursive definition of X and Y

X = 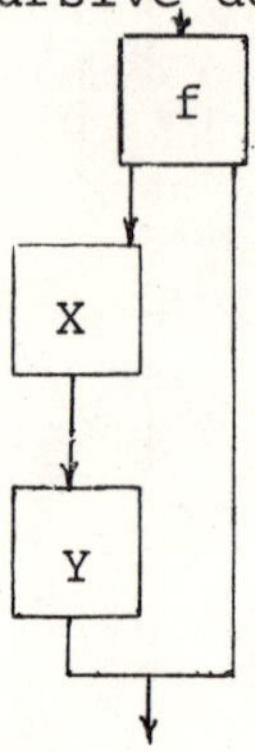

Y = 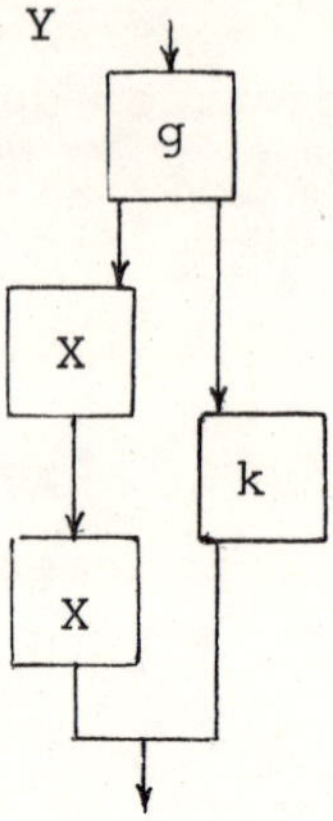

can be written as

$$X = Y \cdot X \cdot f_1 + f_2 \ ; \qquad Y = X \cdot X \cdot g_1 + k \cdot g_2 \ .$$

This defines the abstract recursion scheme (A, Σ, H) with

$A = Pfn(D,D) \times Pfn(D,D)$

Σ componentwise partial addition

and $H_o = (f_2, k \cdot g_2)$

$H_1(a_1, a_2) = (0,0)$

$H_2((a_1,a_2),(a_1',a_2')) = (a_2' \cdot a_1 \cdot f_1, a_1' \cdot a_1 \cdot g_1)$

$H_n \equiv 0$ for $n \geqq 3$.

The recursive definition is then equivalent to the fixpoint equation $(X,Y) = h(X,Y)$.

3. The Abstract Syntax and Semantic Expansion

We do *not* use a syntax that generates a path expression for every path in every flow diagram obtained by repeated substitution. Rather, we use an abstract syntax with one n-ary symbol ω_n for each $n \geqq 0$. ω_n denotes "substitution into, and subsequent evaluation of, all n-substitution paths".

The *abstract syntax* e (where, for reason to become clear below, e is short for *expansion*) comprises the set of trees defined inductively by $\omega_o \in e$.

If $t_1,\ldots,t_n \in e$ then

$\omega_n[t_1,\ldots,t_n] =$ 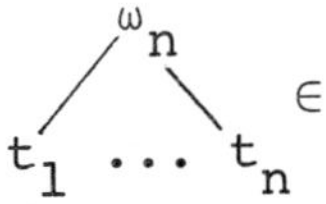 $\in e$.

Given (A,Σ,H) and $s \in e$, the *interpretation* s^h of s is the element of A obtained by "running" H on s:

$$(\omega_o)^h = H_o$$

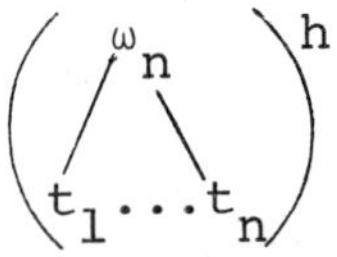 $= H_n(t_1^h,\ldots,t_n^h)$

Since all trees s containing an ω_n for n with $H_n=0$ evaluate to 0 by n-additivity, we may ignore them.

Recall from 3 that for iteration we have $H_o=f_2$, $H_1(a) = a\cdot f_1$ and $H_n=0$ for $n \geq 2$. Thus, in this case we need only consider trees

S_n 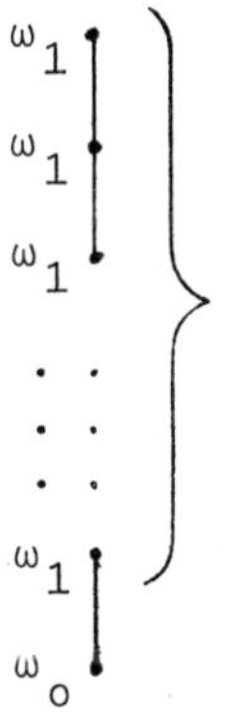 $n \geq 0$ occurrences

Here $s_n^h = f_2 f_1^n$ and the "series expansion"

$$\Sigma(s^h : s\in e) = \Sigma(s_n^h : n \geq 0) = \sum_{n\geq 0} f_2 f_1^n$$

is the desired semantics f^+.

This motivates the main theorem of this paper the proof is outlined in Sections 6 through 8.

THEOREM

Let e *be the abstract syntax as defined above. Let* (A,Σ,H) *be any abstract recursion scheme. For each* s *in* e, *let* s^h *be the element of* A *obtained by running* H *on* s. *Then*

(1) *The expansion* $e^h = \Sigma(s^h : s \in e)$ *is well-defined in* (A,Σ); *and*

(2) e^h *is a fixpoint of* (A,Σ,H): $h(e^h) = e^h$ *and thus the expansion* e^h *provides a semantics for the abstract recursion scheme.*

4. *Correctness Proofs*

We appear to have solved a problem stated in Backus' Turing Lecture:

> The question of the existence of simple expansions that "solve" "quadratic" and higher order equations remains open.

As an axample of how algebraic manipulation may be used to simplify such an expansion (further examples appear in the full paper), we prove that

f(x) := *if* p(x) *then* f(f(g(x))) equals *while* p *do* g.

In sum-of-paths terms, the recursive definition takes the form $f = p_f + f^2 g p_t$ so that we have

$$H_o = p_f \ ; \ H_2(x,y) = yxgp_t .$$

We thus look only at binary trees, where we have

$$\omega_o^h = p_f \ . \ \omega_2[s_1,s_2]^h = s_2^h s_1^h g p_t .$$

Thus

$\omega_2[s_1,s_2]^h = (\ldots p_t)(p_f\ldots)gp_t = 0$ unless $s_2 = \omega_o$, and so we may restrict attention to trees of the form

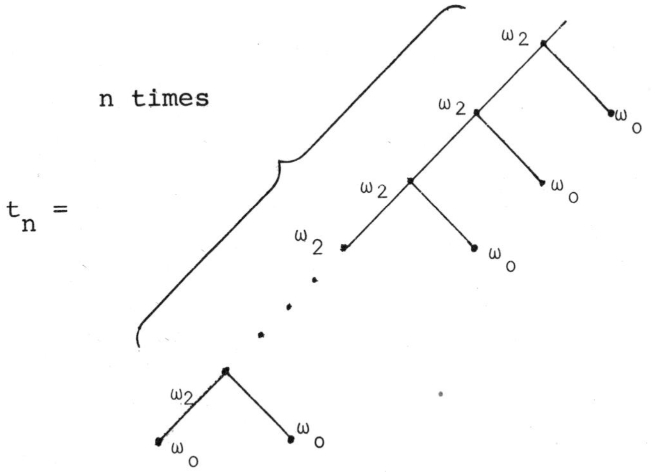

By induction, we see that $t_h^n = p_f(gp_t)^n$: Clearly $t_h^o = \omega_o^h = p_f$. Then

$$t_h^{n+1} = \omega_o^h t_n^h gp_t$$
$$= p_f p_f (gp_t)^n gp_t \cdot$$
$$= p_f(gp_t)^{n+1} \quad \text{since } p_f p_f = p_f.$$

Thus $e^h = \sum_{n\geq 0} t_n^h = \sum_{n\geq 0} p_f(gp_t)^n$

which we recognize as the semantics of the iteration show in 4, i.e. of *while* p *do* g.

4

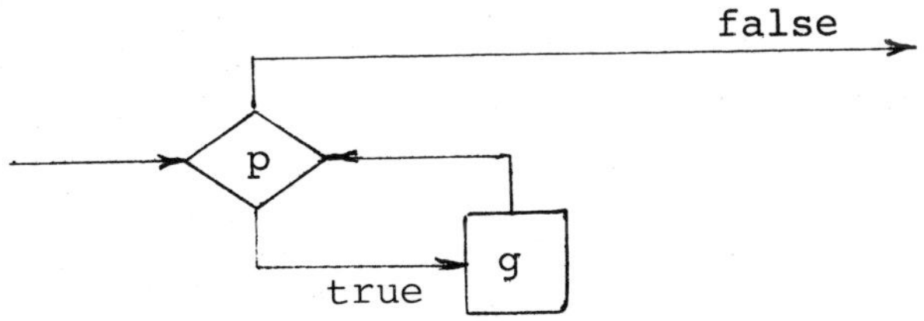

We state three principles for proving properties of programs whose semantics lie in Pfn(D,D). The proofs of these principles are all straightforward.

Tree Induction Rule (Partial Correctness): If for every n with $H_n \neq 0$

$$s_i^h \sqsubseteq g \quad \text{for} \quad 1 \leq i \leq n \Rightarrow H_n(s_1^h,\ldots,s_n^h) \sqsubseteq g$$

then $e^h \sqsubseteq g$ where $\sqsubseteq$ is the usual ordering of $Pfn(D,D)$ as subset of $D \times D$).

Disjointness Lemma: For x in D, $s^h(x)$ is defined for at most one s in e.

Termination Lemma: If for each x in D there exists s in e with $s^h(x)$ defined, then f is total.

5. *Partially-Additive Monoids*

A *partially-additive monoid* is a pair (A,Σ) where A is a set and Σ is a partial operation on countable (i.e. finite or denumerable) sequences in A subject to the following axioms:

Partition-associativity axiom: If the countable set I is partitioned into $(I_j : j \in J)$ then for each family $(a_i : i \in I)$ in A,

$$\Sigma(a_i : i \in I) = \Sigma(\Sigma(a_i : i \in I_j) : j \in J)$$

in the sense that the left side is defined iff the right side is defined and then the values are equal.

Limit axiom: If $(a_i : i \in I)$ is a countable family in A and if $\Sigma(a_i : i \in F)$ is defined for every finite subset F of I, then $\Sigma(a_i : i \in I)$ is defined.

Unary sum axiom: For one-element families, Σa is defined, and $\Sigma a = a$.

In the partition-associativity axiom, I_j may be empty. Since the unary-sum axiom ensures that some sums exist, it follows that the empty sum is defined and provides an additive zero which we denote 0 or 1.

Pfn(A,B) satisfies this with Σ as disjoint union and provides a setting for deterministic program semantics.

Rel(A,B) satisfies this with Σ as arbitrary union and provides a setting for nondeterministic program semantics.

A brief note on the relationship with order--semantics: An ω-cpo (ω-complete partially-ordered set) is a partially-additive monoid with $\Sigma S = \sup(S)$ defined only if S is an ω-chain or finite ordered sequence. On the other hand, given (A,Σ) we may define $a \leq a' \iff a'=a+b$ for some b. This is reflexive and transitive but may not be antisymmetric. There is an upper bound for each ω-chain but it may not be least.

6. *The Summability Theorem*

The summability theorem is the first half of our main theorem. It states that for each abstract recursion scheme (A,Σ,H) the infinite sum

$$e^h = \Sigma(s^h : s \in e)$$

is well-defined as an element of A.

Proof Idea:

Set $S_o=\{\omega_o\}$, $S_{k+1}=\{\omega_n[t_1,\ldots,t_n]:n\geq 0,\ t_i \in S_k\}$. Check by induction that each finite sum

$$T_k = \Sigma(s^h : s \in S_k)$$

is well-defined. Then note that

$$S_o \subset S_1 \subset \ldots \quad \text{with} \quad e = \bigcup_{k \geq 0} S_k$$

and use the limit axiom.

7. *The Canonical Fixpoint*

We now introduce our new concept of canonical fixpoint in a more general setting than that of abstract recursion schemes, i.e. one in which we have any collection A of structures and a collection of homomorphisms, closed under composition.

An *A-recursion scheme* is then a pair (A,h) where $h:A \to A$ need not be a homomorphism. $a \in A$ is a *fixpoint* of (A,h) if $h(a) = a$.

A *scheme homomorphism* $\Phi:(A,h) \to (A',h')$ is a homomorphism $\Phi:A \to A'$ with $h'\Phi = \Phi h$. Such a Φ preserves fixpoints:

$$ha = a \Rightarrow h'\Phi a = \Phi ha = \Phi a.$$

We say (A_o,h_o) is *initial* if for each other A-recursion scheme (A,h) there exists exactly one scheme homomorphism $(A_o,h_o) \to (A,h)$. We denote this by ψ_h.

We say that an assignment $(A,h) \mapsto \alpha_h$ of a fixpoint to each A-recursion scheme is a *canonical fixpoint* if it respects every homomorphis $\Phi : (A,h) \to (A',h')$:

$$\Phi a_h = a_{h'} \ .$$

THEOREM

If there is an initial (A_o,h_o) *with unique fixed point* $a_o = h_o(a_o)$ *then A-recursion schemes have a unique canonical fixpoint given by*

$$(A,h) \mapsto \psi_h(a_o).$$

Proof outline

$\alpha_h = \psi_h \alpha_{h_o} = \psi_h(a_o)$ shows uniqueness; while $\Phi\psi_h(a_o) = \psi_{h'}(\overset{o}{a}_o)$ shows that it is canonical.

Fact:

In the usual order semantics, the structures are ω-cpos, the homomorphisms are continuous maps (preserving suprema of ω-chains) which also preserve $\perp$: while the h of (A,h) is a continuous map which need not preserve $\perp$. In this case, the least fixpoint $\bigvee_{n\geq 0} h^h$ is canonical.

Proof.

Let N be the ω-cpo $N \cup \{\infty\}$ under $\leq$, and let

$$s: N \to N,\ s(n) = n+1\ ,\ s(\infty) = \infty.$$

Then (N,s) has unique fixpoint ∞, and is initial, with $\psi_h : (N,s) \to (A,h)$ uniquely fixed by its homomorphic properties:

$$\psi_h(0) = \perp$$
$$\psi_h(n) = \psi_h(s^n 0) = h^n \perp \qquad \text{for } 0 \leq n \leq \infty$$
$$\psi_h(\infty) = \psi_h(\bigvee_{n\geq 0} s^n 0) = \bigvee_{n\geq 0} (h^n \perp).$$

So the unique canonical fixpoint is

$$(A,h) \mapsto \psi_h(\infty) = \bigvee_{n\geq 0} (h^n \perp).$$

8. *The Canonical Fixpoint Theorem*

Returning now to our setting of abstract recursion schemes, we say that $\Phi : (A,\Sigma,H) \to (A',\Sigma',H')$ is a *scheme homomorphism* if it is:

(i) $H'_n(\Phi a_1,\ldots, a_n) = \Phi H_n(a_1,\ldots,a_n)$; and

(ii) if $\Sigma(a_i : i\in I)$ is defined then $\Sigma'(\Phi a_i : i\in I)$ is defined and is equal to $\Phi\Sigma(a_i : i\in I)$.

These conditions quarante that $\Phi h = h'\Phi$, so that such a Φ is indeed a scheme homomorphism in the sense of the previous section.

Define (A_o, Σ_o, H_o) by

A_o = all subsets of e

$\Sigma_o(S_i : i \in I)$ is defined iff the S_i are *disjoint* subset of e, and then equals $\cup(S_i : i \in I)$

$H_{on} : A_o^n \to A_o$, for $n \geq 0$, is given by $(S_1, \ldots, S_n) \mapsto \{\omega_n[t_1, \ldots, t_n] : \text{each } t_i \text{ in } S_i\}$.

It is then straightforward to prove the following two lemmas.

LEMMA

e *in* A_o *is the unique fixpoint of* (A_o, Σ_o, H_o).

LEMMA

(A_o, Σ_o, H_o) *is initial with*

$$\psi_h(s) = \Sigma(s^h : s \in S).$$

But then, by the theorem of the previous section we immediately have the following canonical fixpoint theorem, which completes and strengthens our "main theorem", stated in Section 3.

THEOREM

$$(A, \Sigma, H) \to e^h = \psi_h(e) = \Sigma(s^h : s \in e)$$

is the unique canonical fixpoint for abstract recursion schemes.

M.A.Arbib
Computer Information Science

E.G.Manes
Mathematics Department

University of Massachusetts
Amherst, Massachusetts 01003, U.S.A

COLLOQUIA MATHEMATICA SOCIETATIS JÁNOS BOLYAI
26. Mathematical Logic in Computer Science,
Salgótarján (HUNGARY), 1978.

AXIOMATIC DESCRIPTION OF MULTI-LEVEL PROGRAM SPECIFICATION

J. Aszalós

There are many efforts in formalizing specifications for multilevel program designs. In practice, nevertheless, it was found that for unskilled persons these methods are sometimes too hard to accept: therefore we attempt to introduce an easy way to produce and to read specifications. First, a short look to its philosophy.

Programming objects (algorithms, data, interfaces, hardware components, programmer, etc.) are sly fellows; they hardly let themselves to formalize for practical purposes. As my main interest is to understand our mental process in composing large programs and to improve the quality of the programs by making this process more explicit, more disciplinized and effective I tried to find the easiest way to induce a *mechanism of thinking* which is close to the common mental operations in practical problem solving and is still apt to formalize in the mathematical sense and is easy to support by a software package. This way of reasoning is outlined in this paper.

My aim is to suggest the chief-programmer an *attitude for the better understanding of the programming problems* and to suggest a way to specify them for the subordinate programmers.

This mental habit should be characterized by some *principles:*

1. The designer of a program *mustn't rely* exclusively on a set of ideas prepared by others. *He must "forget"* such concepts as "list", "cycle", "IFTHENELSE" (except perhaps some basic ones given in the first order logic), and he *must develop his own concepts natural to the problem.* During this process he will "invent" such things as "list", etc. again, but as the components of his own style of thinking. (*The rule of free concept-building*)

2. The problem should be understood in successive steps. The succeeding step presents the preceeding one using more elaborate, refined concepts. (Rule of platonic approach, or *rule of stepwise refinement*)

3. At all steps (level) the whole problem should be totally solved - except the resulting subproblems. (*Spinoza's rule*)

4. The difficulty of understanding the whole problem should be compared to "the sum" of difficulties in understanding it at a certain level and understanding the subproblems at the next level (and not the "product" of the difficulties emerging at the two successive levels). (Rule of the *"dijkstranian" approach*)

5. During the design process the programmer must be aware of his own possiblities, limitations and must regard his own activity too as a "programming object" (*Augustinian rule*, advocated by H. Mills)

6. The formalism used should be easy even for the unskilled programmer or mathematician. Only the fundaments of the mathematical logic are assumed to be well known. Long experience shows that it is hopeless to introduce a formalism of full mathematical rigor for the mass programmers, except at the begining of their carrier.

Following the rules above, specifications should be presented in hierarchically ordered levels. In each level, a problem is to be understood in terms of

- *objects of various types.* In this paper, the names of the types are terminated by a colon, followed perhaps by the names of objects belonging to this type, e.g. list: X,Y; func: length; ... (Operations and functions are regarded too as objects.)

- *assertions* about the objects above, or about the relations between them, e.g. is-emptylist (list), is-equallist (X,Y),... The assertions describe the possible states of the object at this level.

- *limitations;* the limitations, restrictions imposed by the task or by the way of solution are expressed in the first order logic. (These expressions may be regarded axioms.) E.g. $\neg(\exists$ car) (better (car, TRABANT)). The "well-formedness criteria", and the type-descriptions of the operations, functions belong to this group of expressions too.

- *realizations;* the objects, assertions and axioms of the given level are interpreted ("defined") in terms of objects, etc. of the next level. This "mapping" between the levels is expressed by the "realizations". E.g.

unary operation (function at the n. level)	$\Longleftrightarrow$	head, tail, delete (functions at the (n+1). level)

At each level we use two languages to describe the problem: an informal and a more formal one.

The example below presents an extract from a text-editor, in three levels, which has been elaborated for the ANSWER. (ANSWER is a large scale software system supporting the process of program-development, in SZÁMKI, Budapest.)

We use the following notations:

$\longrightarrow$ for mapping,

$\Longrightarrow$ for implication,

$\Rightarrow$, $\Leftrightarrow$, $\Leftarrow$ } for correspondence between two levels.

"lim." stands for "limitation".

A. LEVEL

Formal description:

Objects: textset

func: modify

Assertions: is-empty (textset)

Limitations: 1. $\neg$is-empty (textset)

2. modify: textset$\rightarrow$textset

Informal description:

We want to design a group of procedures operating on a set of text-pieces. The effects of these operations are similar to those of a text-editor, with the restrictions and limitations imposed by the system ANSWER: the text-set contains a built-in knowledge, which can not be deleted.

B. LEVEL

Formal description:

Objects: text: X,Y
func: unary operation,
binary operation

Assertions: is-emptytext (text)
is-equaltext (text, text)

Limitations:

1. $\exists$X (is text(X))
2. binary operation: text x text → text
3. unary operation: text → text
4. is-emptytext(X) ⟹ is-emptytext (unary op.(X))
5. is-emptytext(X) ⟹ is-equaltext (Y, binary op.(X,Y))
6. is-emptytext(Y) ⟹ is-equaltext (X, binary op.(X,Y))
7. is-emptytext (X) & is-emptytext (Y) ⟶ is-equaltext (X,Y)
8. ord (set (text))<∞

Realization:

A LEVEL		B. LEVEL
textset	⟺	set (text)
modify	⟺	binary op., unary op.
lim. 1	⟺	lim. 1
lim. 2	⟺	lim. 2, 3

Informal description

Two kinds of operations are to be developed ("binary" and "unary" operations), both working on objects of the type "text". At least one item of text must exist (lim. 1.), but the number of the text-items must not be unlimited (lim. 8.). Texts may be "empty".

Two texts compared to each other may be "equal".

The effects of the operations on empty text and equal texts are restricted for practical purposes (lim. 4-6.).

Empty texts are equals (lim. 7.).

C. LEVEL

Formal description:

Object: notnegint
charlist: X,Y
char
func: length,
head, tail,
concat, insert, delete

Assertions: is-head (X,Y)
is-tail (X,Y)
is-emptylist (charlist)
is-equallist (X,Y)

Limitations:
1. length: charlist → notnegint
2. is-emptylist(X) ⟹ charlist(X)
3.a is-emptylist(X) ⟹ length(X) = 0
3.b charlist(X) & length(X) = 0 ⟹ is-emptylist(X)
4. ¬is-emptylist(X) ⟹ length(X) > 0
5. head: charlist x notnegint → charlist
6. is-emptylist(X) & charlist(Y) ⟹ is-head(X,Y)
7. charlist(X) & notnegint(N) & N≤length (X) ⟹ charlist (head(X,N)) & is-head (head (X,N), X) & length (head(X,N)) = N
8. charlist(X) ⟹ is-head(X,X)
9. is-equallist(X,Y) & is-head(X,Y) ⟹ is-head(Y,X)

Informal description

The function "length" gives the number of elements contained in a charlist (lim. 1.). The "emptylist" is a charlist with length zero (lim. 2., 3.). The operation "head" displays the first N characters of a charlist as a new charlist (lim. 5., 7.). The emptylist is considered as the head of all (empty or not empty) lists (lim. 6.). Every charlist is the head of itself (lim. 8.). Equallists are the heads of one an other (lim. 9.).

Limitations: (continued for level C)

10. tail: charlist x notnegint → charlist
11. is-emptylist(X) & charlist(Y) ⟹ is-tail(X,Y)
12. charlist(X) & notnegint(N) & length(X) ≥N ⟹ charlist (tail(X,N)) & is-tail (tail (X,N), X) & length (tail(X,N)) = length (X)-N
13. charlist(X)⟹is-tail (X,X)
14. is-equallist(X,Y)⟹is-tail (X,Y) & is-tail (Y,X)

Informal description continued for level C:

The operation "tail" displays the last (length(X) -N) characters of the charlist X (in the same order) as a new list (lim. 10., 12.). The empty list is regarded as the tail of every charlist (lim. 11.). Every charlist is the tail of itself (lim. 13.). Equallists are the tails of each other (lim. 14.).

Limitations: (continued for level C.)

15. concat: charlist x charlist → charlist
16. charlist(X) & charlist(Y) ⟹ charlist (concat(X,Y)) & is-head (X, concat(X,Y)) & is-tail (Y, concat(X,Y)) & length (concat (X,Y)) = length(X)+length(Y)
17. charlist(X) & is-emptylist(Y) ⟹ is-equallist(X, concat(X,Y))
18. is-emptylist(X) & charlist(Y) ⟹ is-equallist(Y, concat(X,Y))

Informal description
(continued for level C.)

The operation "concat" concatenates two charlists in the order given in the parameter-list, and produces a new list (lim. 15., 16.). The concatenation of a list with an emptylist results a list which is equal to the other component (lim. 17., 18.).

Limitations
(continued
for level C.)

19. insert: charlist x charlist x notnegint → charlist

20. charlist(X) & charlist(Y) & notnegint
(N) & N≤length(X) ⟹
charlist (insert(X,Y, N)) &
length (insert(X,Y,N)) = length(X) +
= length(Y) &
is-head (Y, concat(Y, tail(X,N))) &
is-tail (Y, concat(head X,N),Y))

21. charlist(X) & is-emptylist(Y) & not-
negint(N) & N≤length(X) ⟹
charlist (insert (X,Y,N)) &
is-equallist(X, insert(X,Y,N))

22. is-emptylist(X) & charlist(Y) &
⟹ charlist (insert)X,Y,0)) &
is-equallist (Y, insert(X,Y,0))

23. charlist(X) & charlist(Y) & N = length
(X) &
⟹ charlist (insert(X,Y,N)) &
is-tail(Y, insert(X,Y,N)) & is-head
X, insert(X,Y,N))

Informal description
(continued for level C.)

The operation "insert" inserts a charlist Y after the N-th position of the mother-charlist X (lim. 19., 20.). The insertion of an emptylist or into an emptylist results in a new list equal to the other list (lim. 21., 22.). The operations "head" and "tail" are used for making the effect of the operation "insert" more explicit (lim. 23.). (Actually the last three axioms are formally redundant; this kind of redundancy is useful for making ideas more explicit.)

Limitations:
(continued
for level C.) 24. delete: charlist x notnegint x notnegint → charlist

25. charlist(X) & notnegint(N1) & notnegint (N2) & N1≤N2 & N2≤length(X)
⟹ charlist (delete(X,N1,N2)) &
is-head (head (delete(X,N1,N2),N1),X)&
is-tail (tail (delete(X,N1,N2),N2),X)&
length (delete(X,N1,N2)) = N1+length (X)-N2

26. charlist(X) & notnegint(N) & N≤length (X)
⟹ charlist (delete(X,N,N))
& is-equallist(X, delete(X,N,N))

27. charlist(X) & notnegint(N1) &
notnegint(N2) & N1≤N2 & N2≤length(X)⟹
is-equallist (delete(X,N1,N2), concat (head(X,N1), tail(X,N2)))

Informal description
(continued for level C.)

The operation "delete" deletes a sublist from a charlist. The elements to be deleted are specified by means of (not negativ integer) parameters; the first element to be deleted is the subsequent of the one pointed by the first parameter; the last one is specified by the second parameter (lim. 24., 25.). If the parameters are equal, the operation is ineffective (lim. 26.). The ax. 27. has the same meaning as lim. 25. (Both lim. 26. and 27. are redundant.)

Abstractions

B. LEVEL		C. LEVEL
text	⟺	charlist
unary operation	⟺	head, tail, delete
binary operation	⟺	concat, insert
is-emptytext	⟺	is-emptylist
is-equaltext	⟺	is-equallist
lim. 2.	⟸	lim. 5.,7.,10.,12.,12., 24.,25.
lim. 3.	⟸	lim. 15.,16.,19.,20.
lim. 4.	⟸	lim. 7.,12.,24.
lim. 5.,6.	⟸	lim. 16.,20.

Concluding remarks

The example is *not* to show how a text-editor should be generally designed and *not* to show how it was actually designed in the system ANSWER. The purpose of its presentation is only to show the flavour of a style of reasoning about programming objects during the development of a program.

REFERENCES

Dijkstra, E.W.: Notes on Structured Programming. Academic Press, 1972. pp. 1-82.

Hoare, C.A.R.: The Axiomatic Basis of Computer Programming. *Comm. ACM,* Vol. 12, pp. 576-583, 1969.

Hoare, C.A.R.: Proof of Correctness of Data Representations, *Acta Informatica,* Vol. 1. pp. 271-281, 1972.

Jones, C.B.: Formal Development of Programs. IBM Hursly report TR12 117, June 1973.

Jones, C.B.: Program Specifications and Formal Development. International Computing Symposium, pp. 537-553, 1977.

SAM-IV Dömölki B., Farkas Zs., Sántáné-Tóth E.: Software elemek formális leirásának egy módszeréről. SZÁMKI 1969/76. Vol. 3., Budapest, 1976.

Lampson, B.W., Horning, J.J.,London, R.L., Mitchell, J.G.- Popek, G.J.: Report on the Programming Language Euclid. *SIGPLAN Notices.* Vol. 12. No. 2. Febr. 1977.

Makajima, R.,Honda, M.,Nakahara, H.: Describing and Verifying Programs with Abstract Data Types in E.J. Neuhold /ed./. *Formal Description of Programming Concepts*, North-Holland Publishing Company, 1978. pp. 527-556.

Wulf, W.A.,London, R.L.,Shaw, M.: An introduction to the Construction and Verification of Alphard Programs. *IEEE Transactions on Software Engineering*, Vol. SE-2. No. 4. Dec. 1976.

J. Aszalós
Research Institute for
Applied Computer Sciences
H-1536 Budapest, P.O. Box 227.
Hungary

COLLOQUIA MATHEMATICA SOCIETATIS JÁNOS BOLYAI
26. Mathematical Logic in Computer Science
Salgótarján (HUNGARY), 1978.

ON AN INTERACTIVE PROGRAM VERIFIER FOR PROLOG PROGRAMS

K. Balogh

0. INTRODUCTION

The report deals with some problems and solutions connected with an experimental program verification system for proving semantic properties of predicate logic programs.

In order to structure the proof, we trace the property of partial correctness (which has an unusual meaning) back to two simpler properties (see 1.1.).

Since the whole topic would be too broad, my aim is to emphasize two main themes. The first is connected to verification condition generation: it gives some concrete examples, how method of verification and feature of programming language influence each other. The second thomo is about the system and mainly about the working method of the special theorem prover.

The research was influenced and encouraged by similar ones [1] and [2](but there the implementation is based on LISP-systems and the algorithm defining language is

Pascal-oriented instead of PROLOG), on the theoretical basis of [4].

I thank I.Futó and L.Naszvadi for implementing some basic procedures, Mrs.J.Bendl and P.Köves for adding some new built in procedures to the PROLOG system, and P.Szeredi for his remarks on implementation technics and on changes of built-in predicates. The system was completed by a general theorem prover by Mrs.K.Lábadi. I am indebted to my literary adviser, M.Szőts for his remarks and effort to make the report be more readable. The research was financially supported by the KSH OSZI.

1. CONNECTIONS BETWEEN FEATURE OF PROGRAMMING LANGUAGE AND METHOD VERIFICATION IN THE CASE OF PROLOG

The first question would be: what PROLOG is? The semantics of pure PROLOG and the criterion of partial correctness are defined in 1.1. together with an explanation why it is not superfluous to verify programs written in "very high level" languages. In the following section we transform the definition of partial correctness step by step in order to get a practically more usable condition for it. This approach shows also how features of PROLOG influence definition of correctness and method of verification. The definition and the method we got are interesting and applicable also in the case of other (e.g. sequential algorithmic) languages. Finally, in 1.3. on the basis of the realized connections I propose to modify some of the built-in predicates such that they become more adequate to verification.

1.1. *Semantics of pure PROLOG and definition of partial correctness*

The language is based on the Horn-clauses of logic [12, 13]. The main features of it are backtracking and pattern-matching (more exactly first order unification), which are used in a depth-first linear strategy of resolution. This special theorem-proving strategy also gives a useful and clear procedural meaning to PROLOG programs.

The i/o role of the procedure-parameters is not fixed; the procedures can keep a check on the value of their parameters and also can generate values for variables contained by parameters. The implementation is based on structure sharing [11]. Efficiency of a compiler for PROLOG is comparable with that for LISP [9].

On the first approach the *program* is a set of *procedure-declarations* (the so-called *partitions*), each of it being a set of Horn-clauses (alternatives).

For instance declaration of the procedure "rev" to reverse elements of a list (the variables (PROLOG or logical, according to the sense) are marked with upper case letters; they are supposed to be bound universially clausewise) is:

```
rev (nil,nil).                        (first alternative)
rev(X.L1,L3):                         (second alternative;it
  rev (L1,L2),app (L2,X.nil,L3).  has two preconditions)
```

Partition of the used "app" procedure also contains two clauses:

```
app (nil,L,L).
app (X.L1,L2,X.L3):
  app (L1,L2,L3).
```

The meaning of a procedure $p(\bar{X})$ is a relation, denoted by underlining the name: $\underline{p}(\bar{X})$.
E.g. the meaning of the above procedures can be described as the least fixed point of the following formulas [14]:

$\underline{rev}\ (L1', L2') \leftarrow$

$[(L1'=nil \wedge L2'=nil)$ (this corresponds to the first alternative)

$\vee \exists X, L1, L2 (L1'=X.L1 \wedge \underline{rev}(L1,L2) \wedge \underline{app}(L2, X.nil, L2'))$ (this corresponds to the second alternative)

$]$.

$\underline{app}\ (L1', L2', L3') \leftarrow$

$[(L1'=nil\ L2'=L3') \vee$

$\exists X, L1, L3 (L1'=X.L1 \wedge L3'=X.L3 \wedge \underline{app}(L1, L2', L3))]$.

For the least fixed point - in addition to implication - holds equivalence, too.

So, the meaning of a partition is the disjunction of the meaning of its alternatives with the appropriate match, and the meaning of an alternative is the conjunction of the meaning of its preconditions. To derive the proof rules of PROLOG programs, we shall use this fact.

The two first-order clauses express the meaning of the two partitions, if they are considered to be program schemas. The axioms for the data structures (being functions in logic) of the program are:

$nil \neq X.Y.$

$X.Y = U.V \rightarrow (X = U \wedge Y = V).$

They express tha fact, that on the Herbrand universe different ground terms have different values.

The axioms for the data structures of each PROLOG program can be generated automatically.

Let us use the language of first order logic not only to express assertions, but also to describe conditions and rules reffering to such formulas. Let us denote, that the procedure p terminates for $\bar{X}$, by termin $(p,\bar{X})$.

On the first approach, the condition of partial correctness of "rev" with respect to the pre- and post-assertions "i" and "o" is

(O) $[i(L1) \wedge \text{termin}\ (rev,L1,L2)] \rightarrow$
$[\underline{rev}(L1,L2) \rightarrow o(L1,L2)]$,

where i (L1) could be defined to express, that L1 is a list:

$i(L1) \leftrightarrow$
$\quad \text{list}\ (L1);$

$\text{list}\ (L) \leftrightarrow$
$\quad L = \text{nil} \vee \exists X,\ L1\ (L=X.L1 \wedge \text{list}\ (L1))$

and o(L1,L2) could be defined to express, for instance, that each element of L1 occurs in L2 with the same multiplicity, and vica versa

$o(L1,L2) \leftrightarrow$
$\forall X[X \in L1 \vee X \in L2$
$\quad \rightarrow \text{number-of-occurences}\ (X,L1) =$
$\quad \text{number-of-occurences}\ (X,L2)].$

$X \in Y.L \leftrightarrow$
$\quad X = Y \vee X \in L.$

$\text{number-of-occurences}\ (X,L) = N \leftrightarrow$
$\quad (L = \text{nil} \rightarrow N = 0) \wedge$
$\forall Y,L1(L=Y.L1 \rightarrow [(X=Y \rightarrow N=\text{number-of-occurences}\ (X,L1)+1) \wedge$
$\quad (X \neq Y \rightarrow N=\text{number-of-occurences}\ (X,L1))]).$

But the (O) condition is weaker that the usual ones other programming languages. To show it, let us redefine the procedure "rev" with

rev (L1,L2):false, that is for each L1 and L2, the execution of rev terminates with failure.

Therefore

$\underline{\text{rev}}$ (L1,L2) $\leftrightarrow$ false,

and for every one-place relation i and two-place relation o, according to the above definition, rev is partially correct with respect to them. We have to complete the above condition with an other one of form

o1 (L1,L2) $\rightarrow$ $\underline{\text{rev}}$(L1,L2),

where "o1" can be considered playing output role (in the case of fixing the i/o role of the parameters this means, that o1 may contain output parameters of the program rev).

Generally, the definition is the following:

D1 The program "p" is *partially correct* with respect to the pre-assertion "i" and post-assertions "o1" and "o2", iff o1 is sufficient and o2 is necessary condition of the fulfilment of the relation $\underline{p}$ (corresponding to the program) for each input satisfying i on which the program terminates.

That is

$[i(\bar{X}) \wedge \text{termin}(p,\bar{X}) \rightarrow$

(1) $[(o1(X) \rightarrow \underline{p}(\bar{X})) \wedge$

(2) $(\underline{p}(\bar{X}) \rightarrow o2(\bar{X}))]$

or, in Hoare-like notation

(3) $\{i(\bar{X})\}\{o1(\bar{X})\}\ p(\bar{X}) \wedge$

(4) $\{i(\bar{X})\}p(\bar{X})\{o2(\bar{X})\}$.

Designation: According to their meaning, we call o1 *sufficient* and o2 *necessary assertions*, too. (1) and (3) are the *conditions of sufficiency*, (2) and (4) *are the conditions of necessity*.

Remark: The arguments of the program satisfy each property (but not only those) described by (or derivable from) o2, and o1 expresses (or it is derivable from o1) each possible property of them (but not only these).

For a fixed input data the output assertions speak about properties of each possible output corresponding to it and do not predict which of the output will be the actual result after executing the program.

So, supposing a relation $\underline{p}$ assigned to the program p, a language-independent definition of partial correctness is given.

Finally, we have to enlighten the significance of supplying programs with assertions, and of dealing with verification problems.

The next example shows, that even programs written in declarative style may hide some of their properties. These properties can be expressed explicitly by proving the program with assertions (see other examples also in [8]). The procedure into-tree (old--tree, symbol, new-tree) can be used to place symbols into a binary tree. There exists a lexicographic ordering on the domain of symbols. The tree is described by the axioms:

```
tree (nil).
tree (LEFT.ROOT.RIGHT) ↔
        tree (LEFT) ∧ symbol (ROOT) ∧ tree (RIGHT).
```

The partition into-tree is

```
into-tree (nil.X.nil,X.nil).
into-tree (L.X.R,X,L.X.R).
into-tree (L1.Y.R,X,L2.Y.R):
        less (X,Y), into-tree (L1,X,L2).
into-tree (L.Y.R1,X,L.Y.R2):
        greater (X,Y), into-tree (R1,X,R2).
```

During the construction of the tree the following porperties are expressed only implicitly:

- each symbol occurs in the tree at most once;
- the tree (and its subtrees) is ordered in the sense, that each symbol of the left part of the tree

is less than any of the right part of it. Of course, these properties can be expressed in first order logic, too.

The notions, results and methods of verification, both theoretical and implementational, are useful also in the field of the inverse task, in program generation [3,5,6,8]. The value of these themes is given by their correspondance with programming languages, programming methodology and technology: they complete the syntactic requirement of structured programming with semantic ones and give tools to use and handle them.

1.2. A method to verify partial correctness of PROLOG programs

In the preceding section the definition of partial correctness has been introduced. The definition refers to the least fixed point and to the termination of the program, so it is difficult to apply this in practice directly:

- the definition is full of notions of the program (it contains all of them) it is structured only on the abstraction level of the program (which is too low a level to speak about the program); therefore the proof of correctness has to simulate the execution of the program very throughly;

- as lengthes of execution pathes of programs haven't upper bound in general (since they are input-dependent), the user has to apply some kind of induction explicitly during the proof, and it is difficult to guide the mode of the application of that in a universal way [10].

In order to structure the proof, one has to express

and prove general lemmas (verification conditions) stating supposed properties of structured parts of the program. Instead of reproving each application of a program part, one can use the lemma corresponding to that part.

Finding out lemmas is also necessary for applying induction. To make proofs by induction one has to generalize his lemmas suitably and choose ways of applications of the induction rule. Without human inspiration the enormous number of attempts having no success cause "combinatorical explosion". However explicit use of induction makes the proof of termination possible, too [7].

Other methods originated from that of Floyd's. He has shown, that it is possible to give general verification rules to guide construction of verification conditions instead of human inspiration. In accordance with these methods the rule of *induction* is hidden, its use *is implicit* at the expense of demanding to relate some assertions to some points of each (execution-) loop of the program. We shall describe a variant of Floyd's inductive assertions method to prove partial correctness.

To describe a method for sequential languages, it is sufficient to give a method for languages containing only procedures, since in this respect the usual statements can be treated as procedures. We shall derive a verification method only for PROLOG programs, but it can be modified in order to get similar ones for other sequential languages, allowing global variables, calling by value or reference.

Supposing, that *each partition of the program is equipped with input and output assertions*, (so each "loop" of the program has been cut) let us transform the condition for partial correctness step by step in order to get a practically more usable one for it. The aim is to trace property of a partition back to supposed properties of its parts. While generating the verification conditions, we have to simulate execution of the program for concrete data.

But it is a question, how close the simulation would be. During construction of the starting condition (the definition D1 of partial correctness) there was no simulation, as instead of the procedural semantics of PROLOG the declarative one has been explored. In the following subsections we give three conditions according to increasing depth of simulation. Finally, correctness and completeness of the received method is discussed.

1.2.1. Correctness condition on the level of using declerative semantics

Declerative semantics of pure PROLOG does not determine the order of execution of alternatives or preconditions in partitions. It cannot be predicated which of the possible outputs corresponding to a fixed input data will be the actual result, and it is very convenient to the user. The language is nondeterministic. This property is similar of that of output assertions mentioned in the remark of D1.

Applying features of the relation $\underline{p}$ related to the partition of some p and definition of correctness (1), (3) and (2), (4) , we derive properties of the alternatives from properties of that partition.

Let us suppose, that the partition of p consists of alternatives of from

$p(\bar{X}):a_{i1}(\bar{X}),\ldots,a_{in_i}(\bar{X})$ for $i=1,2,\ldots,m$.

That is, the argument expressions from heads of the alternatives are eliminated by generating preconditions with equality (see example of 1.1.).

Let us denote $p(\bar{X})$ by P, $a_{ij}(\bar{X})$ by A_{ij}, and a list of preconditions $A_{i1},A_{i2},\ldots,A_{in_i}$ by the meta variable $\mathcal{A}_i$, then the partition of p is

$$p : \mathcal{A}_i \quad \text{for} \quad i=1,2,\ldots,m.$$

If a list of the preconditions $A_1,A_2,\ldots,A_n$ is not empty ($n \geq 1$), then it can be denoted by "A_1, $\mathcal{A}$", where $\mathcal{A}$ stands for $A_2,\ldots,A_n$. If a list of preconditions is empty, it is denoted by "empty".

We can express the *rules of verification* in the form of Hoare-like axioms

$$(5) \qquad \frac{\{I\}\{O1\}\ P}{\bigvee_{i=1}^{m} [\{I\}\{O1\}\ \mathcal{A}_i]}$$

$$(6) \qquad \frac{\{I\}\ P\ \{O2\}}{\bigwedge_{i=1}^{m} [\{I\}\ \mathcal{A}_i\{O2\}]}$$

Proof of the rules. By the semantics of pure PROLOG $\underline{P} \leftrightarrow \bigvee_{i=1}^{m} \underline{A}_i$.

Substitute it to (1) and (2), and use

for (5) $a \rightarrow (b \vee c) \leftrightarrow (a \rightarrow b) \vee (a \rightarrow c)$

$\forall\bar{X}(q(\bar{X}) \vee r(\bar{X})) \leftarrow \forall\bar{X}q(\bar{X}) \vee \forall\bar{X}r(\bar{X})$

for (6) $(a \vee b) \rightarrow c \leftrightarrow (a \rightarrow c) \wedge (b \rightarrow c)$

$\forall\bar{X}(q(\bar{X}) \wedge r(\bar{X})) \leftrightarrow \forall\bar{X}q(\bar{X}) \wedge \forall\bar{X}r(\bar{X})$.

Finally change the received two conditions into that of a Hoare-like notation.

REMARK

Rule (6) is a transformation producing an equivalent of the condition (4), but rule (5) produces a stronger condition than (3) is. Therefore, under certain conditions one can use rule (5) successfully only after distinguishing cases among values of the input (according to the alternatives of the partition).

And now we compare property of a call-list with the given properties of its elements. If A *is of form* $A_1, A_2, \ldots, A_n$, $\bar{Y}$ *denotes variables of the* A_j*-s not occuring in the heads of the procedure declaration, and for all* j A_j *is partially correct with respect to* I_j $O1_j$ *and* $O2_j$, *the verification rules are*

(7)
$$\frac{\{I\}\{O1\}A}{(I \wedge O1) \rightarrow \exists \bar{Y} \bigwedge_{j=1}^{m} (I_j \wedge O1_j)}$$

(8)
$$\frac{\{I\}\ A\ \{O2\}}{\bigwedge_{k=1}^{n} [I \wedge (\bigwedge_{j=1}^{k-1} O2_{i_j}) \rightarrow I_{i_k}] \wedge [I \wedge (\bigwedge_{j=1}^{n} O2_j) \rightarrow O2]}$$

Empty conjunction is supposed to be true.

Proof

Receall, that $\underline{A} \leftrightarrow \exists \bar{Y} \bigwedge_{j=1}^{n} \underline{A}_j$ by the definition of semantics. By assumption, for all $j=1,2,\ldots,n$ $(I_j \wedge O1_j) \rightarrow \underline{A}_j$ and $I_j \rightarrow (\underline{A}_j \rightarrow O2_j)$. Therefore

$$\bigwedge_{j=1}^{n} (I_j \wedge O1_j) \rightarrow \bigwedge_{j=1}^{n} \underline{A}_j \quad \text{and} \quad \bigwedge_{j=1}^{n} I_j \rightarrow (\bigwedge_{j=1}^{n} \underline{A}_j \rightarrow \bigwedge_{j=1}^{n} O2_j),$$

so

$$\exists\bar{Y} \bigwedge_{j=1}^{n} (I_j \wedge O1_j) \rightarrow \exists\bar{Y} \bigwedge_{j=1}^{n} \underline{A}_j \quad \text{and} \quad \exists\bar{Y} \bigwedge_{j=1}^{n} I_j \rightarrow \exists\bar{Y}(\bigwedge_{j=1}^{n} \underline{A}_j \rightarrow \bigwedge_{j=1}^{n} O2_j)$$

To prove $(I \wedge O1) \rightarrow \underline{A}$, a sufficient condition for the first case is $I \wedge O1 \rightarrow \exists\bar{Y} \bigwedge_{j=1}^{n} (I_j \wedge O1_j)$.

But this flow of thought is a blind alley for the second case. So, let us suppose, that there exists a permutation $i_1, i_2, \ldots, i_n$ of the indices $1,2,\ldots,n$ with the following properties

$$I \wedge (\bigwedge_{j=1}^{k-1} O2_{i_j}) \rightarrow I_{i_k} \quad \text{for } k=1,2,\ldots,n$$

and

$$I \wedge (\bigwedge_{j=1}^{n} O2_j) \rightarrow O2.$$

Then, using the assumption, $I \rightarrow (\underline{A} \rightarrow O2)$ holds, since it is equivalent to $[I \wedge (\bigwedge_{j=1}^{n} \underline{A}_{i_j})] \rightarrow O2$.

One can formulate an analogous condition also for the satisfaction of condition of sufficiency (7), but the former one is more useful.

Applying verification rules (5)-(8) to a program schema one can get a set (conjunction) of first order formulas (verification conditions) stating partial correctness of that automatically. It is another task to prove each element of the set separately with the help of a calculus (logical transformations), using axioms for data structures (data and operations) of the program.

Remarks about arguments of assertions

1./ Assertions are formulae corresponding to some points of the program. They can be structured with references to axiomatically defined notions. They also may contain free variables. The arguments of an assertion and the corresponding partition having the same sequence number are related to each other.

2./ In assertions of programming languages different from PROLOG, other kinds of free variables may occur. Assertions to those languages allowing assignment to change actual value of variables, may contain parameters for transmissing information among assertions. These parameters can be used also to prove properties, from which termination or non-termination can be derived. Assertions of those languages allowing global variables, may contain references to actual values of such variables. Handling of "variable" variables and global variables makes mostly technical difficulties.

If an assertion O2 does not refer to some variables of procedure-parameters, it means that we know nothing about its actual value. But an assertion of type O1 states, that each value of the omitted variable occurs in the output of the procedure. To handle global variables another convention is practical. According to this, we do not speak about global variables, which are supposed to be constant during execution of some program-parts. Verification according to this convention is possible, since it is syntactically controllable, whether all of the changed global variables are referenced by the corresponding assertions.

3./ It would be a too restrictive solution to express the input assertions of partitions by fixing the i/o role of their arguments. One solution would be to construct an input assertion as a disjunction of conditions corresponding of the different variants of i/o role of the arguments. This method is based still on functional and not on relational approach, and instead of being descriptive it reflects execution. The other solution is to speak in an input assertion about each argument, which can play input role in some calls.

Input assertions of this kind are conjunctions of conditions, where one condition speaks about variables, having the same i/o role in each possible call. When using an input assertion to express effect of a call, the conjuncts containing only variables playing output role in that particular call are to be left.

1.2.2. Correctness condition on the level of simulating sequential execution of calls

On this level we use additional assmption to the pure declarative semantics. Procedural semantics of PROLOG states, that the execution of calls (preconditions) of an alternative is sequential. On this level we can use (5) and (6) without change, but verification rules (7) and (8) are to be replaced. We must derive verification rules describing transformations of assertions while executing a precondition of an alternative. Assertions (sets of admissible states) can be transformed both in corresponding and in opposite direction with flow of control.

A transformation is given by creating *the most expressive assertion related to the original one* describing exactly the set of states originated from another one by forward or backward execution of a call. In order to determine the most expressive one, first we characterize the possible assertions. Let us suppose, that $\{I_p\}\ \{O1_p\}\ P$ and $\{I_p\}\ P\{O2_p\}$ is all that is known about the effect of P, and we would like to use it to describe the effect of a particular call of P. The question is: under what conditions we can state partial correctness of that call of P with respect to the new assertions I, O1 and O2, which contain also variables living in the environment of the call in addition to those occuring in the call. Shortly, under what conditions are I, O1 and O2 possible assertions of the call? From the definition of partial correctness, the conditions are the following

(9) $(I \rightarrow I_p) \wedge (O1 \rightarrow I \wedge O1_p)$

(10) $(I \rightarrow I_p) \wedge (I \wedge O2_p \rightarrow O2)$.

On this basis, the most expressive

- sufficient assertion O1 corresponding to I is

(11) $I \wedge O1_p$, if $I \rightarrow I_p$;

- necessary assertion O2 corresponding to I is

(12) $I \wedge O2_p$, if $I \rightarrow I_p$;

- input assertion I corresponding to O1 is

(13) $O1$, if $O1 \rightarrow (I_p \wedge O1_p)$;

- input assertion I corresponding to O2 is

(14) $I_p \wedge (O2_p \rightarrow O2)$.

(11) and (12) correspond to forward simulation of execution, while (13) and (14) correspond to backward simulation of that.

Proof

(11): By (9), if $I \rightarrow I_p$ (which is the consistency condition for the given assertions), $O1 \rightarrow I \wedge O1_p$ characterizes the set of possible sufficient assertions. This set is not empty; e.g. the useless "false" and

$I \wedge O1_p$ satisfy it. The latter is the most expressive sufficient assertion related to I, by definition.

(12): Similar to the proof of (11)

(13): Since (9) is equivalent to $(O1\to I)\wedge(I\to I_p)\wedge \wedge(O1\to O1_p)$, the consistency conditions are $O1 \to I_p$ and $O1 \to O1_p$. Thus by definition, the most expressive input assertion related to O1 is O1.

(14): Use the equivalent from of (10):

$$I \to (I_p \wedge (O2_p \to O2)).$$

Having derived the transformations with the help of the notion of most expressive assertion, we can describe the corresponding verification rules. In order to be short, we give only the forward simulating rules. We use the notations of 1.2.1.

The rules corresponding to (7) are the following

$$(15) \quad \frac{\{I\}\{O1\}\ A,A}{(I\to I_A)\wedge\{I_A\}\{O1_A\}A\wedge\{I\wedge O1_A\}\{O1\}A}$$

$$\frac{\{I\}\{O1\}\ \text{empty}}{O1 \to I}\ .$$

Similarly the rules correspondind to (8) are the following:

$$(16) \quad \frac{\{I\}\ A,A\{O2\}}{(I\to I_A)\wedge\{I_A\}A\{O2_A\}\wedge\{I\wedge O2_A\}A\{O2\}}$$

$$\frac{\{I\}\ \text{empty}\ \{O2\}}{I \to O2}$$

These rules are weaker than the corresponding ones: if one can prove partial correctness of an alternative by the above rules, it can be proved also by (7) and (8). But, we can also simulate the effect of those (e.g. built-in) procedures of the program, i/o role of whose arguments is fixed. So we can prove required definedness of certain arguments (with the help of an especially handled predicate: bound (X)).

1.2.3. Correctness condition on the level of simulating sequential execution of alternatives and calls

As it is unnecessary to speak in assertions about which one of the admissible values will be the actual one, and since backtracking gives all possible solutions (if the program terminates), pure PROLOG with backtrackable built-in predicates can be verified on the previous level. But PROLOG has other built-in predicates too, which modify control: slash, ancestor and slash ancestor. We show in this section, that slash has features corresponding only to the partition it occurs in.

First of all we give the usual procedural semantics of slash.

D2 During execution of a PROLOG program *the parent of a goal* A_1 is that goal A_2, execution of which caused A_1 occurs in the goal sequence. That is, A_2 is the goal matched the head of the clause containing A_1.

D3 *Execution of slash* always succeeds. Backtracking through a slash prohibits control from any new choice till the parent (inclusive) of that slash.

To handle slash we have to take the order of execution of alternatives in some way into account. We use the notion *arrive condition of an alternative*, which is the condition of the fact, that starting from the first alternative (the beginning of the partition) with admissible input, control sometimes can arrive at the head of that alternative. If a partition has pre-assertion I, and none of its alternatives contains slash, then the arrive condition of each alternative is I (because of the possibility of backtracking, supposing termination).

To describe changing of the arrive condition in general, we use the following notions.

D4 *The condition part of an alternative* is
- empty, if the alternative does not contain slash,
- the list of preconditions till the slash (noninclusive).

D5 The *post-assertion* O of a program P is *exacting* with respect to I, if P is partially correct with respect to I, O and O.

It is evident, that the condition part "$A_1,\ldots,A_n$" of an alternative is partially correct with respect to I and the exacting O, if for all i A_i is partially correct with respect to I and the exacting O_i, and $O \leftrightarrow \bigwedge_{i=1}^{n} O_i$. Therefore, according to D4, the *condition formula* C of an alternative is
- false, if the alternative does not contain slash;
- the O above.

In general, the arrive condition of the first alternative is the input assertion, while that of an other alternative is the conjunction of that of the preceding alternative and the negation of the condition formula of the preceding alternative. Only the first slash of an alternative has an effect on arrive condition of the next alternative. So, with the help of the arrive condition we are able to simulate sequential execution of alternatives indirectly, without backtracking.

Verifying an alternative, we have to use assertions related exactly to those parts of the program, which are reached during execution of that alternative. If it does not contain slash, each alternative of the called partitions can be reached (after a sufficiently great number of backtracking, supposing termination).

But verifying alternative containing slash, we have to use assertions of preconditions called by the condition part, which are related exactly to those subpartitions (of the called preconditions), at which control sometimes arrives (for admissible input). (In the condition formula assertions related to whole partitions are used, of curse.)

Contrasted with its effect to the arrive condition, the effect of this feature of slash may not be limited to the partition of its particular occurance in general.

Therefore, slash has a clear meaning (for the programmer) only in special cases.

There are some other built-in predicates in PROLOG, too, which are nonbacktrackable, that is, their effect remains over backtracking (e.g. some of the i/o database-modifying predicates). Using them, the programmer applies his knowledge about sequential execution of alternatives. Therefore, to verify them, one ought to apply backtracking explicitly. In this case the assertions related to partitions could not be used, instead of them assertions connected to alternatives would be necessary. But it seems to be very clumsy and against structured programming.

Moreover, nonbacktrackability is against the basic feature of the language, since in pure PROLOG each solution of the subgoals is hypothetical until solving the whole goal sequence with success.

So instead of even deeper simulation I propose

to change the set of built-in predicates according to verification and structuring.

1.2.4. Problems of correctness and completeness of the method

We have given a method for proving partial correctness: using the verification rules of a level of simulation, one can trace correctness of the program back to that of the elements called by the program. Of course, it may happen, that the given assertions of the elements are too weak for proving the verification conditions. We have proved, that each of the verification rules is correct when applied for carrying out one step. Supposing termination, one can use induction according to the depth of applications of the verification rules (even in the case of proving recursive procedures). With the help of these rules only, total correctness cannot by proved.

The problems with completeness of the method are the same as in the case of algorithmic languages, that are discussed in [15].

1.3. Proposed changes of built-in procedures

In the previous sections we have derived correctness conditions according to the level of simulating execution of the program. Those were the examples of how features of PROLOG effect method of verification. Backwards, I propose to use built-in predicates in a structured way by supplementing new ones and hiding others.

The clarity of PROLOG is reflected by the distance of the verification method from execution. Let us see the classification of existing built-in predicates we have got by this measure.

verification by using		built-in predicates handled
pure declarative semantincs		backtrackable predicates not restricting i/o role of parameters (e.g. insymb (X), interm (X))
simulation of sequential execution of preconditions		backtrackable predicates fixing i/o role of parameters (e.g. plus (N1,N2,X))
simulation of sequential execution of alternatives and preconditions	indirectly	arrive condition modifying feature of slash
	directly	control modifying-predicates (e.g. slash, ancestor)
problematic built-in predicates		non-backtrackable built-in predicates (some i/o and database modifying predicates)

We shall deal with built-in predicates placed to the third and fourth level.

We have seen, that semantics of *slash* is not clear for the programmer in general. The following special case seems to be sufficient for practice. According to this case an occurence of slash in an alternative is allowed, if it is in an alternative which fails for all admissible input, or if the partitions called by the condition part of that alternative contain only one alternative (see 1.2.3). For the first case the meta-predicate not (X) is an example, which is defined by the partition

```
not (X):
    X, slash, fail.
not (X).
```

For the second case the use of the meta-predicate if-1(COND, ACT1, ACT2) is an example

```
if-1(COND,ACT1,ACT2):
    COND, slash, ACT1.
if-1(COND,ACT1,ACT2):
    ACT2.
```

But only those calls of if-1 are allowed, in which the actual value of COND calls only one-alternative partitions.

To refuse the restriction, one solution is to redefine if-1

```
if-2(COND,ACT1,ACT2):
    COND, ACT1.
if-2(COND,ACT1,ACT2):
    not(COND), ACT2.
```

The other solution would be to redefine slash (see D3) by not prohibiting control from any new choice of the partitions called by the condition part of the alternative, while backtracking through a slash. By this solution if-1 and if-2 becomes equivalent, the meaning of not (X) remains the same, but the efficiency of execution of not (X) decreases superflously.

However, its meaning is clearer than the original one, and it is completely verifiable. So, I propose to use the original slash for defining the partition of not (X), and to use not (X) or the redefined slash otherwise.

It would be interesting also for practical reasons to allow backtrackable *output predicates and database modifying predicates*. During (hierarchical) development of the "algorithmic" from less and less descriptive variants (of plans written possibly in the same language), it would be necessary to return to the use of non-backtrackable predicates only in some cases and only on the most algorithmic levels.

Database modifying predicates are used mostly in

- optmizing reexecution of some goalsequence
- optmizing parameter passing
- program modification.

For the sake of efficiency, instead of repeated execution of a long goalsequence, the calculation can be optimized by the help of non-backtrackable database modifying predicates, by placing the alternative solutions into the database of the program. After it the resolution stack can be contracted, too, and solutions can be reached in the database. In order to keep both efficiency, clarity and verifiability, I propose a new built-in predicate, which can be implemented with the help of the non-backtrackable ones.

The meta-predicate is "optimize (GS,N)", where GS is a goal-sequence and N is a (new) partition name given by the user. The axiomatic meaning of this predicate is: execute GS. That is, the assertions for the predicate are the same as the assertions for GS. But the procedural semantics is only equivalent; which is based on the above characterized technic.

So, we have got a procedure, which makes it possible to solve some effeciency problems independently of semantics. Instead of explicit application of non--backtrackable database modifying procedures, if they were used for increasing efficiency, the procedure "optimize" can be applied.

The other application field of the database modifying procedures is to add and inquire about data from database, instead of passing it as a parameter through many procedures. For this kind of application backtrackable version of database modification is necessary. It is important, that if a variable - occuring both in the database and in the goal sequence - gets a value, the value should get also into the occurence in the database. Therefore the backtrackable version of these procedures can be implemented using the resolution stack instead of the area of the original database. In this way the implementation can be more efficient than the original one.

The third (and perhaps last) main application of database handling is program modification. For this application the above kind of backtrackable database modifying predicates is useful.

2. THE VERIFICATION SYSTEM

The system itself is written in PROLOG. It contains three programs.

The first one is a verification condition generator. It needs the algorithm and the supposed effect of it: two kinds of input data. It works on the basis of Hoare-like axioms (15) and (16), and generates the verification conditions expressing partial correctness of the "algorithm" according to the supposed effect.

It handles also the arrive condition modifying feature of slash (see 1.2.3. and 1.3.).

The second program is a special, effective, interactive theorem prover. The inputs of this program are the theorems (e.g. the verification conditions), the definitions and lemmas speaking about the notions of the user, and schemas on a meta-language of first order logic expressing transformations.

The third program of the system is a resolution-based general and complete theorem pover (written by Mrs.K.Lábadi), using connection graphs and set-of-support strategy.

In the followings we deal with the working method of the special theorem prover, which is based on the description of transformations on expressions.

Transformations can be expressed both as data for an interpreter and as procedure. The first solution is more flexible since the user can extend the structure of schemas, while the second is more efficient. In 2.1. we give the main concept of the increactive theorem prover. In 2.2. we describe other fields of applications.

2.1. Idea of the interactive theorem prover

Theorems about computational complexity of complete calculi for first order formulas show, that "combinatorical explosion" is unavoidable in fully automatic systems.Therefore, the user must guide the theorem prover with his imagination. The system can elaborate the next step according to the idea of the user, so the solution is correct. The transformed version is displayed. Then the user has to décide, whether the solution is good for deriving the goal, or not. The interactive theorem prover stores the tree of transformations of the formula, and the user can handle the tree e.g. to make backtrack and choose another alternative when feeling the proof is in blind-alley.

The transformations are structured. There are 40 standard simple transformations, but the user can apply them all in 10 staps by the help of combined ones.

The transformations are clear and the user can add others, since

-(with the expection of 3 transformations) each of them is a data of the program, and not a procedure;

- they are not first-order formulae, but simple schemas of formulae, and the notation is quite usual;

- the schema-interpreter can apply them to each level of the formula automatically, so they are simple and clear.

Their effect, measured by the number of resolution steps is great but clear, so the user can explore interactivity. With the help of meta-theorems the schemas can be simplified to first-order formulae, and so they can be proved by the theorem prover.

The user can select names of simple or combined transformations from the displayed database. The program tries to apply the transformation to the top of the formula, or top down or bottom up (which is a declared property of the schema) to its each level. Only those transformations are implemented as procedures, which are essentially context sensitive (e.g. quantifier elimination, transformation into lexicographic order). The other transformations are expressed as schemas. For instance the transformation to conjunctive normal form can be directly expressed in form of the two schemas describing distribitvity of conjunction and disjunction

$$\frac{X \vee (Y \wedge Z)}{(X \vee Y) \wedge (X \vee Z)} \qquad \frac{(X \wedge Y) \vee Z}{(X \vee Z) \wedge (Y \vee Z)} .$$

Variables of these schemas are simple first order ones (that is first order unification is allowed), and variables of the actual formula (to be transformed) are treated to be constants. (So, in this special case unification is simplified to substitution of appropriate parts of the actual formula to variables of the schema. One can see other cases in 2.2.)

Generammy, the schema can have a third part: the precondition. For instance, the transformation which splits the actual goal two subgoals is of form

$$\text{univvars } (\underset{\approx}{Q}) \vdash \frac{\underset{\approx}{Q}(X \wedge Y)}{\underset{\approx}{Q}(X) \wedge \underset{\approx}{Q}(Y)} .$$

The meta-variable $\underset{\approx}{Q}$ can be bound to any sequence of quantified variables. So the precondition is, that each variable of $\underset{\approx}{Q}$ is universally quantified. In general, the preconditions can be simply composed from some given basic procedures.

The user can direct handling of equality and induction also by schemas. This is important because in general they are much more ineffective than resolution. For instance a schema to apply a kind of induction is the following

$$\frac{\forall X \; \boxed{P} \, (X)}{\boxed{P}\,(0) \wedge \forall X(\boxed{P}\,(X) \rightarrow \boxed{P}\,(X+1))} .$$

In the schema P can be matched with a formula, which depends on the variable bound to X.

A similar example to handle equality is

$$(X \text{ is-not-free-in } A) \wedge \text{univvars } (\underset{\approx}{Q}) \vdash \frac{\forall X \; \underset{\approx}{Q}(X{=}A \rightarrow \boxed{P} \; (X))}{\underset{\approx}{Q}(\boxed{P} \; (A))}$$

This transformation is coarse in cases, when the user wants to substitute only some of the occurences of X. A stronger, but more complicated version is

$$X \text{ is-not-free-in } A \;\vdash\; \frac{\forall X(X=A\rightarrow \text{(P)}(\text{(}\underset{\sim}{AS1}\text{)},X,\text{(}\underset{\sim}{AS2}\text{)}))}{\forall X(X=A\rightarrow \text{(P)}(\text{(}\underset{\sim}{AS1}\text{)},A,\text{(}\underset{\sim}{AS2}\text{)}))} \; ,$$

where $\text{(}\underset{\sim}{AS1}\text{)}$ and $\text{(}\underset{\sim}{AS2}\text{)}$ are (may be empty) sequences of arguments of the actual formula, and (P) is name of a predicate.

By the help of this schema one can substitute occurences of X being complete arguments of (P).

Substituting one arbitrary occurence of X we use a special part of the database of formulas, a stack, and special commands to handle it.

There is a command, with which the user can assign an appropriate part of the actual formula to be the new one. This command - ranging over the original formula - notes down the equalities from the preconditions of the new one into the stack. After some application of this command the occurence of X in question can be obtained as an actual expression, therefore the schema

$$\text{in-database } (X=A) \vdash \frac{X}{A}$$

can be applied. Finally, by the help of an other command, (the transformed version of) the actual formula can be substituted back into the original formula, while deleting the suitable part of the stack.

Schemas and meta-variables of the above type were sufficient to express transformations of formulas into prenex normal form and common simplifications.

2.2. *Other applications of schemas*

We can express a resolution step as a schema, if variables of the actual formula and variables of formulas in the database are treated to be firs-order variables.

The verification rules can be expressed as schemas, too. In this application variables of the actual Hoare--triplet are treated as constants and variables of Hoare-triplets of the database are treated as first--order variables.

By the help of the introduced meta-variables we can describe the meaning - with assertions - of the meta--predicates of PROLOG.

Another application would be to develop programs from non-algorithmic specifications by optimizing transformations [3,5,6,8].

REFERENCES

[1] Good,D.I.,London,R.L.,Bledsoe,W.W.: An iterative program verification system, *IEEE Transactions on Software Engineering*, Vol SE-1, No 1, March, 1975.

[2] Luckham,D.C.: Program verification and verification--oriented programming in Gilchrist,B.(ed) *Information Processing* North Holland, Amsterdam, 1977, pp. 783-793.

[3] Katz,S.M., Manna,Z. Logical analysis of programs, *Communications of ACM*, Vol. 19, No 4, 1976, pp. 188-206.

[4] Manna,Z. *Mathematical Theory of Computation*, McGraw Hill, 1974.

[5] Manna,Z., Waldinger,R.: Synthesis: dreams → programs, Standford Artificial Intelligence Laboratory, Memo AIM-302, Computer Science Department, Report No STAN-CS-77-630, November, 1977.

[6] Burstall,R.M.,Darlington,J.: A transformation system for developing recursive programs, *Journal of ACM*, Vol 24, No 1, 1977, pp.44-67.

[7] Burstall,R.M.: Program proving as hand simulation with a little induction, *Information Processing* 74, North Holland, Amsterdam, 1974.

[8] Clark,K.,Tärnlund,S.A.: A first order theory of data and programs in Gilchrist,B. (ed.) *Information Processing* 77, North Holland, Amsterdam, 1977.

[9] Warren,D.,Pereira,L.: PROLOG - the language and its implementation compared with LISP, *SIGART-SIGPLAN Newsletter, ACM Symposium of AI and Programming Languages*, Rochester,N.Y. August, 1977.

[10] Boyer,R.S., Moore,J.S.: Proving theorems about LISP functions, *Journal of ACM*, Vol.22(1975), No 1, pp. 129-144.

[11] Moore,J.S.: Structure sharing and proof of program properties I, DCL Memo, No 67.

[12] Kowalski,R.: Predicate logic as programming language, *Information Processing* 74, North Holland, Amsterdam, 1974.

[13] Kowalski,R.: Algorithm = logic + control, *Communication of ACM*, Vol 22(1979) pp. 424-436.

[14] Van Emden, M.H.,Kowalski, R.A.: The semantics of Predicate Logic as a Programming Language, *Journal of ACM*, Vol. 23, (1976), pp. 733-742.

[15] Gergely, T.,Ury, L.: Mathematical Theories of Programming Budapest, 1978.

[16] Balogh,K.: On an implemented program verifier (to verify a structured subset of ALGOL 60, implemented in CDL), *Proceedings of 2nd Hungarian Computer Science Conference*, Budapest, 1977, pp. 159-167.

[17] Balogh,K.:Verification of PROLOG programs, NIM IGÜSZI, April, 1977 (in Hungarian).

K.Balogh
NIM IGÜSZI
Markó u.16.
H-1055 Budapest

COLLOQUIA MATHEMATICA SOCIETATIS JÁNOS BOLYAI
26. Mathematical Logic in Computer Science
Salgótarján (HUNGARY), 1978.

ON IMPLEMENTATIONS OF ABSTRACT DATA TYPES

L. Banachowski

Introduction

Some of the problems in the field of computer science are of the following form: we are given a set of instructions. We wish to carry out the sequence of these instructions on a computer. So we look for a data structure that supports the execution of such programs in an efficient way. We also want to know if there exists a data structure that supports the execution of such programs in less than a certain amount of time.

For example, the following instructions are frequently used in computer programs:
insert(x,A) -adds element x to a set A,
delete(x,A) -removes element x from a set A,
member(x,A,p) -p:=x∈A,
find(x,A) -computes the set A containing element x,
union(A,B,C) -combines sets A and B into the new set C.
We would like to implement the sequences of the instructions of the above types on a computer and to prove their correctness.

Simula 67 provides the construction of the class to handle the implementations of collections of instructions within a programming language.

The question about implementations of abstract instructions is also important from the point of view of the methodology of the construction of correct computer programs. It is profitable to divide the process of the construction of a program into (at least) two stages. First we formulate the problem to be solved in general terms reflecting the structure of the problem. In the next step we transform abstract terms into "concrete" ones. In that way also the process of proving program correctness splits into two independent phases: the demonstration of the correctness of an abstract program and its implementation.

This paper is related to the paper of C.A.R.Hoare [5]. The operational point of view on implementations is adopted in contrast with axiomatic ([7],[8]) or categorical ([9]) point of view. The criterion of correctness of an implementation is given, which reduces the question about correctness to the proof of some algorithmic formulas.

1. *Many sorted relational systems and algorithmic languages*

Suppose we are given many-sorted universe of objects $\bigcup_{s \in S} A_s$ with operations and relations forming many-sorted relational system $\mathfrak{A} = \langle \{A_s\}_{s \in S}, \Theta, \rho \rangle$ where Θ is the set of operations, ρ is the set of relations such that for each $o \in \Theta$ there exist $k \geq 0$ and $s_1, \ldots, s_k, s_{k+1} \in S$ such that $o: A_{s_1} x \ldots x A_{s_k} \rightarrow A_{s_{k+1}}$ and for each $r \in \rho$ there exist $k > 0$ and $s_1, \ldots, s_k \in S$ such that $r \subseteq A_{s_1} \times \ldots \times A_{s_k}$.

From this moment on the relational system $\mathfrak{A}$ will be fixed. For a relational system $\mathfrak{A}$ we use formalized algorithmic language denoted by $L(\mathfrak{A})$ (for details see [2] or [6]).

The language $L(\mathfrak{A})$ includes the following basic sets of algorithmic expressions:

1. The set of instructions FS.

These instructions are usually called FS-programs. They are built from the multiple assignment statements and are closed under the formation of compound, branching and while statements.

2. The set of algorithmic terms.

Algorithmic terms include all functional expressions of the first order predicate calculus. An additional algorithmic construction is the following. Given an instruction J and a term τ the expression $J\tau$ is a term. The value of $J\tau$ is the value of τ after performing the instruction J.

3. The set of algorithmic formulas.

This set includes all the formulas of the first order predicate calculus and moreover some constructions binding instructions to formulas. The most important among these is the weakest precondition $wp(J,\alpha)$ of an instruction J with respect to a formula α ([4] and [6]). Instead of the notation $wp(J,\alpha)$, in the papers on algorithmic logic it is customary to write simply $J\alpha$. Roughly speaking the meaning of the formula $J\alpha$ is the value of the formula α after performing the instruction J. More exactly the value of the formula $J\alpha$ at a valuation v of variables is *true* if and only if the computation of J terminates for v and the final valuation of variables statisfies the formula α. In particular the value of the formula J *true* at a valuation v of variable is *true* if and only if the computation of J terminates for v. Similarly, the value of the formula $J_1x=J_2x$ at a valuation v of variable is *true* if and only if the computations of the instructions J_1 and J_2 terminate for v and the final values of the variable x are the same.

By loop we denote the instruction *while true do* x:=x, where x is a fixed variable. The execution of the instruction loop never ends. So, the final valuation of variables for the instruction loop is undefined.

In each instruction J we will distinguish four kinds of variables called as follows:

- input variables,
- output variables,
- data structure variables,
- auxiliary variables.

The only assumptions about this partition of the variables of the instruction J are the following:

- the respective sets of variables of each kind are pairwise disjoint;
- the initial values of output and auxiliary variables do not influence the computation of the instruction J.

Let $K=\{K_o,K_1,\ldots,K_m\}$ be a finite set of instructions each with distinguished four sets of variables and with common data structure variables denoted in the sequel by y. In practice we are interested in executing sequences of instructions starting from a certain initialization of the data structure variables. This initialization is accomplished by the first instruction K_o. We will assume that K_o does not contain and input or output variables and that the unique computation of K_o is finite.

By $\Sigma(K)$ we denote the set of all finite sequences $\sigma = \sigma_o;\sigma_1;\ldots;\sigma_n$ where $n \geq 0$, $\sigma_o = K_o$ and for each $i=1,2,\ldots,n$ $\sigma_i \in K-\{K_o\}$.
The elements of the set $\Sigma(K)$ will be called K-programs.

The intuitive meaning of the above given definitions is as follows. Let us suppose that the instructions of the set $K=\{K_o,K_m,\ldots,K_m\}$ are used in a program P. P will be called the main program. Consider a finite computation of P for some initial valuation of variables. During this computation executions of some instructions from K take place. Suppose that $\sigma = \sigma_o;\sigma_1;\ldots;\sigma_n$ is the sequence of instructions called consecutively by the main program P. First the execution of the instruction $\sigma_o=K_o$ initializes the values of data structure variables. Afterwards some computations take place in the program P, which however do not change the values of data structure variables. The main program P prepares values of input variables for the instruction σ_1 and then σ_1 is called. The execution of the instruction σ_1 can change the values of data structure variables and upon the completion σ_1 communicates the values of its output variables to the main program P. This process is repeated for the instructions $\sigma_2,\sigma_3,\ldots,\sigma_n$, consecutively. At any time the values of data structure variables can be changed only by one of the instructions from K. In this way the values of data structure variables from a certain quality of data distinguished from other data. The main program may use the values of data structure variables only by calling instructions from the set K.

The set K of instructions together with the partitions of variables of each instruction in K into the above described four categories will be called a data type.

With the above presented intuition in mind we adopt the following definition of a computation of a K-program. Namely a computation of a K-program $\sigma = \sigma_o;\sigma_1;\ldots;\sigma_n$ consits of three sequences of values:

- $x_1,x_2,\ldots,x_n$,
- $y_1,y_2,\ldots,y_n,y_{n+1}$,
- $t_1,t_2,\ldots,t_n$,

such that the following conditions hold:

(1) $x_1,x_2,\ldots,x_n$ are values of input variables of the instructions $\sigma_1,\sigma_2,\ldots,\sigma_n$, respectively;

(2) $y_1,y_2,\ldots,y_n,y_{n+1}$ are values of data structure variables;

(3) $t_1,t_2,\ldots,t_n$ are values of output variables of the instructions $\sigma_1,\sigma_2,\ldots,\sigma_n$, respectively;

(4) the computation of the instruction $\sigma_o=K_o$ terminates and the final values of data structure variables are y_1;

(5) for each $i=1,2,\ldots,n$, the computation of the instruction σ_i for the values x_i of input variables and for the values y_i of data structure variables terminates and the final values of data structure variables are y_{i+1} and the final values of output variables are t_i.

It is convenient to present a computation of a K-program $\sigma = \sigma_o;\sigma_1;\ldots;\sigma_n$ by means of the diagram in Fig.1. (assuming the above denotations).

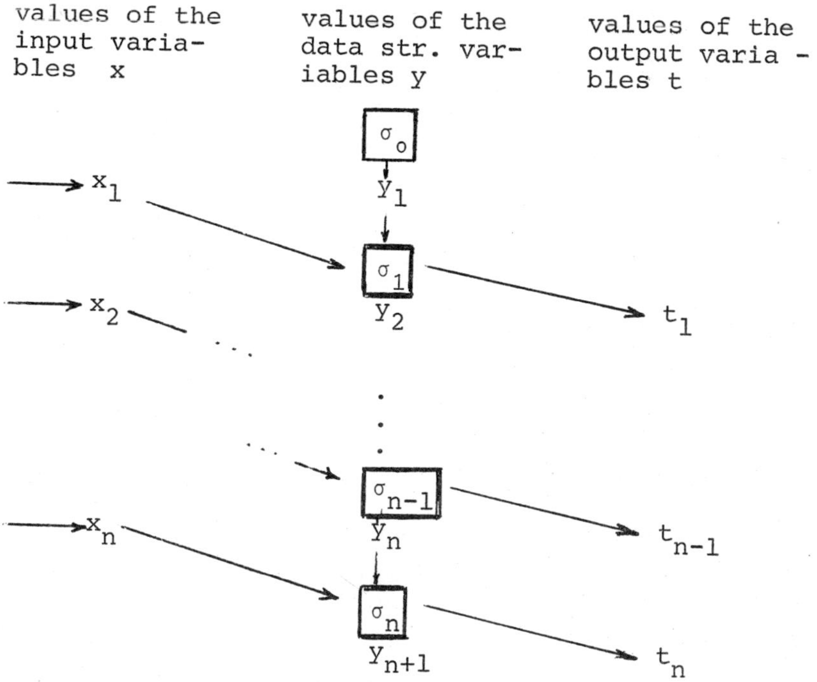

Fig.1.

Example 1.

Let N be a positive integer. Consider the system whose universe includes the set of integers $\{1,2,...,N\}$, the set of Boolean values $\{\textit{false},\textit{true}\}$ and the set $P(N)$ of all subsets of the set $\{1,2,...,N\}$. Assume that the following set operations and relations are in the system under consideration:

- the constant $\emptyset$ (the empty set);
- two-argument operations $\cup$ and $-$ of union and difference os sets, respectively;
- two-argument relation $\in$ of being an element of a set;
- the operation $\{x\}$ of composing one-element set.

Let S be a variable assuming its values from the set P(N), let i be a variable assuming its values from the set {1,2,...,N} and let q be a Boolean variable. Let us consider the following data type DICTIONARY = = {init, insert, delete, member} where

init = (S:=∅);

insert = (S:=S ∪ {i});

delete = (S:=S-{i});

member = (q:=i ∈ S).

We assume that i is an input variable for the instructions insert, delete and member, S is a common data structure variable and q is an output variable for the instruction member. These instructions do not contain any auxiliary variables.

For example the sequence

σ = init; insert; insert; member; delete

is a DICTIONARY-program. If we deliver the values 1,2,2,2 for the input variables of the consecutive instructions in σ, we receive the computation depicted in Fig.2.

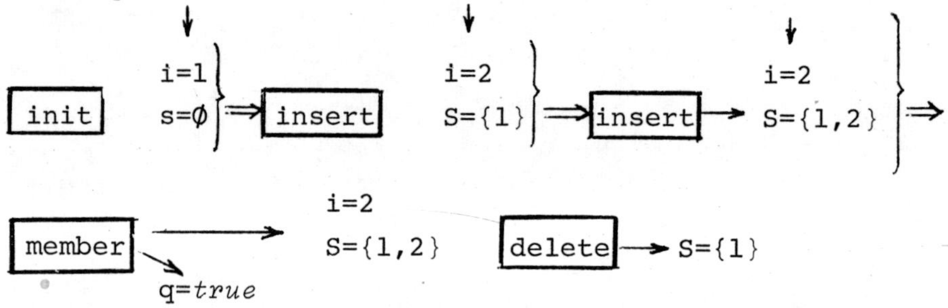

Fig.2.

2. *A notion of an implementation*

By a restriction of many-sorted relational system $\mathfrak{A} = \langle \{A_s\}_{s\in S}, \Theta, \rho \rangle$ we will understand any many-sorted relational system of the form $\mathfrak{B} = \langle \{A_s\}_{s\in S'}, \Theta', \rho' \rangle$ provided $S' \subset S$, $\Theta' \subseteq \Theta$ and $\rho' \subseteq \rho$. In the sequel we will assume that the language $L(\mathfrak{B})$ of the restriction $\mathfrak{B}$ of the system $\mathfrak{A}$ is a restriction of the language $L(\mathfrak{A})$ i.e. $L(\mathfrak{B}) \subset L(\mathfrak{A})$.

Now we will proced to the definition of the notion of an implementation of a data type by other data types. Let $\mathfrak{A}_1$ and $\mathfrak{A}_2$ be restrictions of the system $\mathfrak{A}$ and let $K = \{K_o, K_1, \ldots, K_m\}$, $M = \{M_o, M_1, \ldots, M_m\}$ be two data types such that the instructions of the data type K belong to the language $L(\mathfrak{A}_1)$ and the instructions of the data type M belong to the language $L(\mathfrak{A}_2)$. The system $\mathfrak{A}_1$ will play the role of an "abstract" one, while the system $\mathfrak{A}_2$ will play the role of a "concrete" one. The data type K will be called an abstract data type, on the other hand the data type M will be called a concrete data type. The instructions of the concrete type M will simulate the instructions of the abstract data type. Namely the instruction M_i is to simulate the instruction K_i for $i=0,1,\ldots,m$. If $\sigma = K_{i_o}; K_{i_1}; \ldots$ $\ldots; K_{i_n}$ is a K-program then the following M-program $\bar{\sigma} = M_{i_o}; M_{i_1}; \ldots; M_{i_n}$ is to simulate the K-program σ.

The intuitive meaning of the above given definitions is as follows. Let us suppose that we solved a problem by means of a program P, which uses a data type $K = \{K_o, K_1, \ldots, K_m\}$ and that the data type K includes in its instructions all the operations which cannot be directly carried out by a computer. In the second step we look for a data type $M = \{M_o, M_1, \ldots, M_m\}$, whose instructions can be directly carried out by a computer and such that for each $i=0,1,\ldots,m$ the instruction M_i simulates the action of the instruction K_i. In this way if we replace in the program P each occurrence of an instruction K_i, for $i=0,1,\ldots,m$, by an instruction M_i, we get the program P′ which solves the problem under consideration and which can run on a computer. Given common input data for programs P and P′ the program P calls consecutively some instructions $K_{i_o}, K_{i_1}, \ldots, K_{i_n}$ of the data type K, while at the same time the program P′ calls consecutively the correspondent instructions $M_{i_o}, M_{i_1}, \ldots, M_{i_n}$ of the data type M.

Finally we need to establish the connections between abstract and concrete objects involved in computations of K-program and M-programs. We will use "abstract functions" introduced by Hoare [5]. In contrast with Hoare we will require that these functions are defined by instructions belonging to the language $L(\mathfrak{A})$. The reason of this is to enable to carry out proofs about correctness of implementations of data types in a formal way. Moreover we extend the process of representation on values of input and output variables (Hoare considered the representation only between values of data structure variables). The necessity of this extention will be clear in the following example.

Example 2.

Let N be a positive integer. Consider the system whose universe includes the set of integers $\{1,2,\ldots,N\}$, the set $P(N)$ of all subsets of the set $\{1,2,\ldots,N\}$ and the set $U(N)$ of all partitions of the set $\{1,2,\ldots,N\}$ (i.e. $\pi \in U(N)$ if each element of π is a subset of the set $\{1,2,\ldots,N\}$, $\cup\pi = \{1,2,\ldots,N\}$ and any two non-equal sets A and B in π are non-empty and disjoint).

The list of needed operations and relations for this system is rather long. Since these operations and relations are commonly used in mathematical papers, we do not list them explicitly. From the expressions which use the denotations of these operations and relations the reader will have no trouble in recognizing them.

Let R be a variable assuming its values from the set $U(N)$, let A, B, C, D be variables assuming their values from the set $P(N)$ and let i,q be variables assuming their values from the set $\{1,2,\ldots,N\}$. Let us consider the following data type FU = {init,find,union} where

```
init = begin
          R:=∅; q:=1;
          while q ≤ N do
          begin
            R:=R ∪ {{q}};
            q:=q+1
          end
       end
```

(the instruction init forms the initial partition $R = \{\{1\},\{2\},\ldots,\{N\}\}$);

find = (C:=the unique element of the set ({D∈R:i∈D}))

(the instruction find computes the set C such that the integer i belongs to the set C and the set C belongs to the partition R);

```
union = if  A∈R ∧ B∈R then
        begin C:=A∪B;
              R:=(R-{A,B}) ∪ {C}
        and else loop
```

(if the sets A and B belong to the partition R then the instruction union replaces the sets A and B in the partition R by their union, C = A ∪ B, otherwise the final valuation of variables is undefined). We will present the so-called tree implementation of the type FU (for the details see [1]). Each set belonging to P(N) will be represented as a tree. The whole partition of the set {1,2,...,N} will be represented as a forest of trees, each tree representing one set of the partition. The vertices of this forest are integers 1,2,...,N. The edges of this forest are given by the contents of an integer array T[1:N]. Namely for each i=1,2,...,N, the value T[i] is an integer such that $0 \leq T[i] \leq N$ and if for some $1 \leq i,j \leq N$, T[i]=j then the pair (i,j) forms an edge of the forest under consideration. If T[i] = 0 then this means that the vertex i is a root of some tree. For example the contents of the array T shown in Fig.3.

1	2	3	4	5	6	7	8
0	1	1	3	0	5	1	0

Fig.3.

determines the forest which can be depicted as follows

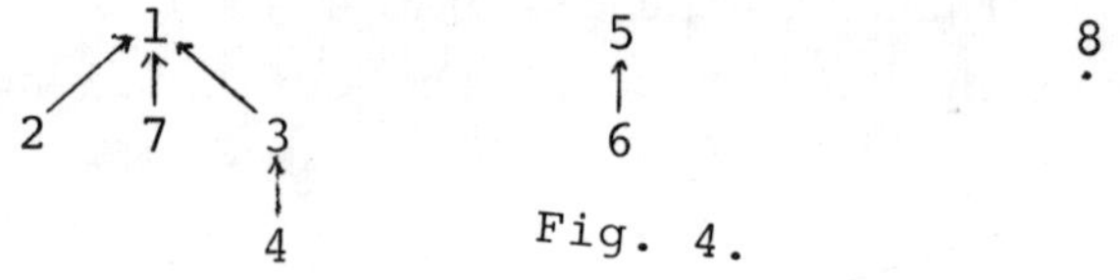

Fig. 4.

(arrows goes from sons to father of vertices).
The forest of Fig.4. represents the partition {{1,2,3,4,7},{5,6},{8}}. Each root of a tree represents the set of all vertices composing this tree. For example the root 1 represents the set {1,2,3,4,7}.
Now we proceed to the definition of the concrete data type FU-tree = (INIT,FIND,UNION). The only data structure variable in the instructions of this data type is the integer array T[1:N]. Let q, i, a, b, c be integer variables. We define

```
INIT = begin
          q:=1;
          while q ≤ N do (q,T[q]):=(q+1,0)
       end
```

(the instruction INIT builds the forest composed of N one element trees);

```
FIND = begin
          (c,q):=(i,T[i]);
          while q ≠0 do (c,q):=(q,T[q])
       end
```

(the instruction FIND begins at the given vertex i and follows the path to the root of the tree. Upon the completion the value of c is the root of the tree);

```
UNION = if T[a]=0 ∧ T[b]=0 then
        begin
          if a ≠ b then T[a]:=b;
          c:=b
        end else loop
```

(the instruction UNION combines the trees given by their roots a and b into one tree. Upon the completion the former root a points now to b and the value of c is the root b of the new tree. If a and b are not roots then the final valuation of variables is undefined).

The variables a,b and i are input variables, c is an output variable, T is a data structure variable and q is an auxiliary variable.

The data structure variables of the types K and M will be denoted by y and z, respectively. To simplify further considerations we will assume that all the instructions of the data type K have common input variables x and common output variables t and similarly, all the instructions of the data type M have common input variables u and common output variables r.

To define an implementation of a data type $K = \{K_0, K_1, \ldots, K_m\}$ by means of a data type $M = \{M_0, M_1, \ldots, M_m\}$, three interpretation instructions must be provided. Namely:

$I_{DS}(z;y)$ - determines the values of data structure variables y that correspond to given values of data structure variables z and the values of the variables z are not altered;

$I_{IN}(z,u;x)$ - determines the values of input variables x that correspond to given values of input variables u and data structure variables z and the values of z and u are not altered;

$I_{OU}(z,r;t)$ - determines the values of output variables t that correspond to given values of output variables r and data structure variables z and the values of r and z are not altered.

The pair $I = (M, I_{DS}, I_{IN}, I_{OU}))$ will be called an implementation of the data type K.

In the sequel we adopt following denotations. For u_o, z_o, r_o being values of input variables u, data structure variables z and output variables r, respectively, we will denote

by $I_{DS}(z_o;y)_{\mathfrak{A}}(y)$ the values of data structure variables y determined by the isntruction I_{DS} for $z=z_o$;

by $I_{IN}(z_o,u_o;x)_{\mathfrak{A}}(x)$ - the values of input variables x determined by the instruction I_{IN} for $z=z_o$ and $u=u_o$;

by $I_{OU}(z_o,r_o;t)$ - the values of output variables t determined by the instruction I_{OU} for $z=z_o$ and $r=r_o$.

Example 3.

Now let's get back to our previous example 2. Let for a natural number h and integer $1 \leq p \leq N$,

$$T^h[p] = \begin{cases} p & \text{if } h=0 \\ T^{h-1}[T[p]] & \text{if } h > 0 \end{cases}$$

Recall that the contents of the array T and an index $1 \leq i \leq N$ determine a subset of the set $\{1,2,\ldots,N\}$ consisting of all vertices $1 \leq j \leq N$ from which the vertex i is reachable in the forest. Denote this set by set (i,T) i.e. set $(i,T)=\{j:\exists n\in\{0,1,\ldots,N\}T^h[j]=i\}$. Now we will define three representation instructions for the tree implementation of the data type FU.Namely:

$I_{DS}(T;R) = (R:=\{set(q,T):1 \leq q \leq N \wedge T[q]=0\})$

(the forest given by T represents the partition consisting of all sets of the form set(q,T), where q is a root of some tree);

$I_{IN}(T,i,a,b;i,A,B) = ((A,B) := (set(a,T),set(b,T))$

(the values a and b represent the sets set(a,T) and

set(b,T), respectively, the interpretation of the value i does not change);

$I_{OU}(T,c;C) = (C:=\mathrm{set}(c,T))$

(the value c represents the set set(c,T)).

3. Correctness of an implementation

The question arises under what conditions M--programs can be regarded as correct simulations of K-programs. We adopt the following definition.

The implementation $I=(M,(I_{IN},I_{DS},I_{OU}))$ of the data type K is said to be correct if for every K-program $\sigma = \sigma_o;\sigma_1;\ldots;\sigma_n$, for every computation of the correspondent M-program $\bar{\sigma} = \mu_o;\mu_1;\ldots;\mu_n$ given in Fig.5.

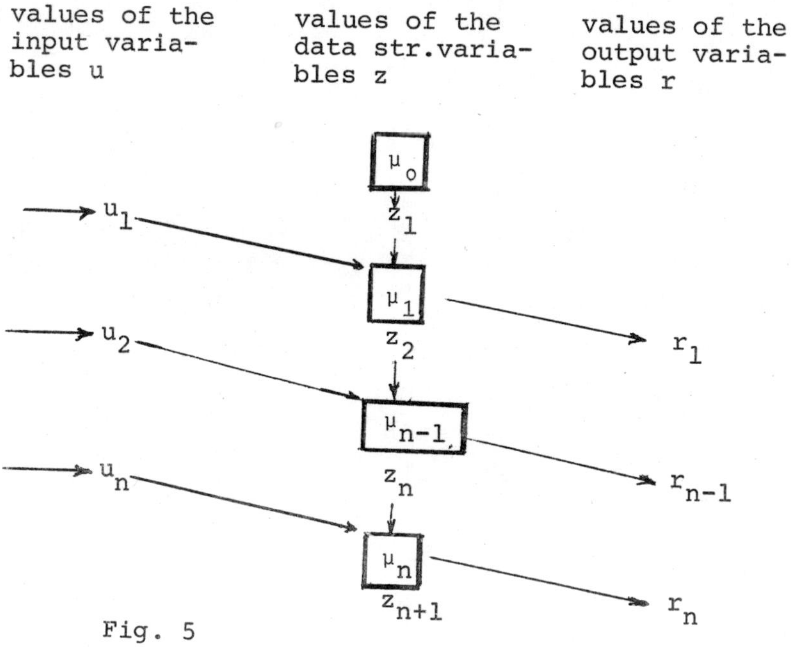

Fig. 5

there exists a computation of the K-program σ of the form given in Fig. 6

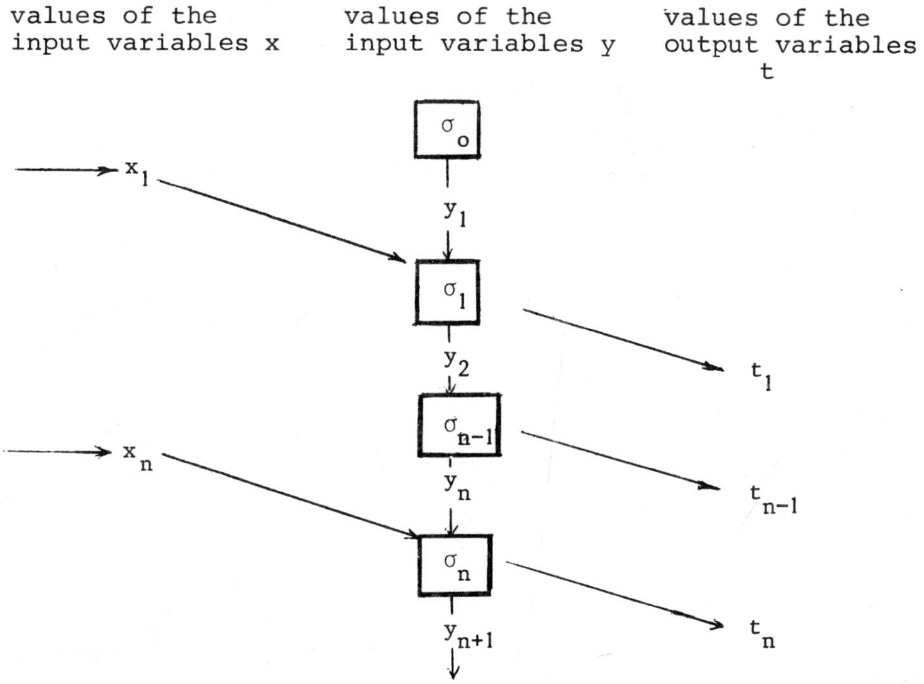

Fig. 6

and such that the following conditions hold:

(1) for each $i=1,2,..,n$ the values $I_{IN}(z_i,u_i;x)_{\mathfrak{A}}(x)$ are defined and equal to x_i;

(2) for each $i=1,2,...,n+1$ the values $I_{DS}(z_i;y)_{\mathfrak{A}}(y)$ are defined and equal to y_i;

(3) for each $i=2,3,...,n+1$ the values $I_{OU}(z_i,r_{i-1};t)_{\mathfrak{A}}(t)$ are defined and equal to t_{i-1}.

That is, by means of the results of the computation of the "concrete" M-program $\bar{\sigma}$ and interpretation instructions we can receive the results of the computation of the "abstract" K-program σ.

It seems that the definition of a correct implementation cannot be immediately written by means of a formula of algorithmic logic. Nevertheless it seems that in practical situations we need not leave behind algorithmic formulas. Our reasoning about sequences of instructions consists of considering each instruction separately.

Below there is a criterion of the correctness of an implementation. It uses the notion of the validity of algorithmic formulas. For the readers who are not familiar with the notation $J\alpha$, an informal explanation of the meaning of particular formulas will be given.

Let $\Delta(z)$ be an arbitrary algorithmic formula which will play the role of an invariant for all the instructions of the concrete data type $M=\{M_o,M_1,\ldots,M_m\}$.

Theorem 1.

If the following formulas are valid in $\mathfrak{A}$

(*a*) $M_o\Delta(z)$,

(*b*) $\bigwedge_{i=1}^{m} (\Delta(z) \wedge M_i \textit{ true} \Rightarrow M_i\Delta(z))$

(*c*) $K_oy = M_oI_{DS}\, y$,

(*d*) $\bigwedge_{i=1}^{m} \{\Delta(z) \wedge M_i \textit{ true} \Rightarrow (M_i\, I_{DS}\, y = I_{IN}\, I_{DS}\, K_i\, y \wedge M_i\, I_{OU}\, t = I_{IN}\, I_{DS}\, K_i\, t)\}$

$I = (M,(I_{DS},I_{IN},I_{OU}))$ *is a correct implementation of the data type* K.

The validity of the consecutive formulas in theorem 1 can be expressed as follows:

($\bar{a}$) *the condition* $\Delta(z)$ *holds after performing the instruction* M_o,

($\bar{b}$) *the condition* $\Delta(z)$ *is preserved under each instruction* M_i *for* $i=1,2,\ldots,m$.

($\bar{c}$) *the computations of all the instructions in the following diagram are finite and this diagram is commutative*

$$\begin{array}{ccc} K_o & \xrightarrow{K_o} & y \\ & & \uparrow {\scriptstyle I_{DS}} \\ M_o & \xrightarrow{M_o} & z \end{array}$$

($\bar{d}$) *for each* $i=1,2,\ldots,m$ *if values of data structure variables* z *satisfy the condition* $\Delta(z)$ *and the computation of the instruction* M_i *terminates then the computations of all the instructions in the following diagram are finite and this diagram is commutative*

$$\begin{array}{ccccc} & y,x & \xrightarrow{K_i} & y',t' & \\ I_{DS},I_{IN} & \uparrow & & \uparrow & I_{DS},I_{OU} \\ & z,u & \xrightarrow{M_i} & z',r' & \end{array}$$

Proof. The validity of the formulas (a) and (b) guarantees that the formula $\Delta(z)$ is an invariant of any M-program $\bar{\sigma}$ for $\sigma \in \Sigma(K)$. That is, during the executinon of any M-program $\bar{\sigma}$ the values of data structure variables z always satisfy the condition $\Delta(z)$.

Let $\sigma = \sigma_o;\sigma_1;\ldots;\sigma_n \in \Sigma(K)$ and $\bar{\sigma} = \mu_o;\mu_1;\ldots;\mu_n \in \Sigma(M)$. Let us consider an arbitrary computation of the M-program $\bar{\sigma}$ given Fig.5. If follows that during the realization of $\bar{\sigma}$ the computation of each instruction μ_i is finite for $i=0,1,\ldots,n$. Because $\Delta(z)$ is an invariant then on account of the validity of the formula (d) all the values $I_{IN}(z_i,u_i;x)_{\mathfrak{A}}(x)$ for $i=1,2,\ldots,n$ are defined. Now carry out the program σ

Boolean array [1:N], i is an input variable assuming its values from the set {1,2,...,N}, q is an output variable of the type Boolean,

I_{IN} = (i:=i)

I_{DS} = S:={j:T[j]}

(the Boolean array T represents a subset S of the set {1,2,...,N} in this way that j ∈ S if and only if T[j] = *true*),

I_{OU} = (i:=i).

In virtue of the corollary 1, in order to prove the correctness of this implementation it is sufficient to prove the validity of the following algorithmic formulas(the conditions concerning the values of the variable q for INSERT and DELETE and T for MEMBER are omitted because the values of these variables are not changed in there):

[S:=∅]S=(*for* j:=1 *to* N *do* T[j]:=*false*;S:={j:T[j]})S ,

(T[i]:=*true*;S:={j:T[j]})S=(S:={j:T[j]};S:=S∪{i})S ∧

(T[i]:=*false*;S:={j:T[j]})S=(S:={j:T[j]};S:=S-{i})S ∧

(q:=T[i])q=(S:={j:T[j]};q:=i ∈ S)q. ◉

4. *A remark on abstract data types defined axiomatically*

The approach presented in the paper can also be applied in the case of abstract data types defined axiomatically. Instead of one abstract relational system we must take a class of abstract relational systems satisfying certain axioms Ax. On the account of the completeness theorem [2] we obtain(using previous denotations):

Corollary 2.
If the formulas (a), (b), (c), (d) are theorems derivable from the set Ax *of axioms then* I *is a correct implementation of the data type* K *in any model of the set* Ax. ◉

References

[1] A.V.Aho, J.E.Hopcroft, J.D.Ullman, *The design and analysis of computer algorithms*, Addison-Wesley, 1974

[2] L.Banachowski, Investigations of properties of programs by means of the extended algorithmic logic, *Fundamenta Informaticae* vol.1, 1977, 93-119

[3] L.Banachowski, A.Kreczmar, M.Mirkowska, H.Rasiowa, A.Salwicki, An introduction to algorithmic logic, in Banach Center Publications, vol.2. 1977

[4] E.W.Dijkstra, *A Discipline of Programming*, Englewood Cliffs, 1976

[5] C.A.R.Hoare, Proof of Correctness of Data Representations, *Acta Informatica*, vol.1. 1972, 271-281

[6] A.Salwicki, Formalized algorithmic languages, *Bull. Acad. Polon. Sci.* vol.18. 1970, 227-232

[7] A.Salwicki, Algorithmic theory of stacks, in J. Winkowski (ed.) *Mathematical Foundations of Computer Science* 1978, Lecture Notes in Computer Science, vol.64, Springer-Verlag, 1978, 452-461

[8] J.V.Guttag, E.Horrowitz, D.R.Musser, Abstract data types and software validation, USC Information Sciences Institute, Rep. ISI/RR-76-48, 1976

[9] J.A.Goguen, J.W.Thatcher, E.G.Wagner, An initial algebra approach to the specification, correctness and implementation of abstract data types, in R.T. Yeh (ed.) *Current trends in programming methodology*, vol.3. Prentice Hall, 1978

Institute of Informatics
Warsaw University
00-901 Warszawa
PKiN pok. 850
Poland

COLLOQUIA MATHEMATICA SOCIETATIS JÁNOS BOLYAI
26. Mathematical Logic in Computer Science,
Salgótarján (HUNGARY), 1978.

PROCESS SEMANTICS OF ALGEBRAIC DATATYPES

J.A.Bergstra and Th.P. van der Weide

0. Introduction

We consider datatypes derived from algebraic structures. Given an algebraic structure we define a process which embodies an operational semantics of the corresponding datatype. In this way we are able to define the operational meaning of axiomatic specifications. Using recursion on processes we can define relative implementability and operational equivalence of algebraic datatypes.

The outline of the paper is as follows. First we introduce the notion of processes. In Section 2 we associate a process with an algebraic structure. The same is done for parametrised families of data structures. The notion of relative implementability is introduced in Section 4. After having introduced the notion of axiomatically specified datatypes we define operational semantics for them. Finally in Section 6 we state some conclusions.

1. Processes

A process P is an entity which can communicate with the outside world W by means of:

- an input alfabet Σ_1
- an output alfabet Σ_o.

After P has been originated, the interaction between P and W will be a permanent alternation of the following actions:

- W sends a message m ($\in\Sigma_1$) to P, and is ready to receive the reply from P
- P computes its reply r ($\in\Sigma_o$) to the received message m, sends is to W, and is ready to receive a new message from W.

The process is completely determined by the function φ, which tells what outputsymbol P will give, after having received inputstring σ from W since its origination (the first letter from σ being the first letter send by W,etc.), for all $\sigma \in \Sigma_1$.

The process P will also be denoted as $<\Sigma_1,\Sigma_o,\varphi>$. For sake of convencience we introduce the set $\{P_\sigma \mid \sigma \in \Sigma_1\}$, referred to as the set of states of P, where P_σ is a function from Σ_1 into Σ_o, defined by:

$$P_\varepsilon = \varphi$$

$$P_\sigma(\tau) = P_\varepsilon(\sigma * \tau)$$

We call P_σ: the state of P after having received inputstring σ from W since its origination.

When we evaluate a function twice for the same argument, but on different moments, the delivered result will be the same. A process however will not have this property in general. Formally:

> whenever $\sigma \neq \tau$, the result delivered by P on argument m, i.e. $P_\sigma(m)$ and $P_\tau(m)$, need not be equal.

So a process is able to learn and forget.

The class of all processes with input alfabet Σ_1 and output alfabet Σ_0 will be denoted by: $PR(\Sigma_1,\Sigma_0)$. Two processes are equal iff their input alfabet, output alfabet and input-output behaviour are equal.

2. *Algebraic Structures and Processes*

We consider algebraic structures

$$\mathfrak{A} = \langle\vec{A},\vec{F},\vec{R},\vec{C}\rangle$$

with finitely many domains ($\vec{A}$), functions ($\vec{F}$), relations ($\vec{R}$) and constants ($\vec{C}$), of several types. We are interested in the use of such algebraic structures as a basis for giving semantics of datatypes (a precise definition of a datatype will follow in section 5). A datatype will essentially be a facility to 'work' with some algebraic structure.

The achieve this, working with an algebraic structure $\mathfrak{A}$ should allow for:

(1) the maintainance of some finite collection $V=\{x,y,\ldots\}$ of variables, each of some fixed type from $\vec{A}$.

(2) communications of the form:

$$x := c$$
$$x := y$$
$$x := f(\vec{v})$$

with $x,y \in \vec{V}$, $c \in \vec{C}$, $f \in \vec{F}$, and $\vec{v}$ a list of variables from V, such that all types are suitable.

(3) questions of the form:

$$r(\vec{v})$$

with $\vec{v}$ a list of variables from V and $r \in R$, such that all types are suitable.

Note that $x=y$ can only be evaluated if equality is one of the relations in $\vec{R}$ (otherwise it has to be computed!)

Now, given $\mathfrak{A}$ and V we define a process $P=P(\mathfrak{A},V)=$ $=\langle\Sigma_1,\Sigma_o,\varphi\rangle$ as follows:

+ Σ_1 is the set of all possible instructions $(x:=c,\ x:=y,\ x:=f(\vec{v}),r(\vec{v}))$ (note that Σ_1 is finite)
+ $\Sigma_o = \{\text{blank, true, false, error}\}$
+ φ is defined by: ($\ddagger$)

$$\varphi(\sigma b)=\begin{cases}\text{blank if } b \text{ is an assignment}\\ \left.\begin{array}{l}\text{true if } r(val(\sigma)(\vec{v}))\\ \text{false if } \neg r(val(\sigma)(\vec{v}))\end{array}\right\}\text{and } \begin{array}{l}b=r(\vec{v})\ \wedge\\ \forall_i[val(\sigma)(v_i)\neq\bot]\end{array}\\ \text{error otherwise}\end{cases}$$

Here $\bot$ is a specially by introduced value, corresponding to 'undefined'.

Furthermore the function

$$val:\ \Sigma_\tau \rightarrow [\{\bot\}\cup\cup\vec{A}]^V$$

is defined by:

$$val(\sigma)(z) = \bot$$

$$val(\sigma\ \text{question})(z) = val(\sigma)(z)$$

$$val(\sigma\ x:=c)(z) = \begin{cases}val(\sigma)(z) & \text{if } x\neq z\\ c & \text{if } x=z\end{cases}$$

$$val(\sigma\ x:=y)(z) = \begin{cases}val(\sigma)(z) & \text{if } x\neq z\\ val(\sigma)(y) & \text{if } x=z\end{cases}$$

$$val(\sigma\ x:Zf(\vec{v}))(z)=\begin{cases}val(\sigma)(z) & \text{if } x\neq z\\ f(val(\sigma)(\vec{v})) & \text{if } x=z\ \wedge\ \forall_i[\mathbf{val}(\sigma)(\mathbf{v}_i)\neq\bot]\\ \bot & \text{otherwise}\end{cases}$$

Remark: It would be closer to reality to choose

$$\Sigma_1 = V\cup\vec{F}\cup\vec{R}\cup\vec{C}\cup\{:,=,(,)\}.$$

However, this approach is not more general, and involves awkward details.

($\ddagger$) $f(val(\sigma)(\vec{v}))$ will stand for $f(val(\sigma)(v_1),\ldots,val(\sigma)(v_k))$, when $v=(v_1,\ldots,v_k)$.

3. *Parametrised Families of Data Structures*

Suppose we are dealing with a family of algebraic structures of the same similarity type T, which are parametrised by sequence $\tau \in \Delta^*$ for some finite alfabet Δ, such that $\mathfrak{A}(\tau)$ is the structure corresponding to τ. Then we will look at processes $Q(T, \mathfrak{A}_i, V)$, which first read a parameterstring $\tau \in \Delta^*$, then a marker #, and thereafter behave like $P(\mathfrak{A}(\tau), V)$ (here V is fixed for Q!). So we have for Q:

+ $\Sigma_1 = \Delta \cup \{\ \} \cup$ ('the set of all possible instructions')

+ Σ_O = {blank, true, false, error}
+ $Q_\tau = P(\mathfrak{A}(\tau),V)_\varepsilon$ for all $\tau \in \Delta^*$

Example:

Let $\Delta = \{0,1\}$. Consider the structure

$$\mathbb{N} = \langle\omega,succ,pred,+,*,=0,=,0,1,c\rangle$$

where ω is the set of natural numbers, and c is some constant in ω. Now we define $\mathbb{N}(\tau)\cdot\mathbb{N}(\tau)$ has the same similarity type as $\mathbb{N}$, but it has a finite domain with $\tau+1$ elements, where τ is seen as a binary written integer. Moreover, succ,+ and * have been adapted to deal with overflow.

$$\mathbb{N}(\tau) = \langle\{0,\ldots,\tau\},succ_\tau,pred_\tau,add_\tau,mult_\tau,zero_\tau, eq_\tau,0,one_\tau,maxint_\tau\rangle$$

where:

- $one_\tau = 1$ if $\tau = 0$, else $one_\tau = 0$
- $maxint_\tau = \tau$
- the operations $pred_\tau, zero_\tau, eq_\tau$ are exactly like pred, = 0 and =.
- the operations $succ_\tau, add_\tau, mult_\tau$ are like succ, + and * when the result is not greater then $maxint_\tau$; in the other case (overflow) the result is undefined, and the output is 'error'.

Notice that overflow conventions are choosen quite similarly to Pascal [7].

In this example the function $Q_{T*}(T,\mathbb{N},V)$ can be realised by a finite automaton. It is in fact a rational data object in the sense of [4].

Notation: when no confusion is likely to arise, we denote $Q(T,\mathfrak{A},\{x_1,\ldots,x_k\})$ as Q^k, and we will write x, y, z, ... for the variables.

4. *Relative Implementability*

In [1] the notion $P \leq_{FC} \vec{Q}$ (P is finite-control reducible to $\vec{Q}$) is introduced for processes P and $\vec{Q}$ with finite input and output alfabet. Processes $P(\mathfrak{A},V)$ and $Q(T,\mathfrak{A},V)$ are of this kind. In [1] it is claimed that $\leq_{FC}$ can be regarded as relative implementability for datatypes. We will now adapt the general formalism in [1] to our special context.

Informally we have $P \leq_{FC} \vec{Q}$ if P, seen as a coroutine which returns outputs in Σ_O^P on inputs in Σ_I^P, can be programmed by means of a simple program M, using given coroutines $\vec{Q}$ (starting in their initial situations), and has only auxiliary datatype finitely many registers (x,y,z,...) for symbols from the input and output alfabets from P and $\vec{Q}$. We say: M computes P from Q, and write: $P = M(\vec{Q})$.

Simple programs are of the form:

INIT(x) ; S

where S is a statement.

Statements are built as follows:

- x:=y , x:=Q(y) (with $Q \in \vec{Q}$), x:=NQA(y) are statements
- if S_1 and S_2 are statements, then so are:

S_1 ; S_2

if BOOL *then* S_1 *else* S_2 *fi*

while BOOL *do* S_2 *od*

Here BOOL stands for a boolean expression built up using $\neg, \wedge, \vee$ from atoms of the form x=y.

The meaning of the instructions is as follows:

- INIT(x) the first input given to P is assigned to x
- x:=NQA(y) the value of y is the output, returned by P, and the next input is assigned to x
- x:=Q(y) the value of y is given as an input to the process Q in $\vec{Q}$, and its reply is assigned to x

The control structures are selfexplaining.

Remarks:

(i) It is easily seen that these programs do not exceed the possibilities of finite control.

(ii) We will often consider the case that Q can give thruth values as outputs. Then one may admit Q(x) as boolean expression. Of course, if Q does not return a thruth value in this case, the execution of P cannot proceed, and P is not totally defined.

(iii) Type errors are possible (for example NQA(x) is only possible if the value of x is an output-symbol from P), and will result similarly in P being undefined.

PROPOSITION

$P \leq_{FC} P,\vec{Q}$

$P \leq_{FC} R,\vec{Q}_1$, and $R \leq_{FC} \vec{Q}_2$ implies $P \leq_{FC} \vec{Q}_1,\vec{Q}_2$

The proofs are easy.

Notice that $P \leq_{FC} Q,Q$ does not imply $P \leq_{FC} Q$ in general.

Definition

$P \equiv_{FC} Q$ iff $P \leq_{FC} Q$ and $Q \leq_{FC} P$

$\equiv_{FC}$ is an equivalence relation. It leads to a degree structure. The bottom degree, **O** , is the degree containing processes which can be realised by the described kind of programs, without the use of auxiliary datatypes (=processes, coroutines). It seems desirable to restrict oneself to computable processes, as these are clearly the only ones which possibly could be implemented.

We have defined $P \leq_{FC} \vec{Q}$ for a list of processes $\vec{Q}$. One might ask how to define $P \leq_{FC} \vec{Q}$, where all processes from $\vec{P}$ use the same processes $\vec{Q}$ simultaneously, exclusively and independently. Therefore we assign to a list $\vec{A}$ of processes a process $\Sigma\vec{A}$ which embodies simultaneos and independent processing of the process from A. The process $\Sigma\vec{A}$ is defined by:

+ input alfabet: the disjoint union of the input alfabets in the processes of $\vec{A}$
+ output alfabet: the disjoint union of the output alfabets in the processes of $\vec{A}$
+ input-output behaviour φ : $\varphi(\omega b) = \psi(\sigma b)$ where ψ is the input-output behaviour of the unique process in $\vec{A}$, which has b as an input symbol, and σ is constructed from ω by deleting all symbols, which are not input symbols to this unique process.

Now we define $\vec{P} \leq_{FC} \vec{Q}$ as: $\Sigma\vec{P} \leq_{FC} Q$.
Notice that we have: $\vec{P} \leq_{FC} \Sigma\vec{P}$,
and so of course: $P \leq_{FC} \vec{Q}$ implies $P \leq_{FC} \Sigma\vec{Q}$.

Examples

1. Let $\mathfrak{A}$ be a given structure, then obviously

   $P(\mathfrak{A}, \{x_1,\dots,x_k\}) \equiv_{FC} P(\mathfrak{A}, \{y_1,\dots,y_k\})$

   So, changing the names of variables does not effect the power of an algebraic datatype.

2. $P(\mathbb{N},\{x_1,x_2\}) \equiv_{FC} P(\mathbb{N},\{x_1,x_2,x_3\})$ (see example from section 3)

Proof: $P' = P(\mathbb{N},\{x_1,x_2\})$ allows for the simulation of two counters simultaneously and independently. Let CO be a process which corresponds to a counter. Then CO, CO $\leq_{FC} P'$, and hence the Turingtape TT (see [1]) can be simulated from P' ($TT \leq_{FC} P'$). As $P'' = P(\mathbb{N},\{x_1,x_2,x_3\})$ is a computable process, it can be computed from TT. Hence $P'' \leq_{FC} P'$. Of course $P' \leq_{FC} P''$ (by just not using x).

Comments: Basically this result states, that two counters and not more are necessary. However the detour along Turingtape is so impractical (i.e. a program M with $P'' = M(P')$ is long and slow) that we almost see this as a negative result, in that $\equiv_{FC}$ does not make sufficiently fine distinstions. Nevertheless we spot the cause of the trouble somewhere else, namely in the fact that P' cannot practically exist at all. There will usually be some upperbound to the integers (overflow). Taking into account that the owerflow can be done within this frame (next example) it makes the tour along Turingtape obsolate.

3. Let $P'' = Q(T,\mathbb{N},\{x_1,\ldots,x_n\})$(see example from Section 3)

Now we have e.g. $P'' \leq_{FC} 2P^2+P^1$.

Proof: We store the four integers from P'' in P^2+P^2. The register from P^1 will be used as a temporary storage, when an operation between registers from different processes P has to be performed.

Comment: Taking into account overflow (as done in this case by looking at a parametrised family of structures) it is quite compatible with comparing the power of algebraic datatypes using $\leq_{FC}$.

Remark: Note that $2P^2+P^1 \leq_{FC} P^4$ does not hold. This can be seen by considering that the simulation should also notice the situation of overflow. But each of the processes in $2P^2+P^1$ may have a different integer capacity. So when the simulation having received the integer capacity of one process, first simulates this process before receiving the integer capacity of an other process, conflicts may result.

5. *Operational Semantics for Axiomatically Specified Datatypes*

We define, according to [1], a datatypes as a triple $<\Sigma_1,\Sigma_o,S>$ where:

- Σ_1 is the input alfabet of the datatype
- Σ_o is the output alfabet of the datatype
- $S \subseteq PR(\Sigma_1,\Sigma_o)$, the operational semantics, is the set of process (called: implementations of the datatype), by which the input-output of the datatype is specified.

So a datatype is something with which we can communicate as specified by one of its implementations.
The datatype is called deterministic, when the communication gives unique answers, i.e. the set S (the operational semantics) contains one element (implementation).

It may be much more convenient to give a description of a datatype, without considering many details, as may be invoked by the specification of the operational semantics. An axiomatic description can be useful in such cases.

An axiomatically specified datatype $\mathbf{D}$ is a pair $<\mathbf{A},V>$, where:

- $\mathbf{A} = \langle T, \Delta \rangle$ is an axiomatisation, i.e.
  + T is a similarity type, called the syntactical description
  + Δ is a theory compatible with T, called the semantical description
- V is a set of variables; each variable is intended to hold values from some sort, as given in T.

This definition is analogous as in [3] and [4], through in both V is not mentioned explicitly. The syntactical description T and the set of variables fix in a straightforward manner the input and output alfabet Σ_I and Σ_O for the associated datatype. The operational semantics $\mathbf{S}(\mathbf{D})$ of this datatype will be defined as the set of all possible implementations of $\mathbf{D}$, i.e.:

$$\mathbf{S}(\mathbf{D}) = \{P(\mathfrak{A}, V) \mid \mathfrak{A} \text{ is a structure for } T, \text{such that } \mathfrak{A} \models \Delta\}$$

Example:
Consider structures of the form $\langle (O,S), \nu, (\Omega, \{e_i\}_{i=1}^{k}) \rangle$, where: the elements of O are called objects,
the elements of S are called selectors,
ν, the assignment operator, is of type $O*S*O \to O$.
The ν-operator assigns to an object in some selector an object as new value. In [5] these structures were introduced, without mentioning the constants $\{e_i\}$. Therefore we must slightly modify the axiomatisation given there.
A datatype, constructed from this similarity type and axiomatisation is obviously deterministic. However it will suffer from the fact that objects can only be built up. It would be convenient if we had some operator ψ to decompose objects, i.e.: $\psi(C) = \langle A, \beta, B \rangle$ such that $\nu(A, \beta, B) = C$. Instead of prescribing how an object can be decomposed, we rather leave this as peculiarity of the chosen implementation. Therefore we now consider

structures of the form:

$$<(O,S,\mathfrak{T}),(\omega,\psi,\pi_1,\pi_2,\pi_3),(\Omega,\{e_i\}_i)>.$$

Here $\mathfrak{T}$ stands for the set of triples $O*S*O$,

π_1 is of type $\mathfrak{T} \to O$, the projection of the first component

π_2 is of type $\mathfrak{T} \to S$, the projection of the second component

π_3 is of type $\mathfrak{T} \to O$, the projection of the third component

ψ is of type $O \to \mathfrak{T}$, the decomposition operator.

The following axioms are added:

$$\nu(\pi_1 \cdot \psi(A), \pi_2 \cdot \psi(A), \pi_3 \cdot \psi(A)) = A$$

$$A \neq \Omega \Rightarrow \pi_3 \cdot \psi(A) \neq \Omega$$

In the datatype constructed along the lines in the preceeding, decomposition is possible. However, decomposition is nondeterministic, i.e. the constructed datatypes are nondeterministic.

We introduce relative implementability for (algebraic) datatypes: we call (algebraic) datatype **D** relatively implementable to the (algebraic) datatype **E**, iff there exists some simple program (see Section 4) M, such that:

$M(\vec{P})$ is an implementation of **D** for all implementations $\vec{P}$ for **E**.

Notation: $\mathbf{D} \leqq_{FC} \mathbf{E}$.

We call two algebraic datatypes $\mathbf{D}_1$ and $\mathbf{D}_2$ equivalent, whenever $\mathbf{D}_1 \leqq_{FC} \mathbf{D}_2 \leqq_{FC} \mathbf{D}_1$ holds.

6. *Conclusions*

1. In [3] we consider structures
   $\mathfrak{A} = <(\vec{A}_{int},\vec{A}_{ext}),(\vec{F}_{int},\vec{F}_{ext}),(\vec{R}_{int},\vec{R}_{ext}),(\vec{C}_{int},\vec{C}_{ext})>.$
   It is clear that operational semantics for such

structures allow only for instructions built from external functions, relations and constants. It may not be possible to give a suitable axiomatisation of the datatype thus obtained, without mentioning internal 'things'.

2. A deterministic datatype will often belong to an initial algebra [6] for the theory Δ. However for our definition it is not necessary that Δ has an initial algebra.
3. Suppose we have constructed a simple program M, in order to show $\mathbf{D} \leq_{FC} \mathbf{E}$ for algebraic **D** and **E**. How can we prove the correctness of M?
   Of course we should have constructed M to be elegant and we should use the information contended in the algebraic datatypes in **E**. However we do not see, what kind of formal proof-system is required here. But surely this can impose demands on the axioms in **E**, which are stronger, that all datatypes in **E** are deterministic.
4. One easily shows, using the completeness of first order logic:
   if Δ is a recursive theory, and $\langle\langle T,\Delta\rangle,V\rangle$ specifies a deterministic datatype **D**
   then **D** can also be specified by $\langle\langle T,\Delta'\rangle,V\rangle$ where Δ' is recursive but also quantorfree.
5. Using our formalism we can define precisely and rather convicingly what it means for two axiomatic specifications of datatypes to be operationally equivalent.

REFERENCES

[1] J.A.Bergstra, "Dynamic Recursion Theory & Degrees of Memory Structures", Report Leiden, May 1978.

[2] J.V.Guttag, "The Specification and Application to Programming of Abstract Datatypes", Technical Report CSRG 59, Dept. of Electrical Engineering and Dept. of Computer Science, University of Toronto, 1975.

[3] J.A.Bergstra, A.Ollongren, Th.P. van der Weide, "An Axiomatisation of the Rational Data Objects" in W.Cowell (ed.) *Portability of Numerical Software. Proceedings 1976.* Lecture Notes in Computer Science Vol. 57, Springer-Verlag 1977.

[4] H.D.Ehrich, "Outline of an Algebraic Theory of Structured Objects", *Automata, Languages and Programming*, Edinburgh University Press, p.508-530,1976.

[5] H.J.M.Goeman, A.Ollongren; Th.P. van der Weide, "Axiomatick van Datastructuren", Mathematical Centre Syllabus 37, Amsterdam, 1978.

[6] J.A.Goguen, J.W.Thatcher, E.G.Wagner, "Initial Algebra Semantics" *Journal of ACM* vol. 24, no 1, January 1977.

[7] K.Jensen, N.E.Wirth, *"PASCAL-User Manual and Report"* Lecture Notes in Computer Science vol. 18, Springer--Verlag, 1974.

J.A.Bergstra, Th.P.van der Weide

Institute of Applied Mathematics and
Computer Science
University of Leiden,
The Netherlands

COLLOQUIA MATHEMATICA SOCIETATIS JÁNOS BOLYAI
26. Mathematical Logic in Computer Science
Salgótarján (HUNGARY), 1978.

MODEL THEORETIC ASPECTS OF ABSTRACT DATA SPECIFICATION[1]

A.Bertoni, G.Mauri, P.A.Miglioli

1.Introduction

The fundamental idea in abstract data specification is to describe a data type independently from every concrete representation of the objects and from every implementation of the operations. In this view, one of the most promising approaches is the algebraic one, as developed by Liskov and Zilles [5],[10], Guttag [2] and ADJ [1], that hinges on the following theses:

a) a data type is a many-sorted equational algebra;

b) an abstract data type is an isomorphism class of initial many-sorted equational algebras.

However, the specification of an abstract data type as initial algebra is often very difficult, and in many cases (see the examples of Majster [7] and ADJ [9], or

[1] This research has been developed in the frame of the Communication and Programming Project of Universitá di Milano and Honeywell Information Systems Italia.

the field of rational numbers) there are no finite equational specifications for the data type. To overcome such technical difficulties, ADJ [9] proposed some methods such as the use of conditional axioms or hidden functions. See also, paragraph 57 of the monograph [13].

There are however some aspects of the data type problem that cannot be clarified only by initiality. In particular, we quote the following points:

a) the initial algebraic approach is not adequate to the analysis of the relations between axioms and their models;
b) in many cases, we have only a partial intuitive knowledge of the data type to be specified, while it needs a complete knowledge of the operations and their properties to give an equational specification; so, the initial algebraic approach cannot serve as a heuristic tool to find new significant operations on a data type;
c) the initial algebraic approach makes no distinction between operations that intuitively have a different role (for example, the "successor" operation in the specification of natural numbers has a constructive role, opposed to the algorithmic role of the "predecessor"); furthermore, it is inadequate to specify data types with relations as well as operations.

For these reasons, we think that a slightly generalized approach, that uses model-theoretic concepts and techniques, would be more adequate. Our claims are that:

a) two kinds of operations are to be distinguished, the "constructive" and the "algorithmic" ones; as the names are pointing out, the operations of the first kind are intended to construct the objects of the data type, those of the second kind to manipulate these objects;

b) the properties of the operations need not be specified by equations, but more generally by first order logical axioms. This in turn allows to consider not only operations, but also relations on the data type.

On this basis, our thesis is that an abstract data type is a model of a set of axioms which is both initial and prime; initiality is to be connected with the constructive operations, primeness with the algorithmic ones. In a sense that will be made more precise below, the prime model gives us a characterization of the maximal set of objects we can describe.

Finally, we introduce the notion of m-completeness of a theory, and show how by requiring this property for the abstract model we can guarantee that all the formulas holding on this model hold also on all the "concrete" representations.

2. *Fundamental Definitions*

The basic notions we need are those of relational structure (generalizing that of algebra) and of first order language.

We start with a *many-sorted alphabet* A, that consists of:

a) a set S of *sorts*;

b) a set Σ of *operation symbols*, together with an *arity-function*

$$\nu_{\Sigma} : \Sigma \to S^{+} x S$$

c) a set R of *relation symbols*, together with an *arity-function*

$$\nu_{R} : R \to S^{+}$$

d) a set C of *constant symbols*, together with a *sort function*

$$\nu_C : C \to S .$$

Definition 2.1 A *structure for* A is a pair $M = \langle M, i_M \rangle$ where $M = \langle M_s | s \in S \rangle$ is a system of non empty sets indexed by S, called the *carriers* of the structure, and i_M is the *interpretation function*, that associates:

i) to every $\sigma \in \Sigma$ with $\nu_\Sigma(\sigma) = \langle s_1 \ldots s_k, s \rangle$ a function

$$i_M(\sigma) = \sigma_M : M_{s_i} \times \ldots \times M_{s_k} \to M_s ;$$

ii) to every $r \in R$ with $\nu_R(r) = s_1 \ldots s_k$ a relation

$$i_M(r) = r_M \subseteq M_{s_i} \times \ldots \times M_{s_k} ;$$

iii) to every $c \in C$ with $\omega_C(c) = s$ an element

$$i_M(c) = c_M \in M_s .$$

Definition 2.2 Let M and M' be two structures for A; a *morphism* $h : M \to M'$ is a system $\langle h_s : M'_s | s \in S \rangle$ of maps such that relations, functions and constants are preserved.

Fact: The class of structures for a given many-sorted alphabet A, together with their morphisms, is a category [6].

Now, we can construct expressions that tell us about the structures by using the symbols of A and:

e) a system $\langle X_s | s \in S \rangle$ of infinite sets of *variables*;

f) the *logical connectives* $\wedge$(and), $\vee$(or), $\sim$(not), $\to$(implies), the *equality symbols* $=$, the *quantifiers* $\forall$(for all) and $\exists$ (there exists).

Definition 2.3 The set T_s of *terms* of sort $s \in S$ is the smallest set containing X_s, all constant

symbols of sort S , and such that if $\sigma \in \Sigma$, $\nu_{\Sigma}(\sigma) =$ $= \langle s_1 \ldots s_k, s \rangle$ and $t_i \in T$, then $\sigma(t_1, \ldots t_k) \in T_s$.
The set AF of *atomic formulas* is the smallest set such that if $t_1, t_2 \in T$ then $t_1 = t_2 \in AF$, and if $r \in R$, $\nu_R(r) = s_1 \ldots s_k$ and $t_i \in T_{s_i}$ then $r(t_1, \ldots, t_k) \in AF$.
The set F of *first order formulas* is the smallest set such that $AF \subseteq F$ and if $\varphi, \psi \in F$, $x \in X_s$, then $\varphi \wedge \psi$, $\sim\varphi$, $\psi \vee \varphi$, $\varphi \rightarrow \psi$, $\forall x \varphi$, $\exists x \varphi \in F$. In the formulas $\forall x \varphi$ and $\exists x \varphi$, the variable x is called *bounded* (by the quantifier). A formula that contains only bounded variables is called a sentence.

Now, given a set T of sentences, let M be a structure such that all the sentences in T hold on M . We say that M is a *model* for T , and write $M \models T$.
The class of models of T form a subcategory of the category of structures.

3. *Initiality in abstract data specification*

Initiality is the property that has been assumed by ADJ [1] to characterize abstract data types, see Section 57 of [13] p.338.

Definition 3.1 An object O in a category C is said to be *initial* iff for every other object O′ there is a unique morphism $h : O \rightarrow O'$.

The fundamental theorem, that guarantees the "abstractness" (independence of any representation) of initial objects is the following [6].

THEOREM 3.1 If O and O' are both initial objects in a category, then they are isomorphic.

In the particular case where C is a category of equational algebras (i.e. of models of a theory without relation symbols and containing only equations) the existence of an initial object is guaranteed by a well known theorem of universal algebra [13],[11]. These theorems are the basis for the ADJ's thesis and abstract data type is the unique (up to isomorphisms) initial object in a category of many sorted equational algebras. For stronger results see Chapter 8 of the standard textbook [13] of universal algebra.

4. Primeness in abstract data specification

The initial algebra approach requires the "a priori" knowledge of allthe operations and their properties; in this section we show how a more flexible approach can be obtained by using the primeness property.

Definition 4.1 Let M and N be two structures on the same alphabet. An *embedding* of M in N is a morphism $h : M \to N$ such that for every atomic formula φ and every assignment s in M

$$M \models \varphi(s) \quad \text{iff} \quad N \models \varphi(h \cdot s)$$

Definition 4.2 A model M of a theory T is called *prime* iff for every model of T , N , there is an embedding $h : M \to N$.

If the model M is initial as well as prime, then the embedding h is unique. In this case, M is in some sense an abstract characterization of the greatest common substructure of the models of T ; the more interesting result, from the point of view of data types, is that every object c of the carrier M can be defined by a logical formula of the type $\exists y \Delta_c(x,y)$ [4], and this definition identifies in an unique way the object corresponding to c in every model of T (concrete representation).

We examine now a well known data type, i.e. the type STACK OF A , with $A = \{a_1,\ldots,a_n\}$, to see how the primeness works.

Let us first build up the set of stacks; this goal can be reached by using the constant λ and the only operation PUSH. Hence, we have the many sorted algebra schematized as follows, in the style of ADJ

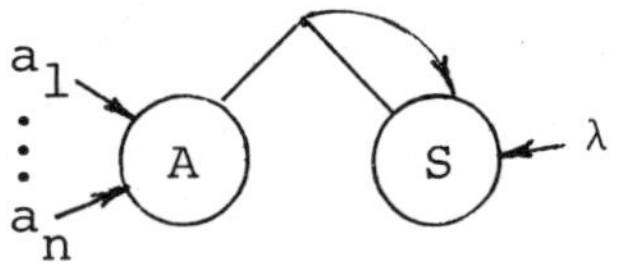

The following model M_o is a representative of the isomorphism class of initial algebras:

$$S_o = A^* \;;\quad PUSH(x,a_i) = \begin{cases} a_i & \text{if } x = \lambda \\ xa_i & \text{if } x \neq \lambda \end{cases}$$

The model M_o does not suffice to capture the intended STACK structure, since two essential operations (TOP and POP) are not defined. Furthermore, M_o is not a prime model. In fact, the structure $\tilde{M}$ with $\tilde{S} = \{\lambda\}$ and $\widetilde{PUSH}(\lambda,a_i) = \lambda$ is similar to M_o , but M_o cannot be embedded in $\tilde{M}$. At this point, we take a second step by finding a set T of axioms such that the category of models of T has yet an initial object, that is also a prime model of the axioms. For STACK, these axioms are:

A1) $\bigvee_{a \in A} x_A = a$;

A2) $\forall z_S \exists! x_A \exists! x_S (z_S = \lambda \;\vee\; z_S = PUSH(x_A,x_S))$;

A3) $\forall x_A, x_S \; PUSH(x_A,x_S) \neq \lambda$.

We remark that the latter axioms can be viewed as a simultaneous definition of the functions:

$$POP : S|\{\lambda\} \to S$$

$$TOP : S|\{\lambda\} \to A$$

Thus our approach points out the intrinsic difference between the "constructive" operation PUSH and the "algorithmic" operations POP and TOP.

5. *Relevance of model theoretic methods*

The above sketched approach allows to apply model theoretic tools to the analysis of abstract data types.

In this section, we show how these techniques could be used to guarantee that all the formulas holding in the initial prime model holds also on every "concrete" representation.

The key concept in this case is that of m-completeness.

Definition 5.1 N is a *substructure* of M, $N \subseteq M$, if $N \subseteq M$ and, for every atomic formula φ and assignment s in N

$$N \models \varphi(s) \quad \text{iff} \quad M \models \varphi(s) \tag{1}$$

N is an *elementary substructure* of M, $N \prec M$, if $N \subset M$ and (1) holds for every formula, not only for the atomic ones.

Definition 5.2 A theory T is *m-complete* if, for every two models N,M of T, if $N \subseteq M$ then $N \prec M$.

Definition 5.3 A theory T is *complete* if for every two models N,M a sentence φ holds in N iff φ holds in M.

THEOREM 5.1 *Let T be an m-complete theory. If T has a prime model, then T is complete.*

It is easy to see that the theory T of the preceding section is not m-complete. For the sake of simplicity, we take A = {a,b} . Let us consider the structure defined by:

$$S_1 = A^*\cup\{A^*b\cup\{\lambda\}\}.\{\lambda,a\}\omega$$
$$PUSH(b,x) = bx$$
$$PUSH(a,x) = \begin{cases} ax & \text{if} \quad x \neq \omega \\ \omega & \text{if} \quad x = \omega \end{cases}$$

This structure is obviously a model, M_1 , of T , and $M_o \subseteq M_1$. However, we have that:

$$M_1 \models \exists x PUSH(a,x) = x \quad \text{and} \quad M_o \models \sim\exists x PUSH(a,x) = x$$

This means that the axioms do not completely characterize their models.
Let now T' be the theory obtained by adding to T the infinite set of axioms

$$\forall x(sx \neq x) \ , \quad s\in\{a,b\}^* \ .$$

THOEREM 5.2 *T' is m-complete.*
Proof Let M,N be two models of T' , and $M \subseteq N$. A sufficient condition for m-completeness [8] is that every formula of kind $\exists y\varphi(y,m)$, where m is a finite vector of elements of M and φ contains no quantifiers, holds in M if it holds in N . A syntactical pre-computation allows in turn to write every φ as a conjunction of atomic or negated atomic formulas; the obvious observation that

$$T \vdash x = \sigma_1\sigma_2\ldots\sigma_s y \leftrightarrow x = y_o \wedge y_o = \sigma_1 y_1 \wedge\ldots\wedge y_{s-1} = \sigma_s y$$

allows then to consider only atoms of one of the following forms:

1) $y_i = \sigma y_k$ 2) $y_k \neq m_j$ 3) $y_i \neq y_k$ 4) $y_k = m_j$

By executing all the substitutions 4), we have only the first three forms left.
We construct now a graph G whose set of vertices is isomorphic to the set of variables y_k and whose edges are labelled with elements of the set A in such a way that $y_k \overset{g}{\to} y_k \subseteq G$ iff the atomic formula $y_i = \sigma y_k$ is contained in $\varphi(y,m)$.
Furthermore, we can identify two vertices y_i , y_j when $y_i \overset{g}{\to} y_k \in G$ and $y_j \overset{g}{\to} y_k \in G$, so obtaining a new graph G' that has the following properties:

i) every vertex has at most one entering edge; in fact, if both $y_i \overset{g}{\to} y_k$ and $y_j \overset{g'}{\to} y_k$ were in G , then $N \models \exists y_k(y_k = \sigma' y_j \wedge y_k = \sigma y_i)$, with $\sigma \neq \sigma$, thus contradicting the axioms of T .

ii) In G' there are no circuits. In fact, the existence of a circuit $y \overset{\sigma_1}{\to} y_1 \overset{\sigma_2}{\to} y_2 \to \cdots \overset{\sigma_s}{\to} y$ would imply $N \models \exists y(y = sy)$, thus contradicting the axioms of T' .

It follows that G' is a forest, with a set $\{x_1,\dots,x_r\}$ of roots, and hence that for every variable y_k there are an unique root x_j and an unique path s such that $y_k = sx_j$.
Let now M be the carrier of M . Since M contains the prime model M_o , we have $M_s \supseteq A^*$. Given the set $M' = \{m_1,\dots,m_\ell\}$ of elements of M which appear in $\varphi(y,m)$, and a k such that $|A|^k > \ell + r$, let us consider the family of languages

$$A = \{A^*a \mid \ell(a) = k, A^*a \cap M' = \emptyset\}$$

Since $|A|^k - \ell > r$, this family contains at least r languages $A^*a_1, A^*a_2,\dots,A^*a_r$. This implies that, for $x_1 = a_1,\dots,x_r = a_r,\dots,y_k = sa_j,\dots,$ we have:

$$\{x_1,x_2,\dots,y_k,\dots\} \subseteq A^* \subseteq M_s$$

$\{x_1, x_2, \ldots, y_k, \ldots\} \cap M' \subseteq \bigcup_{k=1}^{r} A^* a_k \cap M' = \emptyset$, hence $y_k = m_j$ and hence that $\exists y \varphi(y,m)$ holds in M too.

COROLLARY 5.1 T' *is complete.*

COROLLARY 5.2 T' *is decidable.*

So, we have obtained the completion of the theory of STACKS by adding an infinite set of axioms; this set is equivalent to the induction schema:

$$I_\varphi : \varphi(\lambda) \wedge (\varphi(x) \to \bigwedge_\sigma \varphi(\sigma x)) \to \forall x \varphi(x)$$

In fact, by induction on the length of s , it is easy to prove that

$$\varphi I_\varphi \quad \text{implies} \quad \forall s(sx \neq x) .$$

Furthermore, from the completeness of T' , it follows that I_φ or $\sim I_\varphi$ must hold. But I_φ is holding on the initial model, then $\forall s(sx \neq x)$ implies $\forall \varphi I_\varphi$.
Thus a complete axiomatization of STACKS is as follows:

A1) $PUSH(\sigma,x) \neq \lambda$ A2) $\bigvee_\sigma x = \sigma$

A3) $\forall z \exists! \sigma \exists! y(z = \lambda \vee z = PUSH(\sigma,y))$

S1) $[\varphi(\lambda) \wedge (\varphi(x) \to \bigwedge_\sigma \varphi(PUSH(\sigma,x))] \to \forall x \varphi(x)$

We prove now that this result cannot be improved: a complete theory of STACKS cannot be finitely axiomatized (without extending the language).

Let P be a non principal ultrafilter [12] on the set $\{1,2,3,\ldots.n\}$ and $\{M_n | n \geq 1\}$ a set of models such that, for a fixed $a \in A$,

$$M_n \models \exists x(a^n x = x) \wedge \forall x(\bigwedge_{\ell(s)<n} sx \neq x)$$

Such models are easily constructed as the model M_1 .

THEOREM 5.3 $N = \prod_n M_n | P$ *is a model of* T' .
Proof. For $n > l(s)$, we have $M_n \models \forall x(sx = x)$. Thus:

$\{n|M_n \models \sim\exists x(sx = x)\} \supseteq \{n|n>l(s)\} \in P$, being cofinite. It follows that $N \models \forall x(sx \neq x)$ for every s .

THEOREM 5.4 *T' is not finitely axiomatizable.*

Proof. Let T' be finitely axiomatizable; then, there would exist a formula equivalent to the infinite set of axioms $\forall s \forall x(sx \neq x)$.

Then $\sim\varphi \leftrightarrow \exists s \exists x(sx = x)$ and $M_n \models \sim\varphi$. For the well-known properties of ultraproducts, it follows that $N = \prod_n M_n|P \models \sim\varphi$, thus contradicting the preceding theorem. □

Summarizing the content of the last two sections, we can assert that the axiom A2), that allows to define the usual TOP and POP operations, guarantees the existence of a prime model, and it also guarantees that by adding the induction schema the theory becomes complete, with the following consequences:

i) the theory turns out to be decidable:

ii) the functions and the relations (expressible in the language) in any model are those, and only those (expressible in the language) in the initial model.

REFERENCES

[1] Goguen,J.A., Thatcher,J.W., Wagner,E.G., An initial algebra approach to the specification, correctness and implementation of abstract data types, IBM Res.Rep. RC6487, 1976.

[2] Guttag,J.V., Abstract data types and the development of data structures, *SIGPLAN Notices* 8, 1976.

[3] Hardgrave,W.T., A technique for implementing a set processor, *SIGPLAN Notices* 8,1976.

[4] Kreisel,G., Model-theoretic invariants: applications to recursive and hyperarithemtic operations, *Proc.Symp.on the theory of models*, North-Holland, Amsterdam, 1965.

[5] Liskov,B.H., Zilles,S.N., Programming with abstract data types, *SIGPLAN Notices*, 6, 1974.

[6] MacLane,S., *Categories for working mathematician*, Springer,Berlin, 1971.

[7] Majster,M.E., Limits of the "algebraic" specification of abstract data types, *SIGPLAN Notices*, 9, 1977.

[8] Robbinson,A., *Introduction to model theory and to metamathematics of algebra*, North-Holland, Amsterdam, 1963.

[9] Thatcher,J.W., Wagner,E.G., Wright,J.B., Data type specification: parameterization and the power of specification techniques, *Proc. SIGACT 10th Symp. on theory of computing*, 1978.

[10] Zilles,S.N., Algebraic specification of data types, Project MAC progress Report 11, MIT, Cambridge, Mass., 1974.

[11] Cohn,P.M., *Universal Algebra*, Harper and Row, New York, 1965.

[12] Eklof,P.C., Ultraproducts for algebraists, in J.Barwise (ed.) *Handbook of Mathematical Logic*, Norh-Holland, Amsterdam, 1977.

[13] Grätzer,G., *Universal Algebra*, Springer Verlag, 1979.

A.Bertoni, G.Mauri, P.A.Miglioli
Instituto di Cibernetica - Universitá di Milano

COLLOQUIA MATHEMATICA SOCIETATIS JÁNOS BOLYAI
26. Mathematical Logic in Computer Science
Salgótarján (HUNGARY), 1978.

A SEMANTIC THEORY FOR LOGIC PROGRAMMING

F.M. Brown

Abstract

We axiomatize a number of meta theoretic concepts which have been used in Logic Programming, including: meaning, logical truth, nonentailment, assertion and erasure, thus showing that these concepts are logical in nature and need not be defined as they have previously been defined in terms of the operations of any particular interpreter for logic programs.

1. Introduction

One of the basic theses of Logic Programming is that programs written in Logic can be understood merely by reflecting on the intuitive meaning of the programs, without reference to any particular interpreter, or rather automatic theorem prover, that might execute the programs. This thesis would be very nice and have strong implications for programming methodology, if in fact it were true. But unfortunately, if we examine the situation closely we will see that there are features of contemporary logic programming languages which do not seem to be inderstandable without reference to a particular interpreter. Such features are for example:

1. Meaning, such as the universal function of Prolog.
2. Logical Truth.
3. Non-Entailment, such as the Thnot function of Planner and some of the uses of the slash (/) function of Prolog which simulate thnot.
4. Assertion, such as the Thassert function of Planner and the ajout function of Prolog.
5. Erasure, such as the Therase function of Planner, and the suppress function of Prolog.

After describing in Section 2 the logical programming language that we use, we then give in this language a correct axiomatization of each of the above five semantic functions. In particular in Section 3 we axiomatize the concept of meaning, and in Section 4 we axiomatize the modal concept of logical truth which is then used to define the remaining semantic functions: non-entailment, assertion, and erasure. The theory consisting of these axioms is demonstratably consistent relative to 1st order number theory. We mention this fact because the axiomatization of semantic concepts has often fallen into paradox and contradiction.

2. A Logical Programming Language

We describe the syntax of our logical programming language in Section 2.1 and then give a few examples of logic programs in Section 2.2.

2.1. *Notation*

We now explain our notation.

The symbols of classical logic are listed below with their English Translations:

$p \wedge q$	p and q
$p \vee q$	p or q
$p \rightarrow q$	if p then q
$p <-> q$	p iff q
$\sim p$	not p
⧗	true
⧖	false
$\forall X \;\; \phi X$	for all objects X, ϕX holds
$\exists X \;\; \phi X$	for some objects X, ϕX holds
$\forall p \;\; \phi$	for all propositions p, ϕp holds
$\exists p \;\; \phi$	for some propositions p, ϕp holds

Capital letters such as X, Y, Z, S, T, A, B range over objects whereas small letters such as p, q, r range over propositions.

The symbols of modal logic are:

$\vdash p$	p is logically true
$\vdash pq$	p entails q
$\Diamond p$	p is possible
(World p)	p is a World

The last three modal symbols are defined in terms of the first one as follows:

$\vdash pq$	$= \text{df} \vdash (p \rightarrow q)$
$\Diamond p$	$= \text{df} \sim\vdash\sim p$
(World p)	$= \text{df} \; (\Diamond p) \wedge \forall q ((\vdash pq) \vee (\vdash p(\sim q)))$

Equality:

X = Y   X equals Y

where X = Y → (ϕX <-> ϕY) for all sentences ϕ including sentences containing modal connectives such as ⊢.

A data structure of lists formed from:

Nil

(Cons X Y)

where (Cons X Y) is an ordered pair and Nil is not an ordered pair:

(Cons X Y) = (Cons U V) <-> X = U ∧ Y = V

~ (Cons X Y) = Nil

We make three abbreviations as follows:

First for any expression $\alpha_1 \ldots \alpha_n$:

[$\alpha_1 \ldots \alpha_n$] = df (Cons α_1...(Cons α_n Nil)...)

and also:

[α_1 · α_2] = df (Cons $\alpha_1 \alpha_2$)

[ ] = df Nil

Thus [$\alpha_1 \ldots \alpha_n$] may be thought of as being a list whose elements are $\alpha_1, \ldots, \alpha_n$; [$\alpha_1$ · α_2] may be thought of as being an ordered pair; and [] may be thought of as being an empty list.

Finally we include a method of talking about expressions of this logical language, within this language. We do this in our logical language by representing an expression as a list of the names of its symbols. Names of symbols are formed by prefixing to that symbol an accent sign: '. Thus

[['Member 'X '[]] '<-> '⌶]

is a name of the expression (Member X []) <-> ⌶. It is

to be understood that 'X is a constant symbol of our logical language. The apparent visual similarity between a symbol such as: X and its name such as: 'X is merely a pneumonic for the readers convenience which will also allow us to concisely state the criteria for a definition of a meaning function.

2.2. Logic as a Programming Language

This simple language may be thought of as being a programming language. That is, we can implement a system which by making logical inferences can effectively compute various things.

For example a program in this language which could be used to compute whether something is a member of a list would be the three sentences:

```
M1: (Member X[]) <-> ⧖
M2: (Member X (Cons X Y)) <-> ⧗
M3: (~(X=Z)) → ((Member X (Cons Z Y) <-> (Member X Y))
```

If our system were then defined so as to use these sentences to replace the left hand side of an equivalence by the right hand side, checking that any initial conditional sentences were true, then the system could determine that B were a member of the list [A B C] as follows:

```
(Member B [A B C])     :M3
 ↓
(Member B [B C])       :M2
 ↓
 ⧗
```

As a more complex case consider the following program which computes the value of an element V in a list of pairs:

$$[[V_1 \cdot \alpha_1]\ldots[V_n \cdot \alpha_n]]$$

The value of V in such a list is defined to be the first α_i whose V_i equals V.

```
V1: (Val V [])=Nil
V2: (Val V [[V.X].L])=X
V3: (~U=V → (Val V [[U.X].L])=(Val V L)
```

Thus for example the system could determine that the value of C in [[A.B] [C.D]] is D as follows:

```
(Val C [[A.B] [C.D]])   :V3
        ↓
(Val C [[C.D]])         :V2
        ↓
        D
```

More detailed expositions on the use of logic as a programming language are given in [1,2].

3. Meaning

One of the most important features of any reasonably general programming language is the ability to execute a program which has been constructed as a piece of data by another program. For example, a compiler essentially translates one data structure representing a program into another data-structure representing a program which is executable on some particular computer. The key step here is: how does this second data-structure actually become to be turned into a program which can be executed?

In logic, it is easy to represent both a logic program and a data structure which is a name if that logic program along the lines described in Section 2. Furthermore this data structure can easily be manipulated by other logic program, and in particular could be the result of translating a data-structure representing a program of some other language into a data-structure representing a logic program. However, the final key step of translating the data structure representing a logic program into the logic program itself is what appears to be difficult to do in logic.

In the remainder of this section we will show how this may be done in logic, and in particular in Section 3.1 will we give a recursive meaning function M which maps data-structures representing logic programs into the corresponding logic programs. This meaning function M satisfies the criteria that:

$$(M\ \phi) = \Psi$$

if Ψ is obtained from ϕ by

1. eliminating the first accent sign from each symbol beginning with an accent sign.
2. replacing all occurences of [by (.
3. replacing all occurences of] by).

for any expression ϕ consisting solely of the symbols: Cons, Nil, and symbols beginning with an accent sign.

It is worthwhile noting M in this criteria cannot be interpreted as being Tarski's [3] predicate E for Empirical Truth. The reason for this is that whereas Tarski's superficially similar criteria for empirical truth involves bi-implication:

$$(E\ \phi) <-> \Psi$$

our criteria is much stronger and in fact implies synonomity, or rather necessary bi-implication:

$$\vdash((M\ \phi) <-> \Psi)$$

Thus if M in our criteria is interpreted to be empirical truth, then the criteria itself becomes contradictory. We can see this fact by simply letting ϕ be the name of any sentence Ψ which is emprically true but not logically true, such as: "The Morning Star is the Evening Star".

Finally in section 3.2 we discuss some variations of this meaning function, and show why it is not easily represented in clausal form logics.

3.1. Axiomatization of Meaning

A computationally efficient meaning function M satisfying the above criteria is given below. Since all other logical symbols are definable in terms of $\wedge$, ~, $\forall$, and $\vdash$ we have not bothered to list meaning axioms for any logical symbols other than these:

M0: (M S) = (m(closure S) [])
M1: (M[S '$\wedge$ T]A) = (m S A) $\wedge$ (m T A)
M2: (m['~S]A) = ~(m S A)
M3: (m['$\vdash$S]A) = $\vdash$(m S A)
M4: (m['$\phi S_1 \ldots S_n$]A) = (ϕ(m S_1 A)...(m S_n A))
for each non logical symbol ϕ
M5: (m['$\forall$ V S]A) = $\forall$X(m S([V.X].A))

M6: (m V A) = (Val V A)

The variables S, T, $S_1 \ldots S_n$ range over expressions, and the variable V ranges over variables. The closure of a sentence S with free variables $'X_1 \ldots 'X_n$ is defined to be $[\forall 'X_1 \ldots ['\forall 'X_n$ S]...].

An example of the application of this meaning function to the sentence ['∀ 'Z['Y 'movesto 'Z]] is as follows:

```
(M['∀ 'Z['Y 'movesto 'Z]])                          :M0
(m (closure['∀ 'Z['Y 'movesto 'Z]])[])              :closure
(m ['∀ 'Y['∀ 'Z['Y 'movesto 'Z]]][])                :M5
(∀X (m['∀ 'Z['Y  'movesto 'Z]][['Y.X]]))            :rename
```

We cannot immediately apply M5 again because X is a free variable in the subexpression:

```
(m['∀ 'Z['Y 'movesto 'Z]][['Y.X]])
```

and hence be captured by the outlying quantifier: ∀X occurring in 5. Hence one of these variables must first be renamed.

```
(∀Y (m['∀ 'Z['Y 'movesto 'Z]][['Y.Y]]))                                    :M5
(∀Y (∀X (m['Y 'movesto 'Z][['Z.X]['Y.Y]])))                                 :M4
(∀Y (∀X ((m'Y[['Z.X]['Y.Y]]) movesto (m'Z[['Z.X]['Y.Y]]))))                 :M6
(∀Y (∀X ((Val 'Y[['Z.X]['Y.Y]]) movesto (Val 'Z[['Z.X]['Y.Y]])))):Val
(∀Y (∀X (Y movesto X)))                                                     :rename
(∀Y (∀Z (Y movesto Z)))  :universal instantiation and generalization
    (∀Z (Y movesto Z))
```

We note that this meaning function consists only of sentences which recurse through syntactic structure.

Thus this meaning consists entirely of recursive definitions. It is in this sense that we claim that it is a logically true meaning function.

A more detailed description of this meaning function and its philosophical implications is given in [4].

3.2. Variations

Since:

ϕY is equivalent to $\forall X(\phi Z \leftarrow Y = Z)$

and more generally since:

$\phi\ \forall X\ \ (tX)$ is equivalent to $\forall z\ \phi\forall X(zX) \leftarrow \forall Y(tY) = (zY)$

by using these equivalences we can replace the equations of the maning function M by implications to obtain an equivalent meaning function MA:

MA0: $(M\ S)=Z \leftarrow (m(\text{closure}\ S)[\,])=Z$

MA1: $(m[S'\wedge T]A)=(p\wedge q) \leftarrow (m\ S\ A)=p\wedge(m\ T\ A)=q$

MA2: $(m['{\sim}S]A)={\sim}p \leftarrow (m\ S\ A)=p$

MA3: $(m['\vdash S]= p \leftarrow (m\ S\ A)=p$

MA4: $(m['\phi S_1\ldots S_n]A)=(\phi X_1\ldots X_n) \leftarrow (m\ S_1\ A)=X_1\wedge\ldots\wedge(m\ S\ A)=X_n$

MA5: $(m['\forall\ V\ S]A)=\forall X(\phi X) \leftarrow \forall X(m\ S[[V.X].A])=(\phi X)$

MA6: $(m\ V\ A)=Z \leftarrow (Val\ V\ A)=Z$

Note that MA5 when put into Skolem normal form becomes:

MA5': $(m['\forall V\ S]A)=\forall x(\phi X)\quad (m\ S[[V.(Sk\ X\ V\ S\ A(\ X))].A])=$
$=(\phi(Sk\ X\ V\ S\ A(\phi X)))$

where Sk is a Skolem function.

Many comtemporary logical programming languages [1] are to a large degree based on evaluating sentences of a certain form: First they are based on evaluating reverse implications ← somewhat similar to the manner in which the equivalences were evaluated as described in section

2.2. Furthermore, such systems are often defined so as to only match entire atomic propositions such as (Member A B) or (m S A)=Y rather than any embedded terms such as (m S A). And finally, such systems usually require that all sentences be initially skolemized. It will be seen that MA satisfies the first two criteria and if MA5 is replaced by MA5' then all three criteria are satisfied.

There is, however, a problem in that MA5 (and MA5') involves second order unification which usually is not available in such programming systems. Possibly, for this reason, or because MA5 is rather complicated, and therefore difficult to state, or even because of a preference for stating the object language itself in skolem normal form, sentences such as MA5 are not currently used in such programming systems. Thus quantifiers and bound variables are not allowed in the object languages.

Initial free universal object language variables could however be allowed by the simple expedient of replacing MA0 by MA0'.

$$\text{MA0'}: \quad (\text{M S})=\text{Z} \leftarrow (\forall X_1 \ldots \forall X_n (\text{m S a}))=\text{Z}$$

where 'a' here is an association list:

$$[V_1.X_1]\ldots[V_n.X_n]$$

such that $V_1 \ldots V_n$ are all the unbound object language variables in S, and $X_1 \ldots X_n$ are distinct free universal variables. MA0' has essentially the same effect as MA0 since the closure of S is $['\forall V_1 \ldots ['\forall V_n\ S]\ldots]$ which by MA5 becomes: $(\forall X_1 \ldots \forall X_n (\text{m S a}))$.
Note also that M0 has a similar version M0'

$$\text{M0'}: \quad (\text{MS}) = (\forall X_1 \ldots \forall X_n\ (\text{m S a}))$$

We see that MA0', MA1, MA2, MA3, MA4, MA6 constitute a proper meaning function for a quantifier free object language. It should be noted, however, that the replacement of MA0' by MA0" does not constitute a proper meaning function:

MA0": $(M\ S) = Z \leftarrow (m\ S\ a) = Z$

The reason for this that MA0" can, in fact, be false when MA0' is true. To see this first note that MA0" is equivalent to

M0": $(M\ S) = (m\ S\ a)$

We consider now the sentence: $['\phi'X]$ in a universe of two things: By MA0' we get:

$$M['\phi'X] = \forall X(m['\phi'X][['X.X]])$$
$$M['\phi'X] = \forall X\phi X$$
$$M['\phi'X] = \phi_1 \wedge \phi_2$$

whereas by MA0" we get:

$$\forall X(M['\phi'X] =(m['\phi'X][['X.X]])$$
$$\forall X(M['\phi'X] = \phi X)$$
$$M['\phi'X] = \phi_1 \wedge M['\phi'X] = \phi_2$$

But clearly MA0" leads to a contradiction; for example in the case where ϕ_1 is true and ϕ_2 is false it implies that true equals false.

Note however, that a sentence similar to MA0" with the $(m\ S) = X$ replaced by a relation could, however, be consistent if this relation were intended to capture some concept of partial meaning. For example:

(i S A X) means that: X is the meaning in A of some instance of S.

(P S A X) means that: X is implied by the meaning of S in A.

We see that both

(I S X) ← (i S aX) and

(P S X) ← (P S aX)

are probably true.

Note that although I and P are transitive and perhaps reflexive relations, neither is a symmetryc relation. Generally they are related to M as follows:

$$((M\ X)=X \rightarrow (I\ S\ X)) \wedge ((I\ S\ X) \rightarrow (P\ S\ X))$$

A meaning-of-some-instance relation (I) has been implemented in a logical programming system by D. Warren. The sentences of this relation may be obtained from MA0", MA1, MA2, MA3, MA4, MA6 by replacing (m α)=β by (I α β) and (m A α)=β by (i α A β).

4. *Modality*

Another important feature of any reasonably general programming language is the ability to execute a program within a particular context. For example, one might wish to evaluate a particular expression using certain function definitions which are quite different from the function definitions which are active at the top level context. In a logical language the analogy of this would be to evaluate some expressions using a certain database of axioms which could be quite different from the data base of axioms being assumed at the top level.

At first glance, it would appear that it is impossible

to define such a construct in logic because even though one could use the $b \rightarrow p$ concept to created a new context containing additional axioms b which can be used when evaluating p, there does not seem to be any way to stop the use of any axioms a of the top level context such as in $a \rightarrow (b \rightarrow p)$.

However, modal logic offers a way to solve this problem. If we let $\vdash$ be the modal symbol which captures the notion of logical truth, then it is easy to see that $\vdash(b \rightarrow p)$ means that p is to be evaluated using only the axioms b. Thus, for example, if the top level context is a, $a \rightarrow \vdash(b \rightarrow p)$ still only allows p to be evaluated using the axioms b.

The problem is to find a correct axiomatization for the modal concept of logical truth: $\vdash$, for it is certainly clear that there must be special axioms of modality in additon to the normal axioms of classical logic. For example the sentence:

$$\vdash(b \rightarrow (p \wedge q)) \leftrightarrow (\vdash(b \rightarrow p) \wedge \vdash(b \rightarrow q))$$

is intuitively valid although it is not derivable solely from the normal axioms of classical logic.

In section 4.1 we give a correct axiomatization of the modal concept of logical truth. Then in section 4.2 we use this concept to define the semantic functions of non-entailment, assertion, and erasure.

4.1. Axiomatization of Logical Truth

The logical axioms of the modal logic which captures the notion of logical truth includes any complete and reasonable axiomatization of classical quantificational logic, with propositional quantifiers plus the following inference rule and axioms about modality:

RO: form p infer $\vdash p$
A1: $\vdash p \rightarrow p$
A2: $\vdash(p \rightarrow q) \rightarrow (\vdash p \rightarrow \vdash q)$
A3: $\vdash p \vee \vdash \sim \vdash p)$
A4: $(\forall r((\text{World } r) \rightarrow \vdash r\ p)) \rightarrow \vdash p$

R0, A1, A2, and A3 are essentially the inference rule and axioms of S5 modal logic. Axiom A4 which we call Leibniz's postulate expresses his intuition that something is logically true if it is true in all possible worlds.

An efficient sequent calculus proof procedure based on theorems derived from these modal axioms is described in [5].

The consistency problem of modal logic is that from the logical axioms of modal logic we cannot prove certain elementary facts about the possibility of conjunctions of distinct possible negated atomic expressions consisting of non-logical symbols. For example, if we have a theory formulated in our modal logic which contains the non-logical atomic expression (ON A B) then since ~(ON A B) is not logically true, it follows that (ON A B) must be possible. Yet ◇(ON A B) is not a theorem of our modal logic.

Thus, for any theory expressed in modal logic, a certain number of non-logical axioms dealing with possibility should also be added. For example, in the case of the propositional logic, or in the case of the quantificational logic over a finite domain since it reduces to propositional logic, one sufficient but inefficient axiomatization would be to assert the possibility of all consistent disjunctions of conjunctions of literals as additional non-logical axioms:

$$\Diamond(\vee(\wedge \text{Literals}))$$

A more computationally efficient axiomatization which is obtained by noting that the possibility of a disjunction of sentences is implied by the possibility of any one of those sentences:

$$\Diamond p \rightarrow \Diamond(p \vee q)$$

is to assert only the possibility of all consistent conjunctions of literals:

$$\Diamond(\wedge \text{literals})$$

Using our meaning function [4] this may be done in a finite manner by adding the single axiom:

(Conj S) ∧ (Consist S) → ◇(M S)

where Conj and Consist are recursive functions defined as follows:

(Conj S) = df (Lit S) ∨ ∃T ∃R (S=[T' ∧ R] ∧ (Lit T) ∧ (Conj R))

(Lit S) = df (∃T S=['~T] ∧ (Atomicsent T)) ∨ (Atomicsent S)

(Consist [ ]) = df ⧓

(Consist [S,L]) = df (Consist2 S L) ∧ (Consist L)

(Consist2 S [ ]) = df ⧓

(Consist2 S[T.L]) = df ~(Opp S T) ∧ (Consist2 S L)

(Opp S T) = df (∃R S=['~R] ∧ T=R) ∨ (∃R T=['~R] ∧ S=R)

4.2. Non-entailment, Assertion, and Erasure

Having now axiomatized the concept of logical truth, it is easy to define non-entailment and assertion:

D1: (Not-Entail a p) <-> df $\sim \vdash ap$

D2: (Assert a p) <-> df $a \wedge p$

That is p is not entailed by a, iff a implies p is not logically true. And the assertion of p to database a is simply (a and p).

The definition of erasure is, however, slightly more complex, and in general there will be more than one reasonable resulting database b which is obtained by erasing a proposition p from a given database 'a'.

D3: (Erase a p b) <-> $\vdash a\ b \wedge (\vdash b\ p \rightarrow \vdash p)$
$\wedge\ \forall q\ (\vdash a\ q \wedge \vdash q\ b \wedge \sim\vdash bq \rightarrow \vdash q\ p)$

That is b is obtained by erasing p from a iff a entails b, b entails p only if p is logically true, and no proposition stronger that b can be obtained from a by deleting p.

We can see that D3 is indeed a reasonable definition of erasure by noting the following theorem:

T1: $\sim\vdash ap \rightarrow$ ((Erase a p b) <-> $\vdash$(a <-> b))

That is, if p is not entailed by a then erasing p from a merely results in a itself.

This definition of erasure is closely related to Stalnaker's Theory of Conditionals [6] and to Schwind's Theory of Action [7].

5. *Conclusion*

We have axiomatized a number of basic semantic concepts for a logical programming language. An efficient automatic theorem prover based on a sequent calculus [5] derived from these axioms is currently running at Edinburgh, and is being used to prove rather difficult theorem in metamathematics.

Our semantic theory also forms the basis of Brown and Schwind's [8, 9, 10] theory of natural language understanding which is currently being developed.

References

[1] R. Kowalski: Predicate Logic as programming language *Proceedings IFIP*, [1974]

[2] W. Bibel: Predicative Programming Institutsbericht, TU München Abteilung Mathematik, [1974]

[3] A. Tarski: The Concept of Truth in Formalized Languages, (1931) *Logic Semantics, Mathematics*, trans. by J.H. Wooder, Oxford, Clarendon Press, [1956]

[4] F.M. Brown: *The Theory of Meaning* DAI Research Report 35, [1977]

[5] F.M. Brown: A Sequent Calculus for Modal Quantificational Logic *3rd AISB/GI Conference Proceedings, Hamburg*, July [1978]

[6] R.C. Stalnaker: A Theory of Conditionals *Causation and Conditionals*, ed. E. Sora, Oxford University Press, [1975]

[7] C. Schwind: Representing Actions by a Modal Logic
3rd AISB/GI Conference Proceedings, Hamburg,
[1978]

[8] F.M. Brown, C. Schwind: Analysing and Representing
Natural Language in Logic
3rd AISB/GI Conference Proceedings, Hamburgh,
[1978]

[9] F.M. Brown, C. Schwind: Towards an Integrated Theory
of Natural Language Understanding
to appear in *Computational Linguistics Conference*
Proceedings, Bergen, [1978]

[10] F.M. Brown, C. Schwind: Outline of an Integrated
Theory of Natural Language Understanding
to appear as DAI Research Report 50.

F.M. BROWN
Department of Artifical Intelligence
University of Edinburgh

COLLOQUIA MATHEMATICA SOCIETATIS JÁNOS BOLYAI
26. Mathematical Logic in Computer Science,
Salgótarján (HUNGARY), 1978.

INTELLIGENT BACKTRACKING FOR AN INTERPRETER OF HORN CLAUSE LOGIC PROGRAMS

M. Bruynooghe

ABSTRACT

An interpreter of Horn clause logic programs is given a goal statement $<-A_1, A_2, \ldots, A_n$ $(n \geq 1)$ and a set of Horn clauses of the form $B -B_1, \ldots, B_m (m \geq 0)$ and attempts to derive a sequence of goal statements ending in the empty goal statement.

The basic cycle of the interpreter is:

- select a literal A_i in the current goal statement.
- select a Horn clause $B<-B_1, \ldots, B_m$ such that A_i and B have a most general unifier (Θ) and derive a new goal statement:
  $<-(A_1, \ldots, A_{i-1}, B_1, \ldots, B_m, A_{i+1}, \ldots, A_n)\ \Theta$

In general, different Horn clauses match with A_i. This non-determinism can be represented as an or-tree. At the implementation level, this non-determinism is usually handled by backtracking: only one path at a time is explored in the or-tree. A simple backtracking system such as in the current PROLOG respects the total ordering provided by the selection function and returns to the previous goal statement when the current one has no solutions. However, possibly the previous goal statement has no solutions for just the same reasons.

In this paper we define a partial ordering over the different steps in a derivation. This partial ordering allows us to find a goal statement that cannot be unsolvable for the same reasons as the current one ("backtrackpoint").

Moreover, this partial ordering allows us to reorder the steps of the derivation, without losing any possible solutions, such that not all steps executed since the "backtrackpoint" must be undone.

This partial ordering also reduces the amount of computation necessary to derive all solutions when the derivation of a solution contains independent subproblems.

1. *HORN CLAUSE LOGIC PROGRAMS*

1.1. *Syntax and declarative semantics*

A Horn clause program comprises a set of procedure declarations and a "main program" or "goal statement". Procedure declarations are expressions of the form $B\texttt{<-}A_1,\ldots,A_n$ $(n \geq 0)$ where $B,A_1,\ldots,A_n$ are atomic formulas. They can be read as logic statements i.e. for all values of the variables, B is true if $A_1\&\ldots\&A_n$ is true.

An atomic formula is an expression of the form R R(t1,...,tn) where R is a n-adic relation and t1,...,tn are terms, i.e. functional expressions, constants or variables. We distinguish constants from variables by starting the former ones with an upper case letter and the latter ones by a lower case letter.

A declaration $B \leftarrow A_1,\ldots,A_n$ where the atom B has a relation name R is the declaration of a procedure for the relation R.

The main program is given by an expression of the form $\leftarrow A_1,\ldots,A_n$ $(n \geq 1)$ where the "procedure calls" A_i are again atomic formulas. This main program can be interpreted as a request to find a constructive proof of $\exists x1,\ldots,xm\ (A_1 \& \ldots \& A_n)$ where $x1,\ldots,xm$ are the variables in $A_1,\ldots,A_n$.

1.2. The procedural interpretation

The main program $\leftarrow A_1,\ldots,A_n$ is evaluated as follows: we select an arbitrary procedure call $A_i=R(t1,\ldots,tp)$. We invoke a procedure for the relation R of this call, by searching for a declaration $R(t'1,\ldots,t'p)$ $\leftarrow B_1,\ldots,B_n$ such that $R(t1,\ldots,tp)$ and $R(t'1,\ldots,t'p)$ unify with a most general unifier Θ. We transform the main program into a new set of procedure calls $\leftarrow(A_1,\ldots,A_{i-1},B_1,\ldots,B_m,A_{i+1},\ldots A_n)\Theta$ which is the new state of the computation. This process is repeated until we reach an empty set of procedure calls. The composition $\Theta_1 \cdot \Theta_2 \cdot \ldots \cdot \Theta_k$ of the unifying substitutions $\Theta_1,\ldots,\Theta_k$ of this computation, applied to the variables in the initial set of calls, is the output of the computation.

1.3. Non-determinism

The above evaluation mechanism is non-deterministic in two respects. First, there is, in the current set of procedure calls, the choice of the call to be executed in the next step. Control over this selection is of crucial importance for efficiency of the computation and - because the really intelligent scheduler is still unknown and is not likely to be found in the near future - has to be done by the user. In current implementations of the language PROLOG, [1,2,3] selection is done in a strictly left to right order. The user has to adapt his logic programs in such a way that this selection rule gives an adequate efficiency of the computation. Work is under way [4] to provide a more flexible control. This will allow the same overall efficiency with a sometimes much simpler logic program. Notice that the selection rule has no effect on the number of derivable solutions.

The second non-determinism results from the possibility of different procedures matching the selected call. Each of them leads to an alternative branch in the computation. We can explore these branches in parallel, or, we can explore the alternatives sequentially, and find all solutions, using backtracking, if for the strategy used to select the call, each branch terminates with a solution or a failure. Observe that the same restriction on the selection of calls is necessary in order to terminate a parallel execution when searching *all* possible solutions.

Where a naive backtracking system will rigorously explore all alternatives, a more intelligent one will learn from previous failures and successes how to get a faster exploration of the remaining alternatives. In this paper we study such an intelligent backtracking system.

The same subject has been studied by some MSc students at Imperial College in London in a project led by Robert Kowalski, a project that had its origins in the study of Sussman's HACKER. Intelligent backtracking in connection graphs is discussed by Cox and Pietrzykowski [5,6].

2. *NAIVE BACKTRACKING*

An interpreter is given a set S of procedure calls of the form $<-A_1,\ldots,A_n$ (the "goal statement"). Its basic cycle is the following:

- select a literal A_i of S
- select a procedure $B<-B_1,\ldots,B_m$ whose heading B matches with A_i (B and A_i have a most general unifier Θ) and derive the new set $<-A_1,\ldots,A_{i-1},B_1,\ldots,B_m,A_{i+1},\ldots,A_n)\Theta$.

In general, k procedures match the selected literal A_i and thus, k new states are derivable. We can represent these alternative solutions as branches in an *or-tree* e.g.:

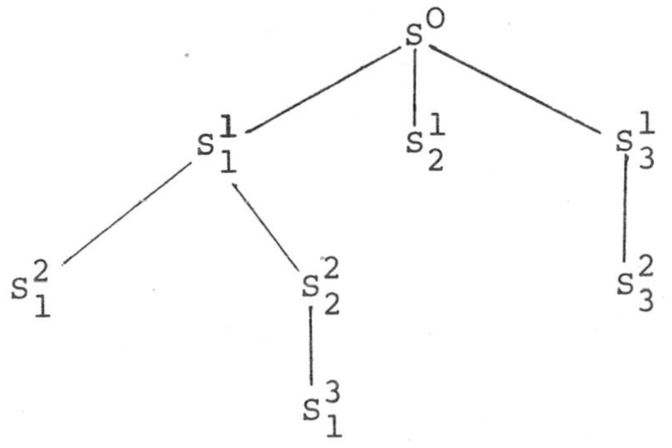

The terminal nodes are either solutions (the empty set of calls) or failure nodes (none of the procedures matches the selected call). Backtracking explores this or-tree using a depth-first, left-to-right search strategy. Once all the solutions to some S_j^i descendant of S_k^{i-1} are found, the system restores S_k^{i-1} and if there are still some untried alternatives to match the literal selected in S_k^{i-1}, one of them is tried and S_{j+1}^i is derived. If there are no more alternatives, it means all solutions to S_k^{i-1} are found and the system backtracks to the previous level.

The shortcoming of this system is that several states occurring in different branches of the or-tree share some procedure calls, and, in executing these calls, the same computations are done in each branch. To be more specific, consider an initial set of calls <-A,B where the literals A and B do not share any variable. Suppose the selection rule is such that the call A is completely executed before the call B is selected. A naive backtrackingsystem will derive a first solution for A characterized by a substitution θ_1 then it will execute the call $B\theta_1$ and will derive all solutions to this call, then it will backtrack to A, deriving a second solution, characterized by θ_2 it will execute the call $B\theta_2, \ldots$ until all solutions are derived. However, because A and B do not share any variables $B\theta_1 = B\theta_2 = \ldots = B\theta_n = B$. The system will execute the same call B n times (n being the number of solutions for A); in other words, the same node will occur in n different branches of the or-tree. A more intelligent system should recognize this independence, should execute both calls A and B only once and should *combine* the n solutions $\theta_1, \ldots, \theta_n$ for A with the m solutions $\sigma_1, \ldots, \sigma_m$ for B to obtain the

$n * m$ solutions $\theta_i \cdot \sigma_j$ for the initial problem -A,B. In the special case where m = O, such an intelligent system will not attempt to remedy the failure of B by deriving another solution for A, but will conclude the unsolvability of the total problem.

The current state of a computation can be characterized by an *and-tree* where the calls of the original goal statement are the branches of the root node. At each step, a leaf is selected and becomes the root of the calls in the body of the applied procedure. The order in which the calls are selected is, in a naive system, also the order in which backtracking is performed and characterizes the whole computation.

Example:

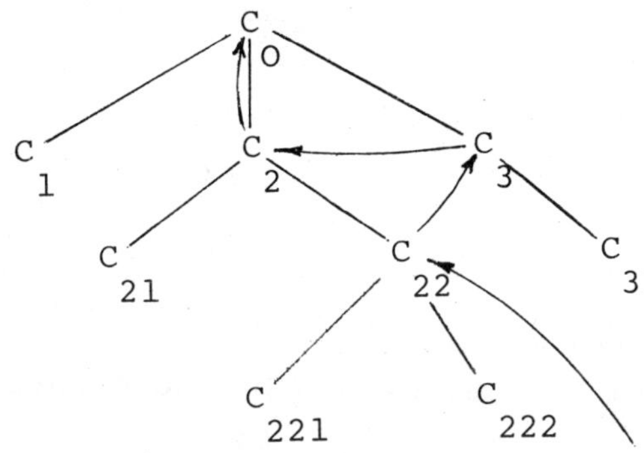

total ordering of selection and backtracking

We can think of the computation as starting with a fictitious goal statement <-C_0. The original goal statement is <-C_1,C_2,C_3. The order of selection is C_0,C_2,C_3,C_{22}. Backtracking occurs in the reverse order. The current goal statement is <-C_1,C_{21},C_{221},C_{222},C_{31}. Once all solutions to the current goal statement are found, the system restores $\leftarrow C_1, C_{21}, C_{22}, C_{31}$. Augmented with, for each call, the list of untried procedures and the applied substitution, this tree defines the whole computation.

The father of a call is the call by which the former is created i.e. C_2 is the father of C_{21} and C_{22}.

The offspring of a call consists of all calls in the subtree with that call as root i.e. the set $\{C_{21}, C_{22}, C_{221}, C_{222}\}$ is the offspring of C_2.

3. A MORE ACCURATE BACKTRACKING ON FAILURE

A call A on some relation R has, when it is executed, a certain pattern (its arguments have a certain value). This pattern (or parts of it) determines the definitions matched by the call. The actual expression of the pattern is determined by some of the calls already executed, especially by its father and other ancestors but, eventually, also by others. In executing a call, the system searches for the first definition matching the call. Assume that the system can determine which of the previously executed calls generate parts of the pattern which are essential for the eventual mismatches and for the first match. Let D_A denote such a set for a call A. We say that a call "depends" on all calls in its D-set. We delay the precise definition of this dependency relation until the next section, but we can already make some observations about it:

1. Including too many calls in the D-set of a call cannot cause the loss of solutions; however, we can expect that a smaller D-set gives more accurate backtracking.

2. Any reasonable definition of the dependency relation has to be transitive. Indeed, when the first definition matching a call A critically depends on the pattern generated during the execution of a call B, and the same relationship exists between the calls B and C, then the call A must also depend on the call C. As a consequence, it is sufficient to keep track of the calls on which each call "depends directly" (those not induced by transitivity).

3. A call is at least dependent on its father (and by transitivity on its other ancestors): indeed, a call only exists because its father has been executed with a particular definition.

Each call in a D-set has a corresponding goal statement, i.e. the goal statement in which that call has been selected.

With all these observations in mind, we can state what the system has to do when it selects some call A which does not match any procedure definition. A naive backtracking system will restore the previous goal statement say G_B and will try another procedure definition for the call B. However, undoing B is insufficient when B does not belong to the set D_A. Indeed, in this case, the system did not need the substitutions generated by the call B to detect the unsolvability of the call A and, thus, changing these substitutions cannot make the call A solvable. We conclude that we have to backtrack further and to restore the goal statement G_C with C the most recently executed call in D_A. Indeed, restoring any of the goal statements between G_C and G_A cannot make the call A solvable, and we do not have to explore the corresponding branches of the or-tree.

Notes:

1. The most recently executed call in D_A is one of those on which A depends directly.
2. Either G_C still contains the call A, eventually with a less specified pattern, or C is the father of (A because the father of A belongs to D_A).
3. The goal statements corresponding to the set D_A are the only points in the or-tree where possibly successful alternatives can start. By alternative, we mean a point which can lead to a different pattern for the call A or no call A at all, in other words a goal statement which can lead to derivations where the same conflict does not occur. On the contrary *all* goal statements in the or-tree corresponding to a *solution* are branching points which can lead to different solutions.

Having rejected the current solution for C, the system searches another definition matching C and, eventually, tries the call A again. To assure that the system - when it, using the other definitions for C, still fails to derive a solution - also tries a new start from the goal statements corresponding to the remainder of the set D_A we have to update the D-set of the call C. We have to replace D_C by the union of D_C and $\{D_A - C\}$. This extended set is still in agreement with the intended definition of the D-sets, indeed it contains all calls which contribute to the generation of patterns essential in the rejection of all tried definitions for the call C. Now, one of these definitions is rejected because it causes a conflict with the call A. Also the pattern of A is essential in this conflict and thus, those calls creating the critical parts of that pattern (those causing the failure of all available definition) must be in the D-set of the call C.

This update of the D-set brings us back to the initial assumption about the D-set (now for the call C instead of A) and thus, the above reasoning can be repeated.

Systematic updating of the D-sets involved - when backtracking and when trying new definitions for a call - assures that the system, in search for a first solution, will try all valuable branching points of the or-tree and will prune all others. As the system has learned from previous failures, these other branches are doomed to fail.

Assuming that the calls in the original goal statement depend on some fictitious node, the direct dependency defines a partial ordering over the nodes of the and-tree. We sketch a simple example (in addition to the and-tree, we indicate the direct dependencies by arrows).

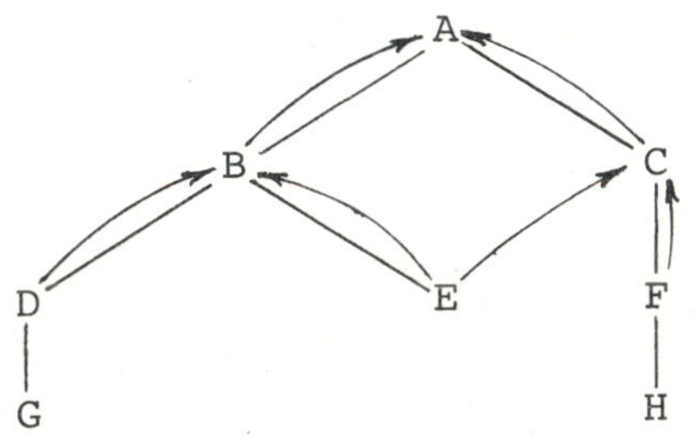

Sequence of goal statements:

G_B=<-B,C

G_C=<-D,E,C

G_D=<-D,E,F

G_F=<-G,E,F

G_E=<-G,E,H

When the call E in G_E fails, the system restores G_C (C being the most recent call in $D_E =$ $= \{B,C\}$)

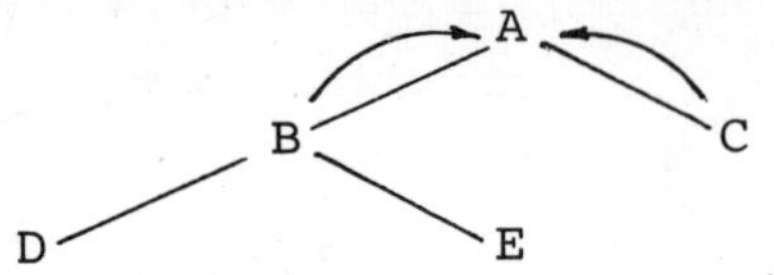

updates D_C: $D_{Cnew} = D_{Cold} \cup \{D_E - C\}$

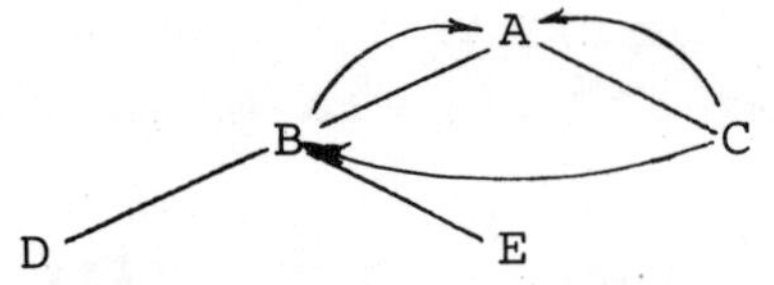

removes direct dependencies induced by transitivity and tries another definition for the call C.

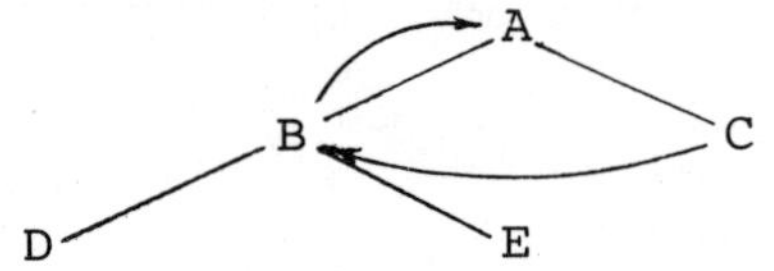

4. *DEFINING DEPENDENCY*

4.1. *A special case: naive backtracking.*

The assumption that every call depends on *all* previously executed calls leads to the total ordering of a naive backtracking system. Indeed, as a consequence of transitivity, each call depends directly on only one call, the one previously executed, and thus backtracking always restores the previous goal statement.

4.2. A definition based on the unification algorithm

It is the unification algorithm which decides whether or not a call matches a certain definition. In doing so, the unification algorithm does not need access to all parts of the pattern. As soon as it finds a mismatch, it does not need the remainder of the pattern. Also in unifying a term with a variable, it does not need the components of the term (as far as the occur-check is unnecessary). The unification algorithm has access to all components generated by calls on which the current call depends. Initially, the current call depends on its father (because as we already said, the call only exists as a consequence of executing the father with a particular procedure definition) and, by transitivity, on those calls on which the father depends and the unification algorithm has access to all substitutions generated by these calls. As soon as the unification algorithm needs access to other substitutions, dependencies are added.

An example:

```
P  (  x  )
      | (1)
      f ( k   ,  l )
          | (2) | (3)
          A     f ( m   ,  n )
                    | (4) | (5)
                    B     f ( p  ,  q )
```

The substitutions x <- f(k,l) have been generated by the call (1), k <- A by (2), l <- f(m,n) by (3), m <- B by (4) and n <- f(p,q) by (5). Suppose the call P initially depends on (1) but not on (2) (3) (4) and

(5). As a consequence, while executing the call P the unification algorithm initially only has access to the substitution x <-f(k,l). It knows k and l are bound but it does not know to which terms.

Assuming that the attempted definition is:

P(f(r,f(s,t))) <- Q(r,s) , P(f(s,t))

the system has to unify the bound variable x with the term f(r,f(s,t)). To do so, it has to know the value of x. Because the call depends on (1) the unification algorithm can access f(k, l) without adding new dependencies. The functors are the same and the system has to unify their arguments. It has to unify the bound variable k with the free variable r. Because this is the first occurrence of r, unification must be possible (whatever the value of k), and the substitution r <- k can be generated without accessing the value of k, thus without making the current call dependent on (2). In the next step the bound variable l has to be unified with the term f(s,t). To do so we have to access the first approximation of l: f(m,n).

The necessity to access this structure makes the current call P dependent on (3). In the next step we have to unify the free variable s with the bound variable m and the free variable t with the bound variable n. Again, because these are the first occurrences of s and t, they cannot occur inside the terms to which m and n are bound, and the substitutions s<-m, t<-n can be generated without accessing the values of m and n, thus without making the current call dependent on (4) and (5).

Giving the substitutions generated by the call P a label (6), the children of P become:

```
Q ( r  ,  s )    and    P ( f ( s  ,  t ))
    | (6) | (6)                 | (6) | (6)
    k     m                     m     n
    | (2) | (6)                 | (4) | (5)
    A     B                     B     f( p , q )
```

They depend initially on their father (6) (the call P) and the calls on which the father depends i.e. (1) and (3).

Notes:

1. An implementor should consider possible optimisations for example, because the children depend on (6), they can as well be replaced by

```
G ( k  ,  m )  and    P ( f ( m  ,  n ))
    | (2) | (4)               | (4) | (5)
    A     B                   B     f ( p  ,  q )
```

2. In unifying a free variable with a bound variable, when the occurcheck is necessary, the current call becomes dependent on all calls generating parts of the term to which that bound variable is bound. As a consequence, to get accurate backtracking, it is necessary to perform the occurcheck only where it really is necessary. Most implementors of languages based on logic do not apply an occurcheck at all, however, some programs have been written where the occurcheck is essential [7]. Further in this paper, we argue that dependencies caused by occurchecks have only to be added when the occurcheck causes the failure of the unification.

In detecting a mismatch, the unification algorithm can have some choice because there can be more than one disagreement between the call and the definition. In such cases it seems preferable to access structures generated by the least recent calls, they will give the deepest backtracking and the highest number of pruned branches in the or-tree.

We illustrate the cacktracking behaviour with the n-queens problem for n=4. The initial goal statement is:

<- Perm (4.3.2.1.Nil, l) Pair (4.3.2.1.Nil, l, q) Safe (q)

The execution order is strictly left to right. The call Perm generates a permutation l of the list 4.3.2.1. Nil, e.g. 1.2.3.4. Nil, the procedure Pair combines its first two argument lists into a queenboard q with one queen in each row and one queen in each column e.g. p(4,1).p(3.2).p(2,3).p(1,4).Nil.

The procedure Safe rejects the queenboard and causes backtracking when it finds two queens on the same diagonal. A naive backtracking system will systematically generate all permutations until some permutation passes all checks created by Safe. In our system, to detect a conflict between for example the first and the second queen, Safe needs no access to the remainder of the permutation, and thus it is not dependent on the nondeterministic procedures generating that remainder. As a consequence it will not try to solve the conflict by generating other remainders of the permutation but will backtrack immediately to the procedure generating the second queen. The interested reader finds a detailed analysis of this behaviour in the remainder of this section.

The complete program is:

```
Perm (Nil, Nil) <-
Perm (x.y, u.v) <- Del (u, x.y, w) Perm (w, v)
Del (x, x.y, y) <-
Del (u, x.y, x.v) <- Del (u, y, v)
Pair (Nil, Nil, Nil) <-
Pair (x.y, u.v, p(x,u).w) <- Pair (y, v, w)
Safe (Nil) <-
Safe (p.q) <- Check (p, q) Safe (q)
Check (p, Nil) <-
Check (p, q.r) <- Diag (p, q) Check (p, r)
```

For simplicity, we assume Diag defined as a set of assertions about queens not on the same diagonal.

```
Diag (p(1,1), p(1,2)) <-
Diag (p(1,1), p(2,3)) <-
        .
        .
        .
```

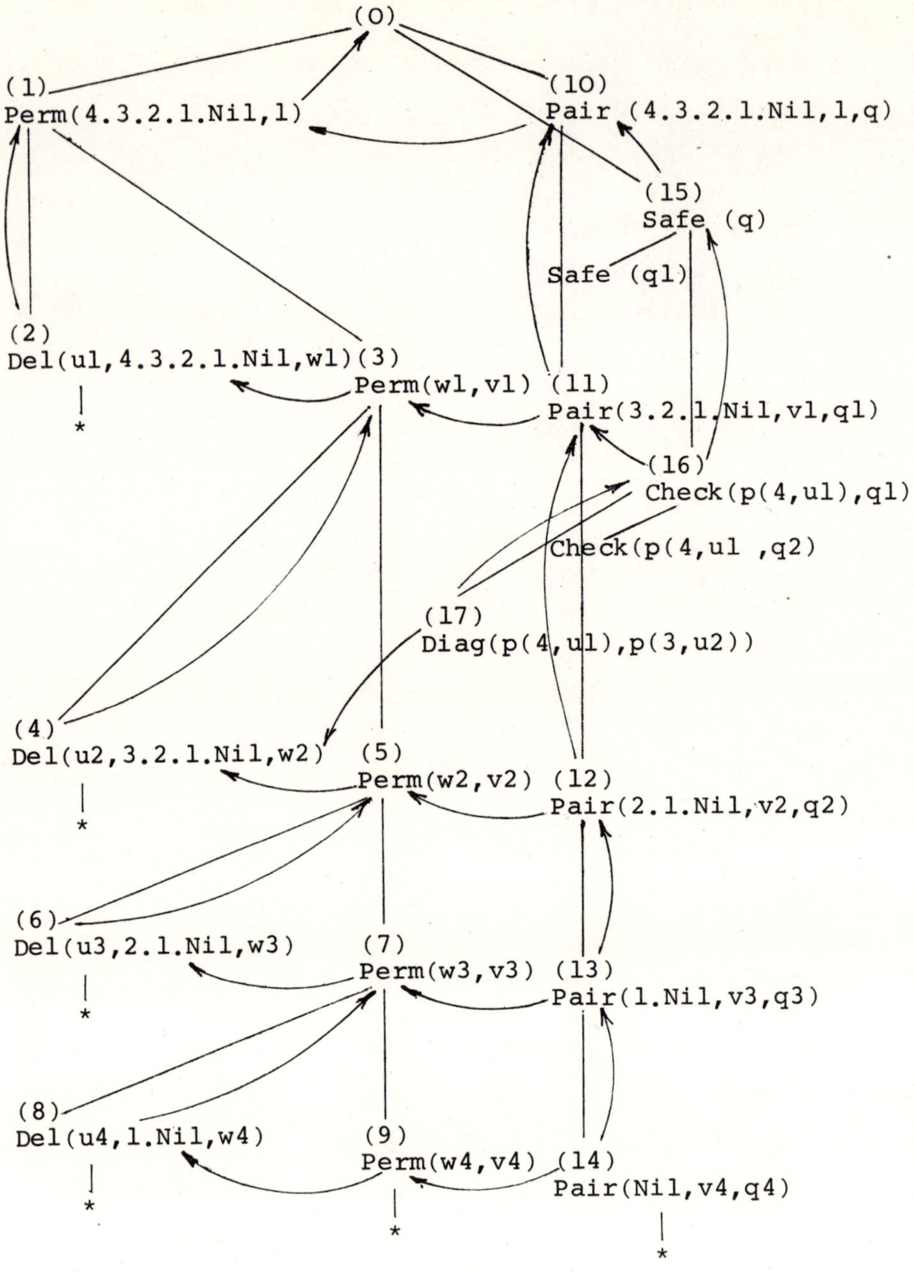

And-tree and dependecy-graph at the point where call (17) causes a failure.

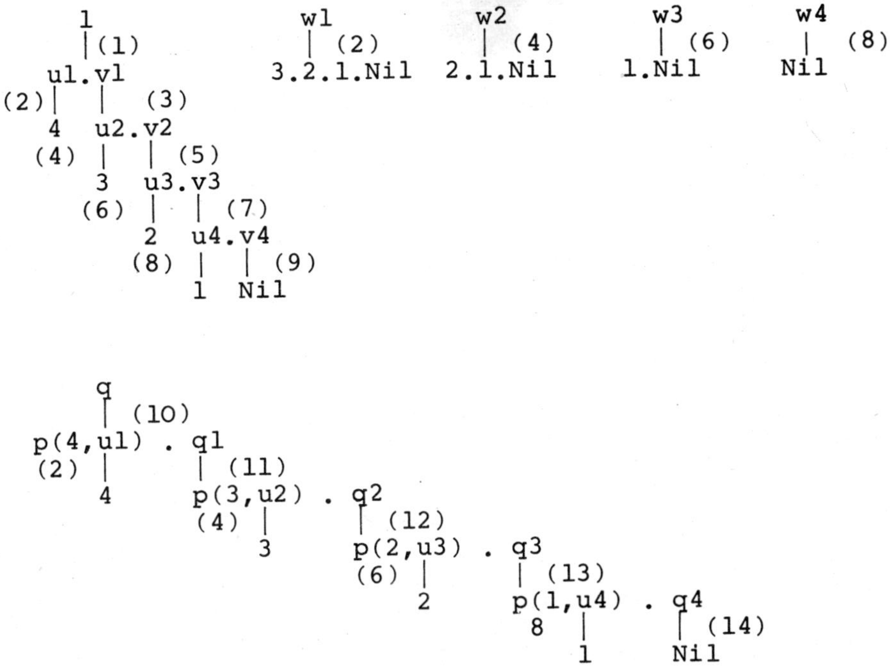

Substitutions generated by the different calls.

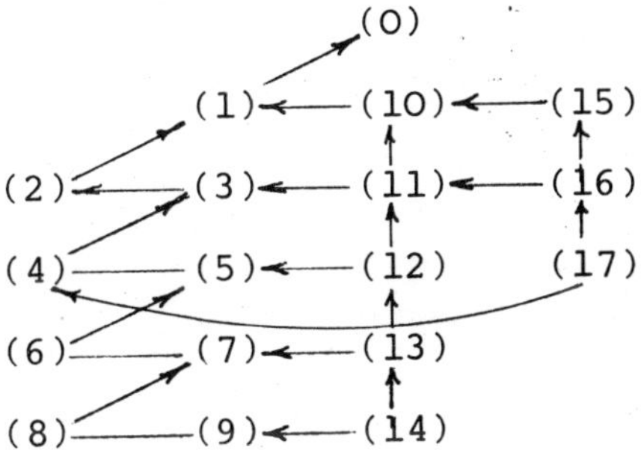

Direct dependency-graph.

(17) fails because of the conflict between the first and the second queen.

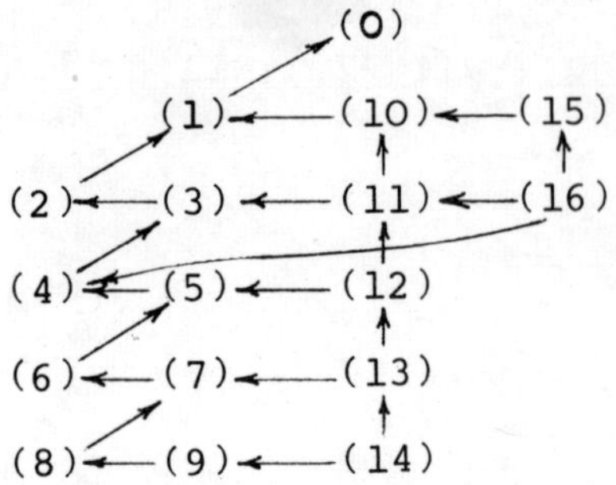

The goal statement prior to the execution of call (16) is restored, and the D-set of (16) is updated. No other definition for (16) is available and the system backtracks further.

- restore the goal statement prior to the execution of (15); update D-set
- restore the goal statement prior to the execution of (11); update D-set

There is still no alternative definition available and the dependency-graph becomes as follows:

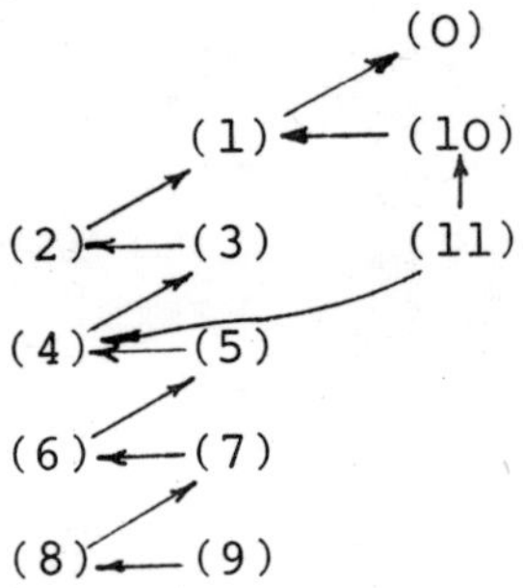

- restore the goal statement prior to the execution of (10); update D-set
- restore the goal statement prior to the execution of (4); update D-set

The dependency-graph is now:

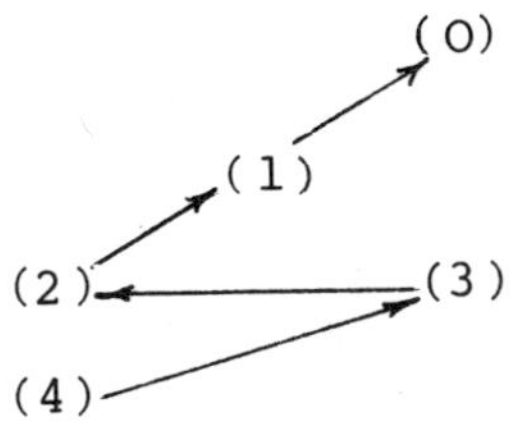

(4) (Del(u2,3.2.1.Nil,w2)) has an alternative definition. The system has backtracked correctly to the point where the second queen was chosen. A naive backtracksystem would try to find other solutions for the last queen, the last but one, ... !

5. *ON THE ORDER OF EXECUTION*

We claim that any order of execution *allowed by the partial ordering* leads to the same current goal statement and the same dependency-graph.

We give a simple example:

```
A( f(x)) <- C(x)
C(x) <- D(g(x))
B(h(y)) <- E(y)
E(y) <- F(k(y))
<- A(x0) B(y0)
```

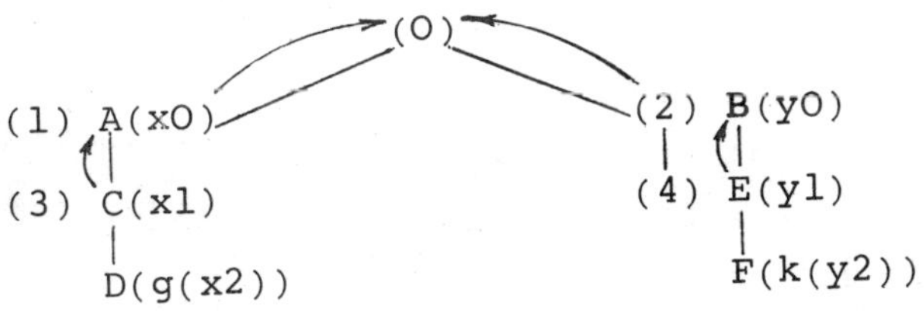

```
x0                y0
| (1)             | (2)
f(x1)             h(y1)
| (3)             | (4)
x2                y2
```

The sequence of calls (1) (2) (3) (4) leads to the goal statement <-D(g(x2)),F(k(y2)).

We claim that any sequence allowed by the partial ordering of the dependency-graph, i.e. (1) (3) (2) (4) and (2) (4) (1) (3), leads to the same goal statement and the same dependency-graph.

We know from resolution theory that the order of different resolution steps does not change the final clause.

We only have to show that the dependencies are also unchanged. We do this by arguing that two successive steps, not dependent on each other, can be interchanged without causing changes in the dependency-graph.

Suppose we execute a call A. It matches the i-th definition and results in a substitution $\sigma = \{x_k <- s_k\}$. The next call executed is a call B. It matches the j-th definition and results in a substitution $\theta = \{y_k <- t_k\}$.

Suppose B is not dependent on A.

Now we execute B before A. Because B was not dependent on A, it did not look at the substitution σ generated by A and consequently it does not matter for B whether or not the substitution σ is present. B will again match the j-th definition with the same substitution θ. Executing A after B, the unification algorithm surely can, as before, reject the first i - 1 definitions without looking at θ and this will result in the same dependencies as before. Because interchanging the resolutionsteps does not change the final clause, the call A must match the i-th definition, but, will the unification algorithm look at the substitution θ? Suppose it does. Then the unification algorithm must find some pair of the form x - s to be unified, with x a variable bound by the substitution θ(x<-t $\in\theta$) and s bound to some term. Indeed, only in such a situation must the unification algorithm look at θ and does A become dependent on B. This is not possible, indeed, in that case executing A before B must have resulted in x<-s$\in\sigma$ and, during the execution of B, a pair x - t to be unified. At that point, either t<-x was generated or the unification algorithm had to look up x and B becomes dependent on A. This contradicts the assumptions that B does not depend on A and that x<-t$\in\theta$. Thus while executing A after B, the unification algorithm does not need θ and it will generate the *same* substitution and the *same* dependencies as before.Because σ and θ are the same as before, all calls executed *after* A and B get the same dependencies and the interchanging of A and B does not effect the dependency-graph.

There is a weak point in the above reasoning: we did not mention occurchecks. Maybe, changing the execution order causes different occurchecks and these occurchecks in turn cause different dependencies.

An example:

```
A(x1,x1) <- RA(x1)
B(f(x2)) <- RB(x2)
<- A(x0,y0) B(x0)
```

Starting with the call A we get x1<-x0 and we have to unify y0 with x1. Before generating the substitution y0<-x1, we have to verify that y0 does not occur inside the substitution for x1. Then we execute the call B and we get the substitution x0<-f(x2).

The dependency-graph is:

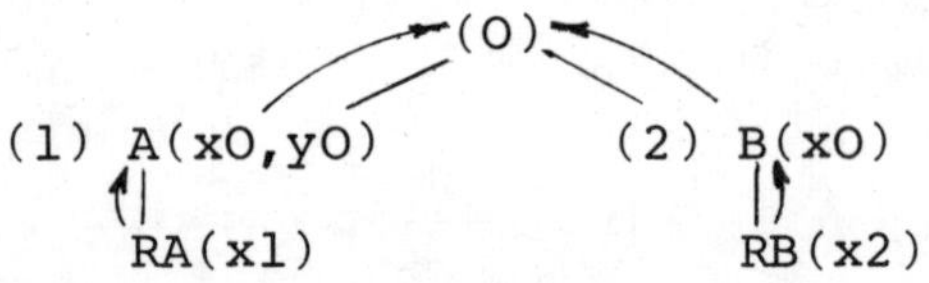

Starting with the call B we get the substitution x0<-f(x2) then while executing A, we have to unify x0 with x1. Because it is the first occurence of x1, we can generate x1<-x0. Then we have to unify y0 with x1: we have to verify that y0 does not occur in the substitution for x1. For this test, we need to access also the substitution x0<-f(x2) thus the call A becomes dependent on B: we get a different dependency-graph.

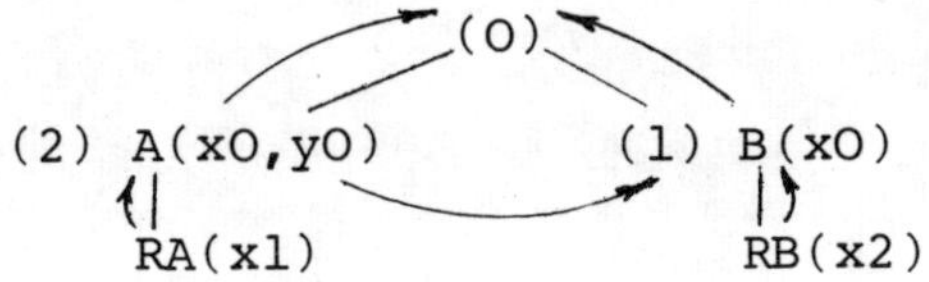

We can solve this problem by some subtle reasoning. As we already mentioned in the previous section, we should only execute the occurchecks which are relly necessary. In fact the only necessary occurchecks are those causing the failure of the unification. Now, when changing the order of execution, we know that the derivation is possible and therefore the occurchecks cannot cause a failure and are unnecessary. Thus the above reasoning was correct and changing the order of execution - respecting the partial ordering - does not change the dependency-graph. We even can apply this observation about the necessity of the occurcheck with retrospecitve effect, i.e. when we detect that it does cause a failure, we conclude we should not have done it and we leave the dependency-graph unchanged. We only add the dependencies caused by an occurcheck when it causes the failure of the unification.

As a consequence of the property that we can change the order of execution, when we backtrack, we can choose between any of the calls on which the failing call depends directly. We undo the selected one, say A, and of the calls executed after A, we have only to undo those dependent on A. Then we make A directly dependent on the other calls on which the failing call depends directly, remove the direct dependencies induced by transitivity of others and try another definition for the call A.

When there is a good selection function available, it seems reasonable to backtrack to the most recent call in the direct dependency relation. If the selection function is rather poor, then it can be useful to analyse the effect on the size of the search space of the different possible "backtrackpoints". Eventually, the selection strategy could learn from this analysis.

In the queensexample at the end of the previous section, the failing call (17) depends directly on (4) and (16). Backtracking to (4) yields the graph:

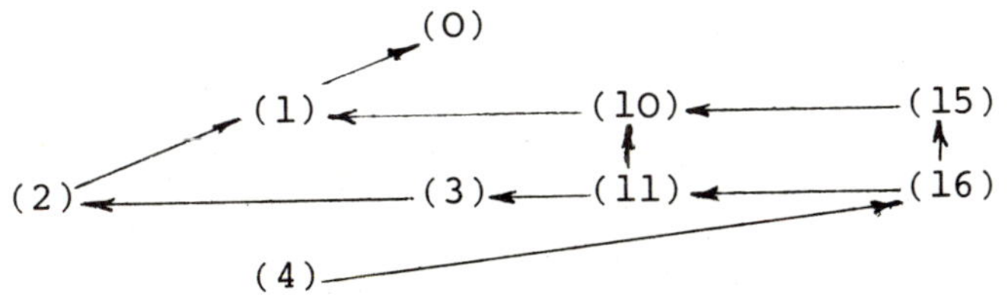

This backtracking is preferable to the one used, because both end with the application of another procedure for call (4), but here, a greater part of the computation has been saved.

6. *FINDING ALL SOLUTIONS*

As we already briefly mentioned in the introductory sections, to find *all* solutions when there are different independent subproblems, each having a set of solutions, a backtracking-system will execute some subproblems several times. The goal of this section is to avoid this inefficiency. We introduce our method by a simple example.

The program:

```
<- AB(x,y)
AB(x,y) <- A(x) B(y)
A(f(x)) <- RA(x)
A(g(x)) <- QA(x)
RA(A) <-
RA(B) <-
QA(C) <-
B(g(y)) <- RB(y)
B(k(y)) <- QB(y)
RB(E) <-
RB(F) <-
QB(G) <-
```

Assume a first solution has been found:

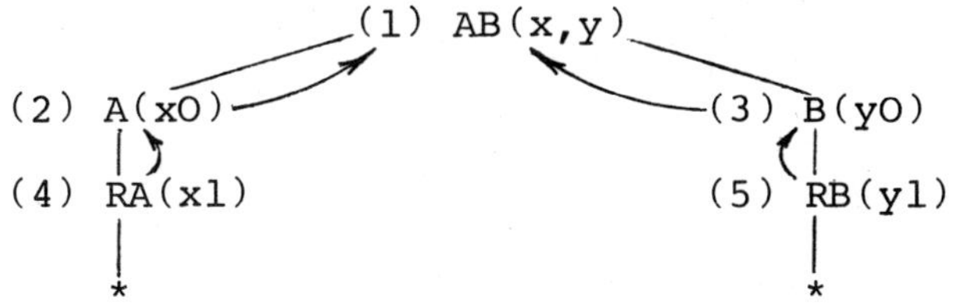

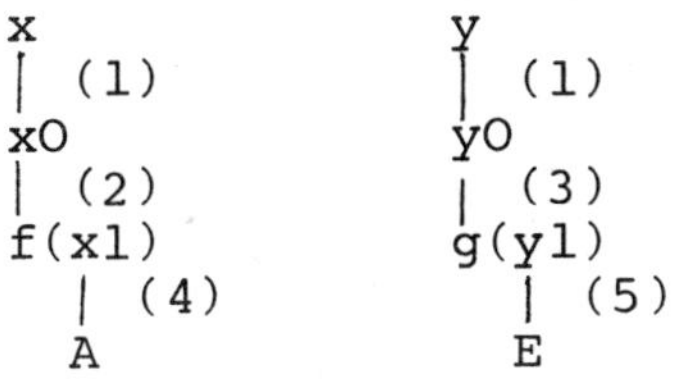

This first solution {x<-f(A), y<-g(E)} can be derived by all sequences allowed by the partial ordering. Let us choose one of these orderings i.e. (1) (2) (3) (4) (5).

The sequence of goal statements in then:

G_0 = AB(x,y)
G_1 = A(xO) B(yO)
G_2 = RA(x1) B(yO)
G_3 = RA(x1) RB(y1)
G_4 = RB(y1)
G_5 = □

To find all solutions, we have to explore all branches of the corresponding or-tree:

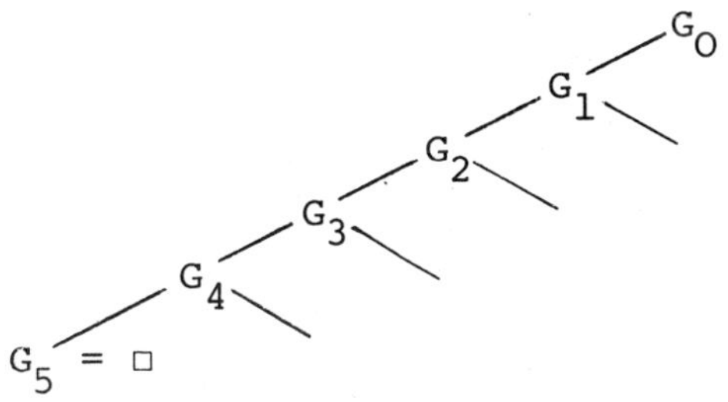

We can explore this or-tree bottom up i.e. find all solutions for G_4 then all solutions for G_3 ,... then all solutions for G_0. To improve the efficiency, we will divide, where possible, these goal statements into *independent* subproblems. We can solve these subproblems independently of each other and obtain the solutions for the complete goal statement by making the cross-product of their solutions. In fact, we will not explicitly compute the set of solutions for each successive goal statement. As the example will illustrate, we only compute a solutionset for well chosen subproblems.(See the "merge procedure" below.)

G_4 (=RB(yl)) consists of a single call, all its solutions can be found simply by bracktracking over the call RB(yl).

By expanding G_4 into G_3, another call RA(xl) enters the goal statement. The dependency-graph shows us that both are independent. All solutions for G_3 could be found by:

- searching for all solutions of each subproblem, using normal backtracking; this yields [{yl<-E}, {yl<-F}] for RB(yl) and [{xl<-a}, {xl<-B}] for RA(xl).
- making the cross-product of both sets; this yields: [{yl<-E, xl<-A}, {yl<-F, xl<-A}, {yl<-E, xl<-B}, {yl<-F, xl<-B}].

In the next step, the subproblem RB(yl) expands into the subproblem B(yO) (the call being replaced by its father) while the other subproblem remains unchanged. Still, the solutions for each of the subproblems can be obtained by backtracking. Similarly, while moving to G_1, the call RA(xl) expands into A(xO) and we have two independent subproblems A(xO) and B(yO): the solutions for each of them can be found by backtracking. The last step is more interesting: here, both independent subproblems merge into the single call AB(x,y). Systematic backtracking is inefficient for this goal statement. We apply the following "merge procedure":

1. Find all solutions for the subproblem A(x0) i.e. [{x0<-f(A)}, {x0<-f(B)}, {x0<-g(C)}]. In general, this is done by recursively applying the "find-all-solutions" methods, which in this example simplifies to systematic backtracking over the nodes (2) and (4). In fact, the system alternates between:
   - Find a new solution: systematic backtracking over all nodes in the current solutions but intelligent backtracking over the new nodes (those entering the graph when it grows again: for these nodes we again search a first solution).
   - Find all solutions for the subproblem.
2. Similarly, find all solutions for B(y0) i.e. [{y0<-g(E)}, {y0<-g(F)}, {y0<-k(G)}].
3. Compute all solutions for A(x0), B(y0) by making the cross-product of both sets: i.e. [{x0<-f(A), y0<-g(E)}, {x0<-f(A), y0<-g(F)}, ...].
4. The "current solution set" for AB(x,y) is obtained by making the composition of the substitution applied on the call ({x<-x0, y<-y0}) with the substitutions of the above set: i.e. [{x<-f(A), y<-g(E)}, ...] (we are only interested in substitutions for the variables in the goal statement AB(x,y)).
5. We can reduce the dependency-graph and summarize the obtained results in a "reduced node". The reduced node 1_* replaces all calls involved ((5) (4) (3) (2) (1)) and has, instead of a single substitution, a set

of substitutions (the "current solution set") characterizing all obtained results.

The graph becomes:

(1_*)
AB (x,y)

A single reduced node with a current solution set and a list of untried procedure definitions (in this example empty).

Now, we determine where the next "merge" operation must be performed, or, as in this example, start to collect the solutions for the initial goal statement:

- take the current solutions set
- find all remaining solutions, by alternating between:
  - find a new solution

  and - find all solutions

Where a backtracking system directly starts collecting all solutions for the initial goal statement, the above procedure collects solutions for well chosen subproblems, and merges these subproblems into bigger ones until finally the original problem with its solutions is obtained.

The merge of subproblems is not always as simple as in the example above. There, in expanding G_1 into G_0, the entering call AB replaced its two independent children A and B. In general G_j consists of an independent subproblem $A_1,\ldots,A_n$ ($n \geq 1$) and another independent subproblem $B_1,\ldots,B_m$ ($m \geq 1$). By extending the goal statement G_j into G_i, a call C enters. C is such that some of the A_i as well as some of the B_j depend on it. The new goal statement consists of the calls $\{A_i\} \cup$

$\{B_j\} \cup \{C\}$ - {children of C}. C becomes the reduced node, its current solution set is obtained as described in the above example. Unlike the simple case, those A_i and B_j which are *not* children of C remain in the graph and depend on C: indeed, they have to be executed again when the backtracking mechanism tries the remaining procedure definitions on C.

When a first solution is found and we start to collect all solutions, we do not have to decide immediately which or-tree we will use in the derivation of all solutions. We can take this decision step by step, we only have to be careful to respect the partial ordering. Respecting the partial ordering means that the call entering the goal statement has no other call depending on it which is not yet in the goal statement. Manipulation is simpler and the alternatives are clearer when we perform an update of the dependency-graph which is similar to the one used in Section 3. Initially, each subproblem consists of a single call with a certain D-set. This call is named the *representative* call of the subproblem. When a subproblem is extended, the entering call becomes the new representative call and its D-set is updated with the D-set of the old representative call; similarly, in the case of a merge, the entering call becomes the representative call and its D-set is updated with the D-sets of the representative calls of the merging subproblems (the representative call becomes the reduced node). Now, the dependencies between the other calls in the subproblems and calls not yet in the goal statement can be ignored and, in each step, to extend the goal statement, we can choose between the calls which are such that the only calls dependent on them are representative calls.

We now give a non-trivial example.

Example:

A program to compute all binary trees having a certain leaf profile.

- x.y represents a list with first element x and remainder y.
- Nil represents an empty list.
- x*y represents a binary tree with x and y subtrees.

```
Tree (x.Nil, x) <-
Tree (x.y.z, u*v)<-Append (s, t, x.y.z) Tree (s, u)
     Tree (t, v)
Append (Nil, x, x) <-
Append (x.u, v, x.w)<-Append (u, v, w)
<- Tree (A.B.C.D.E.F.Nil, z)
```

We give the computation of a first solution in a strictly left to right execution order. As first definition for "Append" we use the one that is best suited for our purposes: dividing the six-element list into two tree-element lists. We only give the complete computation of the left subtree, the right subtree is similar.

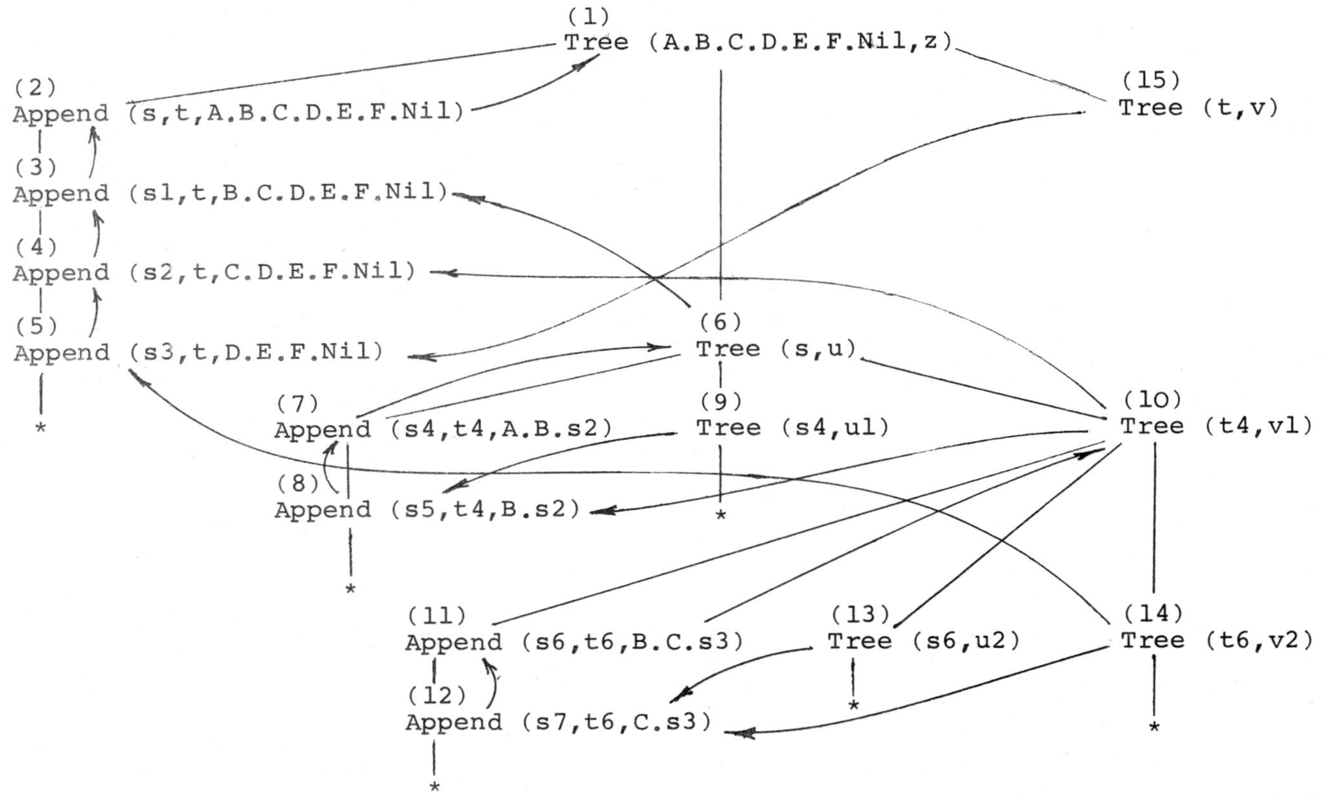

And-tree and dependency-graph.

Substitutions made during the execution of the different calls:

```
 1: {z<-u*v}
 2: {s<-A.s1}
 3: {s1<-B.s2}
 4: {s2<-C.s3}
 5: {s3<-Nil, t<-D.E.F.Nil}
 6: {u<-u1*v1}
 7: {s4<-A.s5}
 8: {s5<-Nil, t4<-B.s2}
 9: {u1<-A}
10: {v1<-u2*v2}
11: {s6<-B.s7}
12: {s7<-Nil, t6<-C.s3}
13: {u2<-B}
14: {v2<-C}
```

We start with the empty goal statement. It can be extended with one of those calls on which nothing depends e.g. (9) (13) or (14). We choose (14). (14) depends on (12) and (5). However, both have other dependent calls not yet in the goal statement. We concentrate on (12). First we extend the goal statement with (13) (now we have two independent subgoals), then we extend it with (12): we have to perform a merge.

Solutions of (14): the current solution is {v2<-C}; backtracking yields no other ones.

Solutions of (13):[{u2<-B}]

Cross-product: [{v2<-C, u2<-B}]

Current solution set of (12) (13) (14):[{s7<-Nil, t6<-C.s3, u2<-B, v2<-C}]

Part of the reduced graph:

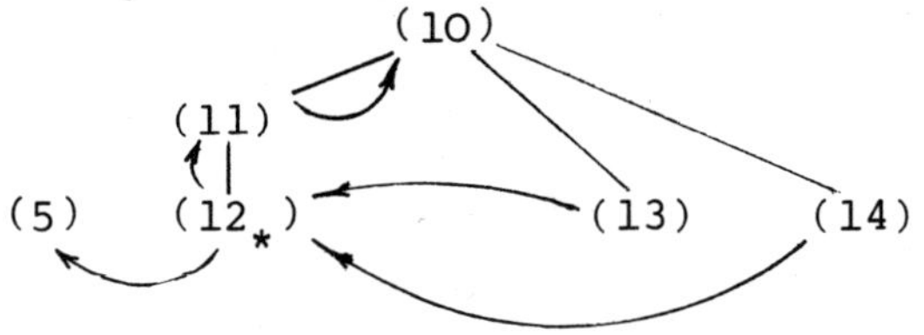

Remark that the calls (13) and (14) stay in the graph and that the "representative" call (12) now depends on (5).

We want to extend the goal statement further with the nodes in the following order (11), (10), (8). However, before extending it with (8), we first have to extend it with (9), which is independent, and then, we have again to perform a merge.

Solutions for (9): [{ul<-A}]

Solutions for (10): current solution [{vl<-B*C}]

We search other solutions by a systematic backtracking over the nodes (12_*) (11) and (10), while a solution is only found when it also solves the calls (13) and (14) dependent on the reduced node (12_*).

In fact, we search a first solution for the subproblem, but starting with one of the untried procedures for call (12). When we find one, we again start to collect all solutions, (still for the subproblem (10)). This subproblem has no other solutions.

Cross-product: [{ul<-A, vl<-B*C}]

Current solution set for (8) (9) (10):
[{ s5<-Nil, t4<-B.s2, ul<-A, vl<-B*C}]

Part of the reduced graph:

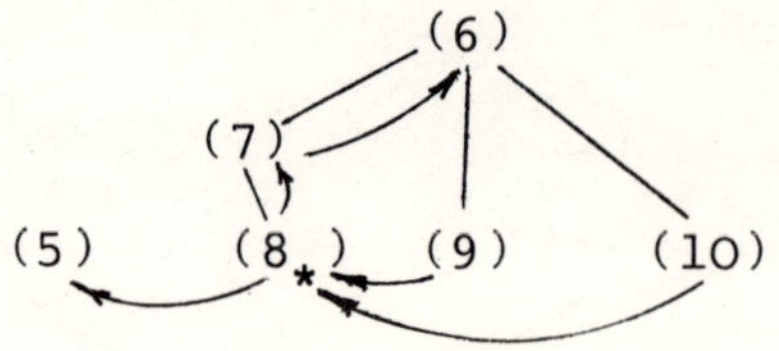

We can further expand the goal statement with (7) and (6). The graph becomes:

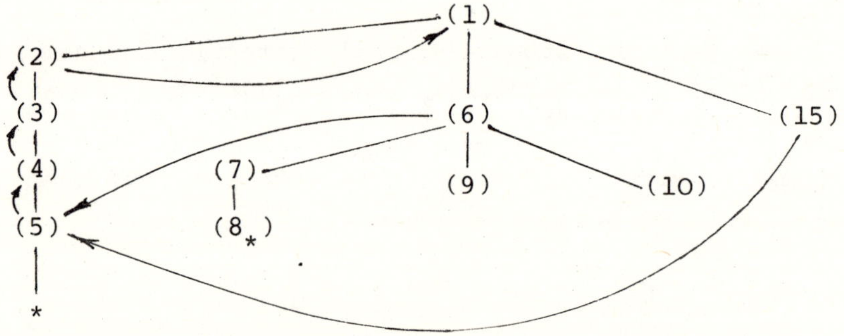

Before extending the goal statement with (5) we have to extend it with the whole subtree of (15), then we have again to perform a merge.

Solutions of (6): current solution set: [{u<-A * (B*C)}]. Other solutions have to be derived by backtracking over the nodes (8*), (7) and (6). This results in a second solution {u<-(A*B) * C}.

Solutions for (15): similar as for the left subtree, we can derive the solution set [{v<-D * (E*F)}, {v<-(D*E) * F}].

After making the cross-product of both sets, we obtain as current solution set for (5) (6) (15):

[{s3<-Nil, t<-D.E.F.Nil, u<-A * (B*C), v<-D * (E*F)},
{s3<-Nil, t<-D.E.F.Nil, u<-A * (B*C), v<-(D*E) * F},
{s3<-Nil, t<-D.E.F.Nil, u<-(A*B) * C, v<-D * (E*F)},
{s3<-Nil, t<-D.E.F.Nil, u<-(B*B) * C, v<-(D*E) * F}]

The reduced graph becomes:

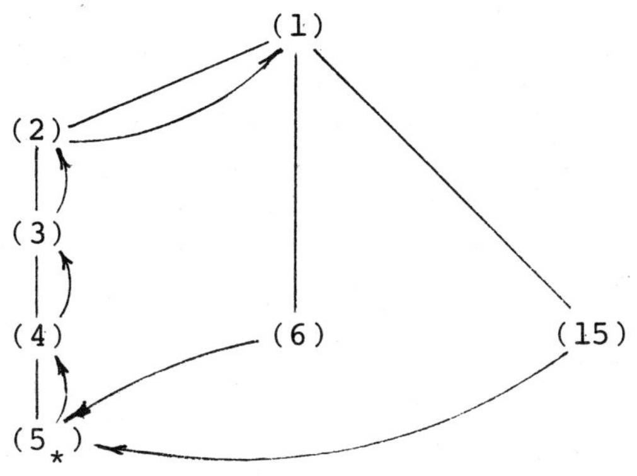

Now the goal statement can be expanded by, in order, (4), (3), (2) and (1) to reach the initial goal statement.

Its current solution set is:

[{(A * (B*C)) * (D * (E*F))},
{(A * (B*C)) * ((D*E) * F)},
{((A*B) * C) * (D * (E*F))},
{((A*B) * C) * ((D*E) * F)}]

Other solutions will be obtained by systematic backtracking over the nodes (5_*) (4) (3) (2) and (1). The process will start with applying the recursive append definition on call (5) and with deriving (using intelligent backtracking) a new solution. Then, applying successive merge operations, this new solution is transformed into a new solution set and the search for another solution in the new reduced graph starts. This process repeats until all solutions are found.

Some subtle problems can appear from time to time. The dependency-graph is based on the procedure definitions already used, but the ones that have not yet been tried can cause some new dependencies. This gives trouble when a call becomes dependent on a reduced node.

We give a simple example:

```
PQ(x,y) <- P(x,y) Q(x)
P(x,y) <- PP(y)
P(A,B) <-
P(B,C) <-
Q(B) <-
Q(C) <-
PP(D) <-
PP(E) <-
```

First solution:

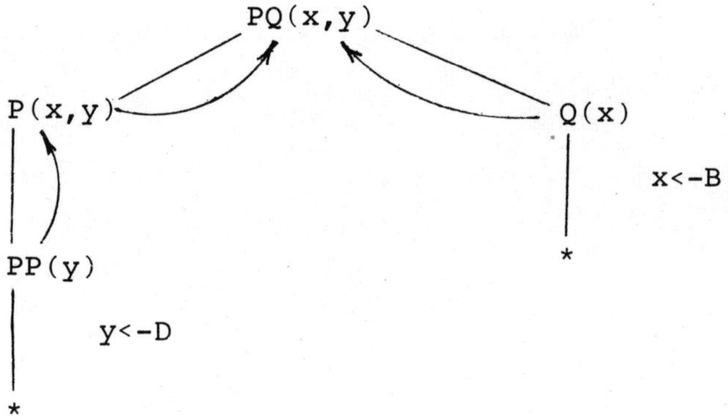

As far as this first solution is concerned, P(x,y) and Q(x) are two independent subproblems which merge into the initial goal statement. To find all solution, we perform a merge.

Solutions of Q(x): [{x<-B}, {x<-C}]

Here we can define an intermediate dependency-graph with node Q reduced and having an empty list of untried procedure definitions.

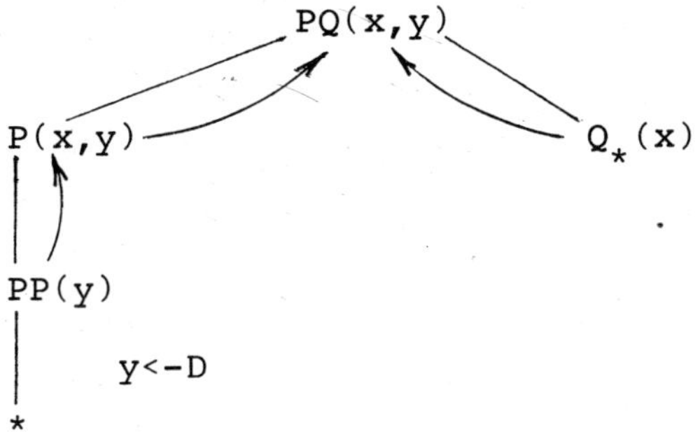

[{x<-B},{x<-C}]

Solutions of P(x,y):

Backtracking over PP results in the solutions {y<-D} and {y<-E}. Trying the other procedure definitions for P makes P dependent on the reduced node Q.

To find all solutions for P(x,y), Q(x), we have to be very careful. A first set of solutions is found by making the cross-product of the solutions already derived for P and Q:

[{x<-B, y<-D}, {x<-B, y<-E} {x<-C, y<-D}, {x<-C, y<-E}].

Others are found by applying the yet untried procedures for P, to each solution of Q. For x<-B, this results in a solution x<-B, y<-C ; x<-C results in a failure. This results in the following reduced graph:

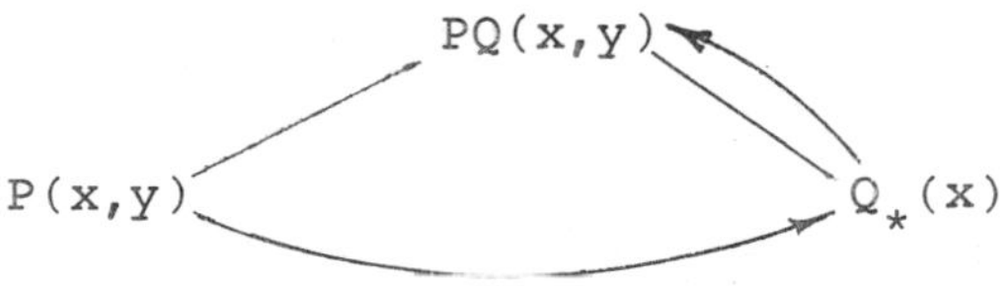

[{x<-B, y<-D}, {x<-B, y<-E}
{x<-C, y<-D}, {x<-C, y<-E},
{x<-B, y<-C}]

Q_* does not change its empty set of untried procedures. Instead of deriving the current solution set for PQ, the additional dependency has forced us to first derive a current solution set for Q(x), P(x,y). The further reduction to derive all solutions for the initial goal statement can be done usual.

7. *EXPLOITING DETERMINISM*

The dependency-graph representing the state of the computation can be simplified by exploiting the present determinism. When we execute a call which depends directly on its father only and there is only one definition between the available definitions matching the call, then we can interpret the execution of this call and the father as a single step in the computation, i.e. obtained by applying a preprocessed procedure on the father. Thus we can remove the node representing the son from the dependency-graph while the list of untried procedures for the father remains unchanged.

A simple example:

```
<- P(B,y) Q(y,z)
P(x,y) <- PA(x,y) PB(x,y)
P(x,y) <- PP(x,y)
PA(A,y) <- PA1(y)
PA(B,y) <- PA2(y)
```

Dependency-graph after executing the calls P and PA:

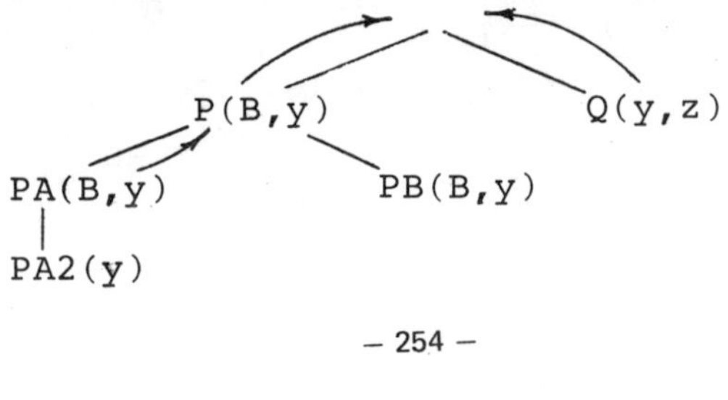

Preprocessing the call PA in the first definition of P leads to two procedures:

P(A,y) <- PA1(y) PB(A,y)
P(B,y) <- PA2(y) PB(B,y)

In general, preprocessing is not interesting: it leads to a different list of untried procedures (for the call P). However, when only one of the preprocessed procedures matches (as in the above example), then the list of untried procedures remains unchanged and we can derive in a single step the following dependency-graph.

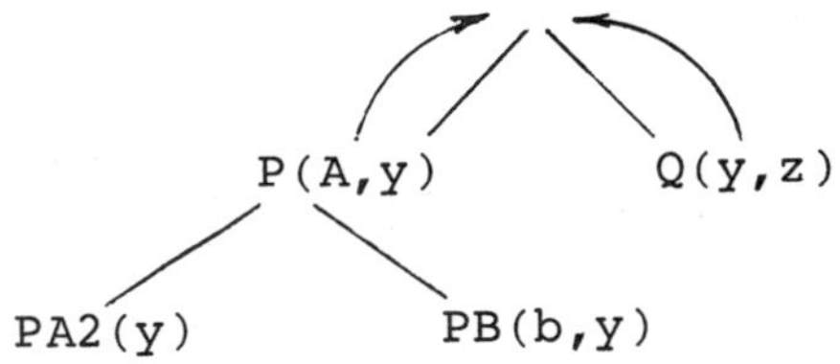

It differs from the first graph only by the fact that the determinate call PA has disappeared. Instead of performing preprocessing, we prefer to perform post-processing, i.e. when we execute the call PA, we observe that it only matches *one* procedure, and only depends on its father; we conclude that we could have applied a preprocessed procedure on the father call without changing its list of untried procedures. We apply the preprocessed procedure afterwards: we remove the determinate call from the graph and we extend the substitution of the father call with the substitutions made during the current step.

In fact, it is not necessary for the call to depend directly on its father only. However, when the son also depends on a third call, then we have to make the father dependent on that third call. This probably will result in a less accurate backtracking: it becomes impossible to backtrack from the son to that third call, also a failure of the father caused by for example another son unnecessarily lead to backtracking towards that third call.

8. CONCLUSIONS

We have defined a partial ordering over the set of calls involved in the execution of a goal statement.

This partial ordering, built by the unification algorithm, allows the accurate determination of all calls which can break a given failure. Moreover, it allows the reordering of the executed calls such that only a minimal number of calls must be undone during backtracking.

This partial ordering also allows the isolation of independent subgoals. A method to find all solutions, which uses a limited form of parallelism to exploit the detected independence of subgoals is given.

Our method still contains non-deterministic steps:

- When detecting that a call does not match a certain definition, the unification algorithm can sometimes choose which substitutions it will access and thus which dependencies it will add.
- When an unsuccessful call triggers backtracking, the system has to choose one of the calls on which the failing call depends directly.
- While searching all solutions, there can be different calls by which the goal statement can be extended.

All these choices effect the size of the search space and thus the efficiency of the computation. Determining (simple) criteria to make the right choices remains an area of further research. Another interesting question is whether we can develop selection strategies which can learn from the behaviour of the backtracking-system. The development of such a learning system could be a major step towards the development of systems which can execute logic programs efficiently without need for any user-specified control information.

9. ACKNOWLEDGEMENTS

I am indebted to Robert Kowalski and Yves Willems for their comments on an earlier draft of the paper. This research is supported by the Belgian "National Fonds voor Wetenschappelijk Onderzoek".

REFERENCES

[1] Battani,G., Meloni,H.: Interpreteur du langage de programmation PROLOG. Groupe d'Intelligence Artificielle, Marseille-Luminy, 1973.

[2] Warren,D., Pereira,M., Pereira,F.: PROLOG - The language and its implementation compared with LISP. *Proceedings ACM Conference on "AI and Programming Languages"*, Rochester, New York, 1977.

[3] Bruynooghe,M.: An interpreter for predicate logic programs. Report CW 10, Applied Maths. & Programming Division. Katholieke Univ. Leuven, Belgium, 1976.

[4] Clark,K., Bruynooghe,M.: A control regime for Horn clause logic programs. (Forthcoming report.)

[5] Cox,P., Pietrzykowski,T.: A Graphical deduction system. Dept. of Computer Science, University of Waterloo, Ontario, Canada, 1976.

[6] Cox,P.: Deduction plans: a graphical proof procedure for the first order predicate calculus. Dept. of Computer Science, Univ. of Waterloo, Ontario, Canada, 1978.

[7] Tarnlund,S.: A logical basis for data bases. Report TRITA - IBADB 1029, Dept. of Computer Science, Royal Institute of Technology, Stockholm, Sweden, 1976.

M. Bruynooghe

Applied Mathematics and Programming Division
Katholieke Universiteit Leuven,
Belgium.

COLLOQUIA MATHEMATICA SOCIETATIS JÁNOS BOLYAI
26. Mathematical Logic in Computer Science,
Salgótarján (HUNGARY), 1978.

OTHER DIRECTIONS FOR AUTOMATIC THEOREM PROVING

D. de Champeaux

1. *WHAT GOES ON IN ATP THESE DAYS?*

In ATP several schools can be distinguished. It is the method employed that divides them. The two leaders are 'resolution' and 'natural deduction'. Resolution has been most thoroughly investigated and is the most attractive from a theoretical point of view, since completeness can be easily verified. Natural deduction however seems closer to the method employed by human beings, and what goes on in a natural deduction proof can be more easily interpreted. Moreover current results of natural deduction are more impressive than those obtained by resolution.

Both schools have in common that finding a proof is seen as a search problem, in which objects in the search space are predicate calculus formulas (clauses in resolution). They only differ in the operators used. To be fair it must be stated that the natural deduction school is quite pragmatic and is not afraid of borrowing from another school, from the procedural one.

For many years these proceduralists have been drumming pervasively. They even developed a special language QA4. Very much was expected from this language since QA3 was a theorem prover (based on resolution). In fact, it was immediately obsolete and it was absorbed by QLISP/Interlisp. Until now theorem provers have not been implemented in these languages (as far as I know). These languages are more geared towards plan generation and automatic porgramming. As stated above they have inspired the natural deduction devotees to add special purpose routines to their programs, which indeed improved their power for a family of problems. However, a theorem prover equipped with these papers is not general anymore, and its speciality is no flexible component.

In the background people are active with higher order logic. Its theoretical problems are unpalatable. Moreover, until now there is no real need for it in practice (excluding, its monadic predicates part and except where higher order language is used for introducing abbreviations or definitions).

To give an impression of current achievements in ATP we give examples of problems that are managable.

(1) In group theory: If G is a group, K is a subgroup and g is an element of G then g is an element of K iff $g \cdot k = K$ (where $g \cdot k$ is a left coset of K). This example has been proved with natural deduction (Nevins, [1]) as well as with resolution (Overbeek, [2]). Nevins did not start from scratch (=definitions and axioms from group theory) but provided a mixture of axioms and lemmas. He also admitted having had restricted interaction between sets of input formulas. Still his result is quite impressive. Overbeek did not specify the input formulas and since the job is formulation dependent one cannot judge whether he did the same.

(2) An example in analysis: If f is a continuous function with the compact set S in its domain then f(S) is compact. This was proved by Ballantyne and Bledsoe [3] using natural deduction.

(3) Again by Bledsoe [4] in analysis: (Intermediate Value Theorem). If f is continuous for $a \leq x \leq b$, $a \leq b$, $f(a) \leq 0$, and $f(b) \geq 0$, then $f(x) = 0$ for some x between a and b.

2. *RESEARCH DIRECTIONS IN THE ATP SCHEME*

For natural deduction as well as for resolution the scheme of theorem provers is simply a program embodying derivation rules operating on axioms and the (negated) theorem with the aim of constructing a confirmation (contradiction).

In resolution people have been busy for many years refining the resolution rule in order to restrict application of that rule, to limit the generation of redundant clauses. A recent addition to the field are connection graphs [5], a data representation for clauses where the label on an edge represents the substitution of unifiable literals. This representation is attractive as no search of unifiable literals is necessary, but connection graph initialization is expensive, while updating the graph - though costly - is necessary at each production of a new clause.

Observe that these improvements fit the ATP scheme. The program working on input formulas becomes more complex, more intelligent. The same holds for other ATP improvement activities. People analysed the equality predicate and they built in special procedures to deal with its occurrence in formulas (paramodulation, simplification rewrite rules, etc. . Others concentrated on associative, idempotent and/or commutative functions and designed procedures to deal with them. These extensions also fit the ATP scheme.

We, however, do not like these extensions since they affect the generality of the ATP and have an ad-hoc flavour. Also we propose to stop working in the resolution refinement paradigm. How then to improve the power of ATP's.

3. *HOW TO IMPROVE THE POWER OF ATP'S?*

In the next paragraphs we discuss three options for getting more powerful ATP's. They are formulated from a resolution point of view but they are also relevant to natural deduction.

3.1. *Preprocessing*

Before turning loose a resolution theorem prover the input - given in predicate calculus formulas - has to be translated into a so called conjunctive normal form. Before doing this it is worthwhile investigating whether it is possible to improve the situation by finding an equivalent but simpler formulation. An example in group theory: Suppose one has axiomatized as follows:

$$(G)[GR1(G) \leftarrow \rightarrow$$

$$\{(x)(y)(x \varepsilon G \; \& \; y \varepsilon G \rightarrow (\;\; z)(z \varepsilon G \; \& \; f(x,y) = z)) \; \&$$

$$(x)(y)(z)(x \varepsilon G \; \& \; y \varepsilon G \; \& \; z \varepsilon G \rightarrow$$

$$f(x,f(y,z)) = f(f(x,y),z)\}]$$

$$(G)(e)[GR2(G,e) \leftrightarrow \{(x)(x \varepsilon G \rightarrow f(x,e) = x) \ \& \ (x)(x \varepsilon G \rightarrow (\ \ y)(y \varepsilon G \ \& \ f(x,y) = e))\}]$$

$$(G)(Group(G) \leftrightarrow Gr1(G) \ \& \ (\ \ e((e \varepsilon G \ \& \ Gr2(G,e)))$$

After deriving 7 theorems and introducing a definition for the inverse function I, one can reformulate a group (provided one does not discuss functions between different groups) thus:

$$(x)(y)(z)\{f(x,f(y,z)) = f(f(x,y),z)\}$$

$$(x)\{f(x,e) = x \ \& \ f(e,x) = x\}$$

$$(x)\{f(x,I(x)) = e \ \& \ f(I(x),x) = e\}$$

Undoubtedly there is no simple algorithm which can do this reformulation activity. Intuitively it requires various procedures such as the use of an existence theorem to intruduce with a definition a predicate or element, the use of a theorem concerning functionality of an argument position in a predicate to introduce a function, the shift of the domain of discourse, etc.

We suspect that without this reformulation capacity any theorem prover information however clever its other components be will remain poor.

The next step which may pay off is to take advantage of the distinction between axioms and definitions. The input of a resolution prover can be simplified considerably by substitution of the definitions so that they do not have to be added in claused form.

We have still not reached the stage where the predicate calculus formula collection obtained might be given to a translator and subsequently submitted to the prover. First it has to be checked whether the theorem to be proved (possibly expanded by the substitution of definitions can be decomposed into independent sub-problems. If so, elimination of aplhabetic variants in the collection of sub-problems may be attempted.

The recognition of independent sub-problems can be easily performed by functions in the predicate-calculus conjunctive-normal-form translator. The basic steps are elimination of $\rightarrow$ and $\leftrightarrow$, the moving inwards of $\sim$ and the moving inwards of the quantifiers (a step often omitted but which can prevent Skolem functions getting spurious arguments). Skipping the case that the leading logical connective is 'or' one finds a first decomposition if and only if the leading connective of the resulting formula is 'and'.

Example: Nevin's coset problem can be formulated in predicate calculus using the reformulation of a group as given above with 3 axioms, 3 definitions and the theorem to be proved. Direct translation in conjunctive normal form produced 30 clauses with together 91 literals. Definition substitution and sub-problem recognition yielded three sub-problems each having 14 clauses and respectively 21, 21 and 18 literals (see also Table 2). A considerable improvement.

3.2. Provide more information to an ATP

In this and the next section we shall undermine the current ATP scheme more rigorously. It is inspired by emerging feelings of pity while observing a theorem prover in action. Without any goal directedness formulas are generated at random, frequently generating duplicates or subsumable clauses which take a lot of effort to recognize as such, generating clauses which are 'obviously' useless, etc. Is the program to be reproached for this stupid behaviour? Or is there any hope that programs can be written that make less stupid choices on the same input?) We do not believe so. This is based on the clumsiness with which people manipulate symbolic structures when they do not have experience with them and/or cannot interpret them. If somebody has any doubt about this he is urged to work on the following example:

Axiom 1 : (x)(y)(u)(v)(w) TER(TER(x,y,u),v,TER(x,y,w)) = = TER(x,y,TER(u,v,w))

2 : (x)(y) TER(y,x,x) = x

3 : (x)(y) TER(x,y,UN(y)) = x.

Theorems: (x)(y) TER(x,x,y) = x

(x)(y) TER(UN(y),y,x) = x

(x)(y) TER(UN(y),x,y) = x.

(This example is taken from one of the Machine Intelligence volumes.) We conjecture therefore that the behaviour of a theorem prover may improve drastically if it is possible to add to it the kind of knowledge that human beings consult while proving a theorem.

We discuss several options to formalize this additional knowledge.

a) Not a very impressive possibility is the use of a domain dependent heuristic function where this function is a composite of syntactic features (in let's say a polynomial construction where the coefficients are domain dependent). If the coefficients convegre in a teaching sequence (in the spirit of Samuel [6]) one may get confidence in the heuristic function. For a collection of syntactic features see Chang and Lee [7] p. 154. Here the additional information derives from a problem family itself. This stresses the necessity to analyse solutions in order to extract aspects that may be of future use.

b) This option is also based on the analysis of solutions and it provides tools to be employed in subsequent efforts. In each branch of mathematics frequently recurring chains of deductions may be observed. E.g. in analysis the chain showing the result that two numbers are arbitrary close when their equality should be shown; the introduction of induction in domains with recursive definitions; the activation of a central theorem or lemma in subsequent theorems, etc. How are we to recognize these chains automatically in order to add them to subsequent input clauses in the form of lemmas, so that they do not have to be derived again?

A candidate for the status of a lemma is the resolvent of two axioms when the decendents of two literals in the axioms are frequently resolved upon in actual proof trees. In case several lemmas are found the same process can be repeated with combinations of them to get more powerful (and more specialized) lemmas.

As in the former option finding a solution is not the real problem but the extracting of tools, strategies, lemmas from that solution (Champeaux [8]).

c) The introduction of external additional knowledge not derived from former solutions is inspired by the set-of-support-strategy. This strategy prescribes the taking of literals for a resolution step only from clauses that do not both belong to the same member of a partition of the clauses. This partition corresponds with an abstract model allowing to assign a truth value to a clause and leading to two disjoint sets. A well known trick to define such an abstract model is to take a syntactic feature of clauses. E.g. A clause is assigned false in a model if all its literals are preceded by 'NOT'. (It is the custom to cancel NOT NOT in clause representation.) Instead we propose to use a real model together with an interpretation function to divide the set of clauses into two disjoint sets. Thus when working on group theory one can pick a model from the integers, the rationals, the reals, complex numbers, a specific finite group, matrices, etc. If the model contains an infinite number of different objects difficulties may arise when a truth value has to be assigned to a clause having a variable ranging over an infinite domain. Calculating the truth value for a fair number of examples (or a 'typical' example, or a 'suspect' example; this information can be a part of the specification of the model under consideration) is a way to get around this difficulty.

d) In contrast with the former options the one discussed now has been actually experimented with. The option is derived from the habit to use a typed predicate calculus. It corresponds with the practice to give constants and/or variables similar names if they belong to or range over the same domain. Domain information attached to identifiers leads to shorter, more compact formulations. One step further is the removal of the restriction that any two different types have to correspond with *disjunct* domain and to allow a type to be the disjunct sum of some other types (and to generalise the unification algorithm to deal with such situations). In (Champeaux [9]) results of a theorem prover working with such a sorted predicate calculus (S.P.C.) are given. We will give two examples: one actually proved (Table 1) and one yet to be proved (Table 2). The latter needs paramodulation to be implemented, but it is already interesting to see how the S.P.C. reduces the input considerably.

problem	subsumption	S.P.C.	# input clauses	# resolvents generated	# total clauses	# clauses in proof	g
Green's sorting problem	NO	NO	10	60	70	22	0.314
		YES	8	24	32	17	0.531
	YES	NO	10	28	38	22	0.597
		YES	8	17	25	17	0.680

Table 1. Showing the difference of using ordinary P.C. versus sorted predicate calculus (S.P.C.). The last column shows that the g-penetrance (= # clauses in proof/total # clauses) improves when using S.P.C.

problem	S.P.C.	input # clauses	 # literals	after subproblem recognition # clauses	 # literals
coset	NO	30	91	14	21
				14	21
				14	18
	YES	13	25	12	16
				12	14
				12	14

Table 2. Showing the difference of using ordinary P.C. versus S.P.C. on the input specification of the coset problem, as well as with and without independent sub-problem recognition. (In each case 3 sub-problems were found.)

3.3. Theorem proving as a multi level search problem

Even when all the above mentioned suggestions for improving the power of ATP's should have been implemented we still expect the resulting program to be weak. The operators, having 'small' consequences where search is involved (for the sake of argument we skip the search while reformulating a problem) are responsible for this weakness. Human beings (having experience in a domain) do not rush off to apply transformations to a conjecture but they first consider it. Is it plausible? Is there a counterexample? Is it a special case of a theorem already available? Is it worthwhile to introduce an abbreviation for a recurring expression? Which theorems/definitions may be useful in a proof? etc. After this preprocessing macro-operators may be activated: apply theorem α to β; expand term α using definition β in occurence γ; split up in cases; simplify terms in equality α; substitute term α for β in occurence γ; etc Subsequently these macro-operators are tentatively executed. And most importantly in case a proof is found, a summary of the proof (e.g. the chain of operators) is attached at the conjecture which gets the status of a theorem. In the preprocessing stage this allows to look for a theorem that is similar to the conjecture and as a consequence it may result in a proof skeleton proposal (Sacerdoti [10]), in which each step has to be refined by the same procedure *or* if this fails by verifying it with a theorem prover embodying all the features described in 3.1 and 3.2. The role of this theorem prover is thus reduced to abridging the gaps in the (refined) skeleton.

3.4. Solving the theorem proving problem by forgetting it

There is a fair chance that the problem of mechanical proving of difficult mathematical conjectures can best be replaced by another problem: how to generate automatically (with respect to a given collection of definitions, axioms, lemmas, theorems, models and similar theories) an interesting conjecture or concept to be defined? Having this capability, at least to some extent, might be essential for generating intermediate stepping stones for a really difficult theorem.

4. SUMMARY

The current ATP paradigm of building intelligence into the program that has to be activated somehow by the axioms and theorem to be proved, is challenged. Arguments are advanced for preprocessing the input, for giving additional information, for introducing multi--level search and finally for giving more attention to the conjecture generation problem.

Since these issue are tightly connected with the central A.I. problem of representation and efficient access to a large amount of complex structured data we wtill do worry slightly about the tendency in the theorem proving community to separate from the A.I. community as voiced at the Workshop on Automatic Deduction, MIT, August 17-19, 1977.

REFERENCES

[1] Nevins, A.J., A Human Oriented Logic for Automatic Theorem-Proving, *Journal of ACM*, 21, (1974), pp. 606-621.

[2] Lusk, E.L., Overbeek, R.A., Experiments with Resolution-Based Theorem-Proving Algorithms, Collected Abstracts, *Workshop on Automatic Deduction*, MIT, Aug. 17-19, 1977.

[3] Ballantyne, A.M., Bledsoe, W.W., Automatic Proofs of Theorems in Analysis Using Nonstandard Techniques, *Journal of ACM*, 24, (1977) pp. 353-376.

[4] Bledsoe, W.W., Set Variables, *IJCAI-77 Proceedings*, MIT, 1977, pp. 501-510.

[5] Kowalski, R.: A Proof Procedure Using Connection Graphs, *Journal of ACM*, 22, (1975) pp. 572-595.

[6] Samuel, A.L.: Some studies in machine learning using the game of checkers, *IBM Journal of Research and Development*, Vol. 3 (1959), pp. 211-229.

[7] Chang, C.L., Lee, R. C.T.: *Symbolic Logic and Mechanical Theorem Proving*, Academic Press, 1973.

[8] Champeaux, D. de: Solutions and their Problems, in Morlet, Ribbens (eds.), *ICS77*, North--Holland, pp. 119-127.

[9] Champeaux, D. de: A theorem prover dating a semantic network, Research Memorandum No 7611, Dept. of Economics, Univ. of Amsterdam.

[10] Sacerdoti, E.: Planning in a Hierarchy of Abstraction Spaces, *Artificial Intelligence*, Vol. 5. (1975), No. 2, pp. 115-135.

D. de CHAMPEAUX

Bedrijfsinformatica
University of Amsterdam
Mirror Co

COLLOQUIA MATHEMATICA SOCIETATIS JÁNOS BOLYAI
26. Mathematical Logic in Computer Science
Salgótarján (HUNGARY), 1978.

ON THE MANY-SORTED LOGICAL APPROACH TO INFORMATION RETRIEVAL SYSTEMS

T. Chkhenkeli

1. Introduction

By the information retrieval system (IRS) we understand here a document retrieval system as opposed to a data retrieval system. A great deal of modern document retrieval systems are based on the principle of the post-coordinate indexing. According to this principle the content of a document and a request can be expressed by a list of words or word combinations in some natural language. These words and word combinations are called lexical units or descriptors of the information retrieval language. V. Finn [1] proposes a logical method to define a descriptor, which permits us to consider document retrieval as a particular case of data retrieval. In papers [2,3,4] which have the common title "Information storage and retrieval system. Mathematical foundations", W. Marek, Z. Pawlak, W. Lipski, M. Jaegerman develop a logical and algebraic theory of a certain class of information retrieval systems containing both document retrieval systems and data retrieval systems . They obtained important results concerning the organization of the computer memory to simplify the implementation of

IRS and the systems with incomplete information.

However, modern document retrieval systems do not entirely satisfy the definitions of IRS given in [2,3,4].

In [2,3] *IRS is defined as a quadruple* $\langle X, A, R_I, U\rangle$, *where*

(i) X *is a set called the set of objects (documents),*

(ii) A *is a nonempty set of descriptors and* R_I *is an equivalence on* A,

(iii) $U: A \to P(X)$ *is a function satisfying the following two conditions:*

(1) *If* $aR_I b$ *and* $a \neq b$ *then* $U(a) \cap U(b) = \emptyset$;
(2) $\cup\{U(b): bR_I a\} = X$ *for each* $a \in A$.

The definitions of the description language L_A, terms and formulas of L_A are given in [2,3]. There are no variables and quantifiers in L_A. The following axioms are assumed for the theory of IRS:

(i) *Substitutions of the axioms of Boolean algebra and equality for terms;*

(ii) *Substitutions of the propositional calculus axioms for formulas;*

(iii) *For each* $a \in A$ *the formula* $a = \sim \sum_{\substack{bR_I a \\ b \neq a}} b$

The *modus ponens* is taken as an inference rule. Following [5] the obtained theory and language L_A may be called the zero order theory and the zero order language while the quadruple $\langle X, A, R_I, U\rangle$ is an interpretation of the language L_A.

We shall consider two IRS of IBM: a simple one the IRMS (Information Retrieval and Management System) and a much more complicated one the DPS (Document Processing System). The equivalence relation R_I, connected with the function U by the conditions (1) and (2), is absent in both systems. Both systems contain the function U which associates a respective set of documents to each descriptor. This function is employed to organize inverted files of both systems and satisfies the condition:

$$(3) \quad \bigcup_{\alpha \in A} \{U(\alpha)\} = X.$$

Each search prescription corresponds to some Boolean expression over the set A, more precisely to some element of the free Boolean algebra $B(A)$ generated by the set A. Let $i:A \to B(A)$ be an embedding, $U:A \to P(X)$ the function satisfying the condition (3); then there exists the homomorphism $h:B(A) \to P(X)$ and h is an extension of the function U. In other words, the diargam

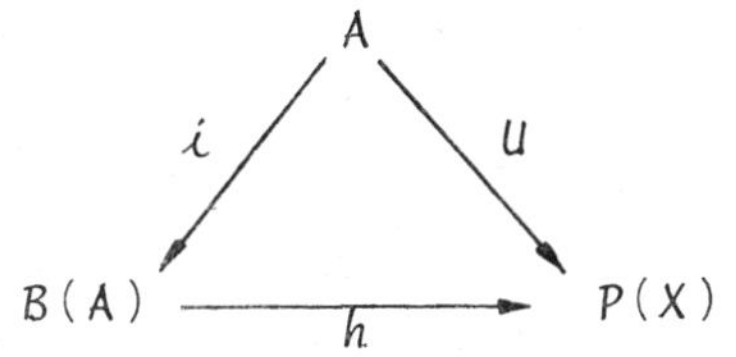

is commutative. Consequently, the homomorphic image $h(a)$ of the element $\alpha \in B(A)$ to which the given search prescription corresponds, will be an answer in both IRMS and DPS. Due to [4] the homomorphism h may be regarded as IRS.

2. The semantics of IRS

The semantic means of IRS constitute an information retrieval language (index language) whose construction includes the following steps:

- the choice of lexical units or descriptors,
- the establishment of paradigmatic relations between lexical units via a conceptual analysis of words,
- the choice of syntagmatic relations and logical relators to connect lexical units or descriptors during the formation of index records and search prescriptions,
- the choice of syntagmatic 1-place relations to establish semantic roles of lexical units or descriptors during the formation of index records and search prescriptions,
- the choice of complementary semantic means to express the content of documents and requests during the formation of index records and search prescriotions,
- the establishment of rules of idexing, i.e. rules to form index records and search prescriptions.

The semantic tools that have been used in Section 1 were the set A of lexical units and Boolean operations applied to them. In order to consider the other semantic tools of IRS by the method of [2,3] we need to extend the description language L_A, to develop an appropriate zero order theory and to modify the definition of IRS. However, we prefer here to follow the thesis formulated in [1] according to which every information retrieval language can be imitated by means of many-sorted predicate calculus. For information on general subject of many-sorted logic the reader is referred to [6].

We start with the consideration of those sets elements of which are used to express the content of documents and requests: A_1 is the set of all lexical units of the given IRS; A_7 is the set of lexical units associated with the given document during the formation of the index record, A_8 is the set of lexical units associated with the given request during the formation of the respective search prescription, A_5 is the set of sets of descriptors such that for each set there exists an index record containing exactly the same descriptors, A_6 is the set of search prescriptions, A_9 is the set of natural numbers.

The elements of set A_1 in IRMS are words or word combinations (of limited length) of a natural language. The set A_1 in IRMS is effectively given by storing it in the computer. In DPS the proper subset B of the set A_1 is given effectively (the elements of this subset will be called descritors below), while the complement B' of the set B to the set A_1 (the elements of which are proper names, abbreviations etc.) is given ineffectively.

The various relations are defined on the sets A_1, A_7, A_8, A_5, A_6, A_9 in the systems IRMS and DPS which are used to express the content of documents and requests. These relations fall into three groups: paradigmatic, syntagmatic and nonlexical relations.

By paradigmatic relations we understand the predicates $P^{i_1,\ldots,i_k}$ defined on the system of sets $A_{i_1},\ldots,A_{i_k}$, where at least one of them A_{i_l} $(1 \leq l \leq k)$ is such that $A_{i_l} = A_1$.

Six paradigmatic relations may be defined in IRMS: $P^{1,1}_{I1}(X^{(1)}_1,X^{(1)}_2)$ means that the descriptor $X^{(1)}_1$ is not the main synonym of the descriptor $X^{(1)}_2$, $P^{1,1}_{I2}(X^{(1)}_1,X^{(1)}_2)$ implies that the descriptor $X^{(1)}_1$ is the main synonym of the descriptor $X^{(1)}_2$, $P^{1,1}_{I3}(X^{(1)}_1,X^{(1)}_2)$ shows that the descriptor $X^{(1)}_1$ expresses a more narrow concept than it does in $X^{(1)}_2$, $P^{1,1}_{I4}(X^{(1)}_1,X^{(1)}_2)$ means that the descriptor $X^{(1)}_1$ expresses a wider concept than it does in $X^{(1)}_2$, $P^{1,1}_{I5}(X^{(1)}_1,X^{(1)}_2)$ implies that $X^{(1)}_1$ and $X^{(1)}_2$ express connected (associative) concepts, $P^{1,1}_{I6}(X^{(1)}_1,X^{(1)}_2)$ shows that $X^{(1)}_1$ and $X^{(1)}_2$ are antonyms.

The following paradigmatic relations may be defined in DPS: $P^{1}_{D1}(X^{(1)})$ means that the lexical unit $X^{(1)}$ is a descriptor, $P^{1,1}_{D2}(X^{(1)}_1,X^{(1)}_2)$ implies that $X^{(1)}_1$ and $X^{(1)}_2$.

are conditional synonyms, $P^{1,1}_{D3}(X^{(1)}_1, X^{(1)}_2)$ shows that $X^{(1)}_1$ and $X^{(1)}_2$ are equivalent lexical units (using the relation $P^{1,1}_{D3}$ it can be set polylingual IRS) and finally

$$P^{1,9}_{D4}(X^{(1)}, X^{(9)}) = \begin{cases} t, & \text{if } P^1_{D1}(X^{(1)}) \text{ and } X^{(9)} \text{ is a frequency of } X^{(1)} \text{ in a given set of documents,} \\ f, & \text{otherwise} \end{cases}$$

By syntagmatic relations we understand the predicates $S^{i_1,\ldots,i_k}$ defined on the system of sets $A_{i_1},\ldots,A_{i_k}$, where for every ℓ $A_{i_\ell} \neq A_1$ and there exists ℓ such that $A_{i_\ell} \subset A_1$ $(1 \leq \ell \leq k)$.

The syntagmatic relations are absent in IRMS. Lexical units in search prescriptions may be provided with weights in DPS, i.e. the following syntagmatic relation may be defined in DPS:

$$S^{8,9}(X^{(8)}, X^{(9)}) = \begin{cases} t, & \text{if } X^{(9)} \text{ is a weight of } X^{(8)}, \\ f, & \text{otherwise.} \end{cases}$$

Many other syntagmatic relations (links) are defined in DPS and may be used for forming only search prescriptions. These are some of them:

$S^{8,8}_{en}(X^{(8)}_1, X^{(8)}_2)$-descriptor $X^{(8)}_2$ and its preceding $X^{(8)}_1$ must belong to the same sentence of the document being searched;

$S^{8,8}_{par}(X^{(8)}_1, X^{(8)}_2)$-descriptor $X^{(8)}_2$ and the preceding one $X^{(8)}_1$ must belong to the same point of the desired document;

$S^{8,8,9}_{+}(X^{(8)}_1, X^{(8)}_2, X^{(9)})$-descriptor $X^{(8)}_2$ must be the $X^{(9)}$-th word after $X^{(8)}_1$ in the desired document;

$S^{8,8,9}_{-}(X^{(8)}_1, X^{(8)}_2, X^{(9)})$-descriptor $X^{(8)}_1$ must be the $X^{(9)}$-th word before the next $X^{(8)}_2$ in the desired document.

The following syntagmatic 1-place relations (roles) are used when forming search prescriptions and index records;

$S^8_{b1}(X^{(8)})(S^7_{b1}(X^{(7)}))$-lexical unit $X^{(8)}$ must be the author's name of the desired document (the lexical unit $X^{(7)}$ is the author's name of the given document);

$S^8_{b2}(X^{(8)})(S^7_{b2}(X^{(7)}))$-lexical unit $X^{(8)}$ must be the name of a publishing house of the desired document (lexical unit $X^{(7)}$ is the name of a publishing house of the given document);

$S^8_{b3}(X^{(8)})(S^7_{b3}(X^{(7)}))$-lexical unit $X^{(8)}$ must be the name of the periodical publication of the document being searched (the lexical unit $X^{(7)}$ is the name of the periodical publication of the given document);

$S^8_{b4}(X^{(8)})(S^7_{b4}(X^{(7)}))$-lexical unit $X^{(8)}$ must belong to the heading of the required document (the lexical unit $X^{(7)}$ belongs to the heading of the given document).

The following syntagmatic relations are also used for forming index records in DPS:

$S^{7,9}_p(X^{(7)},X^{(9)})$ - $X^{(9)}$ is the number of a point, containing the descriptor $X^{(7)}$ of the given document,

$S^{7,9}_s(X^{(7)},X^{(9)})$ - $X^{(9)}$ is the number of the sentence, containing the descriptor $X^{(7)}$ in the given point of the given document,

$S^{7,9}_v(X^{(7)},X^{(9)})$ - $X^{(9)}$ is the number of $X^{(7)}$ in the given sentence of the given document.

By nonlexical relations we understand the predicates $T^{i_1,\ldots,i_k}$ defined on the system of sets $A_{i_1},\ldots,A_{i_k}$, where for every ℓ $A_{i_\ell} \cap A_1 = \emptyset$ and there exists ℓ such that $A_{i_\ell} = A_5$ or $A_{i_\ell} = A_6$ $(1 \le \ell \le k)$.

Below are two examples of nonlexical relations used in DPS:

$T_1^{6,9}(X^{(6)}, X^{(9)})$ - $X^{(9)}$ is the critical weight of the search prescription $X^{(6)}$,

$T_1^{5,9}(X^{(5)}, X^{(9)})$ - $X^{(9)}$ is the publication year of the document index record of which consists of the descriptors from the set $X^{(5)}$.

3. *The model theoretical representation of IRS*

In this section we are going to describe the systems IRMS and DPS omitting details, by using first order many-sorted language with equality and we establish similarity and distinction between these systems from the viewpoint of many-sorted theory.

We start with IRMS. Let us enumerate all the domains required to represent IRMS as a many-sorted (multibasic) algebraic system (model): A_1, A_5, A_6 are defined above, $A_2 = B(A_1)$ is the free Boolean algebra generated by the set A_1, A_3 is the set of documents entered in IRMS, $A_4 = P(A_3)$ is the power set of the set A_3.

The following relations are defined on the domains (basic sets) $A_1, \ldots, A_6$: $P_1^{1,1}$, $P_2^{1,1}$, $P_3^{1,1}$, $P_4^{1,1}$, $P_5^{1,1}$, $P_6^{1,1}$, $\in_7^{1,5}(X^{(1)} \in_7^{1,5} X^{(5)})$ indicates that $X^{(1)}$ is an element of $X^{(5)}$) are paradigmatic relations, $\in_8^{3,4}(X^{(3)} \in_8^{3,4} X^{(4)}$) indicates that $X^{(3)}$ is an element of $X^{(4)}$) is a nonlexical relation. Paradigmatic relations $P_1^{1,1}, \ldots, P_6^{1,1}$ are used in IRMS when forming search prescriptions and, together with the set A_1, they constitute the thesaurus.

Let R_I denote the set of all relations of IRMS. The following operations are defined on the basic sets $A_1, \ldots, A_6$: $\delta_1^{3,5}: A_3 \rightarrow A_5$ is a mapping which associates a certain set of descriptors with each document, $\delta_2^{6,2}: A_6 \rightarrow A_2$ is a function which associates a certain element of the free Boolean algebra with each search prescription, $\delta_3^{1,4}: A_1 \rightarrow A_4$ is a function which associates with each descriptor a certain set of documents the index records of which contain the given descriptor, i.e. $\delta_3^{1,4}(X^{(1)}) = \{X^{(4)} (\forall X^{(3)} (\delta_1^{3,5}(X^{(3)}) = X^{(5)} \& X^{(1)} \in^{1,5} X^{(5)})\}$, $\delta_4^{2,4}: A_2 \rightarrow A_4$ is a homomorphism which is an extension of function $\delta_3^{1,4}$, $\delta_5^{1,2}$ is an embedding of the set A_1 into the set A_2, $\vee_6^{2,2,2}$, $\wedge_7^{2,2,2}$, $\sim_8^{2,2}$ (or simply $\vee$, $\wedge$, $\sim$) are Boolean operations on A_2, $\cup_9^{4,4,4}$, $\cap_{10}^{4,4,4}$, $(\;)_{11}^{';4,4}$ (or simply $\cup$, $\cap$, $'$) are Boolean operations on A_4. Let F_I denote the set of operations of IRMS.

The many-sorted algebraic system

${}^{(6)} = \langle A_1, \ldots, A_6, R_I, F_I \rangle$ represents IRMS if it satisfies the conditions:

1. Axioms of Boolean algebra for $\vee$, $\wedge$, $\sim$;
2. Axioms of Boolean algebra for $\cup$, $\cap$, $'$;
3. $\forall X^{(1)} \forall X^{(2)} (\delta_5^{1,2}(X^{(1)}) = X^{(2)} \rightarrow X^{(1)} = X^{(2)})$

   - this condition implies that $\delta_5^{1,2}$ is the identical embedding;
4. $\forall X^{(6)} \forall Y_1^{(2)} \forall Y_2^{(2)} (\delta_2^{6,2}(X^{(6)}) = Y_1^{(2)}$ &
   & $\delta_2^{6,2}(X^{(6)}) = Y_2^{(2)} \rightarrow Y_1^{(2)} = Y_2^{(2)})$,
5. $\forall X^{(1)} \forall Y_1^{(4)} \forall Y_2^{(4)} (\delta_3^{1,4}(X^{(1)}) = Y_1^{(4)}$ &
   & $\delta_3^{1,4}(X^{(1)}) = Y_2^{(4)} \rightarrow Y_1^{(4)} = Y_2^{(4)})$,

   - the last two conditions show that the mappings $\delta_2^{6,2}$ and $\delta_3^{1,4}$ are functions;
6. $\forall X^{(1)} (\delta_3^{1,4}(X^{(1)}) = \delta_4^{2,4}(\delta_5^{1,2}(X^{(1)})))$

   - the condition means that the diagram

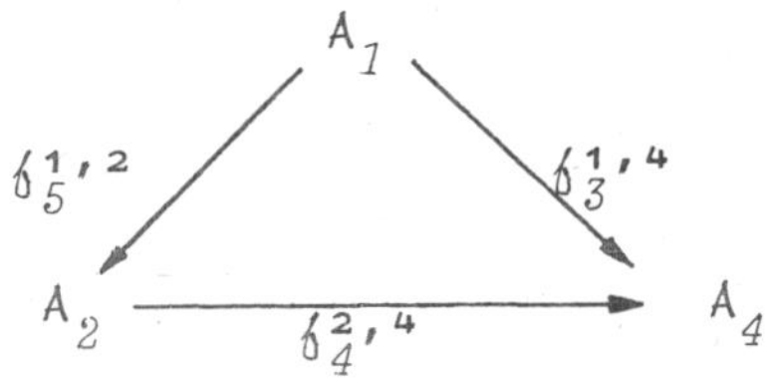

is commutative;

7. $\forall X^{(2)} \forall Y^{(2)} (\delta_4^{2,4}(X^{(2)} \vee Y^{(2)}) = \delta_4^{2,4}(X^{(2)}) \cup \delta_4^{2,4}(Y^{(2)}))$,

8. $\forall X^{(2)} \forall Y^{(2)} (\delta_4^{2,4}(X^{(2)} \wedge Y^{(2)}) = \delta_4^{2,4}(X^{(2)}) \cap \delta_4^{2,4}(Y^{(2)}))$,

9. $\forall X^{(2)} (\delta_4^{2,4}(\sim X^{(2)}) = (\delta_4^{2,4}(X^{(2)}))')$,

- conditions 7,8 and 9 show that $\delta_4^{2,4}$ is a homomorphism;

10. $\exists X^{(3)} \exists X^{(6)} (X^{(3)} \in_8^{2,4} \delta_4^{2,4}(\delta_6^{6,2}(X^{(6)})))$

- the condition shows that a set of outputs of IRS is not empty;

11. $\exists X_1^{(1)} \exists X_2^{(1)} (P_1^{1,1}(X_1^{(1)}, X_2^{(1)}) \vee P_2^{1,1}(X_1^{(1)}, X_2^{(1)}) \vee P_3^{1,1}(X_1^{(1)}, X_2^{(1)}) \vee$
$\vee P_4^{1,1}(X_1^{(1)}, X_2^{(1)}) \vee P_5^{1,1}(X_1^{(1)}, X_2^{(1)}) \vee P_6^{1,1}(X_1^{(1)}, X_2^{(1)}))$

- the condition indicates that at least one of the paradigmatic relations $P_1^{1,1}, \ldots, P_6^{1,1}$ is not empty.

Consider now the first order many-sorted language with equality all relational and function symbols of which are interpreted by the relations of the set R_I and the operations of the set F_I. Let $\mathfrak{F}^{(6)}$ denote a set of closed formulas of this language corresponding to conditions 1-11 stated above. Let $\mathfrak{T}^{(6)}$ close the set $\mathfrak{F}^{(6)}$ under the rules of inference. Then the 6-basic algebraic system $\mathfrak{A}^{(6)} = \langle A_1, \ldots, A_6, R_I, F_I \rangle$ which represents IRMS will be an element of the axiomatizable class of all 6-basic algebraic systems the elementary theory of which is $\mathfrak{T}^{(6)}$.

In addition to the sets $A_1, \ldots, A_6$ other three domains are used in DPS. They are A_7, A_8 and A_9 and have been defined above.

In order to describe DPS by a many-sorted first order language with equality we introduce the relational and function symbols and put into parentheses their interpretations which have also been given above.

The relational symbols for the paradigmatic relations will be denoted by $R_1^1(P_{D1}^1)$, $R_2^{1,1}(P_{D2}^{1,1})$, $R_3^{1,1}(P_{D3}^{1,1})$, $R_4^{1,9}(P_{D4}^{1,9})$, $R_5^{1,5}(\in_7^{1,5})$, $R_6^{1,6}$ ($\in^{1,6}$is the membership of an element of the set A_1 to an element of the set A_6).

The relational symbols for the syntagmatic relations will be written as $R_7^{8,9}(S^{8,9})$, $R_8^{8,8}(S_{en}^{8,8})$, $R_9^{8,8}(S_{par}^{8,8})$, $R_{10}^{8,8,9}(S_+^{8,8,9})$, $R_{11}^{8,8,9}(S_-^{8,8,9})$, $R_{12}^{8}\ (S_{b1}^{8})$, $R_{13}^{7}(S_{b1}^{7})$, $R_{14}^{8}(S_{b2}^{8})$, $R_{15}^{7}(S_{b2}^{7})$, $R_{16}^{8}(S_{b3}^{8})$, $R_{17}^{7}(S_{b3}^{7})$, $R_{18}^{8}(S_{b4}^{8})$, $R_{19}^{7}(S_{b4}^{7})$, $R_{10}^{7,9}(S_p^{7,9})$, $R_{21}^{7,9}(S_s^{7,9})$, $R_{22}^{7,9}(S_v^{7,9})$.

The relational symbols for the nonlexical relations are $R_{23}^{6,9}(T_1^{6,9})$, $R_{24}^{5,9}(T_2^{5,9})$, $R_{25}^{3,4}(\in_8^{3,4})$, $R_{26}^{9,9}$ ($<$ is the inequality on the set A_9).

Let R_D denote the set of all relations of the system DPS. The function symbols for the operations of DPS are: $F_1^{3,5}(\mathfrak{f}_1^{3,5})$, $F_2^{6,2}(\mathfrak{f}_2^{6,2})$, $F_3^{1,4}(\mathfrak{f}_3^{1,4})$, $F_4^{2,4}(\mathfrak{f}_4^{2,4})$, $F_5^{1,2}(\mathfrak{f}_5^{1,2})$, $F_6^{2,2,2}(\vee)$, $F_7^{2,2,2}(\wedge)$, $F_8^{2,2}(\sim)$, $F_9^{4,4,4}(\cup)$, $F_{10}^{4,4,4}(\cap)$, $F_{11}^{4,4}(')$, $F_{12}^{9,9,9}$ ($+$ is the addition on the set A_9).

Let F_D denote the set of all operations of the system DPS. We take as axioms the closed formulas of the set $\mathfrak{F}^{(6)}$ except the last one associated with condition 11 and add to them the sentences characteristic of the basic sets, relations and operations of DPS. These are some of them:

$$\forall X^{(1)} \exists X^{(9)}\ (R_1^1(X^{(1)}) \rightarrow R_4^{1,9}(X^{(1)}, X^{(9)})$$

- if the lexical unit $X^{(1)}$ is a descriptor, then there exists its frequency $X^{(9)}$,

$$\forall X^{(7)} \exists X^{(1)} \exists X^{(9)} \exists X^{(9)} \exists X^{(9)} (X^{(1)} = X^{(7)}\ \& \\ \& R^1(X^{(1)}) \rightarrow R^{7,9}(X^{(7)}, X^{(9)})\ \& R^{7,9}(X^{(7)}, X^{(9)})\ \& \\ \& R^{7,9}(X^{(7)}, X^{(9)}))$$

- if the lexical unit $X^{(7)}$ of the given index record is a descriptor, then $X^{(7)}$ is a number of the point containing $X^{(7)}$, $X_2^{(9)}$ is a number of the sentence containing $X^{(7)}$ in this point and $X_3^{(9)}$ is a number of $X^{(7)}$ in the sentence,

$$\exists X^{(7)} (R_{13}^7(X^{(7)}) \vee R_{15}^7(X^{(7)}) \vee R_{17}^7(X^{(7)}) \vee R_{19}^7(X^{(7)}))$$

- the set of bibliographic data in the given index record will not be empty,

$\forall X^{(8)} \forall X_1^{(9)} \exists X^{(1)} \exists X^{(6)} \exists X_2^{(9)} (R_7^{8,9}(X_1^{(8)}, X^{(9)}) \&$
$\& R_6^{1,6}(X_1^{(1)}, X^{(6)}) \& X^{(1)} = X^{(8)} \rightarrow R_{23}^{6,9}(X_1^{(6)}, X_2^{(9)}))$

- if a lexical unit $X^{(8)}$ of the search prescription $X^{(6)}$ has a weight $X_1^{(9)}$, then $X_2^{(9)}$ is a critical weight of the search prescription $X^{(6)}$.

Let $\mathfrak{F}^{(9)}$ denote the set of all sentences characteristic of the domains, reliatons and operations of DPS. Let $\mathfrak{K}^{(9)} \mathfrak{F}^{(9)}$ denote the axiomatizable class of all 9-basic algebraic systems which satisfy every sentence from $\mathfrak{F}^{(9)}$. The class $\mathfrak{K}^{(9)} \mathfrak{F}^{(9)}$ is regarded as the representation of DPS. The element $\mathfrak{A}^{(9)} = \langle A_1, \ldots, A_9, R_D, F_D \rangle$ of the class $\mathfrak{K}^{(9)} \mathfrak{F}^{(9)}$ characterizes the given search prescription and the given document. If we add to the set $\mathfrak{F}^{(9)}$ the closed formulas which express a certain relevance criterion, we get an axiomatizable subclass of $\mathfrak{K}^{(9)} \mathfrak{F}^{(9)}$. Each member of the obtained class specifies the search prescription and the document which will be obtained in the output in response to the given search prescription. There are two examples of such axioms:

$\forall X_1^{(8)} \forall X_2^{(8)} \exists X_1^{(7)} \exists X_2^{(7)} \exists X_1^{(9)} \exists X_2^{(9)} \exists X_3^{(9)} \exists X_4^{(9)}$
$(R_8^{8,8}(X_1^{(8)}, X_2^{(8)}) \& R_{20}^{7,9}(X_1^{(7)}, X_1^{(9)}) \& R_{21}^{7,9}(X_1^{(7)}, X_2^{(9)}) \&$
$\& R_{20}^{7,9}(X_2^{(7)}, X_3^{(9)}) \& R_{21}^{7,9}(X_2^{(7)}, X_4^{(9)}) \rightarrow X_1^{(8)} = X_1^{(7)} \&$
$\& X_2^{(8)} = X_2^{(7)} \& X_1^{(9)} = X_3^{(9)} \& X_2^{(9)} = X_4^{(9)})$

- if lexical units $X_1^{(8)}$ and $X_2^{(8)}$ of the given search prescription are contained in the desired document, then in the given index record there are lexical units $X_1^{(7)}$ and $X_2^{(7)}$ which coincide with $X_1^{(8)}$ and $X_2^{(8)}$ and have the same numbers of the point and the sentence,

$\forall X^{(8)} \exists X^{(7)} (R_{12}^{8}(X^{(8)}) \rightarrow R_{13}^{7}(X^{(7)}) \& X^{(8)} = X^{(7)})$

- if lexical unit $X^{(8)}$ of the given search prescription is the name of the author of the desired document, then the author's name $X^{(7)}$ of the given document coincides with $X^{(8)}$.

4. *Conclusion*

The IRS without syntagmatic relations (namely IRMS) is represented in this paper as a many-sorted model (6--basic algebraic system) while the IRS with syntagmatic relations (namely DPS) is represented as an axiomatizable class of many-sorted models (9-basic algebraic systems). I have attempted here to present the basic parts of IRS but of course much has been left out.

The tradition of using axioms which speak of various types of entities such as points, lines and planes goes back to the ancient greeks, and was followed by Hilbert in his axiomatization of geometry. Many-sorted models (algebraic systems) can be made into ordinary models (i.e. single-sorted ones) in an obvious way by combining all the domains into a single domain and adding to the language 1-place predicates by which separate domains can be recovered. It seems, however, that many natural concepts become less natural after being translated into terms of the ordinary first order language. The first description of the information retrieval language by means of the many-sorted logic is given in [1].

This paper deals with the document retrieval systems. But it is clear that many-sorted first order language will be convenient for investigating Codd's relational models of data retrieval systems.

Acknowledgements. I would like to express my appreciation to Victor Finn and Leo Esakia for supplying me with valuable information, kind attention and useful suggestions.

References

[1] В.К. Финн: Логийеские проблемы информационного поиска
Москва, "Наука", (1976)

[2] W. Marek, Z. Pawlak: Information storage and retrieval system
Mathematical foundations. Part I, *CC PAS Reports*, No. 149, Warszawa (1974)

[3] W. Lipski, W. Marek: On information storage and retrieval systems
CC PAS Reports, No. 200, Warszawa (1975)

[4] M. Jaegermann: Information storage and retrieval systems
Mathematical foundations, Part IV, *CC PAS Reports*, No. 215, Warszawa (1975)

[5] H. Rasiowa, R.S. Sikorski: *The mathematics of metamathematics*
PWN, Warszawa (1963)

[6] G. Kreisel, G.L. Krivine: *Elements of mathematical logic model theory*
North-Holland, Amsterdam (1967)

T. CHKHENKELI
USSR 380007 Tbilisi
G. Tabidze street 23.

COLLOQUIA MATHEMATICA SOCIETATIS JÁNOS BOLYAI
26. Mathematical Logic in Computer Science
Salgótarján (HUNGARY), 1978.

EQUATIONAL THEORIES AND EQUIVALENCES OF PROGRAMS

B. Courcelle

ABSTRACT

We give conditions on a Church-Rosser term rewriting system S allowing to define S-normal forms for infinite trees (i.e. infinite terms). We obtain in this way a nice characterization of the S-*equivalence* of recursive program schemes (i.e. the equivalence of recursive program schemes in all interpretations which validate S now considered as a set of equations). We also do not use any induction principle (such as "Scott's induction principle") but only consist of checking equations (as in McCarthy [6]).

Introduction

We shall only give the main results and examples. The *recursive program schemes* (rps) that we are to consider are those of [2], a typical example being given in example (3.16). An interpretation is a complete partially ordered set with an algebraic structure (a complete F-magma of [2], a Δ-continuous algebra [1]). If R is a set of equations (see example (3.16)) then C_R is the class of all interpretations where the equations are satisfied, (a variety of ordered algebras in [3]).

For rps Φ and Φ', $\Phi \equiv_R \Phi'$ iff the computed functions Φ_I and Φ'_I are the same for all $I \in C_R$. More generally, $\Phi \leq_R \Phi'$ iff $\Phi_I \leq \Phi'_I$ for all $I \in C_R$. For applications (see [4], example (3.16)) we are mainly interested in characterizing $\leq_R$.

In fact Φ_I can be defined in terms of I and $T(\Phi)$ also denoted $T(\sum,\Phi))$ the infinite tree (infinite term) obtained by unfolding the recursion ad infinitum. We denote by $M^{\infty}_{\Omega}(F,V)$ the set of infinite trees ($CT_F(V)$ in [1]). Hence $\leq_R$ can be defined on $M^{\infty}_{\Omega}(F,V)$ by $T \leq_R T'$ iff $T_I \leq T'_I$ for all $I \in C_R$.

(1.1) Definition

$T \in M^{\infty}_{\Omega}(F,V)$ is R-*maximal* if for $T' \in M^{\infty}_{\Omega}(F,V)$, $T \leq_R T'$ implies $T \equiv_R T'$. This concept has consequences for proofs of programs: let

(1) $\Phi(v_1,\ldots,v_k) = \tau$
be a recursive equation and $\psi(v_1,\ldots,v_k)$ be defined by another rps. Assume now that

(2) $\psi(v_1,\ldots,v_k) \equiv_R \tau\{\psi/\Phi\}$
where the right hand side of (2) is obtained by the substitution of ψ at each occurence of Φ in τ. Since Φ_I is the least fixed-point of the functional Φ_I, we get

(3) $\Phi \leq_R \psi$.
Assuming now that $T(\Phi)$ is R-maximal, we get

(4) $\Phi \equiv_R \psi$

since (3) is equivalent to $T(\Phi) \leq_R T(\psi)$.

This scheme of proof is used in McCarthy [6] where a termination property allows to deduce (4) from (2).

On the other hand, certain transformations of [5] (namely the "folding" and "unfolding" used with the "laws about primitives") produce from ψ a new scheme Φ such that (2) holds. That $T(\Phi)$ is R-maximal insures that the transformation is correct and no inductive proof is necessary. Let us give an example showing that such transformations may fail:

(1.2) Example

$$R : gfv_1 = fgv_1$$

$$\Sigma : \begin{cases} \Theta = f\Theta \\ \psi = g\Theta \end{cases}$$

$$\begin{aligned} \psi &\equiv g\Theta \\ &\equiv gf\Theta \\ &\equiv fg\Theta \\ &\equiv f\psi \qquad \text{(folding)} \end{aligned}$$

let now $\Phi = f\Phi$

then $\Phi \leq_R \psi$ but the equivalence does not hold

since $T(\Phi) = f^\omega = \mathrm{Sup}_n\{f^n\Omega\}$

$T(\psi) = gf^\omega = \mathrm{Sup}_n\{gf^n\Omega\}$

and $T(\psi) \not\leq_R T(\Phi)$ since

$gf^n\Omega \not\leq_R f^m\Omega$ for no $n,m \in N$ (by (1.7) below).

A recursive equation (1) is R-*univocal* if (2) ⇒ (4) for all rps ψ .

We raise the following questions for a given finite $R \subset M_\Omega(F,V) \times M_\Omega(F,V)$:

QUESTIONS: Which equations (and systems of equations) are R-*univocal? Which* $T \in M_\Omega^\infty(F,V)$ *are*

R-maximal; which rps (Σ,Φ) *have a* R-*maximal* $T(\Sigma,\Phi)$?

We answer to these questions (by giving sufficient conditions) when R a confluent noetherian reduction (as in [7]) i.e. a Church-Rosser term rewriting system ([8]) for which any finite term t has a unique R-*normal form* $C_R(t)$. Let us give some more definitions.

(1.3) Notations

$M_\Omega(F,V)$ is the set of finite terms with possible occurrences of Ω (a constant meaning "undefined" or "bottom"), $M(F,V)$ is the set of finite terms without Ω.

$M_\Omega(F,V)$ is ordered by $\prec$ such that $\Omega \prec t$ and $f(t_1,\ldots,t_k) \prec f(t'_1,\ldots,t'_k)$ iff $t_i \prec t'_i$ for $i=1,2,\ldots,k$.

Then $M_\Omega(F,V)$ is the set of *infinite trees*, i.e. of least upper bounds of directed subsets of $M_\Omega(F,V)$, $M^\infty(F,V)$ is the set of infinite trees without Ω, i.e. of maximal elements of $M_\Omega^\infty(F,V)$ with respect to $\prec$.

Letters t, t', u, s,... will usually denote elements of $M_\Omega(F,V)$ and T, T', U elements of $M_\Omega^\infty(F,V)$.

A term t is *linear in* V if each $v_i \in V$ has at most one occurrence in t.

For $t,t_1,\ldots,t_k \in M_\Omega(F,V)$, let $t[t_1/v_{i_1},\ldots,t_k/v_{i_k}]$ be the result of the substitution of t_j to each occurrence of v_{i_j} for $1 \le j \le k$. Let $t[t_1,\ldots,t_k] = t[t_1/v_1,\ldots,t_k/v_k]$.

Let Var(t) be the set of *variables*, i.e. of elements of V occuring in t.

(1.4) Classes of interpretations (see [2] and [9]).

Let C be a class of interpretations; for $T,T' \in M_\Omega^\infty(F,V)$

$T \leq_C T'$ iff $T_I \leq T'_I$ for all $I \in C$.

We slightly modify a definition of [9]:

A mapping $c : M_\Omega(F,V) \to M_\Omega(F,V)$ is a C-*projection* if

$$(1.4.1) : c(t) \equiv_C t$$

$$(1.4.2) : c(t) \prec c(t') \text{ iff } t \leq_C t' \text{ , for all}$$

t, t' in $M_\Omega(F,V)$.

Assuming that C is *algebraic* [2] i.e. that, for all $T, T' \in M_\Omega^\infty(F,V)$:

$$T \leq_C T' \text{ iff } \forall t \prec T, \exists t' \prec T', \text{ s.t. } t \leq_C t' \text{ ,}$$

we can extend c to $M_\Omega^\infty(F,V)$ by $c(T) = \mathrm{Sup}\{c(t)/t \prec T\}$, and (1.4.1) and (1.4.2) hold for $t, t' \in M_\Omega^\infty(F,V)$. Classes C_R are algebraic.

(1.5) It is known from [2] (also [3] for classes C_R) that every class of interpretations C can be represented by a single interpretation H_C . More precisely, one can define an interpretation H_C and values $d_1, \ldots, d_k$ such that, for all k-ary schemes Φ and ψ :

$$\Phi \leq_C \psi \text{ iff } \Phi_{H_C}(d_1, \ldots, d_k) \leq_{H_C} \psi_{H_C}(d_1, \ldots, d_k) \text{ .}$$

It is called the *Herbrand interpretation* of C (it is essentially unique). Unfortunately it is a very complicated object in general. But if C is algebraic and admits a C-projection c, H_C can be more concretely desribed as the set of $c(T)$ for $T \in M_\Omega^\infty(F,V)$, ordered by $\prec$ and with operators so defined:

$$f_{H_C}(T_1, \ldots, T_k) = c(f(T_1, \ldots, T_k)) \text{ .}$$

(1.6) Let now $R \subset M_\Omega(F,V) \times M_\Omega(F,V)$, let $\overset{*}{\underset{R}{\leftrightarrow}}$ as in [7] be the least congruence on $M_\Omega(F,V)$, containing R and such that $t \overset{*}{\underset{R}{\leftrightarrow}} t'$ implies $t[u_1, \ldots, u_k] \overset{*}{\underset{R}{\leftrightarrow}} t'[u_1, \ldots, u_k]$.

Let $C_R = \{I / I$ is an interpretation and $t_I = t'_I$ for all $(t,t') \in R\}$ let $\leq_R$ stand for $\leq_{C_R}$ and $\equiv_R$ for $\equiv_{C_R}$. By [2], $T \leq_R T'$ if and only if $\forall t \prec T, \exists t' \prec T'$ s.t. $t \leq_R t'$. The notation of $\leq_R$ to $M_\Omega(F,V)$ is characterized by:

(1.7) PROPOSITION [2]: $\leq_R$ is the transitive closure of $\prec \cup \overset{*}{\leftrightarrow}$.

Let R be a finite subset of $M_\Omega(F,V) \times M_\Omega(F,V)$, let let $\overset{*}{\underset{R}{\to}}$ be the semi-thue relation associated as in [7].

If R is confluent (Church-Rosser) and noetherian, for each $t \in M_\Omega(F,V)$ there exists a unique t' such that $t \overset{*}{\underset{R}{\to}} t'$ and t' is R-*irreducible* i.e. $t' \underset{R}{\to} t''$ for no t'' ; t' is the R-*normal form of* t and we denote it by $c_R(t)$.

2. *Reductions compatible with* $\prec$.

(2.1) Definition

Let $S \subset M(F,V) \times M(F,V)$ be a (finite left-linear) reduction. It is said *compatible with* $\prec$ *(-or* $\prec$ *-compatible)* if the two following properties are satisfied:

(2.1.1) For all t,t',z such that $t' \prec t \underset{S}{\to} z$ there exists z' such that $t' \overset{*}{\underset{S}{\to}} z' \prec z$.

(2.1.2) For all t, t', z such that $t \prec t'$ and $t \underset{S}{\to} z$ there exists z' such that $t' \overset{*}{\underset{S}{\to}} z'$ and $z \prec z'$.

(2.2) LEMMA: A reduction S *is* $\prec$ *-compatible if and*

only if (2.1.1) and (2.1.2) hold for all t', z *and* t *such that* $\langle t,z\rangle \in S$.

(2.3) Examples

The reduction: $ffv_1 \to f_{v_1}$ is $\prec$ -compatible: the diagrams

$$\begin{array}{ccc} ffv_1 & \to & f_{v_1} \\ \curlyvee & & \curlyvee \\ f\Omega & \overset{*}{\to} & f\Omega \end{array} \quad \text{and} \quad \begin{array}{ccc} ffv_1 & \to & f_{v_1} \\ \curlyvee & & \curlyvee \\ ff\Omega & \to & f\Omega \end{array}$$

show that (2.1.1) holds

(the other cases are trivial); since $ffv_1 \prec t'$ implies $ffv_1 = t'$ (2.1.2) holds trivially.

On the contrary, the reduction $fgv_1 \to gfv_1$ is not $\prec$ -compatible (see examples (1.2) and (2.15)).

(2.4) THEOREM: LET S *be a confluent, noetherian and* $\prec$ *-compatible reduction. Then* c_S *is a* C_S *-projection.*

Hence by (1.4) one can define $c_S(T) = \mathrm{Sup}\{c_S(t)/t<T\}$ for $T\in M_\Omega^\infty(F,V)$ as the S-*normal form of* T . We also obtain $T\leq_S T'$ if and only if $c_S(T) \prec c_S(T')$. Hence, under the hypotheses of (2.4).

(2.5) COROLLARY: An infinite tree T *is* S *-maximal if* Ω *does not occur in* $c_S(T)$.

(2.6) Definition

Assuming that S is confluent, noetherian and $\prec$ -compatible, a recursive scheme (Σ,Φ) is S-*proper* if $c_S(T(\Sigma,\Phi))\in M^\infty(F,V)$ i.e. if Ω does not occur in $c_S(T(\Sigma,\Phi))$.

(2.7) COROLLARY: Let Σ *and* Σ' *be two recursive schemes in the functional variables* $\Phi_1,\ldots,\Phi_n$ *and* $\psi_1,\ldots,\psi_n$ *respectively. Assume that* (Σ,Φ_i) *is* S*-proper for each* i *. If* $\langle \psi_{1_I},\ldots,\psi_{n_I}\rangle$ *is a solution of* Σ *for all* $I\in C_S$ *then* $\Phi_i \equiv_S \psi_i$ *for* $i=1,2,\ldots,n$ *. (*Σ *is called* S*-univocal).*

Let us go back to (3.4) and give sufficient conditions for a reduction S to be $\prec$ -compatible.

(2.8) Definition

Let $W = \{w_1,\ldots,w_\ell\}$ be a set of auxillary variables. Let $s\in M(F,V\ W)$ be linear in W and $t'_1,\ldots,t'_\ell\in M(F,V)$. If $t = s[\Omega/w_1,\ldots,\Omega/w_\ell]$ and $t' = s[t'_1/w_1,\ldots,t'_\ell/w_\ell]$ we say that $t \prec t'$ *with respect to* $(s,t'_1,\ldots,t'_\ell)$.

For every t and $t'\in M_\Omega(F,V)$ such that $t \prec t'$ there exist $W,s,t'_1,\ldots,t'_\ell$ such that $t \prec t'$ w.r.t. $(s,t'_1,\ldots,t'_\ell)$.

(2.9) PROPOSITION: Let S *be the union of two reductions* S_1 *and* S_2 *such that:*

(2.9.1) *If* $\langle \alpha,\beta\rangle\in S_1$ *then* Ω *occurs neither in* α *nor in* β *,*

(2.9.2) *Each element of* S_2 *is of the form* $\langle \alpha,\Omega\rangle$ *,*

(2.9.3) *If* $\langle \alpha,\beta\rangle\in S$ *, if* $\gamma < \alpha$ *w.r.t.* $(\delta,\alpha_1,\ldots,\alpha_\ell)$ *and* $\alpha_1,\ldots,\alpha_\ell\notin V$ *then* $\gamma \overset{*}{\underset{S}{\rightarrow}} \Omega$

Then S *is* $\prec$ *-compatible.*

Sufficient conditions for confluence can be found in [7]. That S is noetherian can be proved with the results of [8] and [10].

We give a sufficient condition for (Σ,Φ) to be S - proper when Σ is non nested. Recall that $\Sigma = \langle \Phi_i = \tau_i \rangle_{1\le i\le n}$ is *non-nested* if each τ_i is of the form:

(2.10) $$\tau_i = \alpha[\Phi_{j_1}(\vec{\beta}_1)/w_1, \ldots, \Phi_{j_\ell}(\vec{\beta}_\ell)/w_\ell]$$

with $\alpha \in M(F, V\cup W)$, $W = \{w_1, \ldots, w_\ell\}$ is a set of auxiliary variables, and $\vec{\beta}_{j_s}$ is a $\rho(\Phi_{j_s})$ -tuple of elments of $M(F,V)$.

(2.11) Let Σ be a non nested system of equations and S be a (noetherian, etc...) reduction such that:

(2.11.1) if $\langle\gamma,\delta\rangle \in S$ and Ω occurs in δ then Ω occurs in γ and $\delta = \Omega$,

(2.11.2) Ω does not occur in τ_i for $i=1,\ldots,n$,

(2.11.3) if $\mu = f(\mu_1,\ldots,\mu_r)$ is a subterm of $c_S(\alpha)$ with α defined in (2.10) for some $i=1,\ldots,n$ and if some w_j occurs in μ then no rule of S is of the from $<f(\gamma_1,\ldots,\gamma_r),\delta>$.

In other words, if f occurs in $c_S(\tau_i)$ "above" some Φ_i (see the diagram) then it can cannot occur as leftmost symbol in the left-hand side of any rule of S .

$c_S(\tau_i)$:

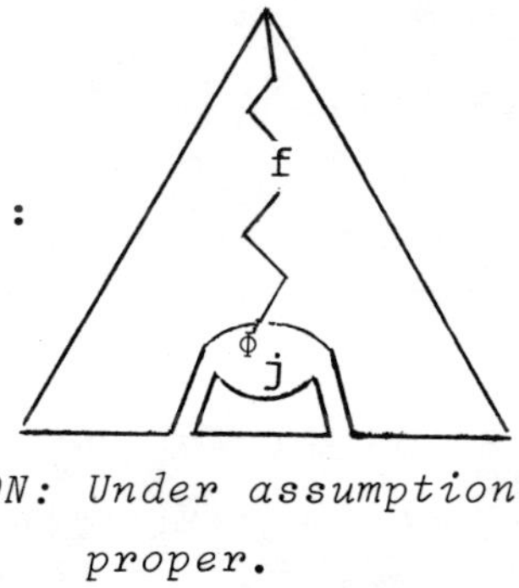

(2.12) PROPOSITION: Under assumptions (2.11), Σ *is* S-*proper.*

Conditions (2.11) are very easy to test, but it would be desirable to enlarge them.

This is only a partial solution to the following:

(2.13) Problems: (1) It is decidable whether (Σ,Φ) is S-proper for a given noetherian, confluent and $-\!\!<$ -compatible reduction S?(my conjecture is yes).

(2) Find easily testable sufficient conditions, weaker than (2.11) for (Σ,Φ) to be S-proper.

We now give some examples and applications of the results of this section.

(2.14) Remark: An example (omitted in this extended abstract) presents a S-univocal equation Σ such that $c_S(T(\Sigma,\Phi))$ is not an algebraic tree, i.e. is not $T(\Sigma',\Phi')$ for any (Σ',Φ'). But proofs using (2.7) do not require $T(\Sigma,\Phi)$ to be in S-normal form.

(2.15) Example: (to be compared with example (1.2)).

Let us consider the following rule:

$gfv_1 \to fgv_1$.

It is not $-\!\!<$ -compatible. Let us add

$g\Omega \to \Omega$.

These two rules form a noetherian, confluent, $-\!\!<$ -compatible reduction S .

The recursive equation $\Phi = f\Phi$ is S-univocal. And gf^{ω} is the same solution as f^{ω} since $c_S(gf^n\Omega) = f^n\Omega$ for every n . It follows that $c_S(gf^{\omega}) = c_S(f^{\omega}) = f^{\omega}$. The equation $\mu = g\mu$ has several solutions, namely Ω and f^{ω} .

(2.16) Example: Let S be the following reduction:

$k(v_1,k(v_2,v_3)) \to k(k(v_1,v_2),v_3)$

$k(v_1,h(v_2,v_3,v_4)) \to h(v_2,k(v_1,v_3),k(v_1,v_4))$

$k(v_1,\Omega) \to \Omega$

$k(\Omega,v_1) \to \Omega$

Conditions (2.9) clearly hold, and it is not difficult to prove that S is confluent and noetherian.

Let us now consider the following recursive schemes:

$$\Sigma_1 : \quad \Phi v_1 = h(v_1, fv_1, k(gv_1, \Phi dv_1))$$

$$\Sigma_2 : \begin{cases} \Theta v_1 = h(v_1, fv_1, \psi(dv_1, gv_1)) \\ \psi(v_1, v_2) = h(v_1, k(v_2, fv_1), \psi(dv_1, k(v_2, gv_1))) \end{cases}$$

CLAIM: $\Phi \equiv_S \Theta$

If h is interpreted as a conditional i.e. as $\lambda x_1 x_2 x_3 \cdot$ *if* πx_1 *then* x_2 *else* x_3 for some continuous predicate π then (Σ_2, Θ) can be considered as an iterative form of (Σ_1, Φ) (via McCarthy's transformation of flowcharts into recursive procedures). The flowchart corresponding to (Σ_2, Θ) is then:

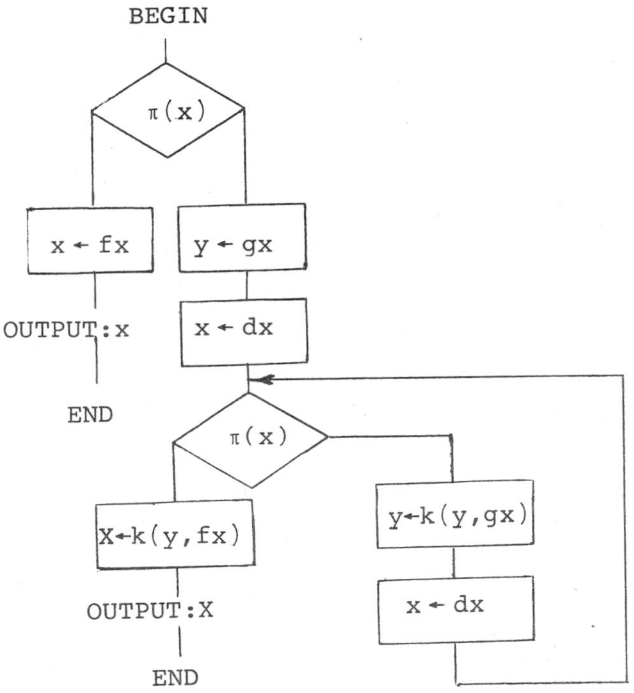

This is essentially the first transformation scheme of [4]. And our claim expresses the correctness of the transformation. Note that our conditions do not use *all* the properties of the conditionals, hence they are weaker than that of [4] where an explicit conditonal operator is used. Hence our transformation scheme is more powerful. Note that replacing flowcharts by recursive procedures allows not to use explicit conditionals.

Let us now prove the claim.

We first prove that $\psi(v_1,v_2) \equiv_S k(v_2,\Phi v_1)$

It is clear by (2.11) that the second equation of Σ_2, $\psi(v_1,v_2) = \tau_\psi$ is S-univocal.

Hence it is sufficient to verify that $k(v_2,\Phi v_1)$ satisfies it w.r.t. C_S.

$$
\begin{aligned}
k(v_2,\Phi v_1) &\equiv_S k(v_2,h(v_1,fv_1,k(gv_1,\Phi dv_1))) \quad (\text{by } \Sigma_1)\\
&\equiv_S h(v_1,k(v_2,fv_1),k(v_2,k(gv_1,\Phi dv_1))) \quad (\text{by } S)\\
&\equiv_S h(v_1,k(v_2,fv_1),k(k(v_2,gv_1),\Phi dv_1))) \quad (\text{by } S)\\
&\equiv_S \tau_\psi\{k(v_2,\Phi v_1)/\psi\} \quad \text{since}
\end{aligned}
$$

$$k(v_2,\Phi v_1)[dv_1,k(v_2,gv_1)] = k(k(v_2,gv_1),\Phi dv_1).$$

Hence now

$$
\begin{aligned}
\Theta v_1 &\equiv_S h(v_1,fv_1,\psi(dv_1,gv_1)) \equiv_S h(v_1,fv_1,k(gv_1,\Phi dv_1))\\
&\equiv_S \Phi v_1 \quad (\text{by } \Sigma_1).
\end{aligned}
$$

And the proof is complete. Notice that we have not used any induction principle. We have only checked equations.

3. *CONCLUSIONS*

This work relates two kinds of investigations about:

(1) classes of interpretations([1],[2],[3],[9])

(2) term rewriting systems ([7],[8],[10])

and produces simple proof methods for the validation of transformations of programs ([5],[4]). It provides a basis for choosing "easy" classes C_R to deal with, and a concrete construction of certain "universal" interpretations H_C... of course it is to be developed in many directions, (improvement of (2.12), R-univocal equations when R is not a reduction) but already provides useful results.

REFERENCES

[1] ADJ, Initial algebra semantics, *Journal of ACM*, Vol. 24,(1977),pp. 68-95.

[2] Courcelle,B., Nivat,M.: Algebraic Families of Interpretations, *Proceedings of 17th IEEE Symposium of Foundations of Computer Science*, 1976, pp. 137-146.

[3] Bloom,S.: Varieties of ordered algebras. *Journal of Computer and System Science*,Vol.13,(1976), pp.200-212.

[4] Burstall,R., Darlington,J.: A system which automatically improves programs, *Proceedings of the 3rd IJCAI*, Stanford,1975, pp.537-542.

[5] Burstall,R., Darlington,J.: A transformation System for developing recursive programs,*Journal of ACM*, Vol.24,(1977).

[6] McCarthy,J.: A basis for a mathematical theory of computation.

[7] Huet,G.: Confluent reductions, *Proceeding of 18th IEEE Symposium of Foundations of Computer Science*, 1977.

[8] Knuth,D., Benedix,P.: Simple word problems in universal algebras, in Leech(ed), *Computational problems in abstract algebra*, Pergamon Press, 1970, pp.263-297.

[9] Berry,G., Courcelle,B.: Canonical forms in stable discrete interpretations, in Michaelson,S., Milner,R. (Eds), *Proceedings of 3rd Colloquium on Automata. Languages and Programming*. Edinburgh University Press, 1976.

[10] Lipton,R., Snyder,L.: On the halting of tree replacement systems, *Proceedings of the Conference on Theoretical Computer Science*, Waterloo, 1977.

B.Courcelle
IRIA
Domaine de Voluceau
Rocquencourt
78150-LE CHESNAY (FRANCE)

COLLOQUIA MATHEMATICA SOCIETATIS JÁNOS BOLYAI
26. Mathematical Logic in Computer Science
Salgótarján (HUNGARY), 1978.

SL:MANY-SORTED LOGIC BASED PROGRAMMING LANGUAGE

G.Dávid, A.Sárközy

0.Introduction

We wished to describe, specify, and prove correctness and to synthetize sequential or parallel programs, especially microprograms. In order to describe a computer system which executes the program to be proved, we need tools to reflect the inhomogeneous data structures of computer systems and the same for programs, as well as for the transformations (i.e. instructions,functions) defined on them. Because the typical data structures are both homogeneous (vectors, arrays) and inhomogeneous ones we had to choose a many-sorted logic as a base for further extension as a programming language. Many-sorted logic [1,2,3] in its form was not suitable so we modified it for logic programming purposes.
Although the use of many-sorted logic is quite natural for abstract data types (see for example VDL [4]) we chose it rather as a programming language than an axiomatic treatment. The logic programming form is much more suitable to

describe and specify programs, to synthetise them because the sentences of this logic and the programming language - i.e.purely syntactical constructs - specify state-transformations, where the states consist of data structures and the transformations are the executions of instructions.

In order to describe these features of a system adequately we extended the many-sorted logic by selectors (as in VDL). Selectors associate substructures which can be manipulated, transformed independently, and the selectors also specify the *sorts* of the selected substructures. Selectors play the roles of "labels" and they are embedded on the logic level, too. Using selectors the typical constructions (from sorts the predicates are built up) in many-sorted logic can be continued: from a given set of sorts the predicates (here they are called *compound predicates*) can be expressed, and further from compound predicates (and maybe from sorts again) new compound predicates can be defined, and so on. This reflects the multilevel hierarchy of data structures. From a compound predicate a finite sequence of selectors points to the lowest level sorts. The logic is called *Structure Logic SL* and in this paper we describe both the mathematical logic form (Section 1) and the programming language form (Section 2) of SL. We shall restrict ourselves to the description and an example only, examples and applications are described in detail elsewhere [5-9].

1. STRUCTURE LOGIC SL

1.1 Syntax

We introduce SL as a classical quantifier-free first

order predicate language. Let the followings be given:

constant symbols
variable symbols
n-ary function symbols
n-ary predicate symbols
logical connectives: $\sim, \vee, \wedge, \rightarrow$
auxiliary symbols: (,)

Unary predicate symbols play the roles of sorts. This set is extended by:

selector-constant symbols
selector-variable symbols
n-ary selector function symbols
type-symbols
structure symbols
composition symbols: $\langle\ ,\ \rangle$, [,], :, '

It should be noted that the word "type" is used as a synonim of predicate for structures but we have to distinguish the n-ary predicate symbols and those ones which are used to describe the properties of data structures. Hence the word "type" is used here as in "abstract data type" and not as the type of the logic.

The elements of SL are constructed in the following way:

Symbols are constant symbols and variable symbols.

Selector symbols are selector-constant symbols and selector-variable symbols, and if $s_1, s_2, \ldots, s_n$ are selector symbols then $[s_1, s_2, \ldots, s_n]$ is a selector symbol.

Selector terms are selector-constant and selector-variable symbols and the selector-functions defined on them.

Terms are the symbols, and if $t_1, t_2, \ldots, t_n$ are terms or selector terms, $f(.,\ldots,.)$ is an n-ary function symbol then $f(t_1, t_2, \ldots, t_n)$ is a term.

There are not any other terms or selector terms but the

repeated application of these rules.

Selector expressions are constructed from selector terms *st* by bracketing: [*st*] and if *se1* and *se2* are selector expressions then *se1.se2* is a selector expression.

(Here and in the following the sign "." represents concatenation).

An *atomic formula* is an arbitrary n-ary predicate symbol $\pi(t_1,t_2,\ldots,t_n)$, where $t_1,t_2,\ldots,t_n$ are terms.

Subtype specification is a pair

$$s:\pi$$

where s is a selector symbol and π is an arbitrary n-ary predicate symbol.

The subtype specification specifies the *sort* of the component selected by the selector s if π is an unary predicate. This is why we did not use sorts explicitly.

Compound predicate is an arbitrary type-symbol Π applied on subtype specifications in the form of

$$\Pi\langle s_1:\pi_1, s_2:\pi_2,\ldots,s_n:\pi_n\rangle \qquad (1)$$

where π_i represents either an n-ary predicate or a compound predicate.

For arbitrary selector expression *se* and compound predicate Π, Π.*se* represents the *subtype* from Π by *se*.

Structures: variable and constant symbols are structures. If S is a structure symbol, and $S_1,S_2,\ldots$ $\ldots,S_n$ are structures, $s_1,s_2,\ldots,s_n$ are selector symbols then

$$S\langle s_1:S_1, s_2:S_2,\ldots,s_n:S_n\rangle \qquad (2)$$

is a structure. If *se* is a selector expression and S is a structure then S.*se* is a structure (and represents a substructure of S selected by *se*).

A special symbol Ω represents the empty structure symbol

Let us denote by $C(S)$ the *carrier* of the structure S, the set of the selector symbols in S:

$$C(S) = \{s_1, s_2, \dots, s_n\}$$

For the empty structure Ω is $s \notin C(S)$ then

$$S[s] = \Omega$$

i.e. the empty structure is always a substructure of an arbitrary structure. For compound predicates like Π in (1) the carrier $C(\Pi)$ is defined in a similar way. The carrier should be finite.

Structure expression: If S is a structure in the form of (2), s^o is an arbitrary selector term, S^o is a structure then

$$S\langle s^o : S^o \rangle$$

is a structure expression. Similarly, for selector terms $s_1^o, s_2^o, \dots, s_k^o$ and structures $S_1^o, S_2^o, \dots, S_k^o$

$$S\langle s_1^o : S_1^o, s_2^o : S_2^o, \dots, s_k^o : S_k^o \rangle$$

is a structure expression and it means the repeated application

$$S\langle \dots \langle\langle s_1^o : S_1^o \rangle, \dots, \rangle, s_k^o : S_k^o \rangle$$

Structure formula: Let a compound predicate Π be given and a structure or structure expression S. The

$$\Pi(S)$$

is a *structure formula*.

Well formed structure formula is constructed from structure formulas $\Pi(S)$, $\Sigma(Z)$ by logical connectives $(\sim\Pi(S))$, $(\Pi(S) \vee \Sigma(Z))$, $(\Pi(S) \wedge \Sigma(Z))$ and $(\Pi(S) \rightarrow \Sigma(Z))$.

These definitions should be explained shorthly by an example: Let the followings be given:

- selector symbols: 0, acc, reg, ins, 7, i, j;
- functions add(.,.), +(.,.) ;

- predicate symbols *bit*, *instruction;*
- type simbols *bit8*, *machine*, *register* and
- structure symbols M, A, R, I .

From these symbols add(i,j) and +(R[i], R[j]) are terms and [i].[O] is a selector expression. Selector-valued functions are not shown in this example. The

bit8 ⟨ [O:7]:*bit* ⟩
register ⟨ [O:7]:*bit8* ⟩
machine ⟨ acc:*bit8*, reg:*register*, ins:*instruction* ⟩

are compound predicates, where [O:7] denotes the range of selectors, i.e. [ϕ:7] = ϕ,1,2,3,4,5,6,7 .

The carrier C(*machine*) = {acc, reg, ins}, and the subtype specifications:

acc: *bit8*
reg: *register* (3)
ins: *instruction*

Let M be a structure

M ⟨ acc:A, reg:R, ins:I ⟩

then the carrier C(M) is the same as for *machine*. The structure expressions

M ⟨ ins:add(i,j) ⟩
M <acc:+(R[i], R[j]) ⟩

change the contents of the components: the first one describes the substructures M[ins], the second one does the same for M[acc]. Then

machine(M)
machine(M⟨ins:add(i,j)⟩) (4)
machine(M⟨acc:+(R[i],R[j])⟩)

are structure formulas and

machine(M⟨ins:add(i,j)⟩) →
→ *machine*(M acc:+ R[i],R[j])⟩) (5)

is a well formed structure formula.

1.2 Semantics

If given a well formed structure formula F, its *interpretation* is defined by a nonempty domain $\mathcal{D}$, by a set $F(\mathcal{D})$ of n-place mappings of $\mathcal{D}^n$ to $\mathcal{D}$; mappings of $\mathcal{D}^n$ to $\{T,F\}$ (true,false) and a set J of integers. The evaluation function ν is defined as follows:

- to each constant ν assigns an element in $\mathcal{D}$ and to each selector constant ν assigns an element in J,
- to each variable symbol x, $\nu(x) \in \mathcal{D}$ and to each selector-variable symbol s, $\nu(s) \in I$.
- for every carrier C : $\nu(C) = \{\nu(c') \mid c' \in C\}$
- to each n-ary function symbol we assign a mapping from $\mathcal{D}^n$ to $\mathcal{D}$, and to each n-ary selector function symbol ν assigns a mapping from I^n to I.
- to each n-ary predicate symbol ν assigns a mapping from $\mathcal{D}^n$ to $\{T,F\}$ (an n-ary relation on $\mathcal{D}$)

For *terms* and *selector terms* in the form of $f(t_1,t_2,\ldots,t_n)$, ν assigns

$$\nu(f(t_1,t_2,\ldots,t_n)) = \nu(f)(\nu(t_1),\ldots,\nu(t_n))$$

For *selector expressions*

$$\nu(se_1 \cdot se_2) = \nu(se_1) \cdot \nu(se_2)$$

and subtype specifications $s : \pi$

$$\nu(s{:}\pi) = \nu(s) : \nu(\pi)$$

For type symbols and structure symbols ν assigns arbitrary names. If Π is a *compound predicate* of (1) then ν assigns

$$\nu(\Pi) = \nu(\Pi)\langle \nu(s_1):\nu(\pi_1),\ldots,\nu(s_n):\nu(\pi_n)\rangle$$

i.e. if $\|\pi\|$ represents the "arity" of a predicate π, then

$$\nu(\Pi) = \underset{s \in \nu(C(\Pi))}{\chi} \quad \mathcal{D}^{\| \nu(\Pi.s) \|}$$

If S is a *structure* of (2) then

$$\nu(S) = \nu(S)\langle \nu(S_1),\dots,\nu(s_n):\nu(S_n)\rangle$$

and ν assigns further relations to π_j and structures to S_i, if they are compound ones.

Structure formulas, like $\Pi(S)$ are interpreted "componentwise": for every $s \in C(\Pi)$

$$\nu(\Pi[s](S[s])) = \nu(\Pi)[\nu(s)](\nu(S)[\nu(s)]) \qquad (6)$$

If *for every* $s \in C(\Pi)$ to (6) ν assigns T then to $\Pi(S)$ we assign T else F .

In (6) $\Pi[s]$ is a compound predicate or n-ary predicate, $S[s]$ is a structure. For structure expressions $S\langle s^o:S^o\rangle$

$$\nu(S\langle s^o:S^o\rangle) = \nu(S)\langle \nu(s^o):\nu(S^o)\rangle$$

If $s^o \in C(\Pi)$, S^o is an arbitrary structure, Π is a compound predicate, then

$$\nu(\Pi(S\langle s^o:S^o\rangle)) =$$
$$= \{\bigwedge_{s_i \in C(\Pi)} \nu(\Pi)[\nu(s_i)](\nu(S)[\nu(s)])\} \wedge$$
$$\wedge\ \nu(\Pi)[\nu(s^o)](\nu(S^o))$$

From $s^o \in C(\Pi)$ it follows that $\Pi(S\langle s^o:S^o\rangle)$ states

$$\Pi[s^o](S^o) \wedge \Pi[s^o](S[s^o])$$

i.e. the new substructure S^o is of the same sort as the original one selected by s^o : $\Pi[s^o]$.

If x is a variable (or s is a selector variable) in a well formed structure formula F, then ν assigns Π to F if and only if for every $\nu(x) \in \mathcal{D}$ (or for every $\nu(s) \in I$) F is true. Well formed structure formulas are evaluated, as usual by truth-table. Models of a formula Σ are those interpretations on which Σ has the value T true. It is easy to prove by (6) that for every formula S

$$\Pi(S) \leftrightarrow \bigwedge_{s\in C(\Pi)} \Pi[s](S[s]) \qquad (7)$$

This provides a lifting mechanism between the layers of structures.
In the example above it is easy to see that a computer consisting of

- an 8-bit accumulator
- an instruction register storing the instruction symbol,
- register bank of eight 8-bit registers and
- an arithmetic unit executing add(i,j) and adds two registers i and j into the accumulator

is a model of (3),(4) and (5).

A repeated application of (7) would transform every well formed structure formula in SL into a first order form

$$\Pi(S) \leftrightarrow \bigwedge_{s'\in EC(\Pi)} \Pi[s'](S[s'])$$

where s' is a selector expression such that $\Pi[s']$ is an n-ary predicate symbol. The set of those selector expressions whose members select only n-ary predicate symbols (and there is not any other selector symbol or selector expression in Π) is called elementary carrier $EC(\Pi)$ of Π . In the example

{[acc].[O],[acc].[1],...,[acc].[7],

[reg].[O].[O],[reg].[O].[1],...

...,[reg].[7].[6],[reg].[7].[7],[ins]}

is the elementary carrier EC(*machine*).

Based on the interpretations we can define substitutions.
A *substitution* σ is a finite set of the form

$$\{t_1/v_1,\ldots,t_n/v_n\ ;\ z_1/s_1,\ldots,z_k/s_k\}$$

where

- every v_i is a variable, s_j is a selector variable,
- every t_i is a term, z_j is a selector term,
- there are different variables after the stroke symbol.

For variables and terms, they are of the same sorts, i.e. for every j, the pair t_j/v_j represents term t_j and variable v_j such that for every unary predicate π

$$\pi(v_j) \rightarrow \pi(t_j)$$

If Σ is a well formed structure formula, then $\Sigma\sigma$ is another formula of SL obtained from Σ by replacing simultaneously each occurence of v_i and s_j by t_i and z_j, respectively.

If σ is a substitution $\{t_i/v_i\}$ of a well formed structure formula Σ and if M is a model of Σ with evaluation ν, which assigns $\nu(v_i)$ is also an element of $\mathcal{D}$, hence this and $\pi(v_i) \rightarrow \pi(t_i)$ imply that M is also a model of $\Sigma\sigma$, where each occurence of v_i is substituted by t_i. The similar is true for the pairs of selector terms and selector variables z_j/s_j. Consequently each model of Σ will be a model of $\Sigma\sigma$, too, hence

$$\Sigma \rightarrow \Sigma\sigma$$

is valid.

The transformation of an arbitrary well formed structure formula F of SL into the first order form based on the elementary carrier defines a translation rule from SL into the quantifier-free first order logic, but variables are assumed to be universally quantified. Consequently the *resolution principle* is described in a similar way as in first order logic. The resolution principle was proved complete directly in SL, i.e. if there is a refutation of a formula F form a set S, of well formed structure formulas then there is a refutation in which only the resolution principle is used. Using resolution a mechanical theorem prover for program synthesis was designed. SL is used for parallel programs, too (9).

2. STRUCTURE LOGIC SL AS A PROGRAMMING LANGUAGE

Earlier we have seen that SL is an appropriate tool to describe data structures and with structure expressions one can specify the transformations executed on these data structures. Having the resolution principle and other mechanical deduction rules (modus ponens, etc), SL can be used as a programming language. But to make both the notations and (primarily) the notions easier for "algorithmic" programmers we modified the logic form and these modifications are described here.

For programming purposes the programmer should "build" up his compound predicates and the structures to form well formed structure formulas. This part is a declaration-part: the user can declare the types of symbols used in the program in SL.

We shall give the syntax of SL in BNF notation but it is slightly modified:

- concatenation is written by dotting "." for example letter.letter
  Set is written as {.....} ; i.e. list of symbols, bracketed by {,}
- symbols, occuring in a special context where they are selectors, are automatically declared as selector and there we shall write "selector" representing an arbitrary symbol.
- it is assumed that a set of unary predicates called

basic-types is given. They are the predicate symbols. Everyone can declare the appropriate set of *basic-types* reflecting the elementary notions of the lowest level from which the structures will be built.

2.1 Symbols

Here the basic symbols are defined, and further used in order syntactical constructions.

letters::=A|...|Z,a|b|...|z;
digits::=0|...|9
externals:=*constant*|*reference*|*formulae*|*basic-types*|
selector|*structure*|*function*|*identities*|
begin|*end*|
<|>|[|]|(|)|:|,|;| logical-connectives;
logical-connectives::=unary-logical-connectives|
binary-logical-connectives;
unary-logical-connectives::= ~ (not)
binary-logical-connectives:: →,∨,∧. (implies,or,and)
alphanumeric::=digit|letter|letter.alphanumeric|
digit.alphanumeric
symbol::=letter|letter.alphanumeric
list::=symbol|symbol.,.list
selector::=*selector*|*sel*|"selector"
selector-domain::=[lower-bound:upper-bound]
lower-bound::="selector"
upper-bound::="selector"
selector-expression::["selector"]|["selector"].
selector-expression
basic-type::={basic-type}|*selector*

2.2 Types and structures

The terminology in Computer Sciences and in the practice is a little bit mixed, so we have to use the word *type* instead of predicate. The programmers should build up compound predicates, i.e. structured data *types* and should declare the data *structures* of the defined types.

So we shall speak about *structures* only. Sometimes we use *struc* for *structure* and *ref* for *reference*.

```
declaration-part::={declarations_
declaration::=type-declaration|structure-declaration|
              constant-declaration|function-declaration
constant-declaration::=constant  list-of-symbols;
type-declarator::=structure|struc
type-name::=symbol
type-declaration::=type-declarator.
                   .type-name{⟨"selector":struc-type-ref}⟩;
struc-type-ref::=type-reference|struc-reference
type-ref::=basic-type|type-name|type-name.
         .selector-expression|type-expression
type-expression::=type-name⟨{"selector":type-reference}⟩;
structure-reference::=structure-name|
                      struc-name.selector expression|
                      struc-expression
structure-declaration::=type-reference|reference(type-name)
struc-declaration::=struc-declarator. list-of-structure-names;
structure-name::=symbol
structure-expression::=structure-name.
                      .⟨{"selector":structure-reference}⟩ ;
predicate-declaration::=predicate.structure-expression;
```

The *predicate-declaration* should be explained. It is obvious from the structure declaration that a type (i.e. a compound predicate) should be associated to every structure symbol. If Π is a type, like *machine* in (3), then it is declared as

structure machine ⟨acc:*bit8*, reg:*register*, ins:*instruction*⟩

and if we need two data structures M and MM of the same type *machine*, then we declare them as:

reference (machine) M,MM

The programmer should write well formed structure formulas (like (5))on M and MM, but it is unnecessary to write the compound predicate "machine" in every formula because M and MM represent implicitly their being a "machine".

Similarly to make the notation simpler in predicate-declaration the structure-expression is used to *name* some substructures, for example

predicate M⟨acc:A, reg:R, ins:I⟩

predicate MM⟨reg:RR⟩

are declaring M as a structure of substructures A,R and I , and MM as MM[reg] is a substructure RR , but the others (namely MM[acc] and MM[ins]) are arbitrary(but they are of the type *machine* [acc] and *machine* [ins], respectively).
With this notation (5) can be written as

M⟨ins:add(i,j)⟩ → M⟨acc:+(R[i],R[j])⟩ (8)

2.3 Functions

In the *declaration-part* the function-symbols to be used should be described. The function-symbols are declared in syntactical sense and they have a *type* which is used in the substitution. Also the formal parameters have types; the specification points to the types of formal parameters. The *identities* describe the semantical equivalencies in syntactical form. For example for the addition +(.,.)

+(x,y) = +(y,x)

The function body may be empty, but if it is written, then it should be evaluated if the actual parameters are constants.

function-declaration::=type-reference.*function*.function symbol.
.specification;identities-declaration;body;
function-symbol::=function-name.(list-of-formal-parameters)
function-name::=symbol

formal-parameters::=symbol

specification::=declaration

identities-declaration::=*identities* list-of-identities;

identity::=term.=.term;

term::=symbol|function name (list-of-terms)

body::=*begin*.local declaration; list-of-structure assignments.*end*

local-declaration::=declaration

structure-assignment::=left-part.:.right-part;

left-part::structure-reference|function-name

right-part::=sturcture-reference|term

From the function-declaration only the type-reference, function symbol and the specification are necessary: they describe what types will be transformed into other ones. The remaining parts are used only during the substitutions.

2.4 Knowledge and problem description

description::=*formulas*{statements};

program-specification::=statements;

statement::=structure expression.binary-logical-connective.
.structure-expression|unary-logical-connective.
.structure-expression

The structures and structure-expressions have associated types. If S is a structure of type Π then

$$\Pi(S) \leftrightarrow \bigwedge_{s\in C(\Pi)} \Pi[s](S[s])$$

This means that for every structure-expression $S\langle s^o:S^o\rangle$

$$\Pi(S\langle s^o:S^o\rangle) \leftrightarrow \tag{9}$$

$$\leftrightarrow \Big(\bigwedge_{\substack{s\in C(\Pi)\\ s\neq s^o}} \Pi[s](S[s])\Big) \wedge \Pi[s^o](S^o)$$

Thus the description becomes simpler and shorter because only the relevant layer and on this layer only the affected substructures should be described. Hence, (8) is a shortened form of the detailed formula

$$M\langle acc:X,\ reg:R,\ ins:add(i,j)\rangle \rightarrow \rightarrow M\langle acc:+(R[i],\ R[j]),\ reg:R,\ ins:U\rangle \tag{10}$$

where X and U are arbitrary symbols representing free variables (of the type *machine*[acc] and *machine* [ins]). Both sides of (10) can be rewritten by (9) in the form of

$$(\mathit{machine}[acc](X) \wedge \mathit{machine}[reg](R) \wedge \mathit{machine}[ins](add(i,j))) \rightarrow$$

$$\rightarrow (\mathit{machine}[acc](+(R[i],\ R[j]) \wedge \mathit{machine}[reg](R) \wedge \mathit{machine}[ins](U))$$

Here *machine*[acc](X), *machine*[ins](U) are T and in both sides we find *machine*[reg](R), i.e. only the formula

$$\mathit{machine}[ins](add(i,j)) \rightarrow \mathit{machine}[acc](+(R[i],\ R[j]))$$

remains.(8) expressed this fact.

The *program specification* is a statement expressing the problem to be solved. A resolution-based theorem prover constructs the proof (if existed).

REFERENCES

[1] Monk,J.D.: *Mathematical Logic*, Springer-Verlag, Berlin 1976.

[2] Feferman,S.: Applications of many sorted interpolation theorems, *Proceedings of the Tarski Symposium*, Providence, R.F. 1974, pp.205-224.

[3] Wang,H.: Logic of many-sorted theories, *Journal of Symbolic Logic*, vol.17,(1972)

[4] Wegner,P.: The Vienna Definition Language, *Computer Surveys*, vol.4,(1972), No.1. pp.5-63.

[5] Dávid,G., Keresztély,S., Sárközy,A.: Microprogram-Synthesis by Theorem Proving, *Proceedings of the II Hungarian Computer Science Conference*, Part 1, Budapest, 1977, pp.293-310.

[6] Dávid,G.: Structured Automatized Design of Microprograms, Lawson,H.W., et al (eds) *Large Scale Integration*, North Holland 1978, pp.241-245.

[7] Dávid,G., Keresztély,S., Losonczy,I., Sárközy,A.: Logic-Based Description of Microcomputers, MTA SZTAKI Tanulmányok 91. Budapest,1978, -p.79-92.

[8] Dávid,G., Keresztély,S., Losonczy,I., Sárközy,A.: Microprogram Synthesis, MTA SZTAKI Tanulmányok 91 Budapest,1978, pp. 187-206.

[9] Dávid,G.: Proving Correctness and Automatic Synthesis of Parallel Programs in J.Miklosko, V.Kotov (eds) *Algorithms, Software and Hardware of Parallel Computers*. VEDA Publ. House, (to appear)

G.Dávid, A.Sárközy
Computer and Automation Institute
of Hungarian Academy of Sciences
Kende u. 13-17. Budapest, Hungary 1111

COLLOQUIA MATHEMATICA SOCIETATIS JÁNOS BOLYAI
26. Mathematical Logic in Computer Science
Salgótarján (HUNGARY), 1978.

DERIVATIVES OF PROGRAMS

J.W. de Bakker and J.I. Zucker

ABSTRACT

The notions of *upper* and *lower derivatives* of a recursive (non-deterministic) program are defines, and used to characterize termination for such a program in terms of the *well-foundedness* of a function with respect to a predicate. This extends earlier work of Hitchcock and Park to the case of *nested recursions*, formulated in terms of a least-fixed-point construct. It is shown how this characterization can be interpreted as stating that a recursive procedure always terminates iff it exhibits neither *global* nor *local* non-termination.

1. INTRODUCTION

The notion of *derivative* of a program was introduced by Hitchcock and Park [H,P] as an aid to investigate properties of program termination. More specifically, they showed how termination of a recursive program scheme may be expressed through the well-foundedness of a relation involving the so-called upper and lower derivatives of the scheme. The framework in which this result is derived is a calculus of binary relations extended with recursion via the least-fixed-point constructs, $\mu X[\ldots]$. However, their main result was proved for : (i) only deterministic programs, (ii) non-nested μ-construct, (iii) some further technical restrictions. In De Bakker [dB1], it was shown how to generalize the theory of [H,P] in the framework of denotational semantics (using the Egli-Milner ordering to deal with nondeterminacy, thus lifting restriction (i)) in such a way that restriction (iii) also disappeared, but maintaining restriction (ii) . The present paper gives the full story in that we now also deal with nested μ- constructs. This needs a non-trivial extension of the definition of upper and lower derivatives (cf. 5. 1c, 5. 3c), and, accordingly, a considerably more intricate proof (surpassing in complexity all proofs in the μ-calculus we have has experience with) of the basic theorem (5.5) connecting these two notions.

Section 2 of this paper describes the syntax, Section 3 provides the necessary backround in denotational semantics, Section 4 introduces a fundamental auxiliary result allowing us to syntactically reduce termination of a program (involving recursion) to termination of its compenents, and in Section 5 we define the upper and lower derivatives of a program and state (without proof) the basic theorem relating the two. Finally, in Section 6 we introduce the notion of a function being well-founded with respect to a predicate, thus refining an idea in [H,P] , and prove as main theorem the announced extension of the result there. The Section closes with an example illustrating how this result may be interpreted as stating that a recursive procedure terminates everywhere iff it exhibits neither *global* nor *local* nontermination.

A fuller exposition of this paper, with detailed proofs, is given in Chapter 8 of [dB2].

2. SYNTAX

The definition in this and the next section, though to some extent variations on familiar themes in denotational semantics, also include some new ideas, e.g., role of $b \in Stat$, of $\mu Z[p]$, and of $f_1 \to f_2$.

Convention. "Let $(\alpha\in)V$ be the set..." is short for "let V be the set..., with variable α ranging over V".

2.1. Definitions

"$\equiv$" denotes identity between syntactic constructs. Let $(n\in)Intc$ be the set of *integer constants*. Let $(x,y\in)Intv$, $(X,Y\in)Stmv$, $(Z\in)Cndv$ be the (infinite, well--ordered) sets of *integer-*, *statement-*, and *condition variables*. Let $(s\in)$ *Iexp* be the set of *integer expressions* defined by

$$s ::= x \mid n \mid s_1+s_2 \mid \textit{if } b \textit{ then } s_1 \textit{ else } s_2 \textit{ fi}$$

Let $(b\in)$ *Bexp* be the set of *boolean expressions* defined by

$$b ::= true \mid s_1=s_2 \mid \neg b \mid b_1 \supset b_2$$

Let $(S\in)$ *Stat* be the set of *statements* defined by

$$S ::= x:=s \mid b \mid S_1;\ S_2 \mid S_1 \cup S_2 \mid X\ \ \mu X[S]$$

Let $(p,q\in)$ *Cond* be the set of *conditions* defined by

$$p ::= true \mid s_1=s_2 \mid \neg p \mid p_1 \supset p_2 \mid \exists x[p] \mid S\{p\} \mid S\langle p\rangle \mid Z \mid \mu Z[p]$$

Let (f∈) *Afor* be the set of *atomic formulae* defined by

$$f ::= p \mid S_1 \sqsubseteq S_2 \mid f_1 \wedge f_1$$

Let (g∈) *Form* be the set of *formulae* defined by

$$g ::= f_1 \rightarrow f_2$$

2.2. *Free and bound variables; substitution*

The variables x, X and Z are *bound* in $\exists x[p]$, $\mu X[S]$ and $\mu Z[p]$ respectively. *intv*(s), *stmv*(S), *cndv* (f), etc., denote the sets of *free* integer-, statement-, and condition variables in s, S, f etc. Constructs which differ at most in their bound (integer, statement or condition) variables are called *congruent* (denoted by "$\cong$").

p[s/x] denotes the result of substituting s for (free occurrences of) x in p; similarly for S[S'/X] and p[q/Z]. The usual precautions to avoid clashes between free and bound variables apply.

2.3. *Remarks*

2.3.1. Integer and boolean expressions are of no concern in our theory - as long as their evaluation always term terminates - and are kept as simple as possible.

2.3.2. Let $S \equiv S(X)$. The $\mu X[S(X)]$ corresponds to a call of the recursive procedure P declared by P <= S(P). The boolean expression b considered as a statement may be understood by the following correspondence with statements in more tradional syntaxes: *if* b *then* S_1 *else* S_2 *fi* $\sim b;S_1 \cup \neg b;S_2$, *while* b *do* S *od* $\sim \mu X[b;S;X \cup \neg b]$ ($X \notin$ *stmv*(X)), and with Dijktra's "guarded commands"

[D]: *if* $b_1 \rightarrow S_1 \square \ldots \square b_n \rightarrow S_n$ *fi* $\sim (b_1;S_1 \cup \ldots \cup b_n;S_n)$, *do* $b_1 \rightarrow S_1 \square \ldots \square b_n \rightarrow S_n$ *od* $\sim \mu X[(b_1;S_1 \cup \ldots \cup b_n;S_n);$ $X \cup \neg b_1 \wedge \ldots \wedge \neg b_n]$ $(X \notin stmv(S_i)$, $i=1,\ldots,n)$.

2.3.3. S{p} and S<p> correspond to the *weakest precondition* for respectively *partial* and *total correctness* of S w.r.t. p.

2.3.4. In $\mu Z[p]$, p is assumed to be *syntactically monotonic* in Z, i.e., Z does not occur in p within the scope of an odd number of $\neg$-symbols (when $p_1 \supset p_2$ is rewritten as $\neg p_1 \vee p_2$). The construct $\mu Z[p]$ allows us to recursively define conditions, which obtain meaning as the usual least fixed point of a suitable operator.

2.3.5. For $f_1 \rightarrow f_2$ cf. Remark 3.6.7 below. A formula *true* $\rightarrow$ f will be abbreviated to f.

3. SEMANTICS

3.1. Complete partial orders and complete lattices

A *complete partial order or cpo* $(x \in)C$ is a partially ordered set with a least element $\perp_C$ such that each (ascending) chain $\langle x_i \rangle_{i=0}^{\infty}$ has a lub $\bigsqcup_i x_i$. A *complete lattice* is a partially ordered set C in which *every* subset X has a lub $\bigsqcup X$ and (hence also) a glb $\bigsqcap X$; thus C is a cpo, with $\perp_C = \bigsqcap C$.

Let C_1 and C_2 be cpo's. A function $f: C_1 \rightarrow C_2$ is *strict* if $f(\perp_{C_1}) = \perp_{C_2}$, *monotonic* if $x_1 \sqsubseteq x_2 \Rightarrow f(x_1) \sqsubseteq f(x_2)$ and *continuous* if it is monotonic and also, for each chain $\langle x_i \rangle_i$ in C_1, $f(\bigsqcup_i x_i) \sqsubseteq \bigsqcup_i f(x_i)$ (or equivalently, $f(\bigsqcup_i x_i) = \bigsqcup_i f(x_i)$). If C_2 is a complete lattice, then $f: C_1 \rightarrow C_2$ is *anticontinuous* if for each chain $\langle x_i \rangle_i$ in C_1, $f(\bigsqcup_i x_i) = \bigsqcap_i f(x_i)$ (which implies that f is anti-monotonic, i.e. $x_1 \sqsubseteq x_2 \Rightarrow f(x_2) \sqsubseteq f(x_1)$).

The sets of all strict, monotonic and continuous functions from C_1 to C_2 are denoted, respectively λ, by $C_1 \rightarrow_s C_2$, $C_1 \rightarrow_m C_2$ and $C_1 \rightarrow_c C_2$. These are all cpo's, when we define $f_1 \sqsubseteq f_2 \overset{df}{\Longleftrightarrow} \forall x \in C_1 (f_1(x) \sqsubseteq f_2(x))$, and $\bot_{C_1 \rightarrow C_2} = \lambda x \in C_1 . \bot_{C_2}$.

A cpo C is *discrete* if for $x_1, x_2 \in C$, $x_1 \sqsubseteq x_2$ iff $x_1 = \bot_C$ or $x_1 = x_2$.

3.2. Least fixed points

If C is a cpo and $f: C \rightarrow_m C$ then the least fixed point of f, μf, may exist. If so, it is given by either of the formulas

$$\mu f = \bigsqcap \{x \mid f(x) = x\}$$

or

$$\mu f = \bigsqcap \{x \mid f(x) \sqsubseteq x\}.$$

The existence of μf is guaranteed by *either* of the following conditions:

(1) f is continuous,

(2) C is a complete lattice (Knaster-Tarski).

In the former case, μf is also given by the formula

$$\mu f = \bigsqcup_{i=0}^{\infty} f^i (\bot_C)$$

where $f^0(\bot_C) = \bot_C$ and $f^{i+1}(\bot_C) = f(f^i(\bot_C))$.

Two useful properties of the least fixed point (for monotonic f), to which we will refer later, are:

fpp ("fixed point property"): $f(\mu f) = \mu f$

lfp ("least fixed point"): $f(x) \sqsubseteq x \Rightarrow \mu f \sqsubseteq x$.

3.3. Some specific cpo's

Let V_0 be the set of integers, and let $(\delta\in)W_0 = \{tt,ff\}$ be the set of truth-values. W_0 is a complete lattice, if we define $\bot_{W_0} = ff$. Let $(\alpha\in)V \overset{df}{=} V_0\cup\{\bot_V\}$ and $(\beta\in)W \overset{df}{=} W_0\cup\{\bot_W\}$. V and W are considered as discrete cpo's. (Note that $\bot_W \neq \bot_{W_0}$!)

For x_1,x_2 in a cpo C, let *if* β *then* x_1 *else* x_2 *fi* $\overset{df}{=} \langle\bot_C$ if $\beta = \bot_W$, x_1 if $\beta = tt$, x_2 if $\beta = ff\rangle$.

Let $(\sigma\in)\Sigma \overset{df}{=} (Intv \to V_0) \cup \{\bot_\Sigma\}$ be the set of *states*. Again, this is a discrete cpo. We will abbreviate $\bot_\Sigma$ to $\bot$. Let $T \overset{df}{=} \{\tau \subset\Sigma \mid \tau$ is finite or $\bot\in\tau\}$. T is a cpo, where we define (Egli-Milner) $\tau_1 \sqsubseteq \tau_2$ iff $\langle\bot\in \tau_1$ and $\tau_1\backslash\{\bot\} \subset \tau_2$, or $\tau_1 = \tau_2\rangle$, and $\bot_T \overset{df}{=} \{\bot\}$.

Let $(\phi\in)M \overset{df}{=} \Sigma \to T$, and $(\pi\in)\Pi \overset{df}{=} \Sigma \to_s W_0$. M is the set of (nondeterministic) *state transformations*, and Π is the set of *predicates* on Σ. Note that Π is a complete lattice (since W is). Let $(\gamma\in)\Gamma \overset{df}{=} (Stmv{\to}M) \cup (Cndv{\to}\Pi)$.

Variants of states etc.: We define $\sigma\{\alpha/x\}$ to be the state σ' such that $\sigma' = \bot$ if $\sigma = \bot$, and otherwise $\sigma'(y) = \langle\sigma(y)$ if $y \not\equiv x$, α if $y \equiv x\rangle$. $\gamma\{\phi/X\}$ and $\gamma\{\pi/Z\}$ are defined similarly.

3.4. Composition of state transformations and predicates

3.4.1. We define:

a) $\phi_1 \circ \phi_2 \stackrel{df}{=} \lambda\sigma . \cup\{\phi_1(\sigma') \mid \sigma' \in \phi_2(\sigma)\}$,

b) $\pi \circ \phi \stackrel{df}{=} \lambda\sigma . \sqcap\{\pi(\sigma') \mid \sigma' \in \phi(\sigma)\}$, and

c) $\pi \square \phi \stackrel{df}{=} \lambda\sigma . (\sigma \neq \bot \wedge \sqcap\{\pi(\sigma') \mid \sigma' \in \phi(\sigma) \setminus \{\bot\}\})$.

The first "o" is used to define the meaning of $S_1;S_2$ (3.5c below), while the second "o" and "□" are used to define the meanings of S<p> and S{p} respectively (3.5d).

3.4.2. Remark

"o" (in both definitions) is monotonic and continuous in both arguments, while "□" is monotonic, but not continuous, in its first argument, and anti--continuous (and hen ce anti-monotonic) in its second.

3.5. Definitions

The functions V: *Iexp* $\to (\Sigma \to V)$, W: *Bexp* $\to (\Sigma \to W)$, M: *Stat* $\to (\Gamma \to M)$, T: *Cond* $\to (\Gamma \to \Pi)$, F: *Afor* $\to (\Gamma \to \Pi)$ are defined by:

a) $V(s)(\bot) = \bot_V$, and for $\sigma \neq \bot$, $V(x)(\sigma) = \sigma(x), \ldots, V(\textit{if } b \textit{ then } s_1 \textit{ else } s_2 \textit{ fi})(\sigma) = \textit{if } W(b)(\sigma) \textit{ then } V(s_1) \textit{ else } V(s_2)(\sigma) \textit{ fi}$

b) $W(b)(\bot) = \bot_W$, and, for $\sigma \neq \bot$, $W(\textit{true})(\sigma) = tt, \ldots, W(b_1 \supset b_2)(\sigma) = (W(b_1)(\sigma) \Rightarrow W(b_2)(\sigma))$

c) $M(\mathbf{x}{:=}\mathbf{s})(\gamma) = \lambda\sigma\cdot\{\sigma\{V(\mathbf{s})(\sigma)/\mathbf{x}\}\}$, $M(\mathbf{b})(\gamma) = \lambda\sigma\ \mathit{if}$ $W(\mathbf{b})(\sigma)\ \mathit{then}\ \{\sigma\}\ \mathit{else}\ \emptyset\ \mathit{fi}$, $M(S_1;S_2((\gamma) = M(S_2)(\gamma)\circ M(S_1)(\gamma)$, $M(S_1 \cup S_2)(\gamma) = M(S_1)(\gamma)\cup M(S_2)(\gamma)$, $M(X)(\gamma) = \gamma(X)$, $M(\mu X[S])(\gamma) = \mu[\lambda\phi\cdot M(S)(\gamma\{\phi/X\})]$.

d) $T(\mathit{true})(\gamma) = \lambda\sigma\cdot(\sigma\neq\bot),\ldots,T(\exists x[p])(\gamma) = \lambda\sigma\cdot\exists\alpha[T(p)(\gamma)(\sigma\{\alpha/x\})],T(S\{p\})(\gamma) = T(p)(\gamma)\square M(S)(\gamma)$, $T(S\langle p\rangle)(\gamma) = T(p)(\gamma)\circ M(S)(\gamma),T(Z)(\gamma) = \gamma(Z),T(\mu Z[p])(\gamma) = \mu[\lambda\pi\cdot T(p)(\gamma\{\pi/Z\})$.

e) $F(p)(\gamma) = T(p)(\gamma),F(S_1\sqsubseteq S_2(\gamma) = \lambda\sigma\cdot((\sigma\neq\bot)\wedge(M(S_1)(\gamma)(\sigma) \sqsubseteq M(S_2)(\gamma)(\sigma)))$, $F(f_1\wedge f_2)(\gamma) = F(f_1)(\gamma)\wedge F(f_2)(\gamma)$.

A formula $g \equiv f_1 \rightarrow f_2$ is called *valid* (denoted by $\models g$) if $\forall\gamma[\ \sigma\neq\bot\ [F(f_1)(\gamma)(\sigma)] = \forall\sigma\neq\bot[F(f_2)(\gamma)(\sigma)]]$, and an *inference* $\dfrac{g_1,\ldots,g_n}{g}$ is called *sound* if $\models g_1,\ldots, \models g_n$ implies $\models g$.

3.6. Remarks

3.6.1. $\Phi \overset{df}{=} \lambda\phi\cdot M(S)(\gamma\{\phi/X\})\in M\rightarrow_c M$, $\Psi \overset{df}{=} \lambda\pi\cdot T(p)(\gamma\{\pi/Z\}) \in\Pi\rightarrow_m\Pi$, hence the least fixed points $\mu\Phi$, $\mu\Psi$ do exist (cf. parts d) and e) of Definition 3.5).

3.6.2. $\models p \supset S\{q\}$ iff S is *partially correct* w.r.t. p,q (often written $\models \{p\}S\{q\}$). $\models p \supset S\langle q\rangle$ iff S is *totally correct* w.r.t. p,q (sometimes written $\models [p]S[q]$).

3.6.3. We have the familiar properties of $S\{q\}$: $= (S_1;S_2)\{q\} = S_1\{S_2\{q\}\}$, $\models S\{q_1\wedge q_2\} = \models S\{q_1\}\wedge S\{q_2\}$, $\models(S_1\cup S_2)\{q\} = S_1\{q\}\wedge S_2\{q\}$, etc, and similarly for $S\langle q\rangle$.

3.6.4. $\models S\langle \mathit{true}\rangle$ holds iff execution of S always terminates (i.e. $\bot \notin M(S)(\gamma)(\sigma)$ for all γ,σ).

3.6.5. Hence $\models S\langle p\rangle = S\langle true\rangle \wedge S\{p\}$.

3.6.6. $S\langle p\rangle$ is monotonic in both S and p, but $S\{p\}$ is anti-monotonic in S (i.e., $\models (S_1 \sqsubseteq S_2) \rightarrow (S_2\{p\} \supset S_1\{p\})$). (Cf. 3.4.2.)

3.6.7. Observe that $\models f_1 \rightarrow f_2$ is a stronger fact than soundness of $\frac{f_1}{f_2}$. The meaning of the former is of the form $\forall\gamma[1 \Rightarrow 2]$, of the latter $\forall\gamma[1] \Rightarrow \forall\gamma[2]$.

3.7. *Fixed point properties for statements and conditions*

We re-state the fixed point properties given above (in 3.2).

fpp $\models \mu X[S] = S[\mu X[S]/X]$

lfp $\models (S[S_1/X] \sqsubseteq S_1) \Rightarrow (\mu X[S] \sqsubseteq S_1)$,

and similarly for $\mu Z[p]$.

3.8. *Continuity and anti-continuity of conditions; Scott's induction rule*

3.8.1. We say that p is *continuous* in X, or *anti-continuous* in X, if $\lambda\phi\cdot\tau(p)(\gamma\{\phi/X\})$ $(\in M \rightarrow \Pi)$ is continuous or anti-continuous respectively.

3.8.2. *Examples.* If X does not occur free in p or q, then (by 3.4.2) $\{X\}p$ is anti-continuous in X, $\langle X\rangle p$ is continuous in X and (hence) $(\langle X\rangle p) \supset q$ is anti-continuous in X.

3.8.3. Below (in 4.3) we will use the following section of Scott's induction rule: The inference

$$\frac{p[\Omega/X],(p\wedge(X \sqsubseteq \mu X[S]))\to p[S/X]}{p[\mu X[S]/X]}$$

is sound, provided p is anti-continuous in X.

4. *TERMINATION*

In this section we study the construct S<*true*>. By remark 3.6.4, we have that the validity of S<*true*> amounts to termination of S (for all γ,σ). We are now interested in a *syntactic* decomposition of S<*true*>, determined by the structure of S. More specifically, we want to define a *condition* $\tilde{S}$ by induction on the complexity of S, such that

$$(*) \qquad \models \tilde{S} = S\langle true\rangle.$$

We will show how to define "~" by induction on the complexity of S, such that (*) is indeed satisfied. Now for $S \equiv X \in Stmv$, there is no possibility of syntactically reducing S, so we *extend* the class of conditions *Cond* with an additional clause $p::=\ldots|\tilde{X}$, and correspondingly extend the definition of T by: $T(\tilde{X})(\sigma) = (\bot\notin\gamma(X)(\sigma))$.

We first give the definition of $\tilde{S}$, and then an explanation of it. (A substitution of the form $p[q/\tilde{X}]$, occurring below, is defined in a natural way; e.g. $\tilde{Y}[q/\tilde{X}] = \langle q$ if $X \equiv Y$, $\tilde{Y}$ otherwise$\rangle$.)

4.1. Definition

a) $(x:=s)^{\sim} \equiv true$, $\tilde{b} \equiv true$

b) $(S_1;S_2)^{\sim} \equiv \tilde{S}_1 \wedge S_1\{\tilde{S}_2\}$, $(S_1 \cup S_2)^{\sim} \equiv \tilde{S}_1 \wedge \tilde{S}_2$

c) $\mu X[S]^{\sim} \equiv \mu Z[\tilde{S}[\mu X[S]/X][Z/\tilde{X}]]$, where Z is (for definiteness) the first condition variable.

Note. One can verify that, for all X and S, $\tilde{S}$ is syntactically monotonic in $\tilde{X}$, and hence clause c is well-formed (cf. 2.3.4).

4.2. Discussion of the above definition

We want to see that (*) holds for $\tilde{S}$ as defined above. This is given by theorem 4.3 below, but a few heuristic remarks on the definition should be helpful now.

Clauses a and b should be clear. (a) Since x:=s and b always terminate, (*) holds for these two types of S. (b) We show that (*) is preserved for these cases: $\models (S_1;S_2)\langle true\rangle = S_1\langle S_2\langle true\rangle\rangle =$ (ind.hyp) $S_1\langle\tilde{S}_2\rangle =$ (by 3.6.5) $S_1\langle true\rangle \wedge S_1\{\tilde{S}_2\} =$ (ind.hyp.) $\tilde{S}_1 \wedge S_1\{\tilde{S}_2\}$. Similarly for the case $S \equiv S_1 \cup S_2$.

Clause c deserves some explanation. We anticipate a result (step b) in the course of proving Theorem 4.3); viz., for each S and S_0,

$$(**) \qquad \models S[S_0/X]^{\sim} = \tilde{S}[S_0/X][\tilde{S}_0/\tilde{X}].$$

(A simpler guess for expressing $S[S_0/X]^{\sim}$ in terms of $\tilde{S}$ and S_0, namely $\models S[S_0/X]^{\sim} = \tilde{S}[\tilde{S}_0/\tilde{X}]$, can be seen to be false by considering e.g. the case $S \equiv X;S_1$ with $X \notin \mathit{stmv}(S_1)$.)

Now taking $S_0 \equiv \mu X[S]$ in (**), and applying *fpp* (3.7), we obtain

$$\models \underbrace{\mu X[S]^{\sim}} = \tilde{S}[\mu X[S]/X][\underbrace{\mu X[S]^{\sim}}/\tilde{X}].$$

Thus $\mu X[S]^{\sim}$ satisfies the above fixed point relationship, making plausible Definition 4.1c (which gives it as the *least* such fixed point).

4.3. THEOREM

$$\models \tilde{S} = S\langle \mathit{true}\rangle.$$

Proof. The proof is fairly involved, and only sketched here. ($i \in \{1,\ldots,n\}$, $n \geq 0$).

a) $S \cong S' \Rightarrow \tilde{S} \cong \tilde{S}'$. This is shown by simultaneously proving, by induction on the complexity of S, that

   (i) $\quad S \cong S' \Rightarrow \tilde{S} \cong \tilde{S}'$

   (ii) $\quad S[X'/X]^{\sim} \cong \tilde{S}[X'/X][\tilde{X}'/\tilde{X}]$

b) $S[S_i/X_i]_i^{\sim} \cong \tilde{S}[S_i/X_i]_i[\tilde{S}_i/\tilde{X}_i]_i$

   Induction on the complexity of S, using part a.

c) $\models \tilde{S}[S_i/X_i]_i[S_i\langle \mathit{true}\rangle/\tilde{X}_i]_i \supset S[S_i/X_i]_i\langle \mathit{true}\rangle$

   (Taking n = 0, we infer that $\models \tilde{S} \supset S\langle \mathit{true}\rangle$)

d) $\models (S'_i \sqsubseteq S''_i)_i \wedge (q'_i \supset q''_i)_i \rightarrow \tilde{S}[S'_i/X_i]_i[q'_i/\tilde{Y}_i]_i \supset \tilde{S}[S''_i/X_i]_i[q''_i/\tilde{Y}_i]_i$

   I.e., $\tilde{S} \equiv \tilde{S}(X,\tilde{Y})$ is monotonic in both X and $\tilde{Y}$. Proved by induction on the complexity of S. The case $S \equiv S_1;S_2$ is not obvious, since then $\tilde{S} \equiv \tilde{S}_1 \wedge S_1\{\tilde{S}_2\}$, and $S_1\{\tilde{S}_2\}$ is not monotonic in S_1 (cf. 3.6.6). But here we use the equivalence $\models \tilde{S}_1 \wedge S_1\{\tilde{S}_2\} = \tilde{S}_1 \wedge S_1\langle\tilde{S}_2\rangle$, (from part c, with n = 0), and note that $S_1<\tilde{S}_2>$ *is* monotonic in S_1.

e) $\models S<true> \supset \tilde{S}$. Induction on the complexity of S. If $S \equiv \mu X[S_0]$, apply Scott's induction rule (3.8.3) with $p \equiv (X<true>) \supset \mu X[S_0]$ (cf. 3.8.2), using the induction hypothesis and parts c), d).

5. *DERIVATIVES*

We will define the upper and lower derivatives of a statement S, and state a fundamental theorem connecting these two notions. Before giving the exact definitions, we make some introductory remarks.

The *upper derivative* of S w.r.t. X, written $\frac{dS}{dX}$, is an element of *Stat*, and has the following intended meaning: Dropping the γ-arguments for simplicity, we have that $\sigma' \in M(\frac{dS}{dX})(\sigma)$ iff execution of S for input state σ leads to σ' as an intermediate state just before execution of X starts. E.G., if $S \equiv S_1;X;S_2;X;S_3 \cup S_4$, $X \notin stmv(S_i)$, i = 1,...,4, then $\frac{dS}{dX} \equiv S_1 \cup S_1;X_1;S_2$. For statements without recursion, we may also briefly say that $\frac{dS}{dX}$ is the union of all prefixes of X in S.

Let $\mathcal{X} \subset Stmv$. The *lower derivative* of S w.r.t. $\mathcal{X}$, written $\delta_{\mathcal{X}}(S)$, is an element of *Cond*, and has the intended meaning: $\delta_{\mathcal{X}}(S)$ is true in a state δ whenever S terminates in σ *provided* that, for each $X \in \mathcal{X}$, execution of X for all states σ' in $M(\frac{dS}{dX})(\sigma)$ terminates. (Hence, $\delta_{\emptyset}(S) \equiv \tilde{S}$.)

(This is essentially the idea as introduced in [H,P] for statements without inner μ-terms. The novelty of our definition lies in clauses c of Definitions 5.1 and 5.3.).

Combining the two intended meanings of $\frac{dS}{dX}$ and $\delta_X(S)$, we except that the following result holds: For each $X \notin \mathcal{X}$,

$$\models \delta_{\mathcal{X}}(S) = \frac{dS}{dX}\{\tilde{X}\} \wedge \delta_{\mathcal{X}\cup\{X\}}(S).$$

Let us give the verbal transliteration of this for the case that $\mathcal{X} = \emptyset$: S terminates in σ iff both (i) and (ii) are satisfied:

(i) Execution of X terminates for all $\sigma'(\neq\bot)$ in $M(\frac{dS}{dX})(\sigma)$,

(ii) S terminates in σ *provided* execution of X for all $\sigma'(\neq\bot)$ in $M(\frac{dS}{dX})(\sigma)$ terminates.

(Note that a more naive equivalence: $\models \tilde{S} = \tilde{X} \wedge \delta_{\{X\}}(S)$ would not work, since termination of X is required for the wrong states.)

5.1. *Definition* (upper derivative)

a) $\frac{dx:=s}{dX} \equiv \mathit{false}$, $\frac{db}{dX} \equiv \mathit{false}$, $\frac{dY}{dX} \equiv \begin{cases} \mathit{true}, & \text{if } X \equiv Y \\ \mathit{false}, & \text{if } X \not\equiv Y \end{cases}$

b) $\frac{dS_1;S_2}{dX} \equiv \frac{dS_1}{dX} \cup S_1; \frac{dS_2}{dX}$, $\frac{d(S_1 \cup S_2)}{dX} \equiv \frac{dS_1}{dX} \cup \frac{dS_2}{dX}$

c) $\frac{d\mu Y[S]}{dX} \equiv \begin{cases} \mathit{false}, \text{ if } X \equiv Y \\ \mu X_1[(\frac{dS}{dX} \cup \frac{dS}{dY}; X_1)[\mu Y[S]/Y]], \text{ if } X \not\equiv Y, \\ \text{where } X_1 \text{ is the first statement variable} \\ \notin \mathit{stmv}(X,Y,S). \end{cases}$

5.2. Remarks

5.2.1. By way of comment to clause 5.1 c) we offer the following: We expect that

$$(*): \quad \models \frac{dS_1[S_2/Y]}{dX} = \frac{dS_1}{dX}[S_2/Y] \cup \frac{dS_1}{dX}[S_2/Y]; \frac{dS_2}{dX}.$$

In words (first forgetting about the substitutions on the right-hand side): Prefixes of X in $S_1[S_2/Y]$ are obtained either as prefixes of X in S_1, or by composing prefixes of Y in S_1 on the right with prefixes of X in S_2. Supplementing this description with the indicated substitutions then explains the plausibility of (*). Taking $S_1 \equiv S$, $S_2 \equiv \mu Y[S]$, and applying *fpp*, we obtain as property of

$$\frac{d\mu Y[S]}{dX} : \models \frac{d\mu Y[S]}{dX} = \frac{dS}{dX}[\mu Y[S]/Y] \cup \frac{dS}{dY}[\mu Y[S]/Y]; \frac{d\mu Y[S]}{dX}.$$

We see that $\frac{d\mu Y[X]}{dX}$ satisfies a fixed point relationship, and, since our fixed points are usually *least*, one may now understand clause 5.1c.

5.2.2. If $Y \notin stmv(S)$ then $\models \frac{dS}{dX} = false$.

5.3 Definition (lower deriviative)

a) $\delta_X(x{:=}s) \equiv true$, $\delta_X(b) \equiv true$, $\delta_X(X) \equiv \begin{cases} true, & X \in X \\ \tilde{X}, & X \notin X \end{cases}$

b) $\delta_X(S_1;S_2) \equiv \delta_X(S_1) \wedge S_1\{\delta_X(S_2)\}$, $\delta_X(S_1 \cup S_2) \equiv \delta_X(S_1) \wedge$
$\wedge\ \delta_X(S_2)$

c) $\delta_X(\mu X[S]) \equiv \mu Z[\delta_{X\setminus\{X\}}(S)[\mu X[S]/X][Z/\tilde{X}]]$, where Z is the first condition variable.

5.4. *Remarks*

5.4.1. The definitions of $\delta_\emptyset(s)$ and $\tilde{S}$ (4.1) coincide.

5.4.2. $\delta_X(S_1;S_2;\ldots;S_n) \equiv \delta_X(S_1)\wedge S_1\{\delta_X(S_2)\} \wedge S_1;$
$S_2\ \{\delta_X(S_3)\}\wedge \ldots \wedge S_1;S_2;\ldots;S_{n-1}\{\delta_X(S_n)\}$.

5.4.3. $X \notin$ *stmv*(S) => $\delta_X(S) \equiv \delta_{X\setminus\{X\}}(S)$.

5.4.4. $\tilde{X}$ free in $\delta_X(S)$ => $X \in stmv(S)\setminus X$.

5.5. *THEOREM*

For $X \notin X$, $\models \delta_X(S) = \frac{dS}{dX}\{\tilde{X}\} \wedge \delta_{X\cup\{X\}}(S)$.

Proof. Induction on the complexity of S. The only interesting case is that $S \equiv \mu Y[S_0]$, $Y \not\equiv X$. We have to show that

$$\mu Z[\delta_{X\setminus\{Y\}}(S_0)[S/Y][Z/\tilde{Y}]]$$

$\models =$

$$\mu X_1[(\frac{dS_0}{dX} \cup \frac{dS_0}{dY}\ ;\ X_1)[S/Y]]\{\tilde{X}\} \wedge \mu Z[\delta_{X\cup\{X\}\setminus\{Y\}}(S_0)[S/Y][Z/\tilde{Y}]].$$

The proof - omitted here - involves fairly complicated manipulations in the μ-calculus, using *fpp* and *lfp* and properties of S{q}(cf. 3.6.3).

5.6. *Corollary*

$$\textit{For } X \notin X, \quad = \delta_X(S) = \frac{dS}{dX} \;<X> \wedge \; \delta_{X\cup\{X\}}(S).$$

Proof. It appears that, in the proof of Theorem 5.5, {p} may be replaced everywhere by <p>.

6. *DERIVATIVES AND TERMINATION*

We express termination of a recursive procedure $\mu X[S]$ in terms of the so-called *well-foundedness* of a function with respect to a predicate (involving $\frac{dS}{dX}$ and $\delta_{\{X\}}(S)$, respectively.)

6.1. *Definition*

ϕ is called well-founded in σ w.r.t. π if

i) There exists no infinite sequence $\sigma_0 = \sigma, \sigma_1, \ldots,$ such that $\sigma_{i+1} \in \phi(\sigma_i)$, $i = 0,1,\ldots$

ii) There exists no finite sequence $\sigma_0 = \sigma, \sigma \;, \ldots, \sigma_k$ such that $\sigma_{i+1} \in \phi(\sigma_i)$, $i = 0,\ldots,k$, $\sigma_k \neq \perp$, and $\pi(\sigma_k) = ff$.

6.2. *Remarks*

6.2.1. By strictness, ϕ is not well-founded in $\perp$ w.r.t. any π.

6.2.2. If, for each $\sigma' \in \phi(\sigma)$ ϕ is well-founded in σ' w.r.t. π, and moreover, $\pi(\sigma) = tt$, then ϕ is well--founded in σ w.r.t. π.

6.3. Lemma

For each ϕ, σ, π

a) $\mu[\lambda\pi'\cdot((\pi' \circ \phi) \wedge \pi)](\sigma) = tt \Rightarrow \phi$ *is well-founded in* σ *w.r.t.* π

b) ϕ *is well-founded in* σ *w.r.t.* $\pi \Rightarrow \mu[\lambda\pi'\cdot((\pi' \circ \phi) \wedge \pi)](\sigma) = tt.$

Proof. a) Let $\pi_1 \overset{df}{=} \mu[\lambda\pi'\cdot((\pi' \circ \phi') \wedge \pi)]$, and let $\pi_{\phi,\pi}$ denote the predicate which, for each σ, expresses that ϕ is well-founded in σ w.r.t. π. We show that $\pi_1 \sqsubseteq \pi_{\phi,\pi}$, or, by *lfp*, that $(\pi_{\phi,\pi} \circ \phi) \wedge \pi \sqsubseteq \pi_{\phi,\pi}$. Now this is immediate by 6.2.2.

b) Let $\pi_2 \overset{df}{=} \mu[\lambda\pi'\cdot((\pi' \circ \phi) \wedge \pi)]$, Assume that ϕ is well-founded in σ w.r.t. π, but $\pi_2(\sigma) = ff$. Clearly, $\sigma \neq \bot$. By *fpp*, then $((\pi_2 \circ \phi) \wedge \pi)(\sigma) = ff$. Thus, either $\pi(\sigma) = ff$, contradicting Definition 6.1 (ii), or there exists $\sigma' \in \phi(\sigma)$, $\sigma' \neq \bot$, such that $\pi_2(\sigma') = ff$. Thus, again by *fpp*, either $\pi(\sigma') = ff$, contradicting 6.1 (ii), or we obtain $\sigma'' \neq \bot$ such that $\sigma'' \in \phi(\sigma')$ and $\pi_2(\sigma'') = ff$. Repeating the argument, either we find a finite sequence $\sigma_0 = \sigma, \ldots, \sigma_k$ $(k \geq 0)$ such that $\sigma_{i+1} \in \phi(\sigma_i), i = 0, \ldots, k-1$, $\sigma_k \neq \bot$, and $\pi(\sigma_k) = ff$, or we obtain an infinite sequence $\sigma_0 = \sigma, \sigma_1, \sigma_2, \ldots,$ such that $\sigma_{i+1} \in \phi(\sigma_i)$, $i = 0, 1, \ldots$. In both cases, we have found a contradiction.

6.4. Definition

S is called well-founded w.r.t. p if for all γ, σ, $M(s)(\gamma)$ is well-founded in σ w.r.t. $T(p)(\gamma)$.

6.5. *Corollary*

a) $\models \; \mu Z[S\langle Z\rangle \wedge p] \Rightarrow S$ *is well-founded w.r.t.* p

b) S *is well-founded w.r.t.* $p \Rightarrow\; \models \mu Z[S\{Z\} \wedge p]$.

6.6. *Definition*

$$\overset{\circ}{S} \equiv \left(\frac{dS}{dX}\right)[\mu X[S]/X],$$

$$\underset{\circ}{S} \equiv \delta_{\{X\}}(S)[\mu X[S]/X].$$

We now come to main theorem of the paper (an intuitive explanation of which is given afterwards).

6.7. *Theorem*

The following two facts are equivalent:

a) $\models \mu X[S]\langle \mathit{true}\rangle$

b) $\overset{\circ}{S}$ *is well-founded w.r.t.* $\underset{\circ}{S}$.

Proof. We have successively:

a) $\models \tilde{S} = \frac{dS}{dX}\{\tilde{X}\} \wedge \delta_{\{X\}}(S)$ (by 5.5 and 5.4.1)

b) $\models \tilde{S}[\mu X[S]/X] = \overset{\circ}{S}\{\tilde{X}\} \wedge \underset{\circ}{S}$ (subst. $\mu X[S]$ for X)

c) $\models \tilde{S}[\mu X[S]/X][Z/\tilde{X}] = \overset{\circ}{S}\{Z\} \wedge \underset{\circ}{S}$ (subst. Z for $\tilde{X}$)

d) $\models \mu Z[\tilde{S}[\mu X[S]/X][Z/\tilde{X}]] = \mu Z[\overset{\circ}{S}\{Z\} \wedge \underset{\circ}{S}]$ (prefixing μZ]

e) $\models \mu X[S]\langle \mathit{true}\rangle = \mu Z[\overset{\circ}{S}\{Z\} \wedge \underset{\circ}{S}]$ (4.1, 4.3)

f) $\models \mu X[S]\langle \mathit{true}\rangle = \mu Z[\overset{\circ}{S}\langle Z\rangle \wedge \underset{\circ}{S}]$ (as a-e, starting from 5.6).

Note: in c, we use that $\tilde{X}$ is not free in $\delta_{\{X\}}(S)$ by 5.4.4, hence also not in $\underset{o}{S}$.)

The theorem now follows from e), f) and Corollary 6.5.

6.8. Discussion

We have derived the following result: A recursive procedure $\mu X[S]$ terminates for all input states $\neq \bot$ iff $\overset{o}{S}$ is well-founded w.r.t. $\underset{o}{S}$. How should one understand this proposition? Let us consider e.g. the procedure $\mu \overset{df}{=} \mu X[S]$, where $S \equiv S_1;X;S_2;X;S_3 \cup S_4$, with $X \notin \mathit{stmv}(S_i)$, $i = 1,\dots,4$. Then $\models \overset{o}{S} = S_1 \cup S_1;\mu;S_2$ (using 5.2.2). Also $\models \delta_{\{X\}}(S) = \tilde{S}_1 \wedge S_1;\ X\ \{\tilde{S}_2\} \wedge S_1;X;S_2;X\ \{\tilde{S}_3\} \wedge \tilde{S}_4$ (using 5.4.2, 5.4.3, 5.4.1), and so

$$\models \underset{o}{S} = \tilde{S}_1 \wedge S_1;\mu\{\ddot{S}_2\} \wedge S_1;\mu;S_2;\mu\{\tilde{S}_3\} \wedge \tilde{S}_4.$$

Forgetting about the γ-arguments, we have that for all σ:

a) There exists no infinite sequence $\sigma_0 = \sigma,\sigma_1,\dots,$ such that $\sigma_{i+1} \in M(S_1 \cup S_1;\mu;S_2)(\sigma_i)$, $i = 0,1,\dots$. Since $\overset{o}{S}$ is nothing but the statement executed between a call of μ at a certain level of recursion depth, and a call at the next deeper level, we see that the non-existence of such an infinite sequence amounts to the absence of infinite recursion, i.e., it is not possible that the procedure goes on calling itself indefinitely.

b) There exists no finite sequence $\sigma_0 = \sigma,\dots,\sigma_k$, such that $\sigma_{i+1} \in M(\overset{o}{S})(\sigma_i)$, $i = 0,\dots,k-1$, $\sigma_k \neq \bot$, and $T(\underset{o}{S})(\sigma_k) = ff$. Assume that, contrariwise, such a sequence would exist. This would mean that, at a certain level of recursion depth, we have obtained an intermediate state $\sigma_k \neq \bot$ such that $T(\underset{o}{S})(\sigma_k) = ff$. By the definition of $\underset{o}{S}$ this means that either

(i) S_1 does not terminate in σ_k, or

(ii) There exists some $\sigma' \neq \perp$ such that $\sigma' \in M(S_1;\mu)(\sigma_k)$ and S_2 does not terminate in σ', or

(iii) There exists some $\sigma'' \neq \perp$ such that $\sigma'' \in M(S_1;\mu;S_2;\mu)(\sigma_k)$ and S_3 does not terminate in σ'', or

(iv) S_4 does not terminate in σ_k.

Altogether, we see that $\underset{\circ}{S}$ is false in $\sigma_k \neq \perp$ precisely when there is some instance of *local* nontermination stemming from σ_k, i.e., nontermination which is not due to infinite recursion of μ, but to nontermination of one of the S_i-components of μ.

Combining results a) and b), we see that $\mu X[S]$ terminates everywhere whenever, for all σ, there is neither the possibility of infinite recursion (global nontermination), nor the possibility of the computation reaching some intermediate state which leads to local nontermination.

References

[dB1] De Bakker, J.W., Semantics and termination of nondeterministic recursive programs in Michaelson, S., Milner, R. (eds.), *Proceedings of 3rd Coll. Automata, Language and Programming*, Edinburgh University Press, 1976, pp. 435-477.

[dB2] De Bakker, J.W., *Mathematical theory of program correctness*, Change Prentice-Hall, Englewood Cliffs, N.J., 1980.

[D] Dijkstra, E.W., *A Discipline of Programming*, Prentice-Hall, Englewood Cliffs, N.J., 1976.

[H,P] Hitchcock, P., Park, D.M.R.: Induction rules and proofs of termination in Nivat, M. (ed.) *Automata, Languages and Programming* North Holland Amsterdam, 1973, pp. 225-251.

J.W. de Bakker
Mathematisch Centrum,
2e Boerhaavestraat 49,
Amsterdam

J.I. Zucker
Dept. of Mathematics and Computer Science,
Bar Ilan University, Ramat Gan, Israel

COLLOQUIA MATHEMATICA SOCIETATIS JÁNOS BOLYAI
26. Mathematical Logic in Computer Science
Salgótarján (HUNGARY), 1978.

*AN ALGORITHMIC MODEL OF STRICT FINITISM**

E. Engeler

Strict finitism does not yet exist as a clearly delineated mathematical theory, all we have to go on are various formulations of its basic tenets[1], which are summarized in section 1 below. In a fashion similar to that of Finitism and Intuitionism it has its roots in an attitude towards the problem of securing the foundations of mathematics. One of the tasks of the logician is to capture such conceptually presented foundational aproaches into a formal logico - mathematical system - as nearly true to the original conceptions as this is possible.

The goal of the present paper is to present, motivate, and discuss a technical Ansatz for a system of strict finitism. We do this with the purpose of exhibiting some of the difficulties in sustaining the original basic

---

* The manuscript of this paper, written in 1971, was lost until recently. We were encouraged to publish it now, because a number of papers have related positions and results have been put forward, and our paper may contribute to this discussion.

[1] For example in Bernays [1,pp.280-281], Wang [5,pp.473-476].

attitudes against the ciriticism that arises once these attitudes are made precise[2].

We do not wish to discuss here in any detail the recent papers[3], which arose from a similar questioning of the role of the arbitrarily large finite in the foundations and in proof theory. As far as we are able to determine, the present approach is the most radically restrictive among these.

In any case, the outcome of our present work is that strict finitism can do no more than produce a mathematical system which may (or may not) be interesting in itself[4]. It is not, in our opinion, a reasonably tenable position in foundation. See the arguments in the last section.

§1 The standpoint

In strict finitism we envision a radicalization of the constructionist viewpoint. This radicalization is motivated by the observation that many of the "constructions" allowed by the constructionists are constructions only in the sense of being potentially executable, indeed, executable only in an idealized world of infinitely patient and gigantic machines. How convincing is such a world for a firm foundation of mathematics? Would it

---

[2] The author is indebted to Bernays, Kreisel, G.H.Müller and others for discussing with him the basic issues involved.

[3] We mention in particular the work of Yessenin-Volpin [7],[8], Geiser [2], Goguen [3], Parikh [4], and Williamsen [6]; see also footnotnes[7] and [8] .

[4] For example, they may be useful for providing a foundation for a strict finitist theory of computation.

not be more realistic to base this foundation on a realm of actually feasible processes? Strict finitism, then, is conceived as treating of the ideas that variously restricted finite beings develop about the concrete mathematical structure which they consider. It thus takes the form of a metatheory, in a classical framework, whose object-theory (or theories) are the sets of ideas of restricted beings referred to above.

Let us therefore consider beings, mathematicians as it were, who try to gain knowledge about a basic mathematical structure, say set theory. The way to gain such knowledge is to perform thought experiments. And, indispensable with scientific experiments, these should be reproducible, hence governed by fixed and communicable programs. Strictly finitist mathematicians operate under restrictions which we could formulate conceptually as follows.

(a) The sets that are considered, i.e. that enter the experiment, are in reality always finite and so are their elements and elements of elements, etc.

(b) Each individual mathematician thinks only during a restricted period of time, and has only restricted imagination.

Our goal now is to construct a series of increasingly intelligent and patient (models of) mathematicians, to investigate what each one's ideas would be about set theory, and on what properties of sets these mathematicians are able to come to a consensus. This consensus is what we call strict finitist truth. Our hope is that the model is realistic enough so that this consensus has a large overlap with classical set theory. For example, our model should explain why, and in what fashion, finite minds can perceive infinite totalities.

§2 *A technical realization of the standpoint*

Let F be the set of *hereditarily finite sets*, i.e. let $F = \bigcup_{i<\omega} R(i)$ where $R(0) = \emptyset$, $R(i+1) =$ power set $R(i)$.

We envision experiments within the relational structure $\mathrm{F} = \langle F, \in \rangle$, conducted according to *programs* that are written in a fixed programming language similar to ALGOL. The exact details of the structure of this language are unimportant here; we shall mostly present programs in form of flow-diagrams (which are self-explanatory).

By a *complexity measure* on programs we understand a function $\mu : \Pi \to N$ from the set Π of all programs to the set N of natural numbers with the property that each set $C_i = \{\pi \in \Pi : \mu(\pi) \leq i\}$ is finite.

A program $\gamma(x)$, containing the variable x as indicated, is called a *generating program* (at x) if there exists a sequence $a_0, a_1, a_2, \ldots$ of sets such that $a_0 = \emptyset$, $F = \{a_0, a_1, a_2, \ldots\}$ and such that $x = a_{i+1}$ is the output of the program for the input $x = a_i$; and $x = a_0$ is the output for input $x = \emptyset$.

For each positive integer i and program π let π^i denote the modified program which arises when each loop in π is allowed to be run through at most i times. The program π^i has one additional exit which is taken if any one of the loops is about to be entered an $i+1$'st time.

The bounding of loops is one of the two main devices by which we implement the inherent restriction of finite minds in our model. Let, for example, the formula $\varphi(x)$ of first-order predicate calculus express some properties of sets. The i-th mathematician will accept $\varphi(x)$ as true for an assignment $x := a$ if the thought-experiment which he associates with φ is successful for input $x := a$. When will he accept $(\forall x)\varphi(x)$?

Assume for the moment that we already know how to associate to the formula φ and any positive integer i such a thought experiment. That is, assume we are given a program 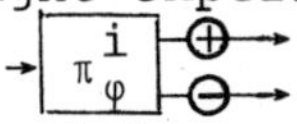 with two exits, and assume that this program takes exit $\oplus$ on input $x := a$ exactly if $\varphi[a]$ holds, (otherwise it terminates in exit $\ominus$).

Consider now the program

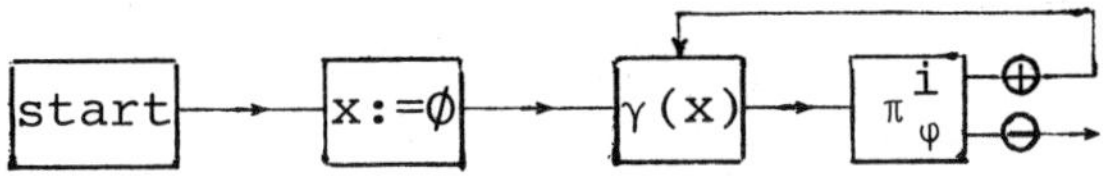

Since $\gamma(x)$ is a generator at x this program does *not* terminate exactly when all sets are such that the i-th mathematician would accept $\varphi[a]$ as true once he gets to test a . However, the i-th mathematician does not have this patience and is willing to accept $(\forall x)\varphi(x)$ already after a limited amount of experimentation. We realize this restriction by letting him accept $(\forall x)\varphi(x)$ iff the i-bounded program

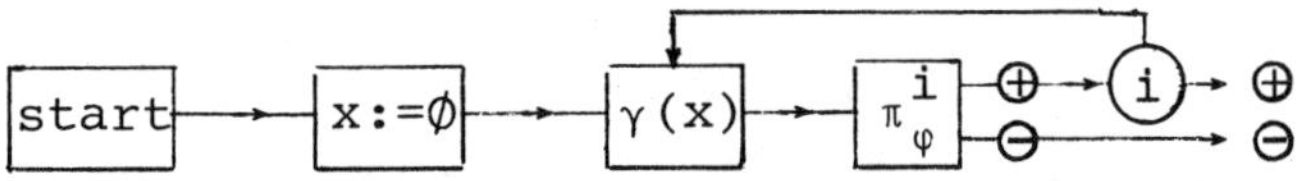

takes exit $\oplus$.

A second device is needed if we wish to implement the restrictions that adhere to strictly finitist statements of existence. Assume as above that we have associated π_φ^i to $\varphi(x)$. When will the i-th mathematician accept $(\exists x)\varphi(x)$? Obviously only if he can think of a program which constructs a set a satisfying φ . Now, the imagination of the i-th mathematician is bounded: he can think up only programs of complexity $\leq i$. For each such construction program $\sigma(x)$ consider the composite program

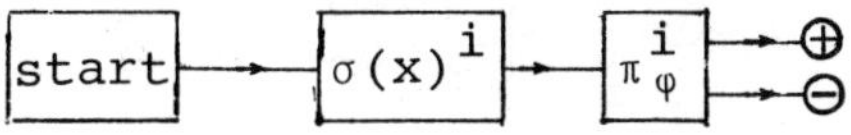

This program terminates in ⊕ exactly if the program $\sigma(x)$ indeed constructs an element such that the i-th mathematician accepts that it satisfies φ . There are only finitely many programs σ of complexity $\leq i$. By combining the above composite programs for each such σ , it is therefore easy to construct a single program 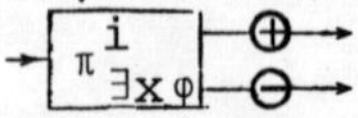 such that the i-th mathematician accepts $(\exists x)\varphi(x)$ as true iff this program terminates in ⊕ .

We still owe the description of the passage between arbitrary first-order formulas $\varphi(x,y,...)$ of set theory and programs 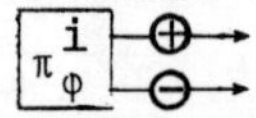 . This procedure is defined recursively as follows:

(1) To the formula $x \in y$ we associate the program

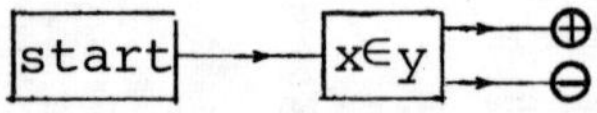

consisting of one conditional instruction

1: *if* $x \in y$ *then go to 2 else go to 3.*

(2) Suppose that there are i-bounded programs 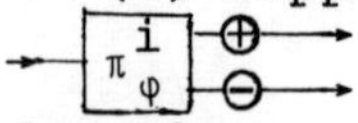 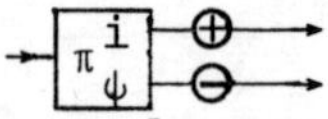 already associated to the formulas φ, ψ and assume that these programs have the following property: If φ has free variables $x_1,...,x_n$ then the values of $x_1,...,x_n$ at termination of π^i_φ are the same as at the start. The programs for $\varphi \wedge x$, $\neg\varphi$, $\varphi \vee x$, $\varphi \rightarrow x$ are found as follows:

$\pi^i_{\varphi\wedge x}$ is 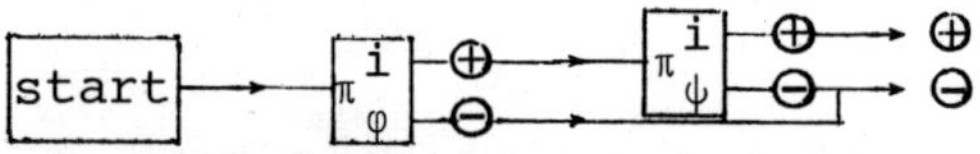

$\pi^i_{\neg\varphi}$ is 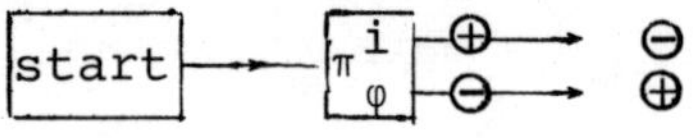

$\pi^i_{\varphi\vee\psi}$ is 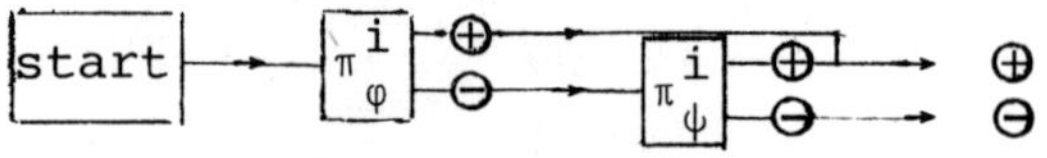

$\pi^i_{\varphi\rightarrow\psi}$ is 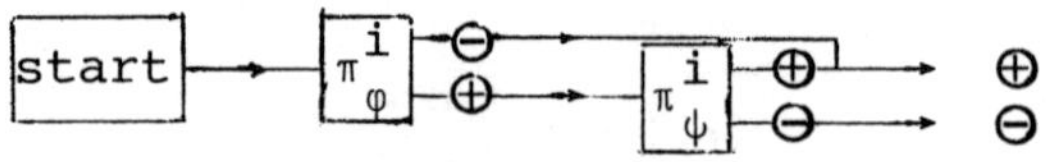

(3) For the formulas $(\forall x)\varphi$ and $(\exists x)\varphi$ we have already described the passage to the corresoponding programs.

The set of *strict finitistically true sentences* can now be defined as the set of all sentences φ such that for all sufficiently large i the i-th mathematician accepts φ. Clearly, the set does not depend on a particular complexity measure that we may have chosen.

Let us now investigate whether some of the familiar statements of set theory are strict finitistically true. First we consider a typical existential axiom, the axiom of pairsets:

$\forall a \forall b \exists x \forall y \, (y \in x \leftrightarrow (y = a \vee y = b))$

Let φ be the quantifier-free part of the above axiom and let π_φ^i be the corresponding verification program of the i-th mathematician. Then

$$\boxed{\text{start}} \to \boxed{\gamma(a)} \to \boxed{\gamma(b)} \to \boxed{x:=\{a,b\}} \to \boxed{\gamma(y)} \to \boxed{\pi_\varphi^i} \begin{array}{c}\oplus\to\textcircled{i}\to\textcircled{i}\to\textcircled{i}\to\oplus\\ \ominus\to\ominus\end{array}$$

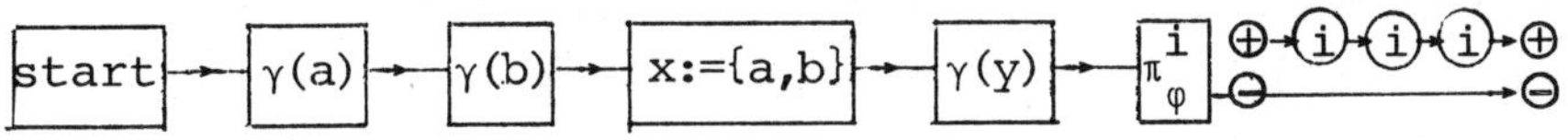

is the program with which the i-th mathematician verifies the pair-set axiom whenever the complexity measure of the program $x := \{a,b\}$ is $\leq i$.

In the case of the *pairset axiom*, and similarly for the powerset and union axioms, the strict finitistical-truth of the statement hinges essentially on the fact that corresponding operations belong to the basic capabilities (i.e. basic instructions) of the programming language. Since we have not yet made a list of these, it is about time to do so:

Operative capabilities:
$x := \{y,z\}, \quad x := P(y), \quad x := \cup y, \quad x := \emptyset, \quad x := y$

Conditional capabilities:
$x \in y, \quad x = y$

For the *axiom of infinity*:

$\exists x(\emptyset \in x \wedge \forall y(y \in x \rightarrow y \cup \{y\} \in x))$,

We need a program that would do something like $x: = \omega$ (which is, of course, not included in the basic capabilities). Consider, then, the following program "$(x:=\omega)^i$":

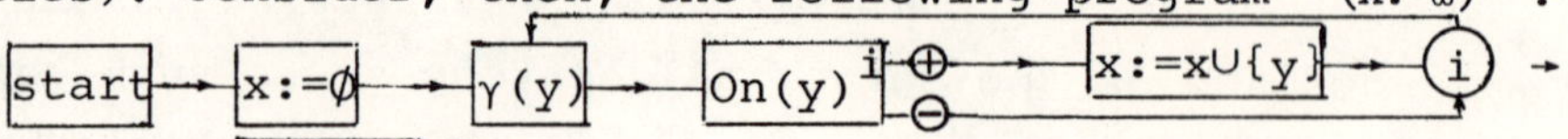

where 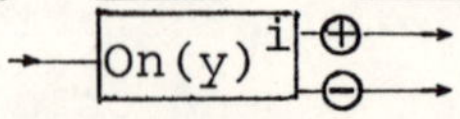

is the obvious program whereby the i-th mathematician checks whether y is an ordinal or not. Let us abbreviate the quantifier-free part of the axiom by $\chi(x)$ and let π^i_χ be the verifying program for $\chi(x)$ then

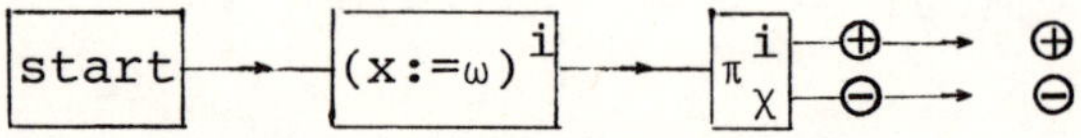

will clearly always terminate in exit $\oplus$, thereby verifying the axiom in question for all mathematicians i for whom $i \geq$ complexity of $(x:=\omega)$.

The verification of the remaining axioms of Zermelo-Fraenkel proceeds along the same lines. But before we investigate what this result means, let us discuss quite generally the properties of the set T of sentences which are strict finitistically true.

A first, obvious, observation is that this set T does not contain φ and $\neg\varphi$ simultaneously for any sentence φ . But T is neither consistent nor complete in the classical sense. Namely, the formula

$$(\exists x)(\emptyset \in x \wedge (\forall y)(y \in x \rightarrow \{y\} \in x)) \wedge (\forall y) \neg (\emptyset \in x \wedge (\forall y) \wedge (y \in x \rightarrow \{y\} \in x))$$

is in T , but is - classically - of the form $\varphi \wedge \neg\varphi$, a contradiction. Incompleteness can be illustrated by the fact that neither the statement that ω is even nor its negation is in T[5].

In order to facilitate the following discussion, let us

[5] This was pointed out to the author by the audience when the present paper was read before the colloquium at Heidelberg (1971). Other examples are easy to be found.

assume that our programming language admits also variables that range over natural numbers, together with some simple capabilities with respect to computations on natural numbers (which we shall introduce as the need arises). Let there be given a Gödel-numbering of sentences, φ_p denoting the sentence with Gödel number p. We have not found a truth-definition of T in T in the sense of Tarski. This would be a formula $\sigma(x)$ such that $(\sigma(p)-\varphi_p)\in T$ for all p. However there is a quasi truth-definition, i.e. there exists a formula $\tau(x)$ such that $\tau(p)\in T$ iff $\varphi_p \in T$.

This can be seen as follows. For each natural number j let $U(j) = \{a : a = \gamma\gamma\ldots(\emptyset)$ for some $i<j\}$, and assume that the operation ${}^{i}x: = U(j)$ belongs to the basic capabilities of the programming language. Furthermore, for each natural numbers p, k,m let $S(p,k,m,) = \{\langle a_1,\ldots,a_m\rangle \in U(k)^m$: 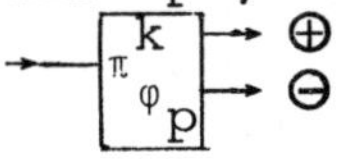 terminates in $+$ on input $\langle a_1,\ldots,a_m\rangle\}$.
Assume again that the operation $y: = S(p,k,m,)$ is among the basic capabilities of the programming language. Now, let $\pi_\tau(z)^i$ be the program

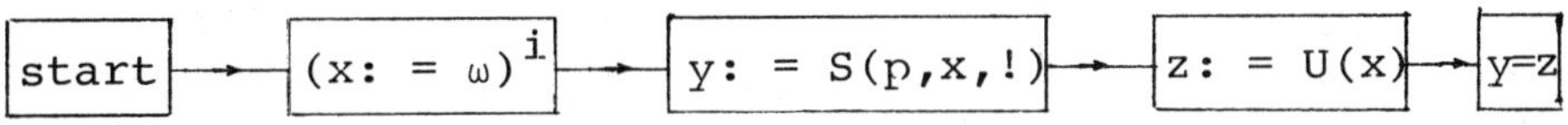

Let p be the Gödel-number of a sentence φ_p. Then
$\tau(p)\in T$
iff $(\exists i_o)(\forall i\le i_o)(\pi_\tau(p)^i$ terminates in exit $\ominus$),
iff $(\exists i_o)(\forall i\ge i_o)(\pi_{\varphi_p}$ terminates in exit $\oplus$),
iff $\varphi \in T$.
Hence $\tau(x)$ is a quasi truth-definition as stated.

If we try-in analogy to the proof of Tarski's theorem - to formulate the Liar's paradox, we obtain again an example of a sentence that is not in T and neither is its negation. Namely, let $d(n)$ be the Gödel-number of the result of substituting the constant symbol for n into the formula φ_n with Gödel-number n (assuming this formula has exactly the free variable x ; for other formulas let $d(n) = 0$). Assume that $x := d(y)$ belongs to the basic operational capabilities of our programming language. Let $\varphi(x)$ be the formula $\neg\tau(d(x))$ whose Gödel-number is m, say. Then neither $\varphi(m)$ nor $\neg\varphi(m)$ are in T, as easily verified.

§3 Conclusion

The fact that T contains both formulas $(\exists x)\varphi(x)$ and $\neg(\forall x)\neg\varphi(x)$ for some φ is surprising. A priori one would expect the axioms and rules of intuitionistic logic (at least) to hold for strict finitistical truth: in particular the intuitionistic theorem $(\exists x)\varphi(x) \rightarrow \neg(\forall x)\neg\varphi(x)$ should be acceptable. Thus, with modus ponens, T would be contradictory - which it is not. Since modus ponens can hardly be disputed by a strict finitist[6], the question arises how plausible the quoted intuitionistic theorem is for him. For the strict finitist it seems reasonable to interpret the above theorem thus: "If there is a construction of an element which all sufficiently large finite minds can perform and such that almost all finite minds are convinced that it has property φ, then it is absurd that almost all finite minds can at the same time be convinced that all elements do not have this property φ". "There is no obvious conceptual reason for a strict finitist not to accept the above statement as true.

Indeed, the above process of interpreting a formula could serve as a way to motivate a formal system of

---

[6] See, however, Goguen [3].

strict finitism (a process that is similar to that which is sometimes used to motivate formal systems of intuitionism).But the point is that the concepts that enter this "interpretation" are too vague. Our more precise interpretation, making sharper use of the restrictions of finite minds expressed in Section 1, thus does in fact diverge from it.

Perhaps it is good at this point to review some arguments *against* the plausibility of strict finitism. We have encountered two examples that illustrated the incompleteness of T . The fact that the statement of liar is neither (strict finitistically) true nor flase may perhaps be pleasing to those who seek the way out of the liars paradox by relegating that statement to a realm of proncuncements which - while grammatical - are of indefinite truth value. More puzzling is the fact that there should be no agreement among finite minds on the parity of ω . Mathematicians, who have finite minds after all, have no trouble in agreeing that ω is ever. How can one explain this? It is obvious that mathematician's minds do not function as naively as our above model makes it out. For example, there seems to exist a sort of interplay between findings based on intuition (or "thought-experiments") and logical deductions from them. We might speculate about a process of educating the intuition which then becomes more acute on questions about the actual infinite[7]. It is not surprising that our naive model of the mind gives only a poor approximation. It would be beautiful, of course, if an exact theory of mathematician's minds were available, and perhaps worthwhile to work towards one. But it is not realistic, in our opinion, to rely upon such an endeavor to "secure the

---

[7] (Added 1979 This idea was made precise by my former colleague Jeroslow [11] and by Hájek [10] as a so-called "experimental logic".

foundations of mathematics". It would be more realistic to leave the theory - or a more appropriate variant of it - where it arose: in computer science, as a theory of feasible processes on a computer[8].

REFERENCES

[1] Bernays, P.: *On Platonism in Mathematics, Benacerraf, Putnam, (eds.) Philosophy of Mathematics*, Prentice-Hall, 1964. pp.274-286.

[2] Geiser, J.R.: A formalization of Essenin-Volpin's proof theory by means of non-standards analysis (abstract), *IVth International Congress for Logic, Methodology, and Philosophy of Science*, Bucharest 1971, p.72.

[3] Goguen, J.A.: The logic of inexact concepts, *Synthese*, Vol.19. (1968), pp. 325-373.

[4] Parikh, R.: Feasibility in number theory, Boston University, 1970. (mimeographed).

[5] Wang, H.: Eighty years of foundational studies, *Dialectica* Vol.12. (1958), pp. 466-497.

[6] Williamsen, J.S.: Induction and artifical intelligence Thesis, Cambridge, 1971.

[7] A.S.Yessenin-Volpin (=A.S.Essenine-Volpine), Le programme ultra-intuitioniste des fondements des mathematiques, in *Infinitistic Methods*, Pergamon Press, 1961, pp. 201-223.

[8] A.S.Yessenin-Volpin, The ultraintuitionistic criticism and the antirational program for the foundations of mathematics, in: Myhill, Kino, Vesley, (eds)

[8] (Added 1979) For notions of "feasible numbers" and "feasible proofs" in computer science cf. the papers by Simon [12] and by Cook [9].

Intuitionism and Proof Theory, North Holland, 1970, pp. 3-45.

added in 1979:

[9] Cook,C.A.: Feasibly constructive proofs and the propositional calculus. Preliminary version. *Proceedings of the 8th ACM Symposium on the Theory of Computing*, 1976, pp. 83-97.

[10] Hájek,P.: Experimental logics and π_3^0 theories. *Journal of Symbolic Logic* Vol.42 (1977), pp. 515-522.

[11] Jeroslow,R.J.: Experimental logics and Δ_2^0 theories. *Journal of Philosophical Logic* Vol.4. (1975), pp. 253-267.

[12] Simon,J.: On feasible numbers. (Preliminary version). *Proceedings of the 8th ACM Symposium on the Theory of Computing*, 1976, pp. 195-207.

E.Engeler
Institut für Informatik
ETH-Zentrum, Hauptgebände
CH-8092 Zürich

COLLOQUIA MATHEMATICA SOCIETATIS JÁNOS BOLYAI
26. Mathematical Logic in Computer Science,
Salgótarján (HUNGARY), 1978.

TIME MODELS FOR PROGRAMMING LOGICS

T. Gergely and L. Úry

ABSTRACT

A class of such models is developed that provides the interpretation of the meaning of programs considering both time and data explicitly. It is shown how different model constructions, developed explicitly or implicitly in different kinds of programming logic, can be expressed in our frame. The proposed model construction is strong enough to become base of such theory of programming that can be built in the frame of classical first order logic and that deals not only with deterministic but with nondeterministic, interactive and parallel programming as well.

To illustrate the usage of the introduced models such kind of descriptive language is considered that handles input-output conditions and which does not use all the possibilities of our models. It is shown that the descriptive language that considers both time and data is complete. Moreover it is shown that if time is considered only in the metalanguage, i.e. the descriptive language can speak only about data then it becomes incomplete.

At the same time some modifications of completeness that seem to be adequate from pragmatical point of view are introduced and with respect to which the descriptive language becomes complete.

1. INTRODUCTION

One of the main directions in the theory of programming is connected with the investigation of semantical aspects of programming languages and programs.

In the above investigation the emphasis is transferred from the programming languages to the so called descriptive ones that characterize programming languages and programs from the semantical point of view.

Investigation in the mentioned area established an independent field, the so called programming logic. Analogously to mathematical logic two main branches, Model Theory and Proof Theory became the main branches of programming logic.

The first is but mathematical theory of semantics while the second deals with questions connected with handling the syntax.

What kind of meaning of programs is interesting for us? Similarly to the saying on the proof of the pudding we believe that the proof of a program is in its execution. This means that the semantical properties of a program is to be characterized by its effect on the environment particularly on its data environment. This principle assumes the operational semantics that will play a significant role in this paper.

Operational semantics presupposes some kind of abstraction of computers. It represents the meaning and semantical properties of programs by "executing" in the chosen abstraction of machines.

In programming logic the abstraction of computers is but the representation of computers in the logic and it is given by certain mathematical objects (models). The model theory of programming logics deals with the construction of these objects with their possible definitions and descriptions and the possible characterisation, etc. The proof theory of programming logic develops and investigates tools that provide, with the power of proof, description and investigation of program properties. Moreover this theory deals with developing and using different exact mathematical notions of proof which consider the specific character and needs of the theory of programming.

Dealing with semantical aspects programming logics may, in principle, consider some concrete programming languages, however for the theoretical investigation it is far more convenient to consider programming languages of abstract syntax. When choosing the syntax, such constructions are taken into account that are essential for many concrete programming languages. (I.e. in the latters the abstract constructions appear in special form.)

Since programming logic handles abstract programming languages together with certain abstractions of computer it is quite natural to ask what this theory can offer to practice, where concrete programming languages and computers are in action.

Programming logic in its theoretical form is not adequate for its use in programming practice directly.

However the results outline a frame within which the usage of both, the concrete computer and programming language of given features can be done at all. Of course in each concrete case the properties by the programming logic should be verified, because the theory states them only in principle and not for some concrete programming situations. Thus e.g. if programming logic states the completeness of its descriptive language, then in the case of concrete programming languages the corresponding complete calculus should be found and for this the theory does not provide any tools.

We believe that the role of programming logic with respect to the programming practice is fairly analogous with that of mathematical logic with respect to mathematical disciplines. Mathematical logic provides the metamathematical foundation of the latters but it has no adequate tools to handle the constructs of a concrete discipline. E.g. while proving a mathematical theorem we do not use the rules provided by mathematical logic because otherwise the proof would become far too large, complicated and hardly understandable. Analogously the main role of programming logic is to provide metamathematical foundation for the theory and practice of programming. It is also believed that programming logic should provide a ground for developing such kind of languages and tools that are adequate to programming siutations of similar level and convenience to those of used in mathematical disciplines (see e.g. Gergely - Vershinin [8]).

In our opinion, modelling computers from the point of view of programming one should consider both the data represented in the computer and the computer's inner clock, i.e. the time given by it. The first is necesarry for the static characterization, the second one is for the dynamic one of the programs.

Thus programming logic aiming at the characterization of different program features should be capable of describing and characterizing the data environment and to describe features and characteristics related to time. There are different possibilities to realize the aboves, e.g.

(i) only data are built into the model construction, time appears only in the metalanguage;

(ii) time appears besides data in the descriptive language but model constructions deal with data only;

(iii) both data and time are handable by the descriptive language and both are built into the model constructions;

(iv) besides aboves even the programming language itself is able to handle time.

Note that most of the versions of programming logic realize possibility (i). There are some exceptions that realize one of the remaining possibilities. (See e.g. Andréka et al [1], Kröger [10], Pnueli [13].)

Explicit time consideration is rather important in the cases where not only sequential and deterministic programs, but nondeterministic, interactive or parallel ones are to be investigated as well.

In the present paper we give a class of such models of computers that explicitly consider both data and time and thus based on it a further research can be done concerning e.g. parallel, nondeterministic and interactive programming. We show that the models used in different programming logics become the special cases of our models.

At the same time we introduce such a kind of descriptive language that handles input-output relations only and thus it does not use all the possibilities of our models. This language is of the spirit of Floyd and Hoare and it is an appropriate version of dynamic logic.

The reason why we decided by this descriptive language is to show the usage of our model construction in a fairly simple way by connecting it to a widely used language.

After having constructed models we investigate the completeness of the introduced descriptive language. Theorem 5.4 gives a positive answer to the question of completeness. However in the theories of programming time is usually considered only in the metalanguage. In our frame this is equivalent to the fact that our model construction considers only standard time (See Definition 2.2). In this case data and time do not play symmetric role in the theory and this leads to some troubles. Here we investigate how the above mentioned positively answered question on completeness turns to negative.

At the same time we introduce some special notions of completeness that seem to be adequate to program synthesis and w.r.t. which the descriptive language becomes complete even in the assymetric cases.

There are several not first order constructions used in programming logic. *Algorithmic logic* is one of the first of such constructions. It uses ω-logic to describe time. (See e.g. Banachowski et al [2]).

The construction of *dynamic logic* (see e.g. Harel [9]) is similar. There the standard arithmetic is "smuggled" into the data and is used (implicitly) as time.

A nonclassical logical construction is proposed in Gabbay et al [6], as model with time structure for a zero-order (propositional) programming logic.

Models introduced in Kröger [10] have unique time structure.

In Andréka et al [1] models with time consideration are presented slightly analogous to ours.

The theory developed here differs from the aboves in that we strictly keep ourselves to the frame of first order logic.

The advantages of using first order logic are:

(i) It is currently used in the research practice so its use is not unfamiliar;

(ii) It has a well developed proof theory that makes one hope to automatize, even if partially, the tools of proofs.

In spite of the fact that first order logic is simpler than the above mentioned ones such a programming logic can be built on it that is at least as expressive as the programming logics based on the above mentioned logics.

The basic notions of first order language which are relevant to our investigation can be found in Appendix.

2. *MODELS WITH INNER TIME*

First we analyse what constituents should be considered in models that abstract computers. Program properties can be described by characterizating the environment under the influence of program. Thus the program properties will be described through changes in program environment resulted by the program execution and through the characterization of time within which the changes occur.

Therefore we have to characterize computers, on which the program execution occur, in an abstract way. We consider the following features of computers.

It is the characteristic feature of a computer what sort of data it can understand and how it can manipulate with them, i.e. what kind of transformation can be done or what kind of properties about data it can check out. Computer functioning is nothing else but data manipulation. The functioning goes on in time namely, in time moments depending on the computer construction and generated or prescribed by the "inner clock".

The data manipulation providing the functioning is the change of data in time. Data change provided by the execution of operators occurs within a time interval characteristic for the computer. More precisely the execution of an operator occurs for a corresponding time cycle. Let us denote the set of disjoint time intervals (that correspond to different operators) by T. From theoretical point of view the time intervals of T can be considered as time moments supposing that change takes place infinitely fast. We also assume that changes cannot be done spontaneously e.g. due to a breakdown but only by a prescribed way. Similarly we assume that a machine works as long as it is needed, i.e. as long as it is required by the program. This means that a machine itself can work infinitely long never stopping due to break-down or power loss. Thus a computer stops only if its overall operations required (by a program) or the stop is required.

The question of which operators should be executed from those allowed by the computer and in what order they should be used for a given set of data is prescribed by a program. Thus computer functioning characterized by data change in time, is connected with the interpretation of the program.

According to the abovesaids it is clear that at least two sets are needed to characterize a computer: the set D of possible data and the set of possible time moments T. Moreover to characterize the functioning on the level of program execution the set of possible data change in time should be taken, i.e. a set I of functions $T \to D$.

It is an obvious wish to use the first order language to describe the computer and its functioning. Since time, data and functions of I are different entities it is advisable to correspond to them different languages or at least different elements of a language. This can be provided by a three sorted language, where the first sort d corresponds to data the second t corresponds to time and the third one i to the set I.

Note that data themselves can be devoted to different subclasses to each of which a different sort should be corresponded (e.g. Boolean, real, integer etc.) Thus to data a set of sorts corresponds. Without restricting the generality we consider only that case when data are only one-sorted.

Let $S \stackrel{d}{=} \{t,d,i\}$ be a set of sorts and let an S-sorted similarity type σ be fixed.

Further on to describe computer and its functioning we use three sorted σ-typed formulas with a new type of models.

Let σ_d, σ_t be the subsignature corresponding only to data and time respectively and let σ_m be the signature containing symbols not occuring in σ_d and σ_t, i.e. σ_m contains the i-sorted homogeneous symbols and also the mixed sorted ones.

Intuitively the symbols of σ_d name those operations and relations that can be used by our computers w.r.t. data. We do not assume any restriction on the symbols of do σ_d though we suppose that $(=,\langle d,d\rangle)\in\sigma_d$.

Signature σ_t is homogeneus as well. We assume that it contains at least such symbols that provide to formulate some natural expectations on time. These symbols are: the moment of starting which is a constant symbol 0, a unary function symbol the successor operation +1 and a binary relation symbol $\leq$ for ordering. I.e. $\{(0,0), (+1;1), (\leq, 2)\} \subset \sigma_t$.

In connection with σ_m we assume only that it contains a heterogeneous function symbol *ext* of type $\langle i,t;d\rangle$ and the relation symbol = of type $\langle i,i\rangle$.

Note that this function serves to concretize the function name f by giving the graph representation, i.e. for a name f *ext*(f,t) is a function that renders to each time moment a d-type entity.
Further on we often do not distinguish between the name f and the function *ext*(f,.) rendered to it. So we write f (t) instead of *ext*(f,t).

Intuitively the symbols the corresponding sort sequence of which contains i provide to formulate some expectations on the name of functions connected with the functioning of our computer.

Now we define the σ-type models as a modification of many sorted models.

2.1 Definition

Let $\sigma = \sigma_t \cup \sigma_d \cup \sigma_m$. Let us fix a σ_t-type (and thus a one - sorted) model $\mathfrak{T}$ which is called *inner time*. Let $T = \mathfrak{T}(t)$ be the universe of the model $\mathfrak{T}$.

A σ-type model $\mathfrak{A}$ (with inner time $\mathfrak{T}$) is a function that satisfies

(i) $\mathrm{do}\,\mathfrak{A} = (\mathrm{do}\,\sigma_d \cup \{d\} \times T \cup \mathrm{do}\sigma_m \cup \{i\} \cup \mathrm{do}\sigma_t \cup \{t\}$;

(ii) $\mathfrak{A}(\ell) = \mathfrak{T}(\ell)$ for any $\ell \in \mathrm{do}\sigma_t$;

(iii) $D_r = \mathfrak{A}(d,r)$ is a non-empty set for any $r \in T$. Intuitively D_r is the set of data existing potentially at moment r.

$I = \mathfrak{A}(i)$ is also a non-empty set for a distinguished set of functions. (See condition (iv) below).

$\mathfrak{A}(t)$ coincides with $\mathfrak{T}(t)$, i.e. $\mathfrak{A}(t) = T$.

(iv) Let $A_t = T$, $A_d = \bigcup_{r \in T} D_r$ and $A_i = I$.

For any relation symbol ρ of type $(s_o, \ldots, s_n)$ belonging to σ we have

$$\mathfrak{A}(\rho) \subset \mathop{\times}_{j=o}^{n} A_{s_j}$$

For any function symbol f of type $(s_o, \ldots, s_{n-1}; s_n)$ belonging to σ we have

$$\mathfrak{A}(f) : \mathop{\times}_{j=o}^{n-1} A_{s_j} \rightarrow A_{s_n}$$

(Here $s_j = i,t,d$).

(v) For the function symbol *ext* we have

a) for arbitrary $f \in I$ and $r \in T$

$$\mathfrak{A}(ext)\,(f,r) \in D_r\ ;$$

b) for any $f_1, f_2 \in I$ if for each $r \in T$

$$\mathfrak{A}(ext)(f_1, r) = \mathfrak{A}(ext)(f_2, r) \text{ then } f_1 = f_2.$$

□

Intuitively the condition a) means that the value of f at t is a data belonging to the r-th set of data.

Condition b) means that if two functions have the same graph then they have the same name in I.

The class of σ-type models is denoted by M_σ. When we would like to emphasize the role of inner time then we use the full notion for $\mathfrak{A}$, i.e. the notion "σ-type model $\mathfrak{A}$ with inner time $\mathfrak{T}$". In this case we use the notation ${}_{\mathfrak{T}}M_\sigma$ for the class of the above models.

For an arbitrary $r \in T$ let us denote by $\mathfrak{D}_r$ the σ_d-type one-sorted model corresponding to r for which

(i) $\mathfrak{D}_r(d) = \mathfrak{A}(d, r)$ and

(ii) $\mathfrak{D}_r(\ell) = \mathfrak{A}(\ell, r)$ for any $\ell \in do\sigma_d$.

Note that a σ-type model with inner time seems to be a three sorted first order model, but it is not the case. Indeed here the model $\mathfrak{T}$ provides the index set for the σ_d-type models $\mathfrak{D}_r$ $(r \in T)$ which makes the difference. If we consider $T \times \bigcup_{r \in T} D_r$ as a universe of a new sort we can obtain a usual three sorted model and the restrictions (iv), (v) can be formulated by the first order language. It means that our theory does not grow up of the frame of first order language.

However if for any $r \in T$ the σ_d-type models are the same, i.e. $\mathfrak{D}_r = \mathfrak{D}$ then σ-type models coincid with the usual three-sorted models, more precisely it is a {t,d,i}-sorted one.

Further on we denote a σ-type model by

$$\mathfrak{A} = (\mathfrak{T}, \{\mathfrak{D}_r | r\in T\}, I).$$

This notation emphasizes that the model $\mathfrak{A}$ has a σ_t-type (one-sorted) model $\mathfrak{T}$ corresponding to time, for each $r\in T$ it has a σ_d-type one-sorted model $\mathfrak{D}_r$ corresponding to data and it has a non-empty set I. Functions and relations of heterogeneous sort or homogeneous in the sort i do not appear explicitly in this notation.

2.2 Definition

Let $\mathfrak{A} = (\mathfrak{T}, \{\mathfrak{D}_r | r\in T\}, I)$ be a σ-type model. The model $\mathfrak{A}$ is said to be *simple* if $\mathfrak{D}_r = \mathfrak{D}$ for any $r\in T$.

If $\mathfrak{D}_r(d) = D$ for any $r\in T$ then $\mathfrak{A}$ is said to be *semisimple*.

Let $\mathfrak{T}$ be a standard model of the Peano arithmetic, i.e. let $\mathfrak{T} = \mathfrak{N} = (N, +, *, 0, 1, =)$. A model $\mathfrak{A}$ with inner time $\mathfrak{N}$ is called *σ-type model standard in time*.

□

We illustrate the notion of model with inner time by different constructions that can be met in literature on programming logic.

2.3 Example

Let us fix a similarity type $\sigma = \sigma_d \cup \sigma_t \cup \{(ext, \langle i, t_j d\rangle)\}$, such that $\eta\subset\sigma_t$ and $\eta\subset\sigma_d$. Remember that η is the (one-sorted) similarity type of arithmetic which is t-sorted in the first case and d-sorted in the second one (see Appendix).

We give three models with inner time which differ only in the set I.

Let the inner time $\mathfrak{I}$ be a copy $\mathfrak{N}'$ of the standard model of Peano axiomatization of arithmetic and let $\mathfrak{D}_n$ for any $n \in N'$ be another copy $\mathfrak{N}$ of the standard model. Though $\mathfrak{N}$ and $\mathfrak{N}'$ are isomorphic for the sake of distinction between time and data a sign " ′ " is given in all components of the model corresponding to inner time that is denoted by $\mathfrak{N}'$.

Moreover let $I_1 = {}^{N'}N$, $I_2 = Def({}^{N'}N)$ and $I_3 = Rec\,({}^{N'}N)$ be the set of all *arbitrary, definable* and *recursive functions* from N′ to N respectively. For each I_i $(i \in [1,3])$ the interpretation of the function symbol *ext* is denoted by ^ and it is defined by $\hat{}(f,n) \stackrel{d}{=} f(n)$, for any $f \in I_j, n \in N', j \in [1,3]$, i.e. $\mathfrak{A}_j\,(ext) = \hat{}$.

Thus our j-th model with inner time is $\mathfrak{A}_j = (\mathfrak{N}', \mathfrak{N}, I_j)$.

These models are said to be *strictly standard* emphasizing that both data and time are considered in standard model of arithmetic.

Note that usually the first model construction $\mathfrak{A}_1$ is used in literature.

However the third model construction $\mathfrak{A}_3$ is quite sufficient to investigate computer programming, because computation realizes recursive functions.

Note that most of the researchers using mathematical logic think in this kind of models while dealing with programming even if this fact is not declared. (See e.g. Manna [12].)

□

2.4 Example

Let σ_d be an arbitrary similarity type and take $\sigma = \sigma_d \cup \eta \cup \{(ext,\langle i,t;d\rangle)\}$. Moreover let $\mathfrak{D} \in M_{\sigma_d}$ be arbitrary. Then a model with standard inner time $\mathfrak{N}$ is $\mathfrak{A} = (\mathfrak{N}, \mathfrak{D}, {}^{N}D)$.

Note that analogous models are considered in *algorithmic logic* (see e.g. Banachowski et al [2]).

This model construction is analysed in details in Gergely-Úry [7].

2.5 Example

Arithmetical universes play a significant role in the investigation of the completeness of dynamic logic (see Harel [9] p.29.).

Arithmetical universes can also be expressed by using our formalism and model construction as follows.

Let $\sigma_d = \{(\cdot,2), (+,2), (\underline{0},1), (\underline{1},1), (R,2)\} \cup \{(nat,1)\}$ and let $\sigma = \sigma_d \cup \eta \cup (ext, \langle i,t,d\rangle)\}$ be a similarity type.

A σ-type model standard in time $\mathfrak{A} = (\mathfrak{N}', \mathfrak{D}, D)$ (where $D = \mathfrak{D}(d)$) is said to be an arithmetical universe if it is of the following properties.

(i) $N \subset D$;

(ii) $\cdot_{\mathfrak{D}} \restriction N'$ and $+_{\mathfrak{D}} \restriction N'$ have the standard meaning;

(iii) $\underline{0}_{\mathfrak{D}} = 0$ and $\underline{1}_{\mathfrak{D}} = 1$;

(iv) $\mathfrak{A} \models nat\,[a]$ iff $a \in N'$;

(v) The function $R_{\mathfrak{D}}$ is but the Gödel function β.

This means that for any $n \in N$ and any sequence $a_1, \ldots, a_n$ $(a_i \in D)$ there exists such a $b \in D$ that $R_{\mathfrak{D}}(b,i) = a_i$ $(i \in [1,n])$.

I.e. by using function $R_{\mathfrak{D}}$ the finite sequences in D can be coded.

(vi) The function symbol *ext* is interpreted in $\mathfrak{A}$ such that for any $n' \in N'$ and $f \in D$ if $\mathfrak{A}(ext)(f,n') = c$ then

$$R_{\mathfrak{D}}(f,n') = c \text{ holds.}$$

It is not difficult to verify that this notion of arithmetical universes coincides with that of used in Harel [9].

2.6 Example

Let the d-sorted similarity type $\sigma_d = \eta \cup \{(\xi,1)\}$, where ξ is a unary relation symbol, which is introduced to describe time. By this we supply our models with inner time supposing that time can be modelled by data.

Let PA be the set of Peano axioms of arithmetic. Take the following additional axioms for ξ:

$$B_o \overset{d}{=} \xi(O)$$

$$B_1 = \xi(x) \rightarrow \xi(x+1)$$

$$B_{2_\phi} = \phi[O/v] \wedge \forall v\ (\phi \wedge \xi(v) \rightarrow \phi[v+1/v]) \rightarrow \forall v(\xi(v) \rightarrow \phi(v)).$$

where $\phi \in F_{\sigma_d}$.

B_o provides O as the starting moment of time B_1 provides that each time moment has a successor and B_2 claimes that the induction works well on time moments, i.e. under the relation symbol ξ.

Let $PA_\xi = \{B_0, B_1\} \cup \{B_{2\phi} | \phi \in F_{\sigma_d}\}$.

Take $PA = PA^* \cup PA_\xi$ which is but the set of Peano axioms with reference to time as well. PA* is consistent if PA is so, because the standard model $\mathfrak{N}$ with $\xi_{\mathfrak{N}} = N$ is a model of PA^*. However there are models on which ξ is not identically true.

Now let $\mathfrak{B}$ be an arbitrary model of PA*. Take $T_{\mathfrak{B}} \overset{d}{=} \{b \in B | \mathfrak{B} \models \xi[b]\}$. Let $\mathfrak{T}_{\mathfrak{B}}$ be the model of similarity type $\{(0,0), (+1,1), (\leq,2)\}$ with the universe $T_{\mathfrak{B}}$. Let $I_{\mathfrak{B}} = \{f | f : T_{\mathfrak{B}} \to B$ and f is parametrically definable in $\mathfrak{B}\}$.

Thus we have the following model construction with inner time: $\mathfrak{A} = (\mathfrak{T}_{\mathfrak{B}}, \mathfrak{B}, I_{\mathfrak{B}})$. Note that if $\mathfrak{T}_{\mathfrak{B}}$ is isomorphic to $\mathfrak{N}$ then we obtain a special arithmetical universe (of 2.5).

This model provides handling of time explicitly without introducing many-sorted signature. Details on the theory of programming that uses such kind of model constructions see in Gergely - Úry [7].

□

The models with inner time given in the previous example are simple. Now we give a semi-simple model construction.

2.7 Example

Let $\sigma_d = \eta \cup \{(Time, 1)\}$, i.e. a new unary relation symbol is added to the similarity type of arithmetic. Moreover let $\eta \subset \sigma_t$ and take $\sigma = \sigma_d \cup \sigma_t \cup \{(ext, \langle i,t;d \rangle)\}$.

Let the inner time be the standard model $\mathfrak{N}$ of arithmetic and for any $n \in N$ let $\mathfrak{D}_n = (\mathfrak{N}, n)$.

Thus we have the following model with standard time

$$\mathfrak{A} = (\mathfrak{N}', \{(\mathfrak{N}, n) \; n\in N\}, I),$$

where I is a non empty set, e.g. $I = {}^{N'}(N\times N)$.

The interpretation of symbols of the signature η in $(\mathfrak{N}, n)$ is the usual, but that of the relation symbol *Time* is n in the model $\mathfrak{D}_n$ $(n\in N')$.

In the signature σ the relation symbol *Time* provides to speak about a clock e.g. in a description of a computer. By using it the exact time of functioning can be asked.

□

2.8 Example

Let $\sigma_t = \{(0,1), (+1, 1), (\leq, 2)\}$ and let σ_d be arbitrary.

Moreover let $\mathfrak{T}$ be such a σ_t-type model that $\mathfrak{T}(t) = T = \alpha$, where α is an ordinal.

A model with inner time $(\mathfrak{T}, \{\mathfrak{D}_t | t\in T\}, I)$ is *monotone* in time w.r.t. data if

(i) for any $t\in T$ $D_{t+1} \subset D_t$;

(ii) for any function symbol $f\in$ do σ_d

$$\mathfrak{A}(f,t+1) = \mathfrak{A}(f,t)\restriction D_{t+1}$$

(iii) for any relation symbol $\rho\in$ do σ_d

$$\mathfrak{A}(\rho,t+1) = \mathfrak{A}(\rho,t)\restriction D_{t+1}$$

This model construction is considered in Kröger [10].

□

Now we give the propositional version of the model construction.

2.9 Example

For a fixed σ-type model $\mathfrak{A} = (\mathfrak{T}, \{\mathfrak{D}_r | r \in T\}, I)$ let us suppose that D_r has exactly one element for any $r \in T$. Hence I has exactly one element as well. If so then the interpretation of function symbols belonging to σ_d is unique and thus we may suppose that σ_d contains only unary relation symbols.

Hence $\mathfrak{A} \restriction_{do\ \sigma_d \times T}$ can be defined as a function

$\hat{\mathfrak{A}} : do\ \sigma_d \to {}^T 2$. Namely for the relation symbol $\rho \in do\ \sigma_d$ let $t \in \hat{\mathfrak{A}}(\rho)$ iff $\mathfrak{A}(\rho, t)$ is true. So in this case our model $\mathfrak{A}$ can be described as a pair $(\mathfrak{T}, \hat{\mathfrak{A}})$ where $\hat{\mathfrak{A}}$ is an arbitrary function of the above form. This model is studied in details in Gabbay et al [6].

□

3. INTERPRETATIONS OF PROGRAMMING LANGUAGE

Now we consider how the above model construction can be used to characterize the meaning of programming languages. Since we are interested in operational semantics first we indroduce the main construction through the changing of which the semantics can be given.

3.1 Definition

Let σ be a similarity type and let $\mathfrak{A} = (\mathfrak{T}, \{\mathfrak{D}_t | t \in T\}, I)$ be an arbitrary σ-type model with inner time.

A *state* in $\mathfrak{A}$ is such a pair $(t, \vec{f})$ that

(i) $t \in T$ or $t = \infty$;

(ii) $\vec{f}$ is a finite vector in I;

(iii) for any $s \geq t$ $\vec{f}(s) = \vec{f}(t)$.

The length of a state $k = (t, \vec{f})$ written as dim(k) is the dimension of the vector $\vec{f}$. Let $\mathfrak{S}_{\mathfrak{A}}$ denote the set of all states in $\mathfrak{A}$.

□

Any state $(t, \vec{f})$ in an arbitrary given σ-type model with inner time represents the computation $\langle \vec{f}(s) | s \leq t \rangle$.

We say that a state or, equivalently, a computation converges iff $t \in T$. Otherwise (in the case $t = \infty$), the computation diverges.

Note that the notion of state provides to consider the flow process of a computation. This permits to describe the computation process occuring in a computer, which is modelled by a model with inner time and not only by the result of the computation process.

States provide to define the semantics of the formulas of F_{σ_d} w.r.t. the model with inner time.

3.2 Definition

Let $\mathfrak{A}$ be as in the previous definition and let

$$\phi \in F_{\sigma_d}, \quad k = (t, \vec{f}) \in \mathfrak{S}_{\mathfrak{A}}.$$

The formula ϕ is valid in $\mathfrak{A}$ w.r.t. the state k (written by $\mathfrak{A} \models \phi[k]$ iff $\dim(k) \geq v(\phi)$ and either $t = \infty$ or $t \neq \infty$ and $\mathfrak{D}_t \models \phi[\vec{f}(t)]$. (Here $v(\phi)$ denotes the number of free variable symbols occuring in the formula ϕ, i.e. $v(\phi) = |\mathrm{Var}\ \phi|$).

□

Now we define the notion of programming language.

3.3 Definition

A programming language $\mathfrak{P}$ is a pair (P,v), where P is an arbitrary set and v is a function $v : P \to N$.

□

Intuitively the elements of P are the programs and $v(p)$ is the number of program variables occuring in $p \in P$.

3.4 Definition

Let $p \in P$ be arbitrary. A σ-type *interpretation* of the programming language $\mathfrak{P} = (P,v)$ is $(\mathfrak{A}, \mathit{Trace})$ iff $\mathfrak{A} \in M_\sigma$ and *Trace* satisfies both

(i) $\mathit{Trace} \subset \mathfrak{S}_{\mathfrak{A}} \times \mathfrak{P} \times \mathfrak{S}_{\mathfrak{A}}$;

(ii) Let $k_j = (t_j, \vec{f}_j)$ $(j = 1,2)$.

If $(k_1,p,k_2) \in \mathit{Trace}$ then

a) $t_1 \leq t_2$,

b) $\vec{f}_1(s) = \vec{f}_2(s)$ for any $s \leq t_1$,

c) $\dim(k_1) = v(p) = \dim(k_2)$.

□

Generally for a σ-type programming language $\mathfrak{P}$ and for a fixed σ-type model $\mathfrak{A}$ there are many different relations *Trace* such that each ($\mathfrak{A}$, *Trace*) is a σ-type interpretation of $\mathfrak{P}$.

Further on if $k_j = (t_j, \vec{f}_j)$ $(j = 1,2)$, then we write $\langle t_1, \vec{f}_1 | p | t_2, \vec{f}_2 \rangle$ instead of $(k_1, p, k_2) \in$ *Trace* in order to emphasize the effect of p on the state $(t_1, \vec{f}_1)$.

Sometimes we use the notation $\langle t_1, \vec{f}_1 | p | t_2, \vec{f}_2 \rangle$ even in that case, when dim $\vec{f}_1 > v(p) = n$. It means that $\langle t_1; (f_{11}, \ldots, f_{1n}) | p | t_2, (f_{21}, \ldots, f_{2n}) \rangle$ and $f_{1i} = f_{2i}$ for any $i > n$.

In other words $\langle t_1, \vec{f}_1 | p | t_2, \vec{f}_2 \rangle$ means that executing the program p after the state $(t_1, \vec{f}_1)$ a new state $(t_2, \vec{f}_2)$ is obtained. Conditions a) and b) in (ii) of 3.4 garantee that the program p is really executed only after the time moment t_1 and the state $(t_1, \vec{f}_1)$ is not changed. An execution of program p on a time interval (t_1, t_2) can be represented by the following sequence $\langle f_2(s) | t_1 \le s \le t_2 \rangle$.

Note that we do not assume that *Trace* is a function. Thus for the same input $(t_1, \vec{f}_1)$ there may be several $(t_2, \vec{f}_2)$ such that $\langle t_1, \vec{f}_1 | p | t_2, \vec{f}_2 \rangle$. Hence the case of non-deterministic program is covered as well.

Further on if $k^o \in \mathfrak{S}_{\mathfrak{A}}$ then its components are denoted by t^o and $\vec{f}^o$ respectively.

3.5 Definition

Let us fix a σ-type interpretation $\mathfrak{A}^* = (\mathfrak{A}, \mathit{Trace})$ of a programming language $\mathfrak{P} = (P, v)$. $\mathfrak{A}^*$ is said to be *deterministic* iff for any $p \in P$ and $k_1 \in \mathfrak{S}_{\mathfrak{A}}$ there is a unique k_2 such that $\langle k_1|p|k_2\rangle$ holds.

$\mathfrak{A}^*$ is said to be *prehistory-independent* iff for any $k_1, k_2, k_1^*, k_2^* \in \mathfrak{S}_{\mathfrak{A}}$ and $p \in P$ if

$$t_1 = t_1^* \text{ and } \vec{f}_1(t_1) = \vec{f}_1^*(t_1^*) \text{ then } t_2 = t_2^*$$

and for any $s \geq t_1$ we have $\vec{f}_2(s) = f_2^*(s)$.

$\mathfrak{A}^*$ is *time-independent* iff for any $p \in P$, $k_1, k_2 \in \mathfrak{S}_{\mathfrak{A}}$ and $t \in T$ there are $\hat{k}_1, \hat{k}_2$ such that

$$\text{if } \langle k_1|p|k_2\rangle \text{ then } \langle \hat{k}_1|p|\hat{k}_2\rangle, \ \hat{t}_1 = t,$$

and the computation sequences of $(\hat{k}_1, \hat{k}_2)$ and (k_1, k_2) are the same. I.e. whenever it starts the computation sequence remains the same.

□

Now we give some examples for programming languages and traces. In the examples many of the model constructions of the previous examples will be used to give interpretation of programs.

3.6 Example

Let $P_t = \{X, G\}$ and let $v(X) = v(G) = 0$, i.e. let $\mathfrak{P}_t = (P_t, v)$ be the programming language. For a σ-type model $\mathfrak{A}$ we define $\mathit{Trace}_{\mathfrak{P}_t}$ in the following way:

$$\langle t_1, \vec{f}_1|X|t_2, \vec{f}_2\rangle \text{ iff } t_2 = t_1+1 \text{ and } \vec{f}_1 = \vec{f}_2$$

$$\langle t_1, \vec{f}_1|G|t_2, \vec{f}_2\rangle \text{ iff } t_2 \geq t_1 \text{ and } \vec{f}_1 = \vec{f}_2 .$$

It is obvious that programs X and G work as "next moment" and "future moment" operations respectively. The σ-type interpretation ($\mathfrak{A}$, $Trace_{\mathfrak{P}_t}$) is nondeterministic and time-independent. Programs X and G do not change the data environment but the time only. □

3.7 Example

We give a programming language with *goto*.

Let us fix a d-sorted similarity type σ_d.

The set U_{σ_d} of σ_d-type commands consists of the following elements:

(i) $\ell : x \leftarrow \tau$

(ii) ℓ : *if* α *then* k

where ℓ, $k \in N$, $\tau \in T_{\sigma_d}$ and $\alpha \in Q_{\sigma_d}$. (See Appendix.)

If $(\ell : u) \in U_{\sigma_d}$ then ℓ is called the *label* of the command. A non-empty sequence $p \in U^+_{\sigma_d}$ of σ_d-type commands is called a σ_d-*type program* iff in the sequence there are no two commands of the same label, i.e. $p = (\ell_o : u_o, \ldots, \ell_n : u_n)$ is a program iff $\ell_k \neq \ell_j$ for any $k \neq j$.

Let $G_{\sigma_d} = \{p | p$ is a σ_d-type program$\}$.

If $p = (\ell_o : u_o, \ldots, \ell_n : u_n)$ then $v(p) = \max \{v(u_k) | k \in [0,n]\}$. Take the programming language $\mathfrak{G} = (G_{\sigma_d}, v)$.

Let $\mathfrak{A}$ be a semi-simple model with inner time $\mathfrak{N}$, i.e. $\mathfrak{A} = (\mathfrak{N}, \{\mathfrak{D}_n | n \in N\}, {}^N D)$ where $\mathfrak{D}_n(d) = D$ for any $n \in N$.

Now we define a σ-type interpretation of the programming language $\mathfrak{G}$.

Fix $(n,\vec{f}) \in \mathfrak{S}_{\mathfrak{A}}$ and let $p = (\ell_o : u_o, \ldots, \ell_n : u_n) \in G_{\sigma_d}$ be such that $k = \dim(\vec{f}) = v\ (p)$.

Now we define a vector of functions $\vec{g} : N \to D^k \times N$ as follows:

(i) $\vec{g}\ (m) = (\vec{f}(m),\ \ell_o)$ for $m \leq n$,

(ii) Let $m > n$. If $(g(\vec{m}))_{k+1} \notin \{\ell_o, \ldots, \ell_n\}$ then $\vec{g}\ (m+1) \overset{d}{=} \vec{g}\ (m)$.

Otherwise there is a j such that $(\vec{g}(m))_{k+1} = \ell_j$.

a) First let u_j be $x_i \leftarrow \tau$. Take

$$(\vec{g}\ (m+1))_s = \begin{cases} (\vec{g}(m))_s & s \neq i,\ k+1; \\ \tau_{\mathfrak{D}_m}[\vec{g}(m)] & s = i, \\ \ell_{j+1} & s = k+1. \end{cases}$$

b) Now let u_j be *if* α *then* ℓ.

If $\mathfrak{A} \models \alpha[\vec{g}(m)]$ then

$$(\vec{g}(m+1))_s = \begin{cases} (\vec{g}(m))_s & s \neq k+1 \\ \ell & s = k+1. \end{cases}$$

Otherwise

$$(\vec{g}(m+1))_s = \begin{cases} (\vec{g}\ (m))_s & s = k+1 \\ \ell_{j+1} & s = k+1 \end{cases}$$

In the end let

$$t = \min \{m \mid (\vec{g}(m))_{k+1} \notin \{\ell_o, \ldots, \ell_n\}\}$$

and take $\vec{f}^* = (g_1, \ldots, g_k)$. By definition $\langle n, \vec{f} \mid p \mid t, \vec{f}^* \rangle \in$ $Trace_{\mathfrak{G}}$.

It is evident that $(\mathfrak{A}, \text{Trace}_{\mathfrak{G}})$ is a σ-type interpretation of the programming language $\mathfrak{G}$.

It is clear that this interpretation is deterministic and prehistory-independent, but generally it is not time-independent, because $\mathfrak{D}_t$' s may differ from one another. If $\mathfrak{A}$ is simple then ($\mathfrak{A}$, $Trace_{\mathfrak{G}}$) is time-independent as well.

4. ACCEPTABLE PROGRAMMING LANGUAGES AND REGULAR MODELS

So far we have illustrated how our model constructions can be used to characterize the meaning of some programming languages. However we would like to get some results concerning certain class of programming languages.

Analogously to real programming situation we consider such class of programming languages that can be interpreted by a given programming language. In order to interpret programming languages we introduce an Algol-like language with a set of "atomic" programs Π.

4.1 Definition

Let us fix a d-sorted similarity type σ_d, a set Π and let $A(\Pi)$ be the minimal set satisfying the followings:

(i) $\Pi \subset A(\Pi)$;

(ii) If $\tau \in T_{\sigma_d}$ then $x_i \leftarrow \tau \in A(\Pi)$;

(iii) If $p_1, p_2 \in A(\Pi)$ and $\alpha \in Q_{\sigma_d}$ then

$p_1; p_2$, *if* α *then* p_1 *else* p_2 *fi* and *while* α *do* p_1 *od* are in $A(\Pi)$ as well.

Moreover let us fix a function $w : \Pi \to N$ and let us define $v : A(\Pi) \to N$ as follows

(i) $v(p) = w(p)$ for any $p \in \Pi$;

(ii) $v(x_j \leftarrow \tau) = \max\{v(x_j), v(\tau)\}$;

(iii) $v(p_1; p_2) = \max\{v(p_1), v(p_2)\}$

(iv) $v(\textit{if } \alpha \textit{ then } p_1 \textit{ else } p_2 \textit{ fi}) =$

$= \max\{v(\alpha), v(p_1), v(p_2)\}$;

(v) $v(\textit{while } \alpha \textit{ do } p \textit{ od}) = \max\{v(\alpha), v(p)\}$.

It is clear that v is well-defined.

Thus the programming language $A_\Pi = (A(\Pi), v)$ has been defined.

□

This programming language is an Algol-like one but it possesses given "atomic" programs, or, in other words, it has program variables, namely the elements of the set Π.

4.2 Definition

Now let $\mathfrak{A}$ be an arbitrary σ-type model with standard time ($\sigma_d \subset \sigma$). Let ($\mathfrak{A}$, $Trace_{\Pi}$) be a σ-type interpretation of (Π, w). Using this interpretation we define an interpretation of the programming language A_{Π}. Indeed let $Trace_{A_{\Pi}}$ be the minimal relation that satisfies the following conditions:

(i) $(k_1,p,k_2) \in Trace_{A_{\Pi}}$ iff $(k_1,p,k_2) \in Trace_{\Pi}$ for any $p \in \Pi$;

(ii) $\langle t_1, \vec{f}_1 | x_j \leftarrow \tau | t_2, \vec{f}_2 \rangle$ iff $t_2 = t_1+1$, $(\vec{f}_2)(t_2) = \tau[\vec{f}_1(t_1)]$ and $(\vec{f}_2)_k(t_2) =$ $= (\vec{f}_1)_k(t_1)$ for any $k \neq j$;

(iii) if $p = p_1; p_2$ then $\langle t_1, \vec{f} | p | t_2, \vec{f}_2 \rangle$ iff there is such a state $(t_3, \vec{f}_3)$ that $\langle t_1, \vec{f}_1 | p_1 | t_3, \vec{f}_3 \rangle$ and $\langle t_3, \vec{f}_3 | p_2 | t_2, \vec{f}_2 \rangle$;

(iv) Now let p = *if* α *then* p_1 *else* p_2 *fi*. $\langle t_1, \vec{f}_1 | p | t_2, \vec{f}_2 \rangle$ iff the following holds. In the case $\mathfrak{A} \models \alpha[t_1, \vec{f}_1]$ $\langle t_1, \vec{f}_1 | p_1 | t_2, \vec{f}_2 \rangle$ holds and in the other case $\langle t_1, \vec{f}_1 | p_2 | t_2, \vec{f}_2 \rangle$ holds;

(v) Now let p = *while* α *do* p_1 *od*. Let us define $\langle t_1, \vec{f}_1 | p | t_2, \vec{f}_2 \rangle$ in such a way that $\langle t_1, \vec{f}_1 | p | t_2, \vec{f}_2 \rangle$ if $\mathfrak{A} \models \neg\alpha[t_1, \vec{f}_1]$ and

$t_2 = t_1, \vec{f}_1 = \vec{f}_2$ or $<t_1, \vec{f}_1|p;$

while α *do* p_1 *od*$|t_2, f_2>$.

Since N is well-founded there is a unique $Trace_{A_\Pi}$ that satisfies the above conditions.

□

Note that in conditions (iv) and (v) it is supposed that the evaluation of the formula α does not take time. Of course, this assumption can be eliminated without any difficulty if needed. If ($\mathfrak{A}$, $Trace_\Pi$) is deterministic, prehistory-independent or time-independent then so is ($\mathfrak{A}$, $Trace_{A_\Pi}$) respectively.

Now we continue investigating programming language A_Π by introducing such tools that enable us to consider models with nonstandard time as well. Further on these tools will play a very important role.

In the case of nonstandard inner time $\mathfrak{T}$ it may happen that the universe T is not well founded.

If so then some problems are affectuated by condition (v). In order to avoid them at least the "interesting" subsets must be supposed to have minimal elements. To formulate this property we introduce the notion of regular models.

4.3 Definition

Let $\downarrow$, $\uparrow$ be two distinct constant symbols in σ_d. A σ-type model $\mathfrak{A} = (\mathfrak{T}, \mathfrak{D}, I)$ is called *regular* iff it satisfies the followings:

1) $\mathfrak{A} \models \uparrow \neq \downarrow$ i.e. D contains at least two definable elements;

2) For any formula α and any t-sorted variable symbol t $\mathfrak{A} \models \alpha[0/t] \wedge \forall t(\alpha \rightarrow \alpha[t+1/t]) \rightarrow \forall t \alpha$ i.e. the induction works well on T.

3) For any α

$$\mathfrak{A} \models \forall t \ \exists! y \alpha \rightarrow \exists f \forall t \ \alpha[ext(f,t)/y],$$

where t is a t-sorted variable symbol, y is a d-sorted and f is an i-sorted one.
I.e. if a function $T \rightarrow D$ is parametrically definable then it is an "intensional" one which can be named by an element of I.

4) The interpretation of the relation symbol $\leq$ (of type (t,t)) in $\mathfrak{A}$ is a linear ordering with minimal element $\mathfrak{A}(0)$ which has a unique successor $\mathfrak{A}(+1)$.
Further on we write 0 instead of $\mathfrak{A}(0)$ and t+1 instead of $\mathfrak{A}(+1)(t)$.

□

Our aim is to prove that in a regular model the interpretation of the programming language A_{Π} can be given. First we list some elementary properties of regular models.

4.4 LEMMA

Let $\mathfrak{A}$ be a σ-type regular model. Then the following holds:

a) $\mathfrak{A} \models \forall t \ (t \neq 0 \rightarrow \exists s \ (s+1=t))$;

b) If R is a parametrically definable subset of T then

(i) if $R \neq \emptyset$ then R has a minimal element;

(ii) *if R is bounded then it has a maximal element;*

(iii) *if $r \in R$ is an intermediate element of R i.e. there are such r', $r'' \in R$ that $r' < r < r''$ then there is a successor and a predecessor of r in R.*

Proof

a) Let $\alpha \overset{d}{=} t \neq 0 \to \exists s(s+1 = t)$. It is clear that $\mathfrak{A} \models \alpha[0/t]$ and $\mathfrak{A} \models \alpha[t+1/t]$ because

$\alpha[0/t]$ is $0 \neq 0 \to \exists s(s+1) = 0$

and $\alpha[t+1/t]$ is $t+1 \neq 0 \to \exists s(s+1 = t+1)$.

Thus by 2) of 4.3 $\mathfrak{A} \models \forall t \alpha$.

b) (i) Let β be such a formula that for an appropriate vector of parameters $\vec{r}$ the set

$R = \{\underset{\sim}{t} | \mathfrak{A} \models \beta\ [\underset{\sim}{t}, \underset{\sim}{\vec{r}}]\}$

is non-empty. Let

$$\alpha_1 = \forall u \exists v(\beta[u/t] \to \beta[v|t] \wedge v<u) \wedge \exists u \beta[u|t]$$

$$\alpha_2 = \forall s(s \leq t \to \neg\beta[s/t]).$$

It is obvious that $\mathfrak{A} \models \alpha_1[\underset{\sim}{\vec{r}}]$ means that $R \neq \emptyset$ and R has no minimal element. It is clear that $\mathfrak{A} \models \alpha_2[\underset{\sim}{t}, \underset{\sim}{\vec{r}}]$ implies $\underset{\sim}{t} \notin R$. To contradict we prove $\mathfrak{A} \models \forall t(\alpha_1 \to \alpha_2)$ which proves that R has a minimal element.

$$(\alpha_1 \to \alpha_2)[0/t] \overset{d}{=} \alpha_1 \to \forall s(s \leq 0 \to \neg\beta[0/t]).$$

By condition 4 of 4.3 0 is the minimal element of T thus α_1 implies $\neg\beta[0/t]$.

Now we turn to prove

$$(\alpha_1 \to \alpha_2) \to (\alpha_1 \to \alpha_2)\ [t+1/t]\ (*)$$

or equivalently if $\mathfrak{A} \models \alpha_1$, $\mathfrak{A} \models \alpha_2$ then $\mathfrak{A} \models \alpha_2[t+1/t]$. If $\mathfrak{A} \not\models \alpha_2[t+1/t] = \forall s(s \leq t+1 \rightarrow \neg\beta[s/t])$ then by $\mathfrak{A} \models \alpha_1$ we get $\mathfrak{A} \models \exists s\ (s \leq t \wedge \beta[s/t])$ which contradicts to $\mathfrak{A} \models \alpha_2$. Thus (*) holds.

The remaining parts of the proof is omitted.

□

4.5 Definition

A binary relation R on T is called *monoton* iff $t_1\ R\ t_2$ implies $t_1 \leq t_2$. (I.e. $R \subset \leq_{\mathfrak{T}}$!)

□

Let us fix a regular σ-type model $\mathfrak{A} = (\mathfrak{T}, \mathfrak{D}, I)$

4.6 LEMMA

Let R be a parametrically definable monoton relation on T. Then there is a binary relation R^* which is the minimal parametrically definable solution of $S = S \circ S$ and $R \subset S$.

Proof

Let $\phi(t_1, t_2, \vec{a})$ be the formula defining R with parameters $\underset{\sim}{\vec{a}}$. Take

$\phi^*(t_1, t_2, \vec{a}) = \quad f_1\ \{ext\ (f_1, t_1) = \uparrow\wedge$

$ext\ (f_1, t_2) = \uparrow\wedge t_1 \leq t_2 \wedge$

$\forall s_1 \forall s_2\ [t_1 \le s_1 \le s_2 \le t_2 \wedge ext(f_1, s_1) = \uparrow \wedge ext\ f_1, s_2) = \uparrow \wedge$

$\forall s (s_1 < s < s_2 \rightarrow\ ext(f_1, s) = \downarrow) \rightarrow \phi(s_1, s_2, a)]\}.$

It is clear by using condition 3) of 4.3 that

$$R^* = \{(\underset{\sim}{t}_1, \underset{\sim}{t}_2) | \mathfrak{A} \models \phi^*[\underset{\sim}{t}_1, \underset{\sim}{t}_2, \underset{\sim}{\vec{a}}]\}$$

satisfies $R^* = R^* \circ R^*$ and $R \subset R^*$.

Now let $\rho(t_1, t_2, \vec{b})$ be such that for an appropriate $\underset{\sim}{\vec{b}}$ if $R_1 = \{(\underset{\sim}{t}_1, \underset{\sim}{t}_2) | \mathfrak{A} \models \rho[\underset{\sim}{t}_1, \underset{\sim}{t}_2, \underset{\sim}{\vec{b}}]\}$ then $R \subset R_1$ and $R_1 = R_1 \circ R_1$. By using induction on t_2 we prove that $R^* \subset R_1$.

First let $\underset{\sim}{t}_2 = 0$. If $(0,0) \in R$ then, of course, $(0,0) \in R^* \cap R_1$.

Let us suppose that for an arbitrary $\underset{\sim}{t} > 0$ holds:

if $\underset{\sim}{t}_1 \le \underset{\sim}{t}_2 \le \underset{\sim}{t}$ and $(\underset{\sim}{t}_1, \underset{\sim}{t}_2) \in R$ then $(\underset{\sim}{t}_1, \underset{\sim}{t}_2) \in R_1$.

If $\underset{\sim}{t}_1 = \underset{\sim}{t}_2 = \underset{\sim}{t}+1$ then $(\underset{\sim}{t}_1, \underset{\sim}{t}_2) \in R$ also holds and thus $(\underset{\sim}{t}_1, \underset{\sim}{t}_2) \in R_1$. Otherwise $\underset{\sim}{t}_1 \le \underset{\sim}{t}$ and by the definition of R^* there is such an $\underset{\sim}{f} \in I$ that for a $\underset{\sim}{t}_3 \in T$ the following holds: $\underset{\sim}{t}_1 \le \underset{\sim}{t}_3 \le \underset{\sim}{t}_2$ and $ext\ (\underset{\sim}{f}, \underset{\sim}{t}_3) = \uparrow$

for every $s \in T$ if $\underset{\sim}{t}_3 < \underset{\sim}{s} < \underset{\sim}{t}_2$ then $ext\ (\underset{\sim}{f}, \underset{\sim}{s}) = \downarrow$.

Now let $\underset{\sim}{f}^*$ be such that

$$ext\ (\underset{\sim}{f}^*, \underset{\sim}{s}) = \begin{cases} ext\ (\underset{\sim}{f}, \underset{\sim}{s}) & \underset{\sim}{s} \ne \underset{\sim}{t}_2 \\ \downarrow & \underset{\sim}{s} = \underset{\sim}{t}_2 \end{cases}$$

Using this $\underset{\sim}{f}^*$ we get $\phi^*\ (\underset{\sim}{t}_1, \underset{\sim}{t}_3, \underset{\sim}{\vec{a}})$ i.e. $(\underset{\sim}{t}_1, \underset{\sim}{t}_3) \in R$ and by our assumption $(\underset{\sim}{t}_1, \underset{\sim}{t}_3) \in R_1$ holds. Since $(\underset{\sim}{t}_1, \underset{\sim}{t}_3) \in R_1, (\underset{\sim}{t}_3, \underset{\sim}{t}_2) \in R$ and $R \subset R_1$ we have $(\underset{\sim}{t}_1, \underset{\sim}{t}_2) \in R_1$.

The abovesaids can be formulated in F_σ and thus we have proved that $R^* \subset R_1$.

□

4.7 Definition

Let $\mathfrak{A}$ be an arbitrary σ-type model. A relation R is said to be *trace-like* in the model $\mathfrak{A}$ if it satisfies both

(i) $R \subset \mathfrak{S}_{\mathfrak{A}} \times \mathfrak{S}_{\mathfrak{A}}$;

(ii) If $((t_1, \vec{f}_1), (t_2, \vec{f}_2)) \in R$.

then a) $t_1 \le t_2$

b) for any $s \le t_1$ $\vec{f}_1 (s) = \vec{f}_2 (s)$.

□

4.8 LEMMA

If R is a parametrically definable trace-like relation in a regular model $\mathfrak{A}$ then there is a relation R^* which is the minimal trace-like parametrically definable solution of $S = S \circ S$ and $R \subset S$.

Proof

The idea of the proof is the same as in the previous lemma but we use the fact, that if $((t_1, \vec{f}_1), (t_2, \vec{f}_2)) \in R$ then $\vec{f}_1$ can be computed from t_1 and $\vec{f}_2$ by taking

$$\vec{f}_1(s) = \begin{cases} \vec{f}_2(s) & s \le t_1, \\ \vec{f}_2(t_1) & s > t_1. \end{cases}$$

□

Relation R^* in the two above lemmas is said to be the *transitive closure* of relation R.

4.9 THEOREM

Let $\mathfrak{A}$ *be a* σ*-type regular model and let* ($\mathfrak{A}$, $Trace_\Pi$) *be such an interpretation of* (Π, w) *that for any* $p \in \Pi$ $Trace_\Pi \restriction p$ *is parametrically definable in* $\mathfrak{A}$. *In this case there is such a unique Trace that*

a) ($\mathfrak{A}$, *Trace) is an interpretation of* A_Π *such that* $Trace \restriction \Pi = Trace_\Pi$*;*

b) $Trace \restriction p$ *is parametrically definable in* $\mathfrak{A}$ *for each* $p \in A(\Pi)$*;*

c) it satisfies conditions (i)-(v) of 4.2.

Proof

The unicity of *Trace* immediately follows from the definiability of $Trace_\Pi \restriction p$(for any $p \in \Pi$) by using induction i.e. condition 2) of 4.3.

We define *Trace* by using induction on the complexity of $p \in A(\Pi)$. If $p \in \Pi$ or $p = x_i \leftarrow \tau$ then *Trace* is given by $Trace_\Pi$ or by condition (ii) of 4.2. Obviously conditions (iii)-(iv) can be verified without any difficulties. By using the previous lemma we give *Trace* for a program p = *while* α *do* p_1 *od*. Let us suppose that for p_1 we have defined *Trace* in such a way that there is a formula ϕ in $\mathfrak{A}$ which parametrically defines $Trace \upharpoonright p_1$. First we construct a new relation $\hat{\phi}$ and prove that $(\hat{\phi})^*$ gives such a relation for which condition (v) holds.

Take $\hat{\phi}(t_1,\vec{f}_1;t_2,\vec{f}_2) \overset{d}{=} (\alpha[\vec{f}_1(t_1)]\wedge\phi(t_1,\vec{f}_1;t_2,\vec{f}_2))\vee$

$$(\neg\alpha[\vec{f}_1(t_1)] \wedge t_1 = t_2 \wedge \vec{f}_1 = \vec{f}_2)$$

Here

$$\alpha[\vec{f}_1(t)] \overset{d}{=} \alpha[\boldsymbol{ext}\ ((\vec{f}_1)_1,t_1)/x_1,\ldots,\boldsymbol{ext}((\vec{f}_1)_n,t_1)/x_n]$$

$\hat{\phi}$ is obviously trace-like.

It is left to the reader that taking $\langle \underset{\sim}{t}_1,\underset{\sim}{\vec{f}}_1|p|\underset{\sim}{t}_2,\underset{\sim}{\vec{f}}_2\rangle$ iff $\mathfrak{A} \models (\hat{\phi})^* [\underset{\sim}{t}_1, \underset{\sim}{\vec{f}}_1, \underset{\sim}{t}_2, \underset{\sim}{\vec{f}}_2]$ then condition (v) of 4.2 is satisfied.

□

From the aboves it follows that the transitive closure of *Trace* plays an important role in interpreting programming languages similar to our Algol fragment. Thus it is quite natural that in the programming language given as the set of regular programs of dynamic logic the transitive closure is provided as a program construction.

4.10 Definition

We define the set of dynamic formulas DL and the set of regular programs RG by simultaneous induction. Let us fix a d-sorted similarity type σ_d. DL and RG are the minimal set for which

$(i)_{RG}$ for any variable x_i and term τ $x_i \leftarrow \tau \in$ RG;

$(ii)_{RG}$ if $p_1, p_2 \in$ RG then

$(p_1; p_2)$ and $(p_1 \cup p_2)$ belong to RG;

$(iii)_{RG}$ if $p \in$ RG then $p^* \in$ RG;

$(i)_{DL}$ $F_{\sigma_d} \subset$ DL;

$(ii)_{DL}$ if $\phi_1, \phi_2 \in$ DL then

$\phi_1 \vee \phi_2 \in$ DL;

$(iii)_{DL}$ if $\phi \in$ DL then $\neg\phi \in$ DL;

(iv) if $\phi \in$ DL, $p \in$ RG then $\langle p\rangle\phi$, $[p]\phi \in$ DL and $(\phi ?) \in$ RG.

□

For the sake of convenience let

$$\textit{if } \alpha \textit{ then } p_1 \textit{ else } p_2 \textit{ fi} \stackrel{d}{=} ((\alpha?);p_1) \cup ((\neg \alpha?); p_2),$$

$$\textit{while } \alpha \textit{ do } p \textit{ od} \stackrel{d}{=} (((\alpha?); p) \cup (\neg \alpha?))^*,$$

$$\cap px \stackrel{d}{=} [p^*]x,$$

$$\cup px \stackrel{d}{=} \langle p^*\rangle x.$$

4.11 Definition

Let $\mathfrak{A}$ be a σ-type regular model. We give a σ-type interpretation of RG in $\mathfrak{A}$.

For this aim we simultaneously define a relation Trace_{RG} in $\mathfrak{A}$ by induction and the validity relation $\models$ for the formulas of DL.

Our induction assumption:

$Trace_{RG}\restriction p$ and $\{k \mid \mathfrak{A} \models \phi[k]\}$ are parametrically definable in $\mathfrak{A}$ for any program $p \in RG$ and formula $\phi \in DL$ the complexity of which is less than that of those to be considered in the followings.

$(i)_{RG}$ For $p = x_j \leftarrow \tau$ let $\langle k_1 | x_j \leftarrow \tau | k_2 \rangle$ iff $(k_1, p, k_2) \in Trace A_{\emptyset}$ (Recall 4.2 and take $\Pi = \emptyset$);

$(ii)_{RG}$ For $p_1, p_2 \in RG$ let $\langle k_1 | p_1; p_2 | k_2 \rangle$ iff there is such a state $k_3 \in \mathfrak{S}_{\mathfrak{A}}$ that $\langle k_1 | p_1 | k_3 \rangle$ and $\langle k_3 | p_2 | k_2 \rangle$ and let $\langle k_1 | p_1 \cup p_2 | k_2 \rangle$ iff $\langle k_1 | p_1 | k_2 \rangle$ or $\langle k_1 | p_2 | k_2 \rangle$;

$(iii)_{RG}$ Let $p \in RG$ be arbitrary. By the induction assumption $Trace_{RG}\restriction p$ is a parametrically definable relation. Since $Trace_{RG}\restriction p$ is obviously a trace-like relation there exists its transitive closure, say R, which is also parametrically definable in $\mathfrak{A}$. Take $Trace_{RG}\restriction p^* \stackrel{d}{=} R$;

$(i)_{DL}$ For $\phi \in F_{\sigma_d}$ and an arbitrary state $k \in \mathfrak{S}_{\mathfrak{A}}$ $\mathfrak{A} \models \phi[k]$ is given by 3.2;

$(ii)_{DL}$ If $\phi_1, \phi_2 \in DL$ then for an arbitrary state $k \in \mathfrak{S}_{\mathfrak{A}}$ $\mathfrak{A} \models (\phi_1 \vee \phi_2)[k]$ iff $\mathfrak{A} \models \phi_1 [k]$ or $\mathfrak{A} \models \phi_2 [k]$;

$(iii)_{DL}$ If $\phi \in DL$ and $k \in \mathfrak{S}_{\mathfrak{A}}$ then $\mathfrak{A} \models \neg \phi[k]$ iff $\mathfrak{A} \not\models \phi[k]$,

(iv) a) $\mathfrak{A} \models \langle p \rangle \phi[k]$ iff there is a $\hat{k} \in \mathfrak{S}_{\mathfrak{A}}$ such that $\langle k|p|\hat{k}\rangle$ and $\mathfrak{A} \models \phi[\hat{k}]$;

b) $\mathfrak{A} \models [p]\phi[k]$ iff for all $\hat{k} \in \mathfrak{S}_{\mathfrak{A}}$ if $\langle k|p|\hat{k}\rangle$ then $\mathfrak{A} \models \phi[\hat{k}]$;

c) $\langle k_1|\phi?|k_2\rangle$ iff $k_1 = k_2$ and $\mathfrak{A} \models \phi[k_1]$.

It is clear that for all programs and formulas obtained by conditions (i)-(iv) of 4.2 the induction assumption also holds. Thus to any σ-type regular model $\mathfrak{A}$ a corresponding σ-type interpretation $(\mathfrak{A}, \mathit{Trace})$ has been defined, which is called a *regular interpretation* (of the programming language RG).

□

Let $\mathfrak{A}$ be a σ-type regular model. The relations Trace_{RG} and Trace_{A} are such that (where $A \overset{d}{=} A_{\emptyset}$):

$$\mathit{Trace}_A \restriction \mathit{if}\ \alpha\ \mathit{then}\ p_1\ \mathit{else}\ p_2\ \mathit{fi} =$$

$$= \mathit{Trace}_{RG} \restriction ((\alpha?);p_1)\cup((\neg\alpha?);p_2)$$

$$\mathit{Trace}_A \restriction \mathit{while}\ \alpha\ \mathit{do}\ p\ \mathit{od} = \mathit{Trace}_{RG} \restriction ((\alpha?);\ p)\cup(\neg\alpha?))^*$$

For a regular model $\mathfrak{A}$, $\mathfrak{A}^*$ denotes either the interpretation $(\mathfrak{A}, \mathit{Trace}_A)$ or $(\mathfrak{A}, \mathit{Trace}_{RG})$ according to the given programming language.

Now we give two subsets of RG and DL.

1) First let RG_1, DL_1 be the same as RG and DL but the condition (iv) now provides programs of the form $(\phi?)$ only for the formulas $\phi \in F_{\sigma_d}$.

The programming language RG_1 obtained this way is the same as the set of regular programs and the language DL_1 is the same as the set of well-formed formulas of *dynamic logic*. (See e.g. in Harel [9]).

2) Now let AL, LP be the minimal subsets of DL and RG, respectively, for which:

(i) $x_i \leftarrow \tau$ is in LP for any variable x_i and term τ;

(ii) if $p_1, p_2 \in$ LP and $\alpha \in Q_{\sigma_d}$ then $(p_1;p_2)$, *if* α *then* p_1, *else* p_2 *fi* and *while* α *do* p_1 *od* are in LP;

(iii) if $\phi_1, \phi_2 \in$ AL then $\phi_1 \vee \phi_2$ and $\neg \phi_1$ are in AL;

(iv) if $p \in$ LP and $\phi \in$AL then $\langle p \rangle\ \phi$, $\cap p \phi$ and $\cup p \phi$ are in AL.

The subsets AL and LP are the same as the corresponding components of *algorithmic logic* introduced and analysed in details by a group of Polish mathematicians (see e.g. Banachowski et al [2]).

In the aboves we introduced the notion of programming language in general as a pair (P,v). To analyse some specific properties of programs this general notion of programming languages should be made concrete. In 3.6 and 3.7 we have seen some concrete programming languages. However the Algol-like language introduced in 4.1 will play an important role below. Namely it will be used to interpret the programs of other programming languages.

First we introduce a programming language the construction of which is general enough to belong to most of the existing programming languages.

4.12 Definition

Let σ_d be a fixed data-sorted similarity type. An acceptable programming language $\mathfrak{P}$ consists of a programming language (P,v) together with the following functions

$\circ : Q_{\sigma_d} \times P \times P \to P,$

$* : Q_{\sigma_d} \times P \to P,$

$; : P \times P \to P$, i.e. $\mathfrak{P} \stackrel{d}{=} ((P,v), \circ, *, ;)$

(Q_{σ_d} see in Appendix).

To give meaning to each program belonging to $\mathfrak{P}$, we interpret the programs in the σ-type interpretations of the Algol-like programming language A ($\stackrel{d}{=} A_{\emptyset}$).

4.13 Definition

Let $\mathfrak{P}$ be an arbitrary acceptable programming language. Let $\mathfrak{A}^* = (\mathfrak{A}, Trace_A)$ be a regular σ-type interpretation of A. Let $Int: P \to A$ be an arbitrary recursive function.

The pair ($\mathfrak{A}^*$, *Int*) is said to be a σ-type *realization* of the acceptable programming language $\mathfrak{P}$ iff it satisfies the conditions below:

(i) $v(p) \leq v(Int\ (p))$ for any $p \in P$;

(ii) $Int\ (\circ(\alpha,p_1,p_2))$ = *if* α *then* $Int\ (p_1)$ *else* $Int\ (p_2)$ *fi* for any $p_1, p_2 \in P$ and $\alpha \in F_{\sigma_d}$;

(iii) $Int\ (p_1;p_2) = (Int(p_1);\ Int\ (p_2))$ for any $p_1, p_2 \in P$;

(iv) $Int\ (*(\alpha,p))$ = *while* α *do* $Int\ (p)$ *od* for any $p \in P$ and $\alpha \in F_{\sigma_d}$.

□

We remark that by using $Trace_A$ we can define an interpretation ($\mathfrak{A}$, *Trace*) of $\mathfrak{P}$ in $\mathfrak{A}$ by taking

$\langle k_1|p|k_2\rangle$ iff $\langle k_1|Int(p)|k_2\rangle$ for any $p \in P$ and $k_1, k_2 \in \mathfrak{S}_{\mathfrak{A}}$.

A class of σ-type realizations $\mathfrak{R}$ of an acceptable programming language $\mathfrak{P}$ is called regular iff all elements of $\mathfrak{R}$ have the same function *Int*.

5. *ON DESCRIPTIVE LANGUAGES*

According to Introduction after having introduced the class of models with inner time and having shown they can be used to represent the meaning in the case of certain programming languages we develope a language for describing properties of programming languages and programs.

Moreover we analyse the completeness of the language that is being defined.

The basic characters of the descriptive language most widely spread were outlined by Naur and developed by Floyd and Hoare.

The main feature of this language is that it provides only comparison of the initial state with the state after the program execution. The language realizes this through the description of the program structure, i.e. by the usage of localization principle.

However this descriptive language tells nothing about the details of execution of the program parts. Though the original versions of this language do not allow any references to time, we allow our version to speak about when a program starts and when it stops.

Note that the dynamic logic which contains the algorithmic logic as a special case is built on the elementary input-output-type statements.

5.1 Definition

Let us fix a σ-type programming language $\mathfrak{P} = (P, v)$

Take $FH \stackrel{d}{=}$

$$\{\phi\langle p\rangle\Psi \mid \phi,\Psi \in F_{\sigma_d}; p \in P\} \cup \{\phi[p]\Psi \mid \phi,\Psi \in F_{\sigma_d}; p \in P\}$$

□

Note that though FH depends on the programming language $\mathfrak{P}$ (more precisely on P) and thus on the similarity type as well we do not indicate this dependency but it is always meant included.

Let ($\mathfrak{A}$, *Trace*) be an arbitrary interpretation of $\mathfrak{P}$, and let $p \in P$, $k_j = (t_j, \vec{f}_j)$ $(j = 1,2)$ be such that $\ell(k_1) = \ell(k_2) \geq v(p) = n$.

Remember that $\langle k_1 | p | k_2\rangle$ means that $\langle t_1, (f_{11}, \ldots, f_{1n}) | p | t_2, (f_{21}, \ldots, f_{2n})\rangle$ and for any index $i \notin [1,n] (\vec{f}_1)_i = (\vec{f}_2)_i$.

Let us fix a σ-type interpretation $\mathfrak{A}^* = (\mathfrak{A},$ *Trace*).

5.2 Definition

Let us fix $p \in P, \phi, \psi \in F_{\sigma_d}$. Let $k \in \mathfrak{S}_{\mathfrak{A}}$ be such that $\ell(k) \geq \max\{v(p), v(\phi), v(\psi)\}$. We say that the program p starting with k is *< >-correct* (read diamond-correct) in $\mathfrak{A}^*$ w.r.t. the input condition ϕ and output condition ψ (written as $\mathfrak{A}^* \models \phi\langle p\rangle\psi[k]$) iff either $\mathfrak{A} \not\models \phi[k]$ or there is such a $k^* \in \mathfrak{S}_{\mathfrak{A}}$ that $\langle k | p | k^*\rangle$ and $\mathfrak{A} \models \psi[k^*]$. If the program p is <>-correct in $\mathfrak{A}^*$ w.r.t. ϕ and ψ for any k then we write $\mathfrak{A}^* \models \phi\langle p\rangle\psi$.

Similarly p starting with k is []-*correct* (read box-correct) in $\mathfrak{A}^*$ w.r.t the input condition ϕ and output condition ψ (written as $\mathfrak{A} \models \phi[p]\psi[k]$) iff either $\mathfrak{A} \not\models \phi[k]$ or for such k^* that $\langle k|p|k^*\rangle$ we have $\mathfrak{A} \models \psi[k^*]$. If the program p is []-correct in $\mathfrak{A}^*$ w.r.t. ϕ and ψ for any state k then we write $\mathfrak{A}^* \models \phi[p]\psi$.

□

Note that if the σ-type interpretation $\mathfrak{A}^*$ is a deterministic one (see 3.5) then $\mathfrak{A}^* \models \phi[p]\psi$ and $\mathfrak{A} \models \phi\langle p\rangle\psi$ coincide.

For any $\chi \in$ FH we have $\mathfrak{A}^* \models \chi$ iff for any state k with $\ell(k) \geq v(p)$

$$\mathfrak{A}^* \models \chi[k] .$$

Let $\mathfrak{R}$ be any class of σ-type interpretations of P. Our descriptive language is

$$\mathfrak{F}^{\mathfrak{R}} \stackrel{d}{=} (\mathrm{FH}, \mathfrak{R}, \models) .$$

Let

$$E_o^{\mathfrak{R}} = \{\phi \in \mathrm{FH} | \mathfrak{R} \models \phi\} .$$

If $\mathfrak{R}$ is an arbitrary class of σ-type regular interpretations, then we can use the formulas of DL, i.e. we have the following descriptive language

$$\mathfrak{D}^{\mathfrak{R}} = (\mathrm{DL}, \mathfrak{R}, \models)$$

Let $D^{\mathfrak{R}} = \{\phi \in \mathrm{DL} | \mathfrak{R} \models \phi\}$

Thus we have defined the descriptive languages $\mathfrak{F}^{\mathfrak{R}}$ and $\mathfrak{D}^{\mathfrak{R}}$ with the set of true formulas $E_o^{\mathfrak{R}}$ and $D^{\mathfrak{R}}$ respectively.

The descriptive language should provide not only the description of program properties, but should also be suitable for proving them. In order to do this many types of calculi can be constructed by varying selection of syntactical rules for derivation of formulas. A calculus does not fulfil its task totally unless it is complete in a certain sense. Completeness means that there is a fixed type calculus by which every true statement or, at least, all the "interesting" statements can be proved. To formulate this, let $L = (F,M,\models)$ be an arbitrary language and let $H \subset Th(L)$ and A be arbitrary.

5.3 Definition

L is complete w.r.t. H and relative to A iff there is a set E such that $H \subset E \subset Th(L)$ and E is recursively enumerable in A. If $E = Th(L)$, $A = \emptyset$ or $E = Th(L)$ we simply say the language L is complete or complete relative to A, respectively. It is clear that the completeness of L means that its true statements are recursively enumerable.

□

Intuitively "E is recursively enumerable in A" means that there is an algorithm (e.g. a Turing machine) enumerating the elements of E such that it has to turn to an oraculum from time to time to learn whether a certain entity belongs to the set A. Thus the enumeration of E is ensured by the algorithm together with the oraculum. The exact definitions see e.g. in Rogers [15].

At the same time the recursive enumerability of the true statements of L means that either $F \subset N$ or at least F can be coded in N. Of course, $F \subset N$ generally does not hold. However, if F is given by a generative grammar then there is a natural embedding of F into N. In this case the above definition works well. Remember that for a class $\mathfrak{R}$ of σ-type models

$$\mathrm{Th}\ (\mathfrak{R}) = \{\phi \mid \phi \in F \ \text{ and } \mathfrak{A} \models \phi \ \text{ for all } \ \mathfrak{A} \in \mathfrak{R}\}.$$

5.4 THEOREM (Completeness)

Let $\mathfrak{R}$ be a class of regular interpretations of a σ-type acceptable programming language $\mathfrak{P}$. The languages $\mathfrak{F}^{\mathfrak{R}} = (FH, \mathfrak{R}, \models)$ and $\mathfrak{D}^{\mathfrak{R}} = (DL, \mathfrak{R}, \models)$ are complete relative to Th $(\mathfrak{R})$.

Proof

By 4.9 and by the regularity of $\mathfrak{R}$ for any $\chi \in$ DL there is a formula $\alpha_\chi \in F_\sigma$ such that for an arbitrary $\mathfrak{A}^* \in \mathfrak{R}$

$$\mathfrak{A} \models \alpha_\chi \quad \text{iff} \quad \mathfrak{A}^* \models \chi \ .$$

Moreover the function $\chi \to \alpha_\chi$ is recursive. This means that $D^{\mathfrak{R}}$ is recursively enumerable in Th $(\mathfrak{R})$. Since $E_0^{\mathfrak{R}}$ is a recursive subset of $D^{\mathfrak{R}}$ it is also recursively enumerable in Th $(\mathfrak{R})$.

□

5.5 COROLLARY

If $\mathfrak{R}$ in the above theorem is recursively axiomatizable then $E_0^{\mathfrak{R}}$, $D^{\mathfrak{R}}$ (and Th $(\mathfrak{R})$ of course) are recursively enumerable too.

□

Note that the notion of ultraproduct can be extended to our model constructions similarly to Gabbay [5]. It is obvious that the ultraproduct of regular models (and interpretations) is also regular.

We give the following theorem without proof.

5.6 THEOREM

Let $\mathfrak{R}$ be a class of σ-type regular interpretations of a σ-type acceptable programming language $\mathfrak{P}$ which is closed under ultraproducts.

The descriptive languages $\mathfrak{F}^{\mathfrak{R}}$ and $\mathfrak{D}^{\mathfrak{R}}$ are compact i.e. a set of formulas $\Sigma \subset FH$ (respectively $\Sigma \subset DL$) is satisfiable in $\mathfrak{R}$ iff each finite subset of Σ is satisfiable in $\mathfrak{R}$.

□

Note that according to the traditions it is convenient to express the program properties by using only the σ_d-type formulas. This means that it is expected to describe time through data. One possibility is shown in 2.6.

In this case such completeness results are needed that provide the enumerability of Th $(\mathfrak{R}_d)$, where $Th(\mathfrak{R}_d) \overset{d}{=} Th\ (\mathfrak{R}) \cap F_{\sigma_d}$. We return to this question in an other paper.

6. MODELS STANDARD IN TIME AND COMPLETENESS

In this section we show that if we restrict ourselves to the models standard in time then the recursive enumerability of true formulas of the descriptive languages gets lost even for the case of Algol-like programming language A. Considering this we try to adjust the notion of completeness such that it becomes adequate to the programming situation namely to program synthesis and with respect to which the descriptive language becomes complete.

From now on let us fix the programming language $\mathfrak{P}$ to be $A_{\emptyset}$, i.e. A, and we consider only interpretations of the form $(\mathfrak{A}, Trace_A)$ where $\mathfrak{A}$ is a model standard in time. Thus a σ-type model $\mathfrak{A}$ standard in time is also used to denote the corresponding interpretation of A.

6.1 Definition

A σ-type model $\mathfrak{A}$ $(=(\mathfrak{N}, \{\mathfrak{D}_n | n \in N\}, I))$ standard in time is called *arithmetic* if

(i) $\eta \subset \sigma_d$;

(ii) $\mathfrak{A}$ is semi-simple;

(iii) for any $n \in N$ the η-type reduct of $\mathfrak{D}_n$ has a submodel isomorphic to N.

A class $\mathfrak{M}$ of models is called *arithmetic* iff each $\mathfrak{A} \in \mathfrak{M}$ is arithmetic.

□

For example if $\mathfrak{D}_n \models PA$ for any $n \in N$ then $\mathfrak{A}$ is arithmetic.

6.2 *THEOREM*

If $\mathfrak{M}$ *is an arithmetic class of* σ*-type models then the language*

$$\mathfrak{F}^{\mathfrak{M}} = (FH, \mathfrak{M}, \models)$$

is not complete (i.e. $E_o^{\mathfrak{M}}$ *is not recursively enumerable)*

Proof

To contradict let us suppose that $E_o^{\mathfrak{M}}$ is complete.

Let $\tau_1, \tau_2 \in T_\eta$ and let k be such that $v\ (t_1) \leq k$ and $v\ (\tau_2) \leq k$.

It is well-known that n-tuples are recursively enumerable (see e.g. Malcev [11]). Let C_k be the program which, for any y, gives the y-th k-tuple $(x_1, \ldots, x_k)$. Take the following program $p_{\tau_1 = \tau_2} \stackrel{d}{=}$

$y \leftarrow 0;\ \ C_k;$ *while* $\tau_1 \neq \tau_2$ *do* $y \leftarrow y+1;$ C *od*.

We prove that

$\mathfrak{M} \models$ *true* $<p_{\tau_1 = \tau_2}>$ *false* iff the Diophantine equation $\tau_1 = \tau_2$ is not solvable.

Indeed let $\mathfrak{A} \in \mathfrak{M}$ be arbitrary and let $u \in \mathfrak{S}_{\mathfrak{A}}$ be any input state. Since $\mathfrak{A}$ is arithmetical the trace of $p_{\tau_1 = \tau_2}$ in $\mathfrak{A}$ starting with the state u computes the k-tuples of natural numbers. These tuples are put into the Diophantine equation $\tau_1 = \tau_2$ one by one.

$p_{\tau_1 = \tau_2}$ stops if it finds a solution of $\tau_1 = \tau_2$ in the subuniverse $D_C \subset D$ (where D_C is the minimal subset of D generated by constants). However $p_{\tau_1 = \tau_2}$

does not satisfy the formula *false*. Hence $\mathfrak{M} \models True$ $\langle p_{\tau_1 = \tau_2} \rangle false$ iff the equation $\tau_1 = \tau_2$ is not solvable.

Let us consider the set of formulas $\{true \langle p_{\tau_1=\tau_2} \rangle false \mid \tau_1, \tau_2 \in T_\eta\}$. Obviously it is recursively enumerable. Hence by our supposition the set

$$\{true \langle p_{\tau_1 = \tau_2} \rangle false \mid \tau_1, \tau_2 \in T_\eta\} \cap \{\chi \in FH \mid \mathfrak{M} \models \chi\} =$$

$$= \{true \langle p_{\tau_1 = \tau_2} \rangle false \mid \tau_1 = \tau_2 \text{ is unsolvable}\}$$

is recursively enumerable too. However this contradicts to the fact that the set of Diophantine equations having no solutions in natural numbers is not recursively enumerable. This fact can be found e.g. in Davis [3].

□

In the literature usually one model is considered instead of a class of models. Namely the model $\hat{\mathfrak{N}} = (\mathfrak{N}', \mathfrak{N}, {}^{N}N)$ which we call *strictly standard*. For this case from the above theorem we have.

6.3 COROLLARY

The language (FH, $\{\hat{\mathfrak{N}}\}$, $\models$) is not complete.

□

Another important result concerning the properties of the language $\mathfrak{F}^{\mathfrak{M}}$ is the following.

6.4 THEOREM

For any arithmetic class $\mathfrak{M}$ of σ-type models the language $\mathfrak{F}^{\mathfrak{M}}$ is not compact.

□

See the proof in De Millo [4].

Theorem 6.2 says that $E_0^{\mathfrak{M}}$ is not recursively enumerable. Since $E_0^{\mathfrak{M}}$ contains e.g. the non-terminating programs too, thus it is obvious that not the whole $E_0^{\mathfrak{M}}$ is interested. However the condition $\chi = \phi\langle p\rangle\psi$ does not contain any restriction within the termination of the program p. So the following question arises: Is it possible to narrow the set $E_0^{\mathfrak{M}}$ so that it becomes recursively enumerable? The narrowing can be done e.g. by using pragmatical considerations of programming to give restriction within the termination.

Thus first let us try the following way. Let us suppose that a fixed arithmetic class $\mathfrak{M}$ of σ-type models is recursively axiomatizable by a set $Ax \subset F_\sigma$. Take

$$E_1^{\mathfrak{M}} = E_1^{Ax} \stackrel{d}{=} \left\{ \phi\langle p\rangle\psi \in FH \;\middle|\; \begin{array}{l} Ax \models \phi\langle p\rangle\psi \text{ and the} \\ \text{program } p \text{ terminates} \\ \text{in } \hat{\mathfrak{N}} \text{ for any such input} \\ k \text{ that } \hat{\mathfrak{N}} \models \phi[k] \end{array} \right\}$$

But the following theorem shows that this narrowing is not sufficient.

6.5 THEOREM

The language $\mathfrak{F}^{Ax}$ is not complete w.r.t. E_1^{Ax}.

Proof

It proceeds as in 6.2. To contradict let $Ax \subset F_\sigma$ be arithmetical and let $H \subset FH$ be recursively enumerable such that $E_1^{Ax} \subset H \subset E_o^{Ax}$.

First we prove that there is a recursively enumerable Ax_1 such that $Md(Ax) = Md(Ax_1)$. Indeed, by our assumptions the set

$$Ax_1 = \{\phi \mid Ax \models \mathit{true} \langle p_\emptyset \rangle \psi\}$$

(where $p_\emptyset$ is the program containing no commands) is recursively enumerable and, of course, this Ax_1 is good.

Thus there is such a Diophantine equation $t_1 = t_2$ that $Ax \not\models \forall \vec{x}_1 \neg(t_1 = t_2)$ and $\mathfrak{N} = \neg(t_1 = t_2)$ see e.g. Davis [3], Theorem 7.7). Now let $\tau_1, \tau_2 \in T_\eta$ such that $v(\tau_1) = v(\tau_2)$.

Let us consider the following program $R_{\tau_1 \tau_2} \stackrel{d}{=}$ $\mathit{if} \neg(t_1 = t_2)\ \mathit{then}\ \ y \leftarrow 0\ \ \mathit{else}\ \ p_{\tau_1 = \tau_2};\ \ y \leftarrow 1\ \mathit{fi}$.

To complete the proof we use the following facts:

(i) for any $\tau_1, \tau_2 \in T_\eta$ for which $v(\tau_1) = v(\tau_2)$ $Ax \models \mathit{true} \langle R_{\tau_1 \tau_2} \rangle\ y = 0$ since $t_1 = t_2$ is not solvable in $\mathfrak{N}$;

(ii) for any $k \in \mathfrak{S}_{\mathfrak{N}}$ the program terminates in $\mathfrak{N}$.

The proof of these facts can be found in Gergely - Úry [7] p. 99.

Let $\Gamma \stackrel{d}{=} \{\mathit{true} \langle R_{\tau_1 \tau_2} \rangle\ y = 0 \mid \tau_1, \tau_2 \in T_\eta\}$
By the above facts it is evident that the set $\Gamma \cap H$ is

isomorphic to the set of Diophantine equations which are unsolvable in natural numbers. Since H is recursively enumerable, thus $\Gamma \cap H$ is recursively enumerable too. This is a contradiction.

□

From the point of view of programming theory it would have been good if the language $\mathfrak{F}^{Ax}$ had proved complete w.r.t. E_1^{Ax} because to demand the termination of the programs in the standard models seems to be quite natural. But it is not the case. The above theorem says that the set of E_1^{Ax} is still far too large. Now we introduce the following narrowing of the set E_o^{Ax} as an extreme case. Take

$$E_2^{Ax} \overset{d}{=} \left\{ \phi\langle p\rangle\psi \in FH \;\middle|\; \begin{array}{l} Ax \;\; \phi\langle p\rangle\psi \text{ and } p \text{ terminates} \\ \text{in every model of } Md(Ax) \text{ for} \\ \text{all such input state } k \text{ that} \\ \mathfrak{A} \models \phi[k] \end{array} \right\}$$

As it can be seen, defining the set E_2^{Ax} we demand the termination of p not only in $\mathfrak{N}$, but in all models of Md(Ax).

6.6 THEOREM

For any recursively enumerable Ax containing PA the language $\mathfrak{F}^{Ax}$ *is complete w.r.t.* E_2^{Ax}.

Proof see in Gergely - Úry [7] p. 102.

□

The last theorem seems to be a strong result, but unfortunately the programs used in practice usually do not terminate in every model of Ax because for such termination the programs should be equivalent with a program without cycles.

A subset E of F_{σ_d} w.r.t. which the language $\mathfrak{F}^{Ax}$ is complete requires far too strong conditions from the termination of programs. So if the descriptive language contains *arithmetic* then it is not complete in the classical sense. One can say that this is the result of the presence of arithmetic and so it is suggested arithmetic be either ruled out or restricted. Suppose we followed these suggestions the applicability of theory would be also narrowed while almost all the applications employ natural numbers. On the other hand if we don't forget the original aim of the programming theory then the completeness of $\mathfrak{F}^{Ax}$ in classical sense is not so important. The aim of the programming theory is not to prove all the possible properties of all programs but to write such programs that provably satisfy the given properties.

Therefore it is reasonable to modify the notion of completeness so that it becomes useful for the programming theory. It is evident to expect to each true formula $\phi\langle p\rangle\psi \in FH$ such a program $p' \in P$ that the formula $\phi\langle p'\rangle\psi$ can already be proved. It is obvious that if we have no further assumptions on program p' then a program, that never terminates, can be appointed to p' and thus it would result in a useless notion. In order to give these a further assumption while having in mind the construction of the sets E_1^{Ax} and E_2^{Ax} consider the following two ways:

(i) p' terminates for all input data that p does for;

(ii) p' terminates for all standard input data that p does for.

Let us see the first way.

6.7 Definition

Let $E \subset E_o^{Ax}$ be arbitrary. $\mathfrak{F}^{Ax}$ is said to be *stop-complete w.r.t. E* iff there is a recursively enumerable set $H \subset E$ such that for every $\phi\langle p\rangle\psi \in E$ there is a p' such that $\phi\langle p'\rangle\psi \in E \cap H$ and for every $\mathfrak{A} \in Md(Ax)$ and for every input state $k \in \mathfrak{S}_{\mathfrak{A}}$ if $\mathfrak{A} \models \phi[k]$ and p terminates starting with k then p' also terminates starting with k.

□

Now we introduce such a subset of E_o^{Ax} w.r.t. which $\mathfrak{F}^{Ax}$ would be stop-complete. Let

$$E_s^{Ax} \stackrel{d}{=} \{\phi\langle p\rangle\psi \in E_o^{Ax} \mid \psi \in Q_{\sigma_d}\}.$$

Let Ax be such a set of axioms that the Gödel coding function is computable.

So far we have not distinguished the variable symbols of input and output conditions, but from the point of view of programming we have to do some distinction. Namely if there is some kind of specification then it is important to carry it out during the execution without any change. This provides the possibility to compare the value obtained at termination with that of the specification. Therefore we shall distinguish two disjoint sets of variable symbols X and Y. X corresponds to the input conditions.

6.8 THEOREM

The language $\mathfrak{F}^{Ax}$ is stop complete w.r.t. E_s^{Ax}.

Proof

Let $\phi \in F^{X}_{\sigma_d}$ and $\psi \in Q^{X\cup Y}_{\sigma_d}$ such that $v(\phi) = n$ and $v(\psi) = N$.

Take the following program $p_\psi \overset{d}{=} y\leftarrow 0;\ C_n;\ P_N;$ *while* $\neg\psi$ *do* $y\leftarrow y+1;\ C_n;\ P_N$ *od*.

Here C_n is a program which for any y gives the y-th n-tuple and P_N is the program that provides for any input $k = (t,\vec{f})$ an appropriate element of $\underbrace{H_k\times\ldots\times H_k}_{N\ \text{times}}$, where H_k is the Herbrand universum generated by $ext(\vec{f},t)$.

The program p_ψ enumerates the elements of H_k^N for any input state k and test whether ψ holds for the actual value of y from $H_k \times\ldots\times H_k$. If ψ holds then p_ψ terminates. It is clear that p_ψ starting in a state $k \in \mathfrak{S}_{\mathfrak{A}}$ terminates in any model $\mathfrak{A} \in Md(Ax)$ (standard in time) iff there is such a state $k^* \in \mathfrak{S}_{\mathfrak{A}}$ that $\mathfrak{A} \models \psi[k, k^*]$. (Remember that ψ refers to both input and output).

Let us take

$$R = \{\phi\langle p_\psi\rangle\psi \mid \phi\in F_{\sigma_d},\ \psi\in Q_{\sigma_d}\}$$

It is evident that this set of formulas is recursively enumerable and from the above consideration follows that R corresponds to 6.7.

□

If E_o^{Ax} were stop complete it would mean that to an arbitrary program p satisfying a specification (ϕ,ψ) there exists such a program which is as good as the previous one in respect to termination and this program can be proved. This kind of completeness would fully serve the programming theory but because of 6.13 this cannot be the case. 6.8 shows that the stop-completeness is acceptable when the specifications can be put down in open formulas. But this seems to be insufficient for programming theory. E.g. such a specification exists for the cases when a system of equations must be solved with given exactness, but no such one exists for the tasks of symbolic manipulation. Now we try the other way, i.e. when we expect p′ to terminate for every standard input data for which p terminates.

Now let us turn to the second possibility of modifying completeness.

6.9 Definition

Let $E \subset E_O^{Ax}$ be arbitrary. $\mathfrak{F}^{Ax}$ is said to be *weak-complete* w.r.t. E iff there is a recursively enumerable set $H \subset E_O^{Ax}$ such that it satisfies both:
(i) for every ϕ,ψ if there is a p such that $\phi\langle p\rangle\psi \in E$ then there exists such a p′ that $\phi\langle p'\rangle\psi \in H\cap E$;
(ii) if p terminates in the standard model then then p′ does as well.
□

We note that in this definition E is usually not recursively enumerable.

6.10 THEOREM

For any $Ax \subset F_\sigma$ *the language* $\mathfrak{F}^{Ax}$ *is weak-complete w.r.t.* $E_O^{Ax^\sigma d}$.

Proof

Consider the following porgram $p' \overset{d}{=}$ *while true do* $y = 1$ *od*.

It is evident that p' never terminates and so $Ax \models \phi\langle p'\rangle\psi$. Hence the set $R = \{\phi\langle p'\rangle\psi \mid \phi \in F_{\sigma_d},\ \psi \in Q_{\sigma_d}\}$ satisfies 6.9.

□

It is clear that this theorem does not satisfy any pragmatical expectations.

Now we introduce such a set E_w^{Ax} which plays an important role in the weak-completeness.

A formula ψ of a fixed type θ (i.e. $\psi \in F_\theta^V$) is said to be cuttable iff there exists such a formula $\psi' \in F_\theta^V$ that $Ax \models \psi \leftrightarrow \psi'$ and ψ' has the form $\exists z_o \mu$ where in μ every quantifier is bounded by a formula $v < \tau$ ($v \in V$, $\tau \in T_\theta^{V\setminus\{v\}}$). Let

$$E_w^{Ax} \overset{d}{=} \left\{ \phi\langle p\rangle\psi \in E_o^{Ax} \;\middle|\; \begin{array}{l} p \text{ terminates in } \mathfrak{N} \text{ for any} \\ \text{such } k \text{ that } \mathfrak{N} \models \phi[k] \text{ and} \\ \psi \text{ is cuttable.} \end{array} \right\}$$

6.11 THEOREM

Let Ax *be recursively enumerable and such that* $PA \subset Ax$. *Then* $\mathfrak{F}^{Ax}$ *is weak-complete w.r.t.* E_w^{Ax}.

□

The proof is trivial by using the fact that to each bounded quantifier an appropriate program can be rendered. For details see Gergely - Úry [7] p. 107.

Let us consider the connection between weak and stop completenesses.

6.12 THEOREM

Let $Ax \supset PA$ *be arbitrary. Then (i)* $\to$ *(ii)* $\to$ *(iii) and (ii)* $\to$ *(iv) where*

(i) $\mathfrak{F}^{Ax}$ *is stop-complete w.r.t.* E_o^{Ax},

(ii) $\mathfrak{F}^{Ax}$ *is stop-complete w.r.t.* E_1^{Ax},

(iii) $\mathfrak{F}^{Ax}$ *is weak-complete w.r.t.* E_1^{Ax},

(iv) $\mathfrak{F}^{Ax}$ *is stop-complete w.r.t.* E_w^{Ax}.

□

The following theorem was proved by L.Pósa [14].

6.13 THEOREM

If $Ax \supset PA$ *then* $\mathfrak{F}^{Ax}$ *is not stop-complete w.r.t. neither* E_o^{Ax} *nor* E_1^{Ax}.

□

Summarizing our considerations of completeness we set the table below.

Let $Ax \supset PA$ be any recursively enumerable set of formulas. The order of reading is as follows "is $\mathfrak{F}^{Ax}$ complete w.r.t. E_o^{Ax}" etc.

is $\mathfrak{F}^{Ax}$ w.r.t.	Complete	Weak complete	Stop complete
E_o^{Ax}	No	Yes	No
E_1^{Ax}	No	No	No
E_w^{Ax}	No	Yes!	?
E_s^{Ax}	No	Yes	Yes
E_2^{Ax}	Yes	Yes	Yes

The above table shows that those sets E w.r.t. which the language $\mathfrak{F}^{Ax}$ is weak-complete are much larger than those w.r.t. which the language is either complete or stop-complete. Therefore it is hoped that the notion of weak-completeness is satisfactory or at least is applicable in programming theory. However we note that the set E_w^{Ax} is not sufficient to cover all the needs of programming practice.

7. CONCLUSION

We wish to emphasize that the aboves are at an intermediate stage of a research on developing meta-methematical tools for a general theory of programming and indirectly for a theory of action.

One of the main points of the attitude of the approach followed here is the explicit and symmetric consideration of both time and data. This attitude is actualized in the model constructions where both time and data are built in. The advantage of these models is that they provide the handling of non-standard time and data. At first this seems to be unnatural in comparison to the so far used attitudes. However, it is but the explicit and correct explanation of our intuition used while programming. The models with inner time are general enough to provide the semantics for programming languages of different "style" such as deterministic, nonterministic, interactive and parallel.

Though our model construction permits to consider different sorts of data, one of its main defects is that data types are not built into it. Only those data types can be handled in our models that can be axiomatized in the language interpretable in our models. However this possibility here is implicit because we use a descriptive language which does not utilize all the possibilities provided by the semantics based on models with inner time. On the other hand this descriptive language beyond illustrating the usage of the models with inner time is strong enough to clarify some wellknown results from a unique pont of view. The descriptive language used here is more advantageous than those of the spirit of Floyd and Hoare. Namely some time dependencies even if only simple ones, can be expressed in it.

We wish to continue this research aiming at covering the defects mentioned above. Our aim for the near future is to show how our tools can be used for dealing with nondeterministic, interactive and parallel programming. For this aim, of course, a more expressive desriptive language should be introduced than the one used here. Developing more expressive programming languages we can describe more about the behaviour of programs. I. e. beyond program and output specification we shall have a powerful tool to specify the behaviour of programs in time and with respect to other interacting computational units (e.g. users, other processors etc.).

Thus following up this attitude we obtain such tools that can contribute to developing higher level programming systems emphasized by Winograd [16].

APPENDIX

Basic logical notions

A *language* in general is represented by a triple $L = (F,M,\models)$, where F is the syntax of the language, the pair $(M,\models)$ is its semantics with M being the class of models and the validity relation $\models \subset M \times F$.

Let S be a non-empty finite set which is called a set of sorts.

An *S-sorted similarity type* (or signature) σ is a pair of functions (σ_R, σ_F) such that

$$\text{(i)} \quad \text{rg } \sigma_R \subset \bigcup_{n=o}^{\omega} S^n, \quad \text{rg } \sigma_F \subset \bigcup_{n=o}^{\omega} S^n,$$

$$\text{(ii)} \quad \text{do } \sigma_R \cap \text{do } \sigma_F = \emptyset,$$

$$\text{(iii)} \quad |\text{do } \sigma_R| \times |\text{do } \sigma_F| \leq \omega.$$

Here ω is the least infinite ordinal number and do f and rg f denote the *domain* and *range* of the function f respectively.

The elements of do σ_R ,and do σ_F are called many sorted relation and function symbols respectively. σ_R and σ_F (i.e. σ) give the corresponding sequence of sorts to each symbol, which is written as follows.

For a relation (function) symbol ρ (respectively f) $\sigma(\rho) = (s_1,\ldots,s_n)$(respectively $\sigma(f) = \langle s_1,\ldots,s_n;s_{n+1}\rangle$) where for all $i \in [1,n]$ $s_i \in S$. (Here $[1,n] \overset{d}{=} \{i \mid 1 \le i \le n\}$.)

We also write $(\rho, \langle s_1,\ldots,s_n\rangle) \in \sigma$ and $(f,\langle s_1,\ldots, s_n;s_{n+1}\rangle)\in\sigma$.

If the sequence of sorts $\langle s_1,\ldots,s_n\rangle$ ($\langle s_1,\ldots,s_n; s_{n+1}\rangle$) consists of the same sort then we say that the corresponding relation(function)symbol is one-sorted or homogeneous.

For homogeneous relation and function symbols we write

(i) $\sigma(\rho) = n$ iff $(\rho,\langle\underbrace{s,\ldots,s}_{n\text{-times}}\rangle) \in \sigma$,

(ii) $\sigma(f) = n$ iff $(f,\langle\underbrace{s,\ldots,s}_{n\text{-times}};s\rangle) \in \sigma$;

i.e. in the homogeneous cases the similarity type renders to the symbols their arity.

If $S = \{s\}$ then the S-sorted similarity type becomes one-sorted and its each relation and function symbol is homogeneous.

Let $S = \{s_1,\ldots,s_k\}$ and let σ be an S-sorted similarity type.

A σ-*type model* is such a function $\mathfrak{A}$ that

(i) $\mathrm{do}\,\mathfrak{A} = \mathrm{do}\,\sigma \cup S$;

(ii) $\mathfrak{A}(s_i) = A_i$ is a nonempty set for any $i \in [1,k]$. ($\bigcup_{i=1}^{k} A_i$ is said to be the universe of the model);

(iii) $\mathfrak{A}(\rho) \subset \times_{j=1}^{n} A_{i_j}$ for any $(\rho,\langle s_{i_1},\ldots,s_{i_n}\rangle) \in \sigma$;

(iv) $\mathfrak{A}(f)\colon \mathop{\times}\limits_{j=1}^{n-1} A_{i_j} \to A_{i_n}$ for any

$(f, \langle s_{i_1}, \ldots, s_{i_{n-1}}; s_{i_n}\rangle) \in \sigma.$

(If $n = 1$ then $\mathop{\times}\limits_{j=1}^{n-1} A_{i_j} = \{\emptyset\}$.)

Now let $S = \{s\}$, then a one-sorted σ-type model is such a function $\mathfrak{A}$ that

(i) $\mathfrak{A}(s) = A$ is a nonempty set;

(ii) $\mathfrak{A}(\rho) \subset {}^{\sigma(\rho)}A$ for any relation symbol $\rho \in \text{do}\,\sigma$;

(iii) $\mathfrak{A}(f)\colon {}^{\sigma(f)}A \to A$ for any function symbol $f \in \text{do}\,\sigma$. (If $\sigma(f) = 0$ then $\mathfrak{A}(f) \in A$).

Note that in the case of one-sorted similarity types usually we omit the adjective "one-sorted". For the sake of convenience sometimes we write $\ell_{\mathfrak{A}}$ instead of $\mathfrak{A}(\ell)$, where $\ell \in \text{do}\,\sigma$.

Let M_σ denote the class of σ-type models.

Let $\mathfrak{A}, \mathfrak{B}$ be S-sorted σ-type models. $\mathfrak{B}$ is said to be a submodel of $\mathfrak{A}$ iff for any $s \in S$ $\mathfrak{B}(s) \subset \mathfrak{A}(s)$, and for any $\ell \in \text{do}\,\sigma$ such that either $(\ell, \langle s_{i_1}, \ldots, s_{i_n}\rangle)$ or $(\ell, \langle s_{i_1}, \ldots, s_{i_{n-1}}; s_{i_n}\rangle)$ we have $\ell_{\mathfrak{A}} \restriction \mathop{\times}\limits_{j=1}^{n} A_{i_j} = \ell_{\mathfrak{B}}$

(Here $\ell \restriction A$ denotes the restriction of the function or relation ℓ to the set A).

Now let σ' be a subsignature of the signature σ $(\sigma' \subset \sigma)$ i.e. $\sigma'_R = \sigma_R \restriction \text{do}\,\sigma'_R$ and $\sigma'_F = \sigma_F \restriction \text{do}\,\sigma'_F$.

Let $\mathfrak{A}$ be an S-sorted σ-type model and let $\mathfrak{B}$ be an S-sorted σ'-type model. $\mathfrak{B}$ is said to be a reduct of

$\mathfrak{A}$ iff for any $s \in S$ $\mathfrak{B}(s) = \mathfrak{A}(s)$, and $\ell_{\mathfrak{A}} = \ell_{\mathfrak{B}}$ for each $\ell \in$ do σ'.

Let $S = \{s_1, \ldots, s_k\}$ be a non-empty set of sorts and let V_i be pairwise disjoint denumerable set of i-sorted variable symbols. Let $V = \bigcup_{i=1}^{k} V_i$.

We denote by $L_\sigma = (F_\sigma^V, M_\sigma, \models)$ the σ-type S-sorted first order language, where F_σ^V is the set of all σ-type formulas with variable symbols belonging to V and $\models \subset M_\sigma X\, F_\sigma^V$ is the validity relation.

We recall that the definition of F_σ^V goes by induction by using the set of σ-type terms denoted by T_σ^V. The meaning of V in T_σ^V is the same as in F_σ^V. Let Q_σ^V denote the set of formulas of F_σ^V not containing any quantifier.

Note that we often omit the indication of V i.e. we often write T_σ, F_σ and Q_σ instead of T_σ^V, F_σ^V and Q_σ^V respectively.

For any $v \in V$, $\tau \in T_\sigma^V$ and $\phi \in F_\sigma^V$ let $\phi[\tau/v]$ denote the collision-free substitution of τ into the free occurences of variable symbol v.

We recall some more notions with respect to language L_σ. Let Var ϕ denote the set of free variable symbols occuring in ϕ. $q : \phi \to A$ is a valuation. $\mathfrak{A} \models \phi[q]$ means that the formula ϕ is valid in the model $\mathfrak{A}$ by the valuation q. $\mathfrak{A} \models \phi$ iff for any valuation $q : \text{Var}\,\phi \to A$ $\mathfrak{A} \models \phi[q]$ holds. Instead of q we often write the concrete values of the function q, e.g. $\mathfrak{A} \models \phi[\vec{a}]$, where $\vec{a} \in A$, i.e. $q\,(\text{Var}\,\phi) = \vec{a}$. Moreover for any variable symbol we often write it underlying by a waved line as its value by a given valuation. E.g. if q is a given valuation then we write $\underset{\sim}{\vec{x}}$ instead of $q(\vec{x})$.

Let $\mathcal{A}$ be an arbitrary class of σ-type models i.e. $\mathcal{A} \subset M_\sigma$ and let $\phi \in F_\sigma^V$ be arbitrary. $\mathcal{A}$ satisfies ϕ iff each model of $\mathcal{A}$ satisfies it, i.e. $\mathcal{A} \models \phi$ iff for any $\mathfrak{A} \in \mathcal{A}$ $\mathfrak{A} \models \phi$ holds.

Let $Th(\mathcal{A}) \overset{d}{=} \{\phi \in F | \mathcal{A} \models \phi\}$ be the set of true formulas in $\mathcal{A}$.

Let $\Sigma \subset F_\sigma$. A model $\mathfrak{A} \in M_\sigma$ satisfies Σ iff it saties-fies each formula of Σ, i.e. $\mathfrak{A} \models \Sigma$ iff for any $\phi \in \Sigma$ $\mathfrak{A} \models \phi$ holds.

Take $Md(\Sigma) \overset{d}{=} \{\mathfrak{A} \in M_\sigma | \mathfrak{A} \models \Sigma\}$ which is the set of models satisfying Σ.

The notion of definability plays a main role among the logical tools used in our approach. Namely, we wonder whether a function or relation of an arbitrary model can be expressed in a fixed language.

Fix a model $\mathfrak{A} \in M_\sigma$ and let $g: \overset{n}{\underset{j=1}{\times}} A_{s_j} \to A_{s_{n+1}}$ and $\rho \subset \overset{n}{\underset{i=1}{\times}} A_{s_i}$ (where each s_i $(i \in [1,n+1])$ belongs to S) be arbi-trary. We recall that the function g (the relation ρ) parametrically definable in $\mathfrak{A}$ if there is a formula

$\phi \in F_\sigma^V$ such that

(i) $Var\ \phi = \{x_1,\dots,x_n,y,a_1,\dots,a_m\}$

$(Var\ \phi = \{x_1,\dots x_n,a_1,\dots,a_m\})$;

(ii) There are $\underset{\sim}{a}_1,\dots,\underset{\sim}{a}_m \in A$ such that for any $\vec{\underset{\sim}{x}} \in A$ and $\underset{\sim}{y} \in A$ (for any $\vec{\underset{\sim}{x}} \in A$)

$$\mathfrak{A} \models \phi\ [\vec{\underset{\sim}{x}},\underset{\sim}{y},\underset{\sim}{a}] \text{ iff } g(\vec{\underset{\sim}{x}}) = \underset{\sim}{y},$$

$$(\mathfrak{A} \models \phi\ [\vec{\underset{\sim}{x}},\underset{\sim}{a}] \text{ iff } \vec{\underset{\sim}{x}} \in \rho).$$

Let η denote the similarity type of aritmetric, i.e. do $\eta_R = \{=\}$ and do $\eta_F = \{0,1,+,\cdot\}$ and $\eta_F(0) = \eta_F(1) = 0$, $\eta_R(=) = \eta_F(+) = \eta_F(\cdot) = 2$.

When η is used as a subsignature of different homogeneous signatures then the sort of η becomes the sort of the corresponding signature.

For the axiomatization of the arithmetic we use the well-known Peano axioms:

$$\alpha_1 \overset{d}{=} \neg(v + 1 = 0)$$

$$\alpha_2 \overset{d}{=} v + 1 = w + 1 \rightarrow v = w$$

$$\alpha_3 \overset{d}{=} v + 0 = 0$$

$$\alpha_4 \overset{d}{=} v + (w + 1) = (v + w) + 1$$

$$\alpha_5 \overset{d}{=} v \cdot 0 = 0$$

$$\alpha_6 \overset{d}{=} v\ (w + 1) = (v \cdot w) + v$$

$$\alpha_{7\phi} \overset{d}{=} \phi[0\ /\ v] \wedge \forall v(\phi \rightarrow \phi[v + 1/v] \rightarrow \forall x \phi$$

where ϕ is a η-type first order formula.

Take

$$PA \overset{d}{=} \{\alpha_i \mid 1 \leq i \leq 6\} \cup \{\alpha_{7\phi} \mid \phi \in F_\sigma^V\}.$$

Note that for convenience's sake we sometimes write $x \not\rho y$ instead of $\neg x\ \rho\ y$ for certain relation symbols ρ.

We use the following abbreviations

true for $x = x$ and

false for $x \neq x$.

If $\vec{f} = (f_1, \ldots, f_n)$ is an arbitrary vector of dimension n then we use the following notation

$$(\vec{f})_i \overset{d}{=} f_i \quad \text{for any } i \in [1,n].$$

REFERENCES

[1] Andréka, H., Németi, J., Sain, I.: Completeness problems in verification of programs and program schemes, in Becvar J. /ed/ *Mathematical Foundations of Computer Science 1977*, Lecture Notes in Computer Science, vol. 74, Springer Verlag, Berlin, 1979, pp. 208-218.

[2] Banachowski, L., Kreczmar, A., Mirkowska, G., Rasiowa, H., Salwicki, A.: An introduction to algorithmic logic, in Mazurkiewich, A., Pawlak, Z. /Eds/, *Mathematical Foundations of Computer Science*, PWN, Warszawa, 1977.

[3] Davis, M., Hilbert tenth problem is unsolvable, *American Mathematical Monthly*, vol. 80. /March 1973/, pp. 233-269.

[4] De Millo, R.A.: Non-definability of certain semantic properties of programs, *Notre Dame Journal of Formal Logic*, vol. XIV /1975/, pp. 583-590.

[5] Gabbay, D.M.: Model Theory for Tense Logics, *Annals of Mathematical Logic*, vol. 8/1975/, pp. 185-236.

[6] Gabbay, D.M., Pnueli, A., Shelah, S., Stavi, Y.: Completeness Results for the Future Fragment of Temporal Logic, 1979 /manuscript/.

[7] Gergely, T., Úry, L.: *Mathematical Theories of Programming*, Budapest, 1978, /manuscript/.

[8] Gergely, T., Vershinin, K.P.: Concept sensitive language for task specification /in this volume/.

[9] Harel,D.: *First-Order Dynamic Logic*. Lecture Notes in Computer Science, vol. 68., Springer Verlag, Berlin, 1979.

[10] Kröger,F., LAR: A logic of algorithmic reasoning. *Acta Informatica*, vol. 8. /1977/, pp. 243-266.

[11] Malcev, A.I. *Algorithms and Recursive Functions*. Nauka, Moscow, 1965. /In Russian/.

[12] Manna, Z.: *Mathematical Theory of Computation*. McGrow-Hill, London, 1974.

[13] Pnueli, A.: The temporal logic of programs. *18th Annual IEEE Symposium on Foundations of Computer Science*, Providence, 1977, pp. 46-57.

[14] Pósa L.: Personal communication, 1979.

[15] Winograd, T.: Beyond programming languages, *Comm. ACM*, vol. 22, N° 7 /1979/, pp. 391-401.

T. Gergely, L. Úry

Research Institute for
Applied Computer Sciences,
P.O. Box 227,
H-1536 BUDAPEST,
Hungary.

COLLOQUIA MATHEMATICA SOCIETATIS JÁNOS BOLYAI
26. Mathematical Logic in Computer Science,
Salgótarján (HUNGARY), 1978.

CONCEPT SENSITIVE FORMAL LANGUAGE FOR TASK SPECIFICATION

T. Gergely and K.P. Vershinin

ABSTRACT

It is convenient to use such a formal language that is very close to natural one in computer-aided problem solving. A formal language of this type is described here which is based on the first order predicate calculus. A special kind of text and its graphical representation, the diagram is introduced. These constructions provide the handling of auxilliary statements during theorem proving and the proof by analogy.

In the end it is shown how the formal language can be used for task specification and how the proof of the corresponding statement can be turned into a correct program directly.

1. *Introduction*

For a very long time computer programs have been considered as functions over natural numbers and the description of different program properties have been provided by formulas of formal arithmetics, the proof of which gives the satisfaction of the properties. This while proper logical considerations have remained in the background. Recently the usage of computers for solving nonnumerical tasks have arisen the need of a special language to describe properties (first of all semantical ones) of different programs of a given programming language. Here mathematical logic and its tools have got into the limelight. A descriptive language should allow the specification of the task to be solved together with that of the main properties of the corresponding program. Moreover, descriptive languages should enable us to describe the main properties of computers that can implement the programming language under consideration.

There are different approaches towards the introduction of such descriptive languages. In our opinion detailed in [1], the mathematical logical approach seems to be most perspective. There are two possibilities to develop a theory of programming within the frame of mathematical logic. The first one focuses on the programming. Here an appropriate descriptive language of some kind of logic is introduced to a given programming language. In this approach the question of the completeness of the calculus corresponding to the descriptive language plays a significant role.

The other approach is task-oriented, i.e. the main role is to specify the task to be solved in a formal language and to construct a corresponding proof by using a calculus corresponding to the language. Then a program can be released from the proof.

Note that constructive and intuitionistic logics are applicable in this second approach while in the first one the main role is played by the classical first order one. See details about the second approach in [2]. In this approach the specification of task is essential. Here the questions of automatic theorem proving become significant. Moreover the field of using such an approach is practically unlimited. Here the language of formal arithmetic is insufficient for task specification. To make search in proof more effective as it was mentioned e.g. in [3] it is necessary to have a language more powerful and expressive than the language of first order logic.

In this work we describe a language which is convenient to specify task conditions and to represent data necessary to solve the task.

This language is very close to the natural one though it is formal, i.e. it can be traced back to the classical first order logic. This closeness is ensured by *concept sensitivity* that means that the language contains concepts which can be introduced and eliminated according to the requirements.

We note that the first order logic itself can be applied for specifying tasks and describing the necessary data but the description can hardly be handled and understood by users. The fact that the language used here is based on the first order classical logic provide a detailed or contracted representation according to need. From other point of view this permits using different kinds of calculus adequate to the different contraction levels of the descriptions.

Details of the language can be found in [4]. We show how the language named TL (text language) provides formalizing reasoning by analogy and introducing auxilliary statements into the proofs. In [5] some preliminary ideas on the proof by analogy based on concept sensitive language can be found.

2. *Informal description of the language*

Let us first explain some basic ideas which have played an important role in developing the language TL. Suppose that we would like to formulate and prove a statement $\mathfrak{P}$ of the group theory. If $\mathfrak{P}$ can be formalised within an elementary theory then it is not important to leave the frames of classical first order logic. It is even more so when $\mathfrak{P}$ can be formulated only within the category of groups (e.g. see theorem on homomorphisms). Then the description can be done by introducing auxilliary predicates and by raising the order of the language, though it would be more rational to use the language of categories. Moreover if the statement $\mathfrak{P}$ speaks about objects of several categories, then the language of first order logic becomes quite inadequate. In such cases some new concepts (e.g. subgroup, annulator, order, etc.) are introduced in natural mathematical languages, and these would be used as terms but when we turn to the predicate logic these terms should be replaced by predicates.

We prefer to leave these units of language as terms and prefer interpreting them as class of objects (e.g. class of all groups, class of all subgroups of a given group, class of all annulators of a given element, etc.)

and we call them *constructions*. We introduce symbol BE in its appropriate conjugation to denote belonging to something.

To reduce concepts in natural languages adjectives are very often used such that e.g. Abelian, prime, equal to 1 modulo p etc. These adjectives are always verbal. We also introduce language units called attributes. Besides these concepts can be restricted by concerning only those elements of the corresponding class on which a given function obtains a given value (this is the set of level of the given function). E.g. "groups" $\xrightarrow{\text{order}}$ "natural numbers". Then group of order 2 or group of prime order denotes the class of all groups of given property. The units (order 2, cardinality ω_o, etc.) of the language TL are called right attributes. In natural languages bounded quantifiers are usually used, i.e. the domain of our reasoning consists of a few universes. The following figure illustrates this:

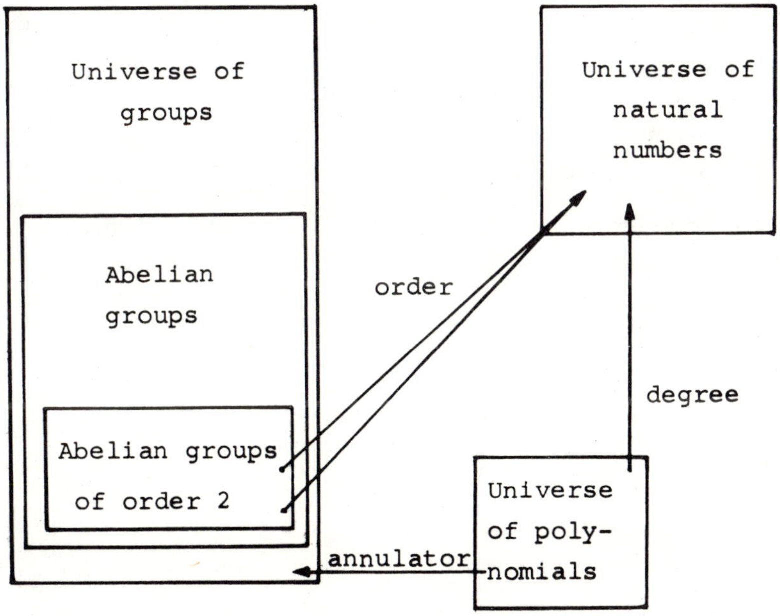

We note that the "natural" proof is neither a proof in Hilbert's sense nor is it in Gentzen's, but it is a fairly complicatedly organised text which is divided into parts of type "case", "subpoint", "auxilliary proof" etc. Certain methods of proof need specially organised texts even within a part. Such methods are e.g. proof by induction, by indirect methods. The language TL considers the possibility to structure texts. We emphasize that this language is connected with none of the special notions of derivation and the language is to be considered neither as a language of proof nor as a programming language.

3. *Formal description of* TL

Now we describe the language more formally. The sentence is the main unit of TL . Sentences are of two kinds: statements and suppositions. The latters are of the following forms: LET <statement> or SUPPOSE-THAT <statement>. TL-units not used independently are constructions, attributes and symbolic terms. All units of TL are obtained either by substitution into the so called formalizers or by concatenation. The formalizers are analogous to the usual predicate and function symbols.

According to the type of received units we distinguish:

c - formalizers that are of constructions,

al - formalizers that are of left attributes,

ar_1 - formalizers that are of right attributes of first type,

ar_2 - formalizers that are of right attributes of second type,

r - formalizers that are of atomic statements (relations),

sf - formalizers that are usual function symbols,

sp - formalizers that are usual predicate symbols.

We connect with every n-ary formalizer f its sort $\sigma(f)$ the n+1 tuple $<s_1,...,s_n,s>$, where $s_1,...,s_n$ are sorts of objects, which may be placed into the corresponding argument places of f and s is the sort of obtained object. E.g. for f ≖ HOMOMORPHISM_IN _* $\sigma(f)$ = <GROUP, GROUP, FUNCTION>. Semantically a sort is a class of objects (e.g. sets, natural numbers). Syntactically it corresponds to 0-ary c-formalizers SET, GROUP, REAL.

Note that the sort of an attribute is the sort of that construction to which the attribute can be attached. Semantically the sort of attribute is the class of objects which can possess the property expressed by the attribute.

We also distinguish logical and non-logical formalizers. The latter may be definable or undefinable.

Logical formalizers are:

IF _ THEN _ ,
_ AND _ ,
_ OR _ , IT-IS-WRONG-THAT _ , _ IFF _, EVERY_, SOME _ , THERE-EXISTS _ , _ IS-UNIQUE, _ SUCH-THAT _ , LET _ , SUPPOSE _ , _ WILL-BE-STAND-FOR _.

*Here ≖ denotes the "graphical equality".

Non-logical undefinable formalizers are as follows:
r - formalizers: _ ∊ _,

_ IS _ , _ = _ , _ BELONGS-TO _ , _ CONSISTS-OF _,
c - formalizers: CLASS, SET, ELEMENT-OF _ ,
sf- formalizers: {_}, {_ | _}.

Formalizers of attributes are obtained from r-formalizers and unary c-formalizers in the following way.

From unary r-formalizers we obtain al-formalizers. E.g.

f_1 × _ IS-ABELIAN is r-formalizer,

f_2 × ABELIAN is al-formalizer,

f_1 × _ IS-MEROMORPHIC is r-formalizer,

f_2 × MEROMORPHIC is al-formalizer.

From the r-formalizers of arity more than 1 we obtain ar_1 - formalizers.

f_1 × _ DIVIDES _ is r-formalizer,

f_2 × DIVIDING _ is ar_1-formalizer,

f_1 × _ IS-COMPARED-WITH _ MODULO _ is r-formalizer,

f_2 × COMPARED-WITH _ MODULO _ is ar_1-formalizer.

We obtain ar_2-formalizers from unary c-formalizers by the permutation of components in sort tuple. E.g.

f_1 × LENGTH-OF _ is c-formalizer,
$\sigma(f_1) = \langle WORD, NUMBER \rangle$;

f_2 × OF-THE-LENGTH _ is ar_2-formalizer,
$\sigma(f_2) = \langle NUMBER, WORD \rangle$;

$f_1 \doteq$ ORDER-OF _ is e-formalizer,
$\sigma(f_1) = \langle$GROUP, NUMBER$\rangle$;

$f_2 \doteq$ OF-ORDER _ ar_2-formalizer,
$\sigma(f_2) = \langle$NUMBER, GROUP$\rangle$.

Besides substitution into formalizers there is one more way of forming attributes. We explain it by an example: from ar_2-formalizer OF-THE-ORDER _ and al-formalizer FINITE we may obtain the right attribute OF-THE FINITE ORDER.

Attributes are used as follows. Left attributes are attached to the left side of a whole construction e.g. SUBGROUP OF G , FINITE SUBGROUP OF G. Right attributes are attached to the right side of the first "meaningful" word of c-formalizers. This word is called the leader of the given formalizer. E.g. SUBGROUP OF INDEX 2 OF G is a construction obtained in this way; SUBGROUP is the leader of its formalizer SUBGROUP OF_ ; OF INDEX 2 is the right attribute obtained from the ar_2-formalizer OF INDEX _ .

We call the units of TL, obtained by attaching attributes to constructions, attributed constructions (a-constructions).

Other non-logical formalizers are introduced by definitions. Admitted types of definitions are strictly fixed. One can select the defined formalizers and their sort from the text of a definition. It is easy to formulate rules of application of definitions. During the parsing of TL-text we have a possibility to deal with some restricted fragments of TL-grammar and this fragment may be automatically changed after having analysed the corresponding definitions in the text.

Let us see some examples of definitions:

D1. LET M BE SET.
P IS PARTITION OF M IFF P IS SET SUCH-THAT (IF $X \in P$ THEN X IS SUBSET OF M) AND (IF $X,Y \in P$ THEN $X \cap Y = \emptyset$) AND IF $X \in M$ THEN X BELONGS-TO SOME ELEMENT OF P).

D2. LET M, N BE SETS. LET P BE PARTITION OF M. P IS-COMPATIBLE-WITH N IFF CARDINALITY OF EVERY ELEMENT OF P IS-EQUAL-TO CARDINALITY OF N.

D3. LET M BE SET. LET R BE EQUIVALENCE-RELATION ON M.
SUPPOSE S(R) IS SET SUCH-THAT IF $X \in S(R)$ THEN THERE-EXISTS ELEMENT Y OF M SUCH-THAT $X = \{Z | Z\ R\ Y\}$.

Here D1 is the definition of the c-formalizer PARTITION OF _ its sort is <SET,SET>; D2 is the definition of the r-formalizer _ IS- COMPATIBLE-WITH _ of the sort <SET,SET,BOOLEAN>; D3 is the definition of the sf-formalizer S(_), its sort is <SET,SET>.

Let us detail such an object as a construction. We noted above that a construction may be treated as a designation of a class. It follows that we should indicate the meaning of every occurence of a construction that can be:

(i) arbitrary element of the corresponding class;

(ii) some elements of the class;

(iii) earlier mentioned element of the class.

In the first case we add the word EVERY to the left side of the construction. In the second case we use the word SOME in the same way. We call these objects *bounded constructions*. In the third case we add to leader of the corresponding c-formalizer the string ◊N, where N is the designation (or name) of the given object (e.g. PARTITION ◊S (R) OF SET ◊M IS-COMPATIBLE-WITH H). The objects obtained in such a way are called *named constructions*.

We consider bounded and named constructions and the usual symbolic terms to be TL-terms, and we permit to place them into the argument places of non-logical formalizers.

Finally let us make some remarks on the quantification. Remember that we use only restricted quantifiers in TL and the quantifier on each variable X must be restricted by the formula of the type X IS T or X∈T where T is a TL-term. The universal quantifier being restricted may be introduced either by using implication with X IS T (X∈T) as an antecedent or by using bounded construction (of the type EVERY K). The existential quantifier may be introduced either by means of bounded constructions (of the type SOME K) or by means of substitution in formalizer THERE-EXISTS _ .

4. *Representation of TL-statements in predicate calculus*

Let some sets of c-, r-, sf-, sp-formalizers be fixed.

We show in this section how to translate a TL-statement into a formula of ordinary predicate language of the following type:

a) functional symbols coincide with c-formalizers and sf-formalizers;

b) predicate symbols are $=$, ε (both are binary) and all r-formalizers and sp-formalizers;

c) $\wedge$, $\vee$, $\supset$, iff, $\forall$, $\exists$ are the usual logical connectives and quantifiers;

d) variable symbols coincide with those of TL.

Terms, atomic formulae, formulae are defined as usually.

Let $\mathfrak{P}$ be a statement of TL. Let $T_{\mathfrak{P}}$ be the parse tree of $\mathfrak{P}$. Now we are interested in that case when the "main formalizer" f of $\mathfrak{P}$ is non-logical.

We call the node N of $T_{\mathfrak{P}}$ *term-like (t-node)* iff N corresponds to some occurence of a construction in $\mathfrak{P}$. We connect a tree $T^i_{\mathfrak{P}}$ with i-th argument place of f. It is "growing" from this argument place and consists of t-nodes only. (Note, that some $T^i_{\mathfrak{P}}$ might be empty).

We call an occurence of a construction in $\mathfrak{P}$ *term-like (t-occurence)* iff it corresponds to a t-node in some $T^i_{\mathfrak{P}}$ i.e. occurences in attributes are not term-like. We call the t-occurence of a construction in $\mathfrak{P}$ *unblocked* iff it corresponds to a "leaf" in some $T^i_{\mathfrak{P}}$.

The nodes of the tree $T_{\mathfrak{P}}$ are to be ordered by depth first from left to right.

We shall write $\mathfrak{P} = \mathfrak{P}(K)$ iff K is the first (according to the order mentioned above) unblocked t-occurence of the construction in $\mathfrak{P}$.

Let $\forall X_K \mathfrak{S}$ ($\exists X_K \mathfrak{S}$) be an abbreviation for $\forall X(X \in K \supset \mathfrak{S})$ ($\exists X(X \in K \supset \mathfrak{S})$ accordingly).

Let K_E denote the named construction K with the name E.

Now we form rules for the elimination of bounded and named constructions (we call them *B-rules*).

B_1. $\mathfrak{P}$ (EVERY K) $\to$ IF X IS K THEN $\mathfrak{P}$(X), where X is a new variable symbol;

B_2. $\mathfrak{P}$ (SOME K) $\to$ THERE-EXISTS K_X SUCH-THAT $\mathfrak{P}$(X), where X is a new variable symbol;

B_3. $\mathfrak{P}$ (K_X SUCH-THAT $\mathfrak{G}$) $\to$ IF X IS K AND $\mathfrak{G}$ THEN $\mathfrak{P}$(X), where X was not mentioned earlier;

B_4. $\mathfrak{P}$(K_X) $\to$ IF X IS K THEN , where X was not mentioned earlier;

B_5. $\mathfrak{P}$(K_E) $\to$ $\mathfrak{P}$(E), where E is a symbolic term or variable symbol mentioned earlier.

Next we show how to eliminate the t-occurences of a-constructions. Note that after applying B-rules a-constructions may have t-occurence only in substatement of the form of "t IS K", where t is a symbolic term. Then it is sufficient to eliminate a-constructions from such a statement. The rules T_1-T_4 below represent this.

T_1. Let K be an a-construction with the left attribute A, i.e. K ≡ A K′.
t IS K $\to$ t IS K′ AND t IS-A.

Example

G IS FINITE GROUP $\to$ G IS GROUP AND G IS-FINITE.

□

T_2. Let K be an a-construction with the right attribute A of the first type. Let L be the leader of K, i.e.
K ≡ LAK′.
t IS K $\to$ t IS L K′ AND t IS-A.

Example

P IS PARTITION COMPATIBLE-WITH M OF G → P IS PARTITION OF G AND P IS-COMPATIBLE-WITH M.

□

T_3. Let K be an a-construction with the right attribute A of the second type. Let L be the leader of K.
Let A be obtained by the substitution of term t in ar_2-formalizer $F\_$,
i.e. A ≖ F t_1; K ≖ L OF-Ft_1 K'.
t IS K → t IS LK' AND t_1 IS Ft.

□

Example

H IS SUBGROUP OF-INDEX 2 OF G → H IS SUBGROUP OF G AND 2 IS INDEX OF H.

□

T_4. Let a right attribute A be obtained from a left attribute A_1 by concatenating it to the leader L of ar_2-attribute R(i.e. R ≖ L_).
Let K be an a-construction with the attribute A (i.e. K ≖ L_1 OF AK' ≖ L_1 OF A_1 LK', where L_1 is the leader of K).
t IS K → t is L_1 K' AND L OF t IS-A_1.

□

Example

H IS $\underbrace{\text{SUBGROUP}}_{L_1}$, OF $\underbrace{\text{FINITE}}_{A_1}$ $\underbrace{\text{ORDER}}_{L}$ $\underbrace{\text{OF G}}_{K'}$ →

→ H IS SUBGROUP OF G AND ORDER OF H IS-FINITE.

□

Evidently after applying B-and T-rules we obtain a statement of TL when it is possible and this does not contain bounded constructions and attributes.

A construction is called elementary if it does not contain occurences of attributes and bounded constructions. We call a statement literal that is obtained by substituting symbolic terms into a non-logical r-formalizer or a statement of the form "t IS K" where t is a symbolic term, K is an elementary construction

Finally we form a number of rules, application of which completes the translation. (Here $^*[\mathfrak{P}]$ will denote the result of translation of a statement $\mathfrak{P}$).

* 1. If $\mathfrak{D}$ is a literal then $^*[\mathfrak{D}] \bowtie \mathfrak{D}$.

* 2. $^*[t \text{ IS } K] \bowtie t \varepsilon K$, where K is an elementary construction .

* 3. $^*[t_1 \epsilon t_2] \bowtie t_1 \varepsilon t_2$.

* 4. $^*[t_1 \text{ IS-EQUAL-TO } t_2] \bowtie t_1 = t_2$.

* 5. $^*[\mathfrak{D}_1 \text{ AND } \mathfrak{D}_2] \bowtie {}^*[\mathfrak{D}_1] \,\&\, {}^*[\mathfrak{D}_2]$.

* 6. $^*[\mathfrak{D}_1 \text{ OR } \mathfrak{D}_2] \bowtie {}^*[\mathfrak{D}_1] \vee {}^*[\mathfrak{D}_2]$.

* 7. $^*[\text{IT-IS-WRONG-THAT } \mathfrak{D}] \bowtie \neg {}^*[\mathfrak{D}]$.

* 8. If X is a variable symbol which occures only in $\mathfrak{D}$, $\mathfrak{D}_1$ then
$^*[\text{IF X IS K THEN } \mathfrak{D}] \bowtie \forall X_K {}^*[\mathfrak{D}]$.

* 9. Under the same condition
$^*[\text{IF X IS K AND } \mathfrak{D}_1 \text{ THEN } \mathfrak{D}] \bowtie \forall X_K {}^*[\text{IF } \mathfrak{D}_1 \text{ THEN } \mathfrak{D}]$

*10. Under the same condition
$^*[\text{IF } X \in t \text{ THEN } \mathfrak{D}] \bowtie \forall X_t {}^*[\mathfrak{D}]$.

*11. Under the same condition
$^*[\text{IF } X \in t \text{ AND } \mathfrak{D}_1 \text{ THEN } \mathfrak{D}] \bowtie \forall X_t {}^*[\text{IF } \mathfrak{D}_1 \text{ THEN } \mathfrak{D}]$.

*12. $^*[\text{IF } \mathfrak{D}_1 \text{ THEN } \mathfrak{D}_2] \asymp {}^*[\mathfrak{D}_1] \supset {}^*[\mathfrak{D}_2]$

*13. Under the condition as in * 8.
$^*[\text{THERE-EXISTS } K_X \text{ SUCH-THAT } \mathfrak{D}] \asymp \exists X_K {}^*[\mathfrak{D}]$

*14. $^*[\text{THERE-EXISTS } K] \asymp \exists X_{\sigma(K)} {}^*[\text{ X IS K}]$, where $\sigma(K)$ is the sort of K, X is a new variable symbol.

Example

$\mathfrak{P} \asymp$ EVERY SUBGROUP OF SOME GROUP IS-ABELIAN.

$\mathfrak{P} \xrightarrow{B2}$ THERE-EXISTS GROUP ◊X SUCH-THAT EVERY SUBGROUP OF X IS-ABELIAN $\xrightarrow{B1}$ THERE-EXISTS GROUP ◊X SUCH THAT IF Y IS SUBGROUP OF X THEN Y IS-ABELIAN $\xrightarrow{*13}$ ∃X(X IS GROUP AND IF Y IS SUBGROUP OF X THEN Y IS-ABELIAN) $\xrightarrow{*2,*5}$ ∃X (Xε GROUP & IF Y IS SUBGROUP OF X THEN Y IS-ABELIAN) $\xrightarrow{*8,*12}$ ∃X(Xε GROUP & ∀Y(Yε SUBGROUP OF X ⊃ Y IS-ABELIAN)).

[ranslation is completed.

]

. TL-*texts*

A TL-text is a sequence of sentences and separators.

a) Syntactical separators are used for punctuational purposes:

_ (blank), . (dot), , (comma), (,) .

b) Structural separators point out the structure of the text such as division in chapters, sections, paragraphs and analogous items (e.g. CHAPTER 3.1, 2.7.4 etc.).

c) Logical separators point out the logical structure of the text. We devide them into three groups:

I. DEFINITION
THEOREM
LEMMA
PROOF
QED
IT IS EVIDENT
LET US SHOW THAT
THIS ITEM IS COMPLETED
CASE
THE CASE IS CONSIDERED
THE CASE IS IMPOSSIBLE

II. BY THE OPPOSITE
CONTRADICTION
INDUCTION ON
BASE
STEP
THE INDUCTION IS COMPLETED
LET US CONSTRUCT
THE CONSTRUCTION IS COMPLETED

III. ACCORDING TO
AS THE RESULT OF
SINCE
THEN

Separators of the third group serve for reference. Separators of the second group mark the beginning and the end of text which corresponds to a certain proof method. Some of the separators of the first and second groups are left and some of them are right. They play the role of a left and a right bracket.

Structural and logical separators turn TL-text into an ordered set of sections. It turns out that many arguments are localised in their sections. For example every supposition "acts" in the section where it is introduced and in all "subordinate" sections. It gives possibility to restrict the possible "field of discourse" during logical analysis of a given text. Say, if we are searching for "logical predecessors" of a given statement in a structured text then we can restrict ourselves to sections which are, in a sense, "relevant" to the statement. The above is analoguous with the *localization* of variables in programming languages.

6. *Situations and their transformations*

In a "natural" mathematical text usually every sentence has some kind of a neighbourhood or environment consisting of other sentences. If we try to check the correctness of some statements we must use some elements of its environment as auxilliary statements or lemmas. But the environment may contain, and it usually does, too many elements and we never use all of them but only the "suitable" or "relevant" ones. How do we select them? That is a question of great importance. Another question connected with it is how we search and apply alalogy in proofs.

Certainly we do not pretend to answer these questions in full but we present some approach to the formalization of these problems.

We call TL-sentences of the form $t_1 \varepsilon t_2$, ε-sentences, where t_1 is a symbolic term, t_2 is a TL-term, ε is ∈ or BE. t_1 is called the *subject* of ε-sentence $t_1 \varepsilon t_2$.

A set S of ε-statements is called *situation* iff for each element $t_1 \varepsilon t_2$ of S and for each subterm t of t_2 there exists an element $\mathfrak{P}$ in S such that t is the subject of $\mathfrak{P}$.

Let $\mathfrak{P} = t_1 \varepsilon t_2$, $\mathfrak{P}' = t_1' \varepsilon t_2'$ be ε-sentences. We write $\mathfrak{P} < \mathfrak{P}'$ iff t_1' occurs in t_2.

It is easy to see that the transitive closure ≤ of the relation < turns a situation S into poset. The diagram of the poset is called the *diagram of S* and is denoted by *D(S)*.

Now we define some ordering on the set of all c-formalizers. Let f_1^r, f_2^s be c-formalizers of arity r and s respectively. Suppose $f_1^r \lesssim f_2^s$ iff the followings are true:

1. if $r \geq s$ then $\forall X \forall X_1 \ldots \forall X_r (X \varepsilon f_1^r (X_1 \ldots, X_r) \supset$ $\supset X \varepsilon f_2^s (X_{i_1}, \ldots X_{i_s}))$
   (a permutation of arguments is possible);
2. if $r < s$ then $\forall X \forall X_1 \ldots \forall X_r \exists X_{r+1} \ldots \exists X_s (X \varepsilon$ $\varepsilon f_1^r (X_1, \ldots, X_r) \supset X \varepsilon f_2^s (X_{i_1}, \ldots, X_{i_s}))$
   (a permutation of arguments is possible).

Examples

Since ∀X∀Y (X IS SUBGROUP OF Y ⊃ X IS SUBSET OF Y) then SUBGROUP OF_ $\lesssim$ SUBSET OF _.
Since ∀X ∀Y (X IS PARTITION OF Y ⊃ X IS SET) then PARTITION OF _ $\lesssim$ SET.

Since $\forall X \forall Y \exists \phi$ (X IS NORMAL-SUBGROUP OF Y $\supset$ X IS KERNEL OF ϕ IN Y) then

NORMAL-SUBGROUP OF _ $\lesssim$ KERNEL OF _ IN _.

□

Note, that 0-ary c-formalizers (they correspond to sorts) are maximal under this ordering. Sorts are themselves ordered by $\lesssim$. E.g.

RING $\lesssim$ GROUP $\lesssim$ SET.

We call a situtation S *pure* iff subjects of all elements of S are variables. If ρ is a substitution then $S\rho = \{\mathfrak{P}\rho \mid \mathfrak{P} \in S\}$

Now we show how to represent situations in TL.

Let $S = \{t_i \varepsilon K^i \mid i = 1, \ldots, n\}$ be a situation. For each $t_i \varepsilon K^i$ we consider the construction $K^i_{t_i}$ named by t_i. There exist 0-ary constructions among them. Replace in each $K^i_{t_i}$ its symbolic term arguments by those $K^j_{t_j}$, which are named by these terms. Repeat this process as long as it is possible. We obtain a set of named constructions. None of them contains symbolic terms as arguments, but named constructions only. Choose those elements of the set which do not occur in the other ones. It is easy to see that the set consisting of parse trees of these elements coincides with D(S).

Example

Let S = {G IS GROUP, H IS SUBGROUP OF G, A IS ABELIAN GROUP, ϕ IS HOMOMORPHISM FROM A INTO H}. The corresponding set of named constructions is {GROUP ◇G, SUBGROUP ◇H OF G, ABELIAN GROUP ◇A, HOMOMORPHISM ◇ϕ FROM A INTO H}. After all possible replacements we obtain {GROUP ◇G, SUBGROUP ◇H OF GROUP ◇G, ABELIAN GROUP

◊A, HOMOMORPHISM ◊ϕ FROM ABELIAN GROUP A INTO SUBGROUP ◊H OF GROUP ◊G}. We choose a unique element HOMOMORPHISM ◊ϕ FROM ABELIAN GROUP ◊A INTO SUBGROUP ◊H OF GROUP ◊G. Its parse tree is as follows

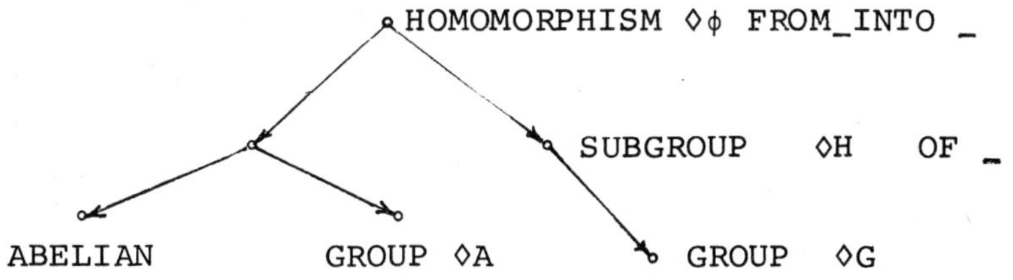

and this evidently coincides with D(S).

□

Now we see that if S contains a-constructions then D(S) contains nodes which correspond to attributes. Let us call them *a-nodes* and the others *c-nodes*. Let us describe now some operations on situations.

Furtheron we shall not differ in between a situation and its diagram.

1. *Coupling*

   Let D(S) contain two c-nodes N_1, N_2 with subjects t_1, t_2 and c-formalizers f_1, f_2 respectively. If $t_1 = t_2$ and $f_1 \lesssim f_2$ then we may remove the node N_2 and all arcs emanated from it and we can emanate all these arcs directly from the node N_1.

Example

S = {G IS GROUP, H IS SUBGROUP OF G, H IS SUBSET OF G }.

D(S) is

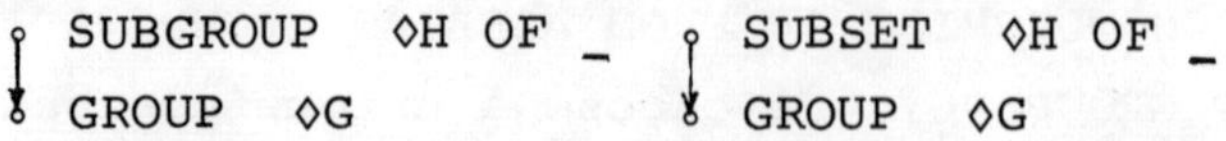

Since SUBGROUP OF _ $\lesssim$ SUBSET OF _ and H = H
we obtain after coupling

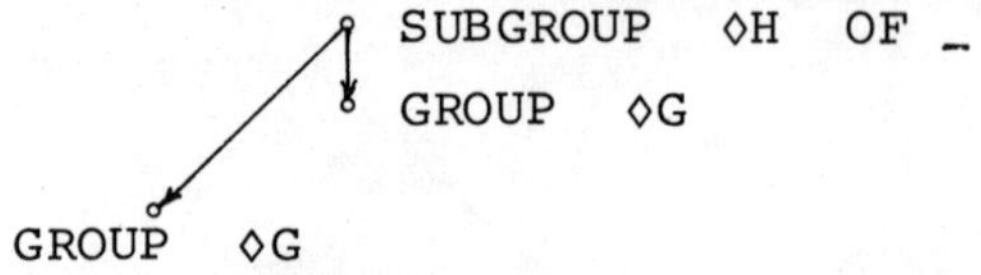

Applying coupling once more we obtain

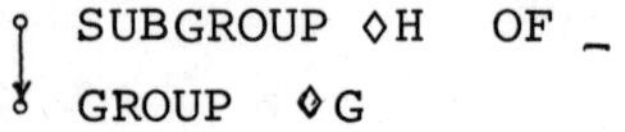

□

2 Roughing

If D S contains a fragment of the form

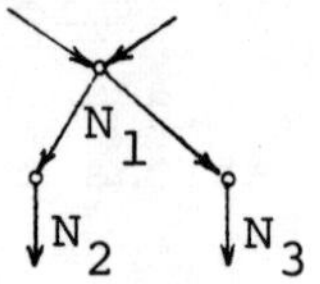

where N_1, N_2 are c-nodes, N_3 is a-node. We can remove N_1 and all arcs emanated from it and we can emanate these arcs directly from the node N_2.

Example

S = {G IS FINITE GROUP, H IS SUBGROUP OF G}

D(S) is

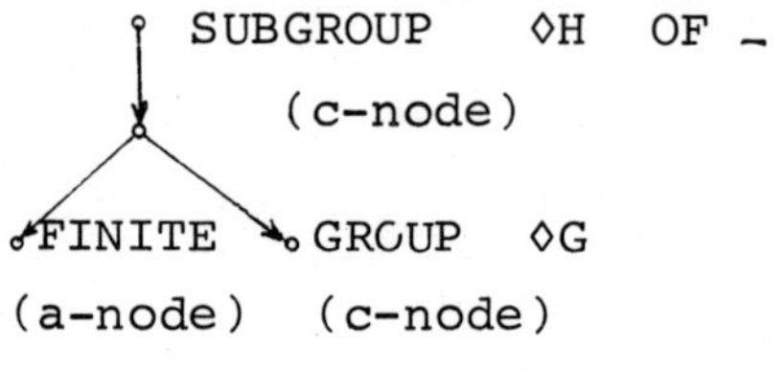

After roughing we obtain

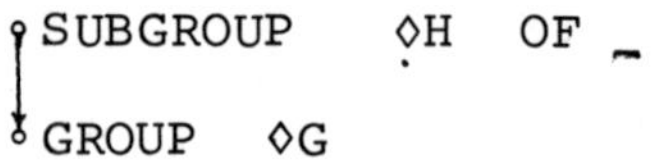

□

3. Refining

If a statement $\mathfrak{P}$ of the form tr $(t_1,\ldots,t_n)$ (where r_ is its formelizer) and situation S (containing a node N with subject t) is given then we may add to D(S) an a-node which corresponds to the attribute r $(t_1,\ldots,t_n)$.

Example

S = {G IS GROUP, H IS SUBGROUP OF G, S(R(H)) IS PARTITION OF G}; $\mathfrak{P}$ = S(R(H)) IS-COMPATIBLE-WITH H. D(S) is

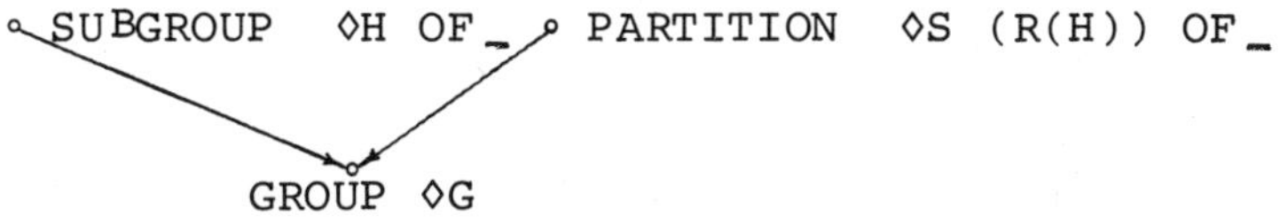

After refining we obtain

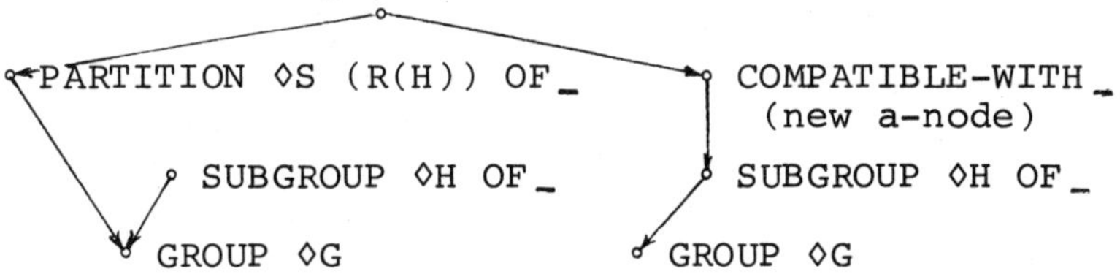

and then after two couplings we obtain

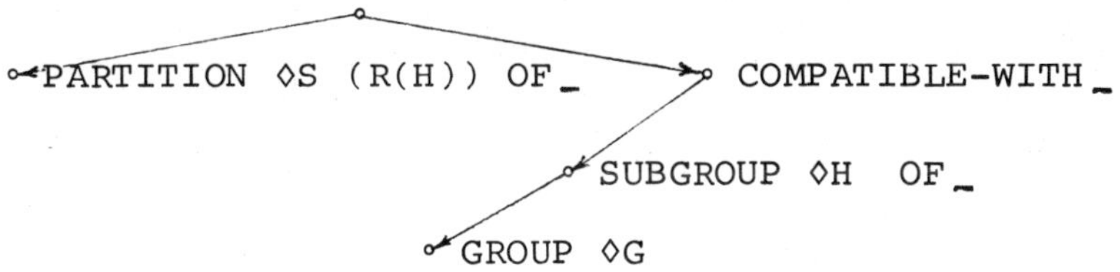

□

4. Completion

If a statement of the form t IS k $(t_1,\ldots,t_n)$ and a situation S (containing nodes $N_1,\ldots,N_m$ with subjects $t_1, \ldots t_m$ accordingly) are given then we can add to S the node N with subject t and direct arcs from N into $N_1, \ldots, N_m$.

Example

S = {G IS GROUP, H IS SUBGROUP OF G};

$\mathfrak{P}$ * R (H) IS EQUIVALENCE-RELATION ON G.

D(S) is

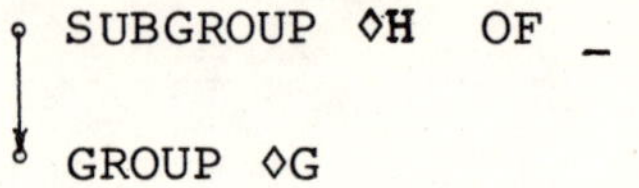

After completion we obtain

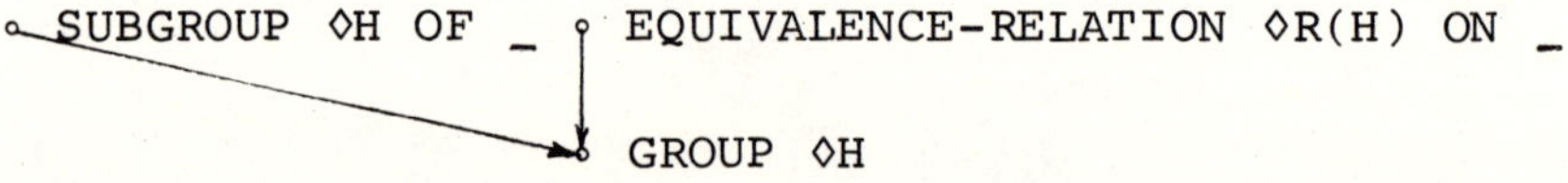

□

Finally we note that situations may be transformed by applying definition of corresponding c-formalizers.

7. *Situations and selection of auxilliary statements*

We now consider one aspect of using situations.

Let S_1, S_2 be pure situations. We say that S_1 is *more general* than S_2 ($S_1 \sqsubseteq S_2$ iff there exists an injection ϕ of $D(S_1)$ into $D(S_2)$ such that if $t\varepsilon k(t_1, \ldots, t_n) \epsilon S$ and $\phi(t\varepsilon k\ (t_1, \ldots, t_n)) = t'\varepsilon k'\ (t'_1, \ldots, t'_m)$ then $k' \lesssim k$, where k, k' are corresponding c-formalizers.

Example

S_1 = {M IS SET, R IS EQUIVALENCE-RELATION ON M},
S_2 = {G IS GROUP, H IS SUBGROUP OF G, R(H) IS EQUIVALENCE-RELATION ON G}.

$D(S_1)$ is

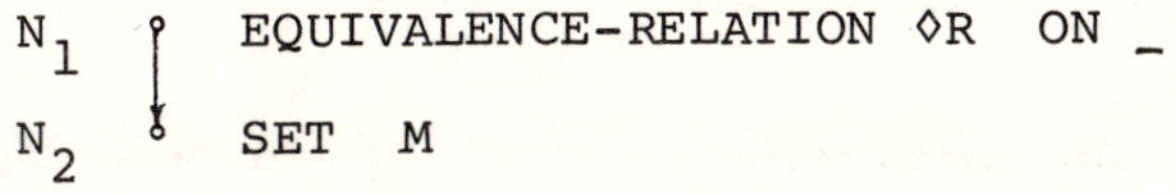

$D(S_2)$ is

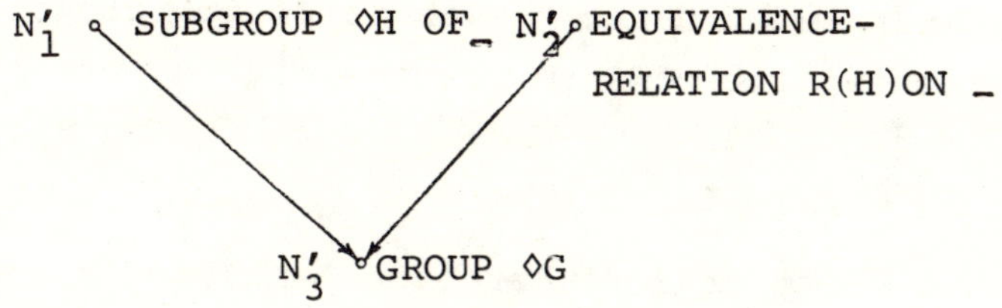

Let $\phi(N_1) = N_2'$, $\phi(N_2) = N_3'$. This shows that $S_1 \sqsubseteq S_2$.

□

Recall that we use only statements with restricted quantifiers in TL. Let $\mathfrak{P}$ be a TL-statement and $X_1,\ldots,X_n$ be variable symbols of $\mathfrak{P}$ which are bounded by universal quantifiers. Then $\mathfrak{P}$ may be represented in the form of $\forall X_1 \ldots \forall X_n (X_1 \varepsilon K_1 \wedge \ldots \wedge X_n \varepsilon K_n \supset \mathfrak{P}')$ (*). It is easy to show that $S = \{X_1 \varepsilon K_1, \ldots, X_n \varepsilon K_n\}$ is a situation. Then $\mathfrak{P}$ has a form of $\forall X_1 \ldots \forall X_n (\wedge S \supset \mathfrak{P}')$, where $\wedge S$ denotes the conjunction of all components of S.

Let now $\mathfrak{P}$ be a statement with free variables $X_1,\ldots,X_n$ and S be a situation. We say that $\mathfrak{Q}$ is *valid w.r.t.* S iff there exists a substitution ρ such that $\forall X_1 \ldots \forall X_n (\wedge S\rho \supset \mathfrak{Q})$.

Finally we say that $\mathfrak{Q}$ *is applicable for proving* $\mathfrak{P}$ if $\mathfrak{Q}$ is valid under those premises, which are presupposed in proof of $\mathfrak{P}$.

Our considerations are based mainly on the following

PROPOSITION

Let 𝔓 has the form (*). Then every statement, which is valid under a situation more general than S, is applicable for proving 𝔓.

The proof of the proposition is quite easy and we omit it.

We consider the proof of *Lagrange's Theorem* as an example of using the above proposition. Note first that no resolution-like methods give us the proof.

Example

Let us formulate the Langrange's Theorem.

LET G BE FINITE GROUP. LET H BE SUBGROUP OF G. THEN CARDINALITY OF H DIVIDES CARDINALITY OF G.

Suppose that there are the following definition and lemmas among many others in the environment of the theorem.

LET M BE SET. LET R BE EQUIVALENCE-RELATION ON M.
DEFINITION.
SUPPOSE S(R) BE SET SUCH-THAT $K \in S(R)$ IFF K IS SUBSET OF M AND THERE-EXISTS ELEMENT ◊X OF M SUCH THAT $Y \in K$ IFF X R Y.
LET G BE GROUP, LET H BE SUBSET OF G.
DEFINITION.
SUPPOSE R(H) IS RELATION ON G SUCH-THAT IF $X \in G$ AND $Y \in G$ THEN X R(H) Y IFF $X \cdot Y^{-1} \in H$.
LET M, N BE SETS, LET S BE PARTITION OF M.
DEFINITION.
S IS-COMPATIBLE-WITH N IFF IF $K \in S$ THEN CARDINALITY OF K = CARDINALITY OF N.

It follows from the above shown definition that conditions of any theorem (lemma) may be represented by a situation. We shall form lemmas by using that.

L_1.

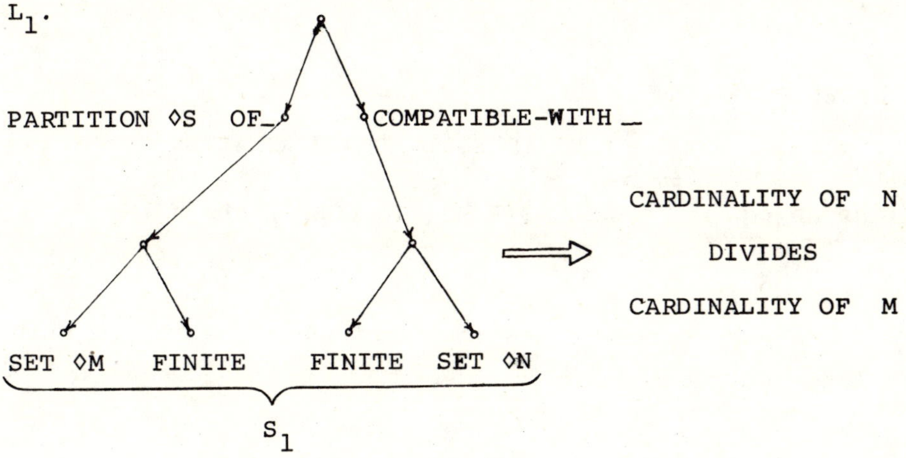

PARTITION ◊S OF_
COMPATIBLE-WITH _
SET ◊M
FINITE
FINITE
SET ◊N
S_1
CARDINALITY OF N
DIVIDES
CARDINALITY OF M

L_2.

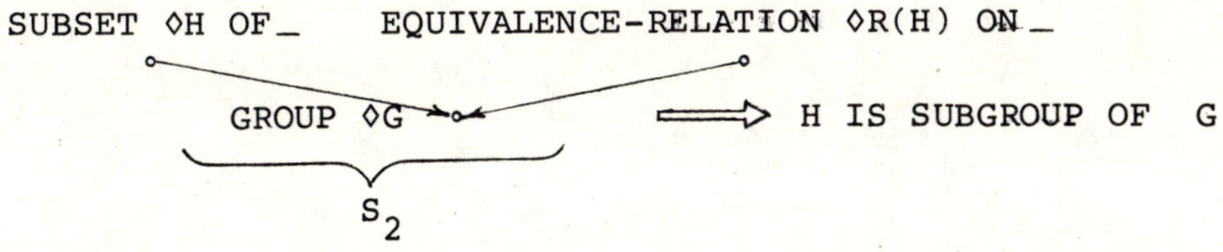

SUBSET ◊H OF _
EQUIVALENCE-RELATION ◊R(H) ON _
GROUP ◊G
S_2
H IS SUBGROUP OF G

L_3.

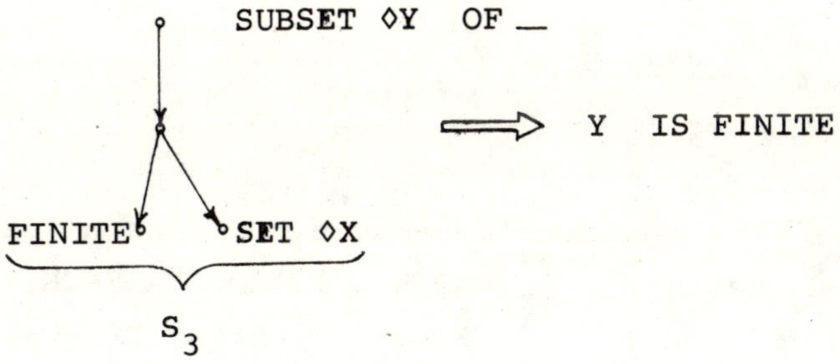

SUBSET ◊Y OF _
FINITE
SET ◊X
S_3
Y IS FINITE

L_4.

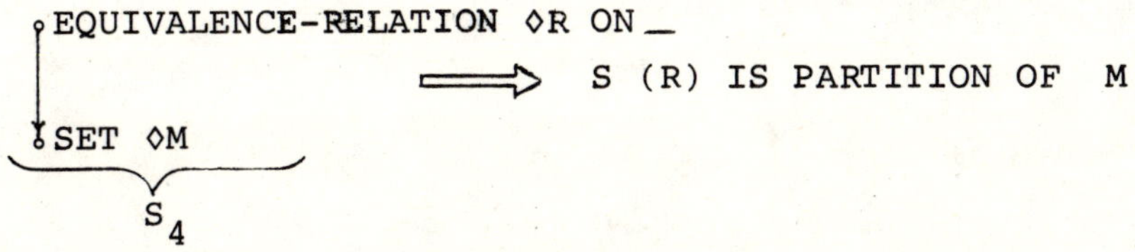

EQUIVALENCE-RELATION ◊R ON _
SET ◊M
S_4
S (R) IS PARTITION OF M

L_5.

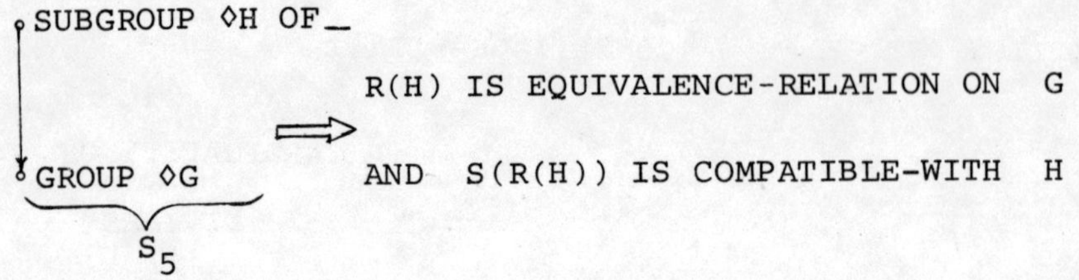

Let us represent now the Lagrange's Theorem in such a form too.

T.

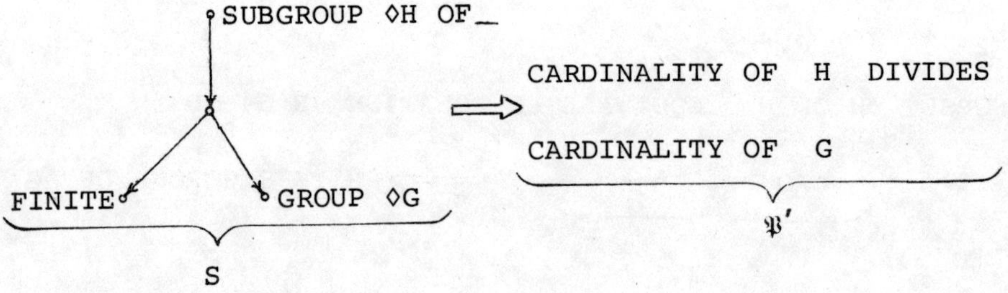

$S_3 \subseteq S$. Only L_3 is applicable. We obtain $\mathfrak{P}_1$ = H IS FINITE.

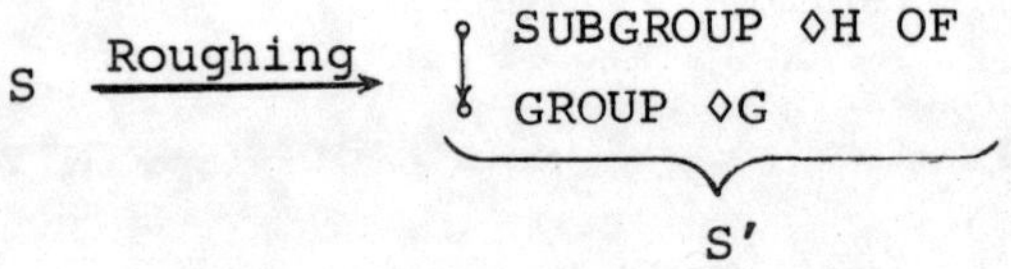

Only L_5 is applicable because L_3 has already been applied. We obtain $\mathfrak{P}_2$ = R(H) IS EQUIVALENCE-RELATION ON G and $\mathfrak{P}_3$ = S (R(H)) IS COMPATIBLE-WITH H.

Nothing is applicable that provides new result. We can use completion by $\mathfrak{P}_2$.

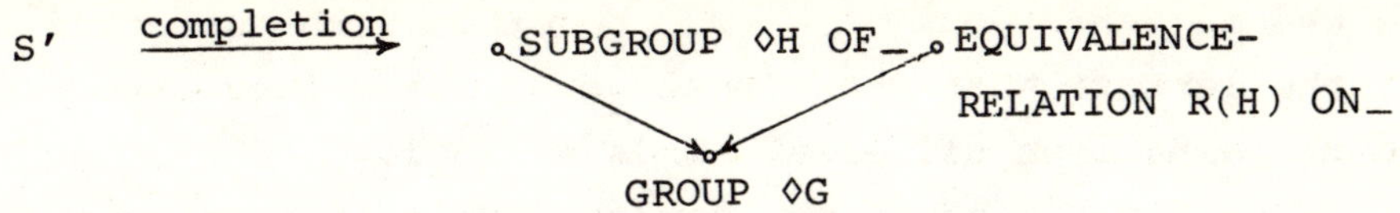

L_2 is applicable but the result is not new. L_4 is applicable and gives a new result: $\mathfrak{P}_4$ = S(R(H)) IS PARTITION OF G.

Nothing is applicable that provides new result. We use completion and refining by $\mathfrak{P}_4$, $\mathfrak{P}_1$, $\mathfrak{P}_3$ and obtain

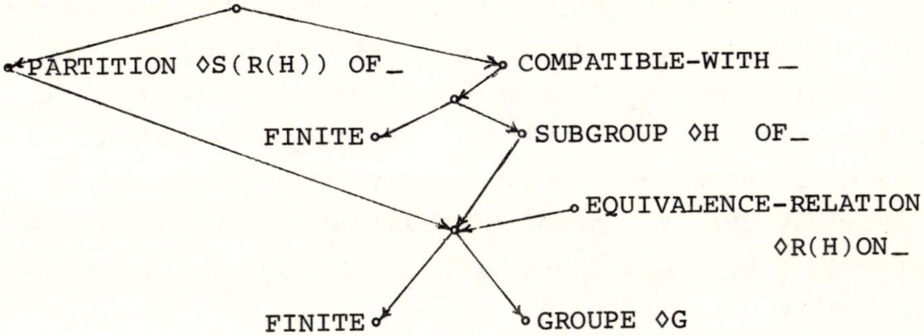

L_1 is applicable. We obtain $\mathfrak{P}'$= CARDINALITY OF H DIVIDES CARDINALITY OF G. This completes the proof of Lagrange's Theorem.

Note that it is not difficult to transform such inference into a quite "natural" proof.

8. Situations and proof by analogy

In nontrivial problem solving activity and so in theorem proving as well the method of analogical reasoning plays a very important role. The variety of analogies used in theorem proving activity is large enough.

However the common sense of proving by different kinds of analogy is that while proving a theorem T_1

we use an "analogous" one T_2 together with the proof of the latter. What does "analogous" mean? Intuitively we can establish different kinds of analogy.

The first kind of analogy occurs in the case when the theorems T_1 and T_2 consist of the same concepts and T_2 can be transformed into T_1 by renaming the variables. E.g. it can be met very often in proofs "let us prove the statement for the case $a \leq b$. For the case $b < a$ the proof is analogous."

The second kind of analogy occurs when the concepts of theorems T_1 and T_2, even after having renamed the variables of T_2, do not coincide, but some kind of symmetry (e.g. duality) can still be established between them. In this case, which is evidently "deeper" than the previous one, the proof of theorem T_2 can be transformed into a proof of T_1 by using the established symmetry.

The third kind of analogy is connected with situations such that a situation S' considered earlier is being found to be analogous with the situation S under consideration. Here the analogy means that some correspondence between the concepts of the situations S and S' can be established. Moreover this correspondence is attempted to be extended on to other relevant concepts. Finally to prove the statement T_1 the scheme of the given proof of T_2 would be transformed into a scheme of possible proof of T_1 and according to this proof the concrete proof theorem T_1 can be attempted to be constructed.

To turn the aboves more concrete first let us make the concept of "proof scheme" precise.

Let us consider some proof text T in TL. This text T is divided into sections according to the definition (see Section 5). Every section is indicated by a separator or by a pair of separators (left and right ones).

Some of the sections are proofs in turn. (Often they begin with the separator LET-US-SHOW-THAT and ends by the separator THE-ITEM-IS-COMPLETED.)Every one of them contains the statement to be proved. Such a statement is called the goal of a given section and section which have the goal - *g-section*.

Let us remove from the text T all sections, which are not g-sections. We obtain a text T_o containing only g-sections. The text T_o is to be the O-level scheme of T.

Let then replace in T_o each g-section by its goal. We obtain a text T_1 and call it 1-level scheme of T and so on. Thus different level of schemes of a given proof text, i.e. proof schemes of different levels can be obtained. For the proof one of these levels may become adequate by analogy.

To be more precise about the above mentioned kinds of analogy we return to the notion "situation".

Let S_1, S_2 be situations, T be a TL-text. The situation S_1 is said to be *analogous* to S_2 w.r.t. T iff

(i) $D(S_1)$ is isomorphic to $D(S_2)$ and

(ii) the definition of every formalizer f_1, which is defined in S_1 and occurs in T, may be transformed into the definition of some formalizer f_2 occuring in S_2.

Note that the isomorphism between $D(S_1)$ and $D(S_2)$ determines a correspondence between formalizers occuring in S_1 and S_2. ii means that the correspondence may be extended on to all formalizers occuring in T.

Depending on the type of transformation that can establish the connection between formalizers occuring in the text and in a situation given erlier we obtain analogies of different type.

Now we illustrate the usage of all three kinds of analogy.

Example 8.1

Here the usage of the first kind of analogy is connected with an item of the proof in the proof text below. At the same time the proof of THEOREM 2 uses the second kind of analogy.

THEOREM 1.

LET G BE GROUP. LET H BE SUBGROUP OF G.LET A,B BE LEFT COSETS OF G W.R.T. H. THEN $A = B$ OR $A \cap B = \emptyset$.
PROOF. LET-IT-IS-WRONG-THAT $A \cap B = \emptyset$.
LET $X \in A$ AND $X \in B$. $A = a * H$ WHERE $a \in G$.
$B = b * H$ WHERE $b \in G$. $X = a \cdot h_1$ WHERE $h_1 \in H$.
$X = b \cdot h_2$ WHERE $h_2 \in H$. $a \cdot h_1 = b \cdot h_2$. $a = b \cdot h_2 \cdot h_1^{-1}$.
$h_2 \cdot h_1^{-1} \in H$. $A = b \cdot h_2 \cdot h_1^{-1} * H$. $h_2 \cdot h_1 * H \subset H$. $A \subset b * H$.
$A \subset B$. BY ANALOGY $B \subset A$. (This is an analogy of the first kind.) THEN $A = B$. Q E D.

THEOREM 2.

LET G BE GROUP. LET H BÉ SUBGROUP OF G.LET A,B BE RI T COSETS OF G W.R.T. H.THEN $A = B$ OR $A \cap B = \emptyset$.
PROOF. SYMMETRIC TO THE PROOF OF THEOREM 1.

This symmetry of the proof is the replacement of each product of the form $a_1 \cdot a_2 \cdot \ldots \cdot a_k$ to the "symmetric" product $a_k \cdot a_{k-1} \cdot \ldots \cdot a_1$.

□

Example 8.2

Here we use the well-known analogy between normal subgroups and ideals in rings.

THEOREM 3.

LET G BE GROUP. LET H BE NORMAL-SUBGROUP OF G. THEN G/H IS GROUP.

PROOF. LET $A,B \in G/H$. $A = a * H$, WHERE $a \in G$. $B = b * H$, WHERE $b \in G$. LET-US-SHOW-THAT $A*B = a \cdot b * H$. LET-US-SHOW-THAT $A * B \subset a \cdot b * H$. LET $X \in A*B$. $X = a \cdot h_1 \cdot b \cdot h_2$, WHERE $h_1, h_2 \in H$. SINCE H IS NORMAL SUBGROUP THEN $h_1 \cdot b = b \cdot h_1'$, WHERE $h_1' \in H$. $X = a \cdot b \cdot h_1' \cdot h_2$. $h_1' \cdot h_2 \in H$. $X \in (a \cdot b) * H$. $A*B \subset (a \cdot b)*H$. THIS-ITEM-IS-COMPLETED. LET-US-SHOW-THAT $(a \cdot b)*H \subset A*B$ SYMMETRIC TO THE PRECEDING. (The symmetry now consists of "reading proof from the end to the beginning".)

THE-ITEM-IS-COMPLETED.

LET-US-SHOW-THAT $A = a^{-1} * H$. $A^{-1} = (a * H)^{-1}$

$= \{(a \cdot h)^{-1} | h \in H\} = \{h^{-1} \cdot a^{-1} | h \in H\} = \{h' \cdot a^{-1} | h' \in H\} =$

$= a^{-1} \cdot h'' | h'' \in H\} = a^{-1} * H$. THE-ITEM-IS-COMPLETED.

QED.

The situation is S_1:

NORMAL SUBGROUP ◊H OF _

↓

GROUP ◊G

The scheme is:

LET A,B $\in$ G/H. A = a $*$ H, WHERE a $\in$ G. B=b$*$H, WHERE b $\in$ G. A $*$ B = (a$\cdot$b) $*$ H. $A^{-1} = a^{-1} * H$.

Let us consider

THEOREM 4.

LET R BE RING. LET S BE IDEAL OF R. THEN R/S IS RING.

The situation is S_2 :

IDEAL ◇S OF_

RING ◇R

It is evident that S_1 is analogous to S_2 according to the text of the proof of THEOREM 3. Then it is easy to see how to transform the given scheme into a new one and to complete the proof of THEOREM 4.

□

It follows from the aboves that if a statement T has to be proven then it can be done by using analogy as follows.

(i) first its situational form S should be established;

(ii) the corresponding diagram D(S) should be constructed;

(iii) one should find an analogous theorem T′ proven earlier for which

a) transformation should be found between the formalisers of S_T and $S_{T'}$;

b) the isomorphism of $D(S_T)$ and $D(S_{T'})$ should be analysed.

9. Program design from TL *texts*

To obtain a program (a text of programs, to be more precise) from the text of a proof of statement with existential quantifiers seems to be a promising approach to program design.

It is quite natural to consider only constructive proofs of existence, i.e. those proofs where terms corresponding to the variables bounded by existential quantifiers shoud be constructed explicitly. Thus we ignore the indirect proofs. For the constructivity we suppose that

(i) all Skolem functions are computable with respect to the functions given in the theorem to be proved;

(ii) every disjunction occuring in the text of the proof is constructively valid, i.e. for each value of free variables it is indicated which part of the disjunction is valid.

Furtheron we show that under the above suppositions how a program can be obtained from a TL-text.

Program design means (i) the specification of data type used in the program;
(ii) the interpretation of operations on the specified data types;
(iii) the sequence of operators that forms the program body.

Data types can be specified by analysing the definition of notions occuring within the text of the theorem and that of the proof. (More precisely, these notions occur in the corresponding situation.)

The sequence of operators can be obtained by analysing the text of the proof.

Our aim here is to present some ideas about the program design from TL texts without providing exact instruction for this.

For example let us consider the following definitions.

DEFINITION 1 .

LET A BE COMMUTATIVE RING WITH UNIT ELEMENT. f IS A POLYNOMIAL OVER A IFF f IS FUNCTION FROM N INTO A SUCH THAT $\{n | n \in N$ AND $f(n) \neq 0_A\}$ IS FINITE.

DEFINITION 2 .

LET A BE COMMUTATIVE RING WITH UNIT ELEMENT. LET F BE POLYNOMIAL OVER A. d IS DEGREE OF f IFF d IS NATURAL NUMBER AND $f(d) \neq 0_A$ AND IF k IS NATURAL NUMBER AND $k>d$ THEN $f(k) = 0_A$.

By analysing the first definition one can see that ARRAY is the data type corresponding to polynomials since an array consisting of elements of a set A is a function from an initial segment of natural numbers into A. It is also obvious that the operations on polynomials can be easily transformed into the operators on the elements of the corresponding arrays.

The analysis of DEFINITION 2 shows that the degree of polynomials is of type INTEGER.

Note that the language units corresponding to the notions of natural language presented in TL provide such an analysis of definitions. When we turn to the language of predicate calculus the details of formalism prevent us to see it as a whole.

Now we take a theorem together with its proof and show how to get the corresponding program from the text.

LET A BE COMMUNICATIVE RING WITH UNIT ELEMENT.

LET f BE POLYNOMIAL OVER A. deg (f) DENOTES DEGREE OF f.

LET d = deg (f).

$f(d)\ X^{d} + f(d-1)\ X^{d-1} + \ldots + f(1)\ X + f(0)$ DENOTES f.

THEOREM.

LET f,g BE POLYNOMIALS OVER A.
LET DEG (f) ≥ 0 AND deg (g) ≥ 0.
LET g (deg(g)) = 1_A.
THEN THERE ARE POLYNOMIALS ◊q, r OVER A SUCH THAT

f = g q + r AND deg (r) < deg (g).

PROOF.

SUPPOSE n = deg (f) AND m = deg (g).
LET $f = f(n)\ X^{n} + \ldots + f(0)$.
LET $g = g(m)\ X^{m} + \ldots + g(0)$.
g(m) IS THE UNIT ELEMENT OF A. f (n) ≠ 0.
INDUCTION ON n.
BASIS. SUPPOSE n = 0.
deg (g) > deg (f) OR deg (g) = 0.

CASE 1.

TAKE q = 0 AND r = f. CASE 1 HAS BEEN INVESTIGATED.

CASE 2.

TAKE $q = f(n)/g(m)$ AND $r = 0$. CASE 2 HAS BEEN INVESTIGATED.

STEP. LET US SUPPOSE THAT IF k IS NATURAL NUMBER THEN IF $k < n$ AND h IS POLYNOMIALS OVER A AND $\deg(h)=k$ THEN THERE ARE POLYNOMIALS $\Diamond u,v$ OVER A SUCH THAT $h = g\cdot u + v$ AND $\deg(v) < \deg(g)$. SHOW THAT IF $\deg(f) = n$ THEN THERE EXIST POLYNOMIALS $\Diamond q,r$ OVER A SUCH THAT $f = g\cdot q + r$ AND $\deg(r) < \deg(g)$.
$\deg(g) > \deg(f)$ OR $\deg(g) \leq \deg(f)$.

CASE 1.

TAKE $q = 0$ AND $r = f$. CASE 1 HAS BEEN INVESTIGATED.

CASE 2.

TAKE $h = f-(f(n)/g(m))X^{n-m}\cdot g$. $\deg(h) < n$.
THERE ARE POLYNOMIALS $\Diamond u,v$ OVER A SUCH THAT
$h = u\cdot g + v$. $\deg(v) < \deg(g)$.
$f = (f(n)/g(m))\cdot X^{n-m}\cdot g + u\cdot g + v$.

TAKE $q = (f(n)/g(m))\cdot X^{n-m} + u$.
TAKE $r = v$.
CASE 2 HAS BEEN INVESTIGATED.
THUS INDUCTION IS OVER.

During the analysis of the above text of proof we follow the forthcoming ideas. A sequence of conditional commands corresponds either to statement of the form "$\mathfrak{S}_1$ OR $\mathfrak{S}_2$" or to succéssive separation of cases.

Namely we correspond to separators CASE 1,..., CASE n labels L1, ... Ln and the operators take the following form:

if $\mathfrak{S}_1$ *then* L1

if $\mathfrak{S}_2$ *then* L2

.
.
.

if $\mathfrak{S}_n$ *then* Ln .

The assignment operator X: = T corresponds to statement of the form "TAKE X = T", where T is a TL-term Operator brackets e.g. *do ... od* correspond to the beginning and the end of TL-sections.

Loop operator corresponds to section INDUCTION where it is important that the basis of induction plays the role of the "degenerated case", i.e. such a case the loop for which should not really be executed. Consequently operators corresponding to the basis of induction are extracted from the body of the loop operator.

The statement ensuring the usage of the inductive hypothesis in the step of induction becomes the loop predicate that allows the continuation of the loop. In our example deg (g) ≤ deg (f) is such a statement.

The TL-variable to which the inductive hypothesis (in our case f) can be used becomes the loop variable. All the variables changing within the proof of the inductive steps are identified by variables that play the same role in the inductive hypothesis (in our case h is identified with f, q with u and r with v), All other statements of TL-texts can be either neglected or considered as conditions the non-satisfiability of which leads to the immediate stopping of the program. In our case, for example "g (m) IS UNIT ELEMENT OF A", and "f (n) ≠ O" are such statements.

Variables can be described by corresponding them to the appropriate data structure as it has been mentioned above.

Considering all the aboves we can transform the text of the example into the following program (we omit the description of variables).

```
        n: = deg(f); m: = deg(g);
        if g(m) ≠ 1_A then L4
        if f(n) = 0_A then L4
        if m>n = 0    then L1
        if m = n = 0 then L2
        if m>n ≠ 0    then L1
        if m≤n ≠ 0    then L3
L1:     q: = 0; r:= f; goto L4;
L2:     q: = f(n)/g(m); r: = 0; goto L4;
L3:     q: = 0; while  n≥m  do
        d: = deg(f);   q: =(f(d)/g(m))·X^(d-m)+ q;
        f: = f- (f(d)/g(m))·X^(d-m)· g;   r: = f od;
L4:     stop .
```

f, g are the input parameters of the program and q, r are the output ones. It is evident that the text from which the program has been constructed is at the same time the proof of the program correctness. Thus the tools of logical derivation can be considered as tools for the verification of the program extracted from the appropriate TL texts.

10. Conclusion

Switching to the high level programming languages has provided us with a significant progress in the technology of designing very complex programs and program systems. Analogously we believe that in the logical constructions the usage of languages being more powerful than the language of predicate logic may result in some progress in the investigations of the tasks connected with theorem proving.

We hope that TL language proves to be a milestone on this way. As it has been shown this language permits the usage of such proof technics that e.g. handle lemmas, analogies etc. and which turn the proving procedures more effective and closer to human intuition.

Moreover the languages of TL type make the man-machine communication easier. In our opinion they ensure a more "intimate" connection between man and computers in the process of problem solving.

Thus the language TL turned out to be quite a convenient tool of task specification. This was illustrated by mathematical tasks and it was shown how a proof described by the use of this language can be turned into a correct program that ensure this solution of the tasks the corresponding TL statement of which is proved.

REFERENCES

[1] Gergely, T., Ury, L., Mathematical Theories of Programming, Budapest, 1978, (manuscript).

[2] Непейвода Н.Н., Применение теории доказательств к задаче построения правильных программ, Кибернетика, №2, 1979, стр. 43-48.

[3] Глушков В.М., Капитонова Ю.В., Автоматизация поиска доказательств теорем математических теорий и интеллектуальные машины, Кибернетика, №5, 1972, стр. 2-6.

[4] Глушков В.М., Вершинин К.П., Летичевский А.А., Малеваный Н.П., Костырко В.Ф., О формальном языке для записи математических текстов, в сборнике Автоматизация поиска доказательств теорем в математике, издательство ИК АН УССР, Киев, 1974, стр. 3-36.

[5] Вершинин К.П., Гергей Т., Основы доказательства по аналогии, KFKI-75-79, Budapest, 1975.

T.Gergely
Research Institute for
Applied Computer Science
P.O.Box 227, H-1536 Budapest
Hungary

K.P.Vershinin
Institute of Cybernetics of
Ukrainian Academy of Sciences
Prospekt 40-letija Oktjabrja 142/144
Kiev-207, USSR

COLLOQUIA MATHEMATICA SOCIETATIS JÁNOS BOLYAI
26. Mathematical Logic in Computer Science
Salgótarján (HUNGARY), 1978

FULL WEAK SECOND-ORDER LOGIC VERSUS ALGORITHMIC LOGIC

M. Grabowski

1. *Introduction*

Let me start with an explanation: we obtain the full weak second order logic (abbreviated to $\mathbf{L}^{f}_{w}$) from the ordinary first-order one by adding quantifiers: "for every finite relation" and "there exists a finite relation".

It is well known that sentences about programs cannot be formalized in the first order logic; in order to formalize them, many formal logical system were constructed in recent years. Classical model-theoretical and logical problems are connected with these systems, for instance: completeness, position in Kleene-Mostowski's hierarchy, expressive power, Löwenheim's and Hanf's numbers, etc. .

Let us make an evident remark: the sentences about programs may be formalized in the full weak second order logic as well as in $\mathbf{L}_{\omega_1\omega}$-logic or in the second order logic. The following questions arise:

1. How much does the power of L^f_w exceed the requirements of formalization of sentences about programs?

2. Is the difference between general logical and model-theoretical properties of L^f_w and of logic of programs essentially large?

I hope, that the paper gives a partial answer to the questions 1 and 2.

In the paper we compare properties of L^f_w and of algorithmic logic [2] (abbreviated to L_a). We observe a certain uniformity of properties of L_a and L^f_w and we shall obtain this uniformity in a certain general and simple way - we shall prove that L^f_w *and* L_a *are mutualy interpretable*. This result is a simple corollary from Salwicki's result ([4]) on representability of stacks (finite sequences) in L_a. The method of proof of the mutual interpretability gives the following corollaries:

1./ The sets of tautologies of L_a and L^f_w are recursively isomorphic.

2./ Logics L_a, L^f_w have the same Löwenheim's numbers and the same Hanf's number.

3./ The interpolation theorem holds for L_a iff it holds for L^f_w. (I remark here - that we can establish separately - that the interpolation theorem does not hold for L_a and nor does it for L^f_w; our proof of validity of corollary 3 is different and illustrated a certain general way of establishing a uniformity of properties L_a and L^f_w.)

In the final section we make some observations on models, in which L_a and L^f_w are equivalent. In this direction our results are very incomplete and the subject seems to deserve a further study.

2. *Translatability and Interpretability*

In this section we determine basic notions and denotations, then we prove the validity of the properties of L_a and L_w^f formulated in the introduction.

L_1 - denotes the ordinary first-order logic.

L_w - weak second-order logic - it is obtained from L_1 by adding variables denoting finite sets, atomic formulas of the form "$z \in X$" (where z is a term and X is a set variable) and by adding quantifiers:

"$(\forall X)$" - for every finite set X the following holds ...

"$(\exists X)$" - there exists a finite set X such that ...

L_w^f - full weak second-order logic; we obtain it form L_1 by adding variables denoting finite relations (for every new infinite number of relational variables of arity n), atomic formulas of the form $R(t_1,\dots,t_n)$ where $t_1,\dots,t_n$ are terms and R is a n-ary relational variable and by adding quantifiers:

"$(\forall R)$" - for every finite n-ary relation R

"$(\exists R)$" - there exists a finite n-ary relation R

} for R being a n-ary relational variable

L_a - algorithmic logic (simplified) which is defined as follows:

(I) every formula from L_1 is a formula of L_a

(II) programs:

substitutions of the form <variable>:=<term> are programs; if K,M are programs and α is an open formula, then

begin K;M *end*

if α *then* K *else* M

while α *do* K

are programs

(III) if α,β are formulas from L_a and K is a program, then $(\alpha \vee \beta)$, $(\alpha \wedge \beta)$, $(\alpha \Rightarrow \beta)$, $\neg\alpha$, $(\forall x)\alpha$, $(\exists x)\alpha$, $K\alpha$, $\bigcup K\alpha$, $\bigcap K\alpha$ are formulas of L_a.

The meaning:

$K\alpha$ - after executing of K, formula α is satisfied on the output valuation; if K diverges, then by definition the value of $K\alpha$ is false.

$\bigcup K\alpha$ - means infinite disjunction $\alpha \vee K\alpha \vee KK\alpha \vee \ldots$ $\ldots \vee K^n\alpha \vee \ldots$ i.e. there exists $n \in \omega$ such that n-times. $K \ldots K\alpha$ holds.

$\bigcap K\alpha$ - means infinite conjunction $\alpha \wedge K\alpha \wedge \ldots \wedge$ $\ldots \wedge K^n\alpha \wedge \ldots$ i.e. for every $n \in \omega$, $\underbrace{K \ldots K}_{n\text{-times}}\alpha$ holds.

We omit the natural definitions of free variables and of satisfiability. The symbol $\mathfrak{A} \models \alpha[v]$ denotes that formula α is satisfied in a relational system $\mathfrak{A}$ by a valuation v - an exact formalism is superfluous here. $Ax \models \alpha$ denotes the semantic consequence: formula α is valid in every model of the set Ax of formulas.

Let L , L' be in $\{L_1, L_w, L_w^f, L_a\}$

DEFINITION 2.1.

By an *interpretation of logic* L *in logic* L' we mean an effective mapping $I : L \rightarrow L'$ from the set of formulas L into the set of formulas of L' and from the power-set of the set of formulas of into the power set of the set of formulas of L'; $I : 2^L \rightarrow 2^{L'}$ such that for every set Ax of formulas from L and for every formula $\alpha \in L$ the following equivalence holds:

$$Ax \models \alpha \quad \text{iff} \quad I(Ax) \models I(\alpha)$$

In particular, a formula $\alpha \in L$ is a tautology iff $I(\emptyset) \models I(\alpha)$.

DEFINITION 2.2.

By a *translation of* $\mathbf{L}$ *to* $\mathbf{L}'$ we mean an effective one-one mapping: t: $\mathbf{L} \to \mathbf{L}'$ from the set of formulas of $\mathbf{L}$ into the set of formulas of $\mathbf{L}'$ such that: for every formula $\alpha \in \mathbf{L}$, if α possesses only individual variables as free ones then the same holds for $t(\alpha)$; and for every relational system $\mathfrak{A}$, for every formula $\alpha \in \mathbf{L}$ with only individual variables as free ones, and for every valuation ν of individual variables the following holds:

$$\mathfrak{A} \models \alpha[\nu] \quad \text{iff} \quad \mathfrak{A} \models t(\alpha)[\nu]$$

Let us notice, that a translation can be treated as an interpretation, if we assume $t(Ax) = \{t(\alpha) \mid \alpha \in Ax\}$, of course, there exists a translation of $\mathbf{L}_1$ to $\mathbf{L}_w$, to $\mathbf{L}_w^f$ and to $\mathbf{L}_a$. We have the following theorem on translatability and interpretability between logics $\mathbf{L}_w$, $\mathbf{L}_w^f$, $\mathbf{L}_a$.

THEOREM 2.1.

(I) *There exists a translation of* $\mathbf{L}_a$ *into* $\mathbf{L}_w^f$

(II) *There does not exist a translation of* $\mathbf{L}_a$ *into* $\mathbf{L}_w$

(III) *There exists an interpretation of* $\mathbf{L}_w^f$ *in* $\mathbf{L}_a$

(IV) *There does not exist a translation of* $\mathbf{L}_w^f$ *into* $\mathbf{L}_a$

PROOF.

(I) Quite simple; it is sufficient to notice that:

1. If 2n-ary relation $r(\vec{x},\vec{y})$ is definiable in a system $\mathfrak{A}$ by a formula φ from $\mathbf{L}_w^f$, then the transitive closure of r, $r^*(\vec{x},\vec{y})$, is also definiable in by a formula $\varphi^*(\vec{x},\vec{y})$ from $\mathbf{L}_w^f$: $(\vec{x}=\vec{y}) \vee (\exists R)$ (2n-ary finite relation R is compatible with $\varphi \wedge R$ is a linear order with $\vec{x}$ as the first element and $\vec{y}$ as the last element).

2. If the transition relation of programs K,M ($Trans_K$,$Trans_M$) are definiable in L_w^f, then the transition relations of programs *begin* K;M *end*, *if* α *then* K *else* M, *while* α *do* K are also definable in L_w^f.

3. $(\cup K\alpha)(\vec{x})$ means that there exists $\vec{y}$ such that $Trans_K^*(\vec{x},\vec{y})$ and $\alpha(\vec{y})$; similarly for $\cap K\alpha$.

(II) We can define the addition and multiplication in the system $\langle\omega;0,S,=\rangle$ by the formulas from L_a, so the existence of a translation of L_a into L_w is inconsistent with Büchi's result (cf. [1]) on decidability of the weak second order theory of system $\langle\omega;0,S,=\rangle$.

(III) Let E^* denote the set of all finite sequences of elements from a set E and let Fin(E) denote the family of all finite subsets of a set E.

Salwicki (cf. [4]) gave a collection of axioms of ATS (algorithmic theory of stacks) written in L_a, such that every model of ATS, proper for identity, is isomorphic to an algebra of finite sequences.

ATS: A1/ S (empty)

A2/ S(s) ⇒ *begin while* ¬ empty (s) *do* s := pop(s) *end* true

A3/ S(s) ⇒ (¬empty(s) ⇒ s = push(top(s),pop(s)))

A4/ (E(e) ∧ (S(s) ⇒ e = top(push(e,s))

A5/ (E(e) ∧ (S(s) ⇒ s = pop(push(e,s))

A6/ (E(e) ∧ (S(s) ⇒ ¬empty(push(e,s))

A7/ (S(s) ∧ (S(s) ⇒ ($s=s^o$ ⇔ *begin* s_1:=s; s_2:=s^o;

```
bool := true;
        while bool ∧ ¬empty(s1) ∧ ¬empty(s2) do
        begin bool:=bool ∧ top(s1)=top(s2);
              s1 := pop(s1); s2 := pop(s2)
        end
              end bool ∧ empty(s1) ∧ empty(s2)
```

A8/ $(E(e) \wedge S(s)) \Rightarrow$ (member(e,s) $\Leftrightarrow$

$\Leftrightarrow$ *begin* bool:=false; s_1:=s; x:=top(s);

while $\neg$ empty(s_1) *do*

begin *if* x=e *then* bool:=true;

s_1:=pop(s_1); x:=top(s_1)

end

end bool)

A9/ axioms for equality.

The axioms for ATS possess (cf. [4]) the following property: if ATS $\models$ <X;E,S,=,member,...> then $E \cup S$ is a subsystem of <X;E,S,...> isomorphic to the algebra of finite sequences $<E \cup E^*;E,E^*,...>$ and $<E \cup S;member>$ is isomorphic to $<E \cup Fin(E);\in>$.

Let us define, for n=1,2,... sets ATS_n:

$A1_n$/ S_n (empty)

$A2_n$/ $S_n(s) \Rightarrow$ *begin while* $\neg$empty(s) *do* s:=pop_n(s)

end true

$A3_n$/ $S_n(s) \Rightarrow$ (empty(s) $\Rightarrow$ s=$push_n(Top_n^1(s),...$

$...,Top_n^n(s),pop_n(s)))$

$A4_n$/ $(E(e_1) \wedge ... \wedge E(e_n) \wedge S_n(s)) \Rightarrow$

$\Rightarrow e_1=Top_n^1(push_n(e_1,...,e_n,s)) \wedge ...$

$... \wedge e_n=Top_n^n(push_n(e_1,...,e_n,s)))$

$A5_n$/ $(E(e_1) \wedge ... \wedge E(e_n) \wedge S_n(s) \Rightarrow$

$\Rightarrow s=pop_n(push_n(e_1,...,e_n,s))$

$A6_n$/ $(E(e_1) \wedge ... \wedge E(e_n) \wedge S_n(s)) \Rightarrow$

$\Rightarrow \neg empty(push_n(e_1,...,e_n,s))$

$A7_n$/ $(E(e_1) \wedge ... \wedge E(e_n) \wedge S_n(s)) \Rightarrow$

$\Rightarrow (member_n(e_1,...,e_n,s) \Leftrightarrow$

$\Leftrightarrow$ *begin* bool:=false; s_1:=s;

$x_1:=Top_n^1(s);...;x_n:=Top_n^n(s)$;

while $\neg$empty(s_1) *do*

begin if $x_1=e_1 \wedge \ldots \wedge x_n=e_n$ *then* bool:=true;

$s_1:=pop_n(s_1)$;

$x_1:=Top_n^1(s_1);\ldots;x_n:=Top_n^n(s_1)$;

end

end bool)

$A8_n$/ $(S_n(s)\wedge S_n(s)) \Rightarrow (s=s^o \Leftrightarrow$

$\Leftrightarrow$ *begin* $s_1:=s$; $s_2:=s^o$; bool:=true;

while bool $\wedge$ $\neg$empty(s_1) $\wedge$ $\neg$empty(s_2) *do*

begin bool:=bool$\wedge Top_n^1(s_1)=Top_n^1(s_2)\wedge\ldots$

$\ldots \wedge Top_n^n(s_1)=Top_n^n(s_2)$;

$s_1:=pop_n(s_1);s_2:=pop_n(s_2)$

end

end bool $\wedge$ empty(s_1) $\wedge$ empty(s_2))

$A9_n$/ Axioms for equality for the symbols:

$push_n$, pop_n,...

We can prove analogously as in [4], that

if $ATS_n \models \langle E\cup S_n;E,S_n,\ldots\rangle$ then $\langle E^n\cup S_n;member_n\rangle$ is isomorphic to $\langle E^n\cup Fin(E^n);\in\rangle$ (*)

Let $ATS^* = \bigcup_{n=1}^{\infty} ATS_n \cup \{(\forall x)_{x\neq empty}\neg(S_i(x)\wedge S_j(x))\}_{i\neq j}\cup$

$\cup\ \{(\forall x)\neg(E(x)\wedge S_i(x))\}_{i=1,2,\ldots}$

From the property (*) we derive

if $ATS^* \models \langle E\cup S_1\cup S_2\cup\ldots;\ldots\rangle$ then $\langle E^i\cup S_i;member_i\rangle$ is isomorphic to $\langle E^i\cup Fin(E^i);\in\rangle$ (**)

Now we are ready to define an interpretation I of L_w^f in L_a.

(1) if $\alpha\in L_1$, then $T(\alpha)$ is obtained by relativisation of all individual variables in α to the predicate E

(2) $I(R(x_1,\ldots,x_n)) = E(x_n) \wedge \ldots$

$\ldots \wedge E(x_n) \wedge S_n(R) \wedge member_n(x_1,\ldots,x_n,R)$

(3) $I((\forall R)\alpha) = (\forall R)(S_n(R) \Rightarrow I(\alpha))$

$I((\exists R)\alpha) = (\exists R)(S_n(R) \wedge I(\alpha))$

(4) $I(Ax) = ATS^* \cup \{I(\alpha)\mid\alpha\in Ax\}$

In (2) and (3) R is a n-ary relational variable. From the property (**) we obtain:

for every set Ax of formulas from $\mathbf{L}^{f}_{w}$ and for every formula α from $\mathbf{L}^{f}_{w}$, $Ax \models \alpha$ iff $I(Ax) \models I(\alpha)$. This end the proof of proposition III.

(IV) It is sufficient to consider a relational system (Danko [2]), in which stop-property of recursive procedures is not definable in $\mathbf{L}_a$ and to porve that the transition relation of recursive procedures is definable in $\mathbf{L}^{f}_{w}$. ◉

COROLLARY 2.2.

Let I be the interpretation defined in the proof of the theorem 2.2.

If a set Ax *of formulas from* $\mathbf{L}^{f}_{w}$ *is consistent then the set* I(Ax) *is consistent also and for every relational system* $\mathfrak{A} = \langle A, \ldots \rangle$, *if* $\mathfrak{A} \models Ax$ *then* $\langle A \cup A^*; A, A^*, \ldots \rangle = I(Ax)$. ◉

By *Löwenheim's number* of logic $\mathbf{L}$ we mean the least cardinal number π such that for every set Ax of formulas from $\mathbf{L}$, if Ax has a model, then Ax has a model of cardinality less than or equal to π.

By *Hanf's number* of logic $\mathbf{L}$ we mean the least cardinal number π such that for every set Ax of formulas from $\mathbf{L}$ if Ax has a model of cardinality π, then Ax has models of arbitrarily high cardinality.

As it is well known, Löwenheim's number of logics $\mathbf{L}_a$, $\mathbf{L}^{f}_{w}$ are equal to ω. Theorem 2.1. (I) and corollary 2.2. give together the following:

COROLLARY 2.3.

Logics $\mathbf{L}_a$ *and* $\mathbf{L}^{f}_{w}$ *have the same Hanf's numbers.*

COROLLARY 2.4.

The sets of tautologies of $\mathbf{L}_a$ *and of* $\mathbf{L}^f_w$ *are recursively isomorphic.*

PROOF.

Let us denote by the Symb($\mathbf{L}$) the set of all functional and relational symbols of logic $\mathbf{L} \in \{\mathbf{L}_a, \mathbf{L}^f_w\}$. Let us assume,

$$\mathrm{Symb} = \mathrm{Symb}(\ ^{f}_{w}) \cup \{S_1, S_2, \ldots, E, push_1, \ldots\} ,$$

(the second component of union contains symbols from ATS*), and that for every $n \in \omega$ there are ω functional and ω relational symbols of arity n in Symb($\mathbf{L}^f_w$);

then we can construct effective one-one mappings:

(1) translation $t: \mathbf{L}_a \to \mathbf{L}^f_w$ such that for every formula $\alpha \in \mathbf{L}_a$, $\models \alpha$ iff $\models t(\alpha)$

(2) mapping $I_1 : \mathbf{L}^f_w \to \mathbf{L}_a$ defined as follows:

$$I_1(\alpha) \overset{df}{=\!=} \bigwedge_{\beta \in \bigcup_{n=1}^{k} ATS_n} \beta \wedge I(\alpha)$$

where k is a natural number such that all the symbols from the set $\{E, S_1, S_2, \ldots, push_1, push_2, \ldots, \ldots\}$ (i.e. from ATS*) occuring in α are in a finite set $\{E, S_1, \ldots, S_k, push_1, \ldots, push_k, \ldots, \ldots\}$.

The mapping I_1 is one-one and possesses the property that for every formula $\beta \in \mathbf{L}^f_w$ $\models \beta$ iff $\models I_1(\beta)$.

The points (1) and (2) give together the thesis of corollary 2.4. ●

Similarly, with the help of interpretation and translation defined in the proof of theorem 2.1, we state the following

COROLLARY 2.5.

The interpolation theorem holds for $\mathbf{L}_a$ *iff it holds for* $\mathbf{L}^f_w$.

III. Some examples of models in which L_a and L_w^f are equivalent

Let K be a class of relational systems and let $L, L' \in \{L_w^f, L_a, L_1\}$. We say, that L and L' are equivalent in K iff for every $\mathfrak{A} \in K$ the following holds: every relation definable in $\mathfrak{A}$ by a formula from L is also definable in $\mathfrak{A}$ by a formula from L' and conversely. We say that logics L, L' are equivalent in a system $\mathfrak{A}$ iff they are equivalent in the class $\{\mathfrak{A}\}$.

For instance, in the system $N=\langle\omega;0,S,+,\cdot,=\rangle$ logics L_1, L_a, L_w^f are equivalent because of well known parametization of finite sequences of natural numbers by natural numbers. In order to generalize an idea of parametrization, we introduce the definition:

DEFINITION

Let $L \in \{L_1, L_w^f, L_a\}$ and let $\mathfrak{A}$ be a relational system with a carrier A.

The family of finite relations in A is parametrizable in $\mathfrak{A}$ by L iff there exists a sequence $\beta_1,\alpha_1,\beta_2,\alpha_2,\ldots$ (β_i possesses one free variable, α_i possesses $i+1$ free variables), such that

(1) for every $n\in\omega$ and every $a\in A$ the set

$$\{\langle a_1,\ldots,a_n\rangle\in A^n \mid \mathfrak{A}\models\alpha_n[a,a_1,\ldots,a_n] \text{ and } \mathfrak{A}\models {}_n[a]\}$$

is finite

(2) for every finite relation $R \subseteq A^n$ there exists $a\in A$ such that

$$R=\{\langle a_1,\ldots,a_n\rangle\ \mathfrak{A}\models\alpha_n[a,a_1,\ldots,a_n] \text{ and } \mathfrak{A}\models\beta_n[a]\}.$$

Of course, if in a system $\mathfrak{A}$ the family of finite relations is parametrizable by L_a (L_1) then L_w^f is equivalent in $\mathfrak{A}$ to L_a (L_1).

The field Q of rational numbers forms another example of a system in which L_a, L_w^f, L_1 are equivalent - it follows from Julia Robinson's result [3] that the set of natural numbers is definable in Q by a formula from L_1, so we can parametrize the family of finite relations in Q by L_1, because rational numbers can be represented as pairs of natural numbers. The set of natural number is definable in the field R of real numbers by a formula from L_a. Every real $x \in R$ can be represented in the only one way as a continued fraction:

$$x = a_o + a_1^{-1} + a_2^{-1} + \dots + a_n^{-1} + \dots$$

where $a_o = \text{entire}(x)$, a_n is a natural number for any $n=1,2,\dots$. The function $f(x,n) \overset{df}{=} a_n$ is definable in R by some formulas from L_a, so we can parametrize the family of finite relations in R by L_a with the help of the following formulas:

$Seq_R(x)$: $(\forall n\in\omega)$ $(f(x,n)$ is a number of a sequence of natural numbers $\wedge$ $(\forall i,j\in\omega)(f(x,i),f(x,j)$ are numbers of sequences of equal length$))$)

$lh_R(x) = n$: $Seq_R(x) \wedge$ $(\forall i\in\omega)$ $(f(x,i)$ is number of sequence of length $n)$

$(x)_i = y$: $i\in\omega \wedge Seq_R(x) \wedge i \leq lh_R(x) \wedge$ $(\forall n\in\omega)$ equals to the i-th element of the sequence with number $f(x,n))$

So L_a and L_w^f are equivalent in R.

The following problem seems to contain the essence of connections between L_a and L_w^f:

1. Is it true, that for every relational system $\mathfrak{A}$, if L_a and L_w^f are equivalent in $\mathfrak{A}$ then the family of finite relations in the carrier of $\mathfrak{A}$ is parametrizable in $\mathfrak{A}$ by L_a?

2. Does there esists a system in which L_a is equivalent to L_1 but L_w^f is not equivalent to L_1?

3. Does there exist a class of relational systems axiomatizable in L_1, in which L_a and L_w^f are equivalent?

REFRERENCES

[1] Buchi,J.R., Weak second order arithmetic and finite automata, *Zaitchs. Math. Logik Grund. Math.*, 1960.

[2] Danko,W., Definiability in algorithmic logic, *Foundamenta Informaticae* /to appear/.

[3] Robinson,J. Definiability and decision problems in arithmetic, *Journal of Symbolic Logic*, vol.14, No 2, 1949.

[4] Salwicki,A., Algorithmic theory of stacks, *Foundamenta Informaticae* /to appear/

M. GRABOWSKI
University of Warsaw

COLLOQUIA MATHEMATICA SOCIETATIS JÁNOS BOLYAI
26. Mathematical Logic in Computer Science,
Salgótarján (HUNGARY), 1978.

ON K-RECURSIVE DEFINITIONS AND BICATEGORIES

H.J. Hoehnke

In Budach-Hoehnke [1] it was show that the C-automata in a category C with finite products form a bicategory in the sense of J. Bénabou. Here we replace C by any strict dht-symmetric category K in the sense of the author [2-3]; but our considerations carry over to arbitrary dht-symmetric categories.

The standard model for K is the category Par of all sets and all partial mappings between sets. K-automata are the same as K-recursive definitions (K-r.d.) in the sense of the author [2]. Par-r.d. are also called recursive partial definitions. Let Def_K be the class of all K-r.d. i.e. of all quintuples

$$\langle A,B,S,l,m\rangle = S \xleftarrow{m} S\otimes A \xrightarrow{l} B = A \xrightarrow[l,m]{S} B$$

where $A,B,S \in |K|$ (the class of all objects of K) and $l,m \in K$. The composition

$$(1) \qquad (A \xrightarrow[l,m]{S} B)(B \xrightarrow[l',m']{S'} C) = A \xrightarrow[l'',m'']{S'\otimes S} C$$

of two K-r.d. $A \xrightarrow[l,m]{S} B$ and $B \xrightarrow[l',m']{S'} C$ is defined as the serial connection according to

$$1'' = (1_{S'} \otimes 1)1',\ m'' = (1_{S'} \otimes d_{S\otimes A}(1\otimes m))(m'\otimes 1_S)$$

and is again a K-r.d. The K-r.d. $I_A = A \xrightarrow[1_A, t_A]{I} A$ is called the identity of Def_K; here $1_A : A \to A$ is the identity of $A \in |K|$ and $t_A : A \to I$ a distinguished morphism of the signature of the enriched category K; $d_A : A \to A\otimes A$ denotes the diagonal morphism of $A \in |K|$.

As easily can be seen I_A is a left identity of any K-r.d. $A \xrightarrow[1,m]{S} B$ with respect to the composition (1). But in general I_B is not a right identity of $A \xrightarrow[1,m]{S} B$. Indeed, it holds

(2) $$(A \xrightarrow[1,m]{S} B)(B \xrightarrow[1_B, t_B]{I} B) = A \xrightarrow[1,m_o]{S} B$$

with

$$m_o = d_{S\otimes A}(1\otimes m)(t_B \otimes 1_S) = d_{S\otimes A}(1t_B \otimes 1_{S\otimes A})m.$$

If $1t_B = t_{S\otimes A}$, then $m_o = m$; but in general $1t_B = t_{S\otimes A}$ does not hold and in Par this equality is equivalent with 1 to be (not partial but) full. This is H. Lugowski's observation. Thus we have the fact that contrary to the case of C where Def_C is a category (or at least determines a bicategory) with respect to the composition (1) and with the identities $I_A (A \in |K|)$ this is not true if we replace C by an arbitrary K although this was claimed by the author [2], Theorem 8.1. Since in general $A \xrightarrow[1,m_o]{S} B$ in (2) is not isomorphic to $A \xrightarrow[1,m]{S} B$ and in order to obtain some structure similar to a bicategory we modify the concepts of a homomorphism and of a bicategory accordingly.

We introduce the concept of a pseudohomomorphism of K-r.d. in accordance with the concept of a pseudohomomorphism in the sence of the author [3]. A pseudohomomorphism

(3) $\langle s_A, s_B, s_S\rangle : A \xrightarrow[1,m]{S} B \quad A' \xrightarrow[1',m']{S'} B'$

is a triple $\langle s_A, s_B, s_S\rangle$ of morphisms

$$s_A : A \to A', \quad s_B : B \to B', \quad s_S : S \to S' \in K$$

such that the following diagram (4) is commutative

$$\begin{array}{ccccccccc}
S' & \xleftarrow{s_S} & S & \xleftarrow{m} & S\otimes A & \xrightarrow{1} & B & \xrightarrow{s_B} & B' \\
{\scriptstyle p_1^{S'S'}}\uparrow\uparrow{\scriptstyle p_2^{S'S'}} & & & & \downarrow{\scriptstyle d} & & & & {\scriptstyle p_1^{B'B'}}\uparrow\uparrow{\scriptstyle p_2^{B'B'}} \\
S'\otimes S' & \xleftarrow[m'\otimes s_S]{} & S'\otimes A'\otimes S & \xleftarrow[s_A\otimes s_B\otimes m]{} & S\otimes A\otimes S\otimes A & \xrightarrow[s_A\otimes s_B\otimes 1]{} & S'\otimes A'\otimes B & \xrightarrow[1'\otimes s_B]{} & B'\otimes B'
\end{array}$$

with

$$p_1^{AB} = 1_A \otimes t_B, \quad p_2^{AB} = t_A \otimes 1_B$$

The K-r.d. considered as heterogeneous algebras in K together with their pseudohomomorphisms form a category. This category together with the composition of K-r.d. according to (1) and with the identities I_A yield a slight modification of a bicategory in the sense of J.Bénabou which is also denoted by Def_K. The details are left to the reader and will be published elsewhere. We only mention:

LEMMA 1.

$$\langle 1_A, 1_B, 1_S\rangle : \quad A \xrightarrow[1,m]{S} B \quad B \xrightarrow[1_B, t_B]{I} B \longrightarrow A \xrightarrow[1,m]{S} B$$

is a pseudohomomorphism of K-r.d.

A pseudohomomorphism (3) will be called pseudoisomorphism if the morphisms s_A, s_B, s_S are all invertible. Then Lemma 1 expresses the fact that the right composition of $A \xrightarrow[1,m]{S} B$ with I_B is not equal to $A \xrightarrow[1,m]{S} B$ but at least pseudohomomorphic with $A \xrightarrow[1,m]{S} B$.

The mentioned modification of the concept of a bicategory concerns with the nature of the components

$$r(A \overset{S}{\underset{1,m}{}} B) = r(A,B) : (A \overset{S}{\underset{1,m}{}} B) I_B \to A \overset{S}{\underset{1,m}{}} B$$

of the natural transformation

$$r(A,B):c(A,B,B)(-,I_B) \to Id$$

occuring in the axioms of a bicategory which in our modification need not be isomorphisms but only pseudoisomorphisms.

Proof of lemma 1: The comutativity of the right part of the diagram (4) is obvious. The commutativity of the left part of (4) is equivalent with the commutativity of the following diagram

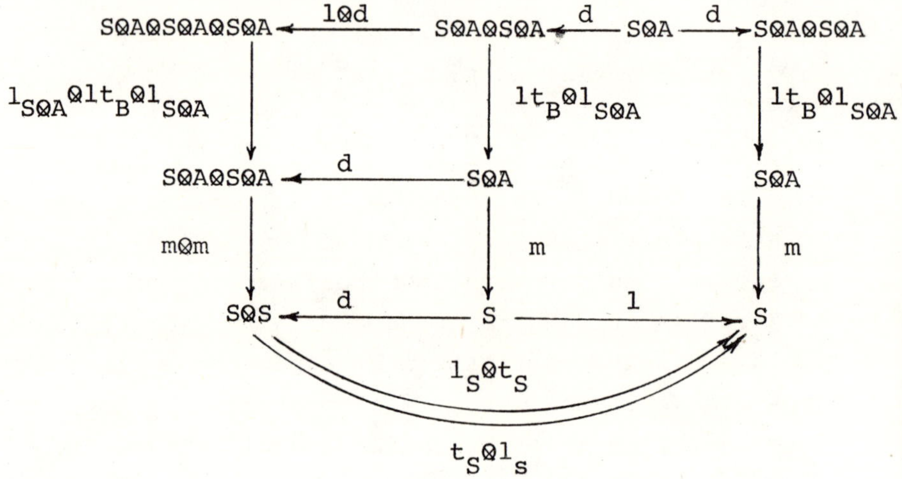

The difference between pseudoisomorphisms and isomorphisms is elucidated by

LEMMA 2.

Let (3) be a pseudoisomorphism which is invertible. Then it is an isomorphism (in the ordinary sense) between K-r.d.

Proof.

Since (3) is invertible, the components s_A, s_B, s_S have to be invertible in K. By assumption also

$$(3^{-1})\quad \langle s_A^{-1}, s_B^{-1}, s_S^{-1}\rangle : A' \xrightarrow[1',m']{S'} B' \to A \xrightarrow[1,m]{S} B$$

is a pseudoisomorphism. We have to show that (3) and

(3^{-1}) are (ordinary) homomorphisms, i.e. that the following diagrams commute:

$$\begin{array}{ccccc}
S & \xleftarrow{m} & S\otimes A & \xrightarrow{l} & B \\
\downarrow{\scriptstyle s_S} & & \downarrow{\scriptstyle s_S\otimes s_A} & & \downarrow{\scriptstyle s_B} \\
S' & \xleftarrow[m']{} & S'\otimes A' & \xrightarrow[l']{} & B'
\end{array}$$

$$\begin{array}{ccccc}
S' & \xleftarrow{m'} & S'\otimes A' & \xrightarrow{l'} & B' \\
\downarrow{\scriptstyle s_B^{-1}} & & \downarrow{\scriptstyle s_S^{-1}\otimes s_A^{-1}} & & \downarrow{\scriptstyle s_B^{-1}} \\
S & \xleftarrow[m]{} & S\otimes A & \xrightarrow[l]{} & B
\end{array}$$

Since s_A, s_B, s_S are invertible it is sufficient to prove the commutativity of the left one of these two diagrams. By (3) and (3^{-1}) we get

$$\begin{aligned}
ls_B &= d((s_S\otimes s_A)l't_B\otimes ls_B) = d((s_S\otimes s_A)l't_{B'}\otimes ls_B)c_{IB} \\
&= dc_{S\otimes A\ S\otimes A}(ls_B\otimes(s_S\otimes s_A)l't_{B'}) \\
&= d(ls_B\otimes(s_S\otimes s_A)l't_{B'}) \\
&= d((s_S\otimes s_A)(s_S^{-1}\otimes s_A^{-1})ls_B\otimes(s_S\otimes s_A)l's_B^{-1}t_B) \\
&= (s_S\otimes s_A)d'((s_S^{-1}\otimes s_A^{-1})ls_B\otimes l's_B^{-1}t_B) \\
&= (s_S\otimes s_A)l'.
\end{aligned}$$

Similarly it holds $ms_S=(s_S\otimes s_A)m'$. Here $c_{AB}:A\otimes B \to B\otimes A$ is the symmetry morphism of K belonging to $A,B \in |K|$.

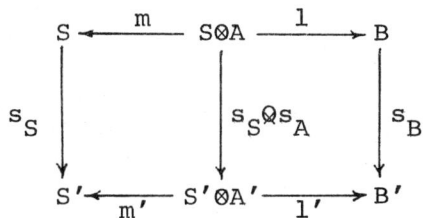

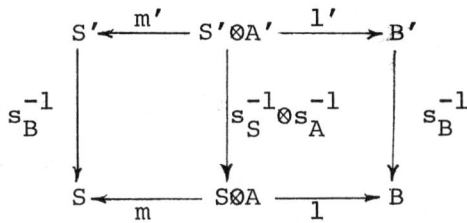

REFERENCES

[1] Budach,L., Hoehnke,H.J., *Automaten und Funktoren* Berlin, 1975.

[2] Hoehnke,H.J., *On partial recursive definitions and programs*, Fundamentals of Computation Theory, Proc. of the 1977. International FCT-Conference, Poznan--Kornik, Poland, September 19-23, Berlin-Heidelberg--New York, 1977.

[3] Hoehnke,H.J. On partial algebras, *Proc. Conf. Univ. Algebra*, Esztergom (HUNGARY) 1977, to appear.

Hans-Jürgen HOEHNKE
Zentralinstitut für Mathematik und Mechanik
Akademie der Wissenschaften der DDR
Mohrenstr. 39
108 Berlin
DDR

COLLOQUIA MATHEMATICA SOCIETATIS JÁNOS BOLYAI
26. Mathematical Logic in Computer Science,
Salgótarján (HUNGARY), 1978.

ON SEMANTICS OF PROGRAMMING LANGUAGES DEFINED BY UNIVERSAL ALGEBRAIC TOOLS

Z. Márkusz, M. Szőts

Background

In this paper we introduce a way of providing semantics for a programming language. Our approach is based on Universal Grammar created by R.Montague [1]. This kind of semantics is an algebraic one. In the case of algebraic semantics [2] the syntax is considered as a heterogenous word algebra. The carrier of this algebra is the set of the expressions of the language. This includes the basic expressions, which are not constructed from other expressions. Other elements are generated by the rules of the syntax and these rules are considered as the operators of the algebra (later on these are called "syntactical operators"). The word algebra is a free algebra generated by the set of basic expressions. So if there is a function from the set of basic expressions to the carrier of any algebra similar to the word algebra, it can be extended to a homomorphism between the two algebras. This unique homomorphism is considered to assign meaning to any expression of the language. So a model is the pair of an algebra and a function from the basic expressions into the algebra. This can lead to undesirable cases, such every numbering of the language can be considered as a model. On the other hand the usage of Universal Grammar forces us to characterize the acceptable interpretations. The most important idea is that the denotation of an expression depends on the point of reference. The point of reference is the pair of a possible world and the context of use. The first is usually some kind of time scale, the second is the assignment of variables. So the meaning of an expression is a binary function rendering a denotation to any possible world and context of use (Fig.1). Using Universal Grammar, after defining the syntax one needs to characterize:

- the structure of the possible worlds and the context of use, that is the set of points of reference;
- the sets of possible denotations, with the set of the points of references this gives the possible carriers of the algebras;
- the possible operations of the meaning algebras corresponding to the syntactical operations, these will be called "semantical operations";
- the possible functions from the set of basic expressions to the carriers of meaning algebras.

We shall define a language speaking *about* programs, which includes programs as expressions. We use first order classical logic to speak about programs. The definition of this language will be done in three steps. First we define the first order logic, this will serve also as an example for the use of Universal Grammar. The most important section is the definition of the programming language, namely giving meaning to programs.

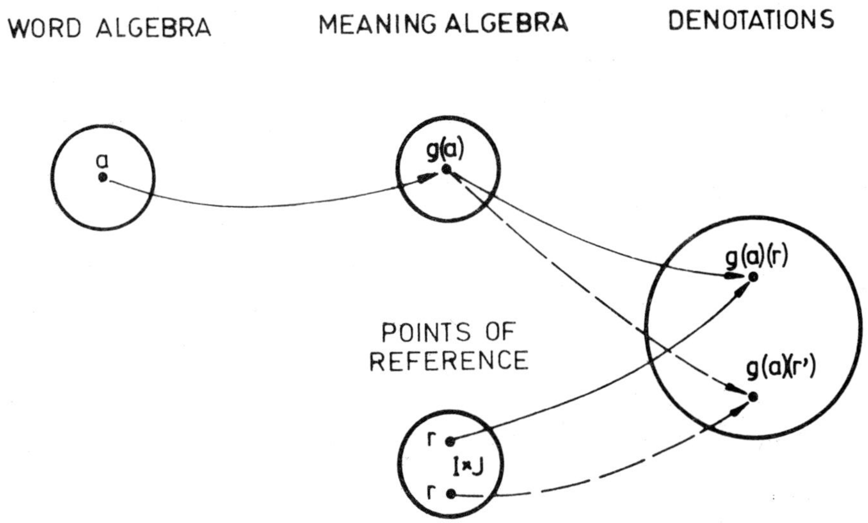

Figure 1.

Finally we define the expressions speaking about programs. However, before these three steps we fix the set of points of reference. In the case of first order logic the set of possible worlds include only one element, so the set of the points of reference is $\{i_o\} \times J_2$, where J_2 is the set of the assignments of variables. In the case of programming language the possible worlds represent the states of the computer during the execution of programs, the context of use is the input assignment. So the set of points of reference is $I \times J_1$, where I is a set isomoroph with ω that is, I is a time scale, and J_1 is the set of the possible input assignments. The set of reference points of our language is $I \times J_1 \times J_2$.

First order classical logic

The types of the word algebra representing the syntax of the first order classical logic are:

$$T_1 = \{e, t, \langle e^n, e\rangle, \langle e^n, t\rangle : 0<n\},$$

where the symbols stand for:

- e - type of entities;
- t - type of truth-values;
- $\langle e^n, e\rangle$ - type of n-ary functions;
- $\langle e^n, t\rangle$ - type of n-ary relations.

The primitive symbols which shall be used:

$Var = \{v_n : 0<n<\omega\}$ - variable symbols;

$Con = \{c_n : 0<n<\omega\}$ - constant symbols;

$\{f_{n,i} : 0<n<\omega,\ 0<i<\omega\}$ - function symbols, n is for the arity of the function;

$\{p_{n,i} : 0<n<\omega,\ 0<i<\omega\}$ - relation symbols;

$\uparrow$ - symbol of "TRUE";

$\neg, \vee, \exists, (,)$ - special symbols.

Now we start to define the syntax of the language. First let us see the basic expressions according to the types:

$X_e = Var \cup Con$

$X_t = \{\uparrow\}$

$X_{\langle e^n, e\rangle} = \{f_{n,i} : 0<n<\omega,\ 0<i<\omega\}$

$X_{\langle e^n, t\rangle} = \{p_{n,i} : 0<n<\omega,\ 0<i<\omega\}$

Now we have to define the syntactical operations. An operation is represented by a finite series of symbols, where the first is the operation symbol, the second one is the arity, and the followings, one after the other, are the types of arguments respectively and the last one is the type of the value of the function. The following operations will be defined:

$F_{<e,n>}$ for forming the terms, $F_{<t,n>}$ for the atomic formulas, and $F_{\neg}$, $F_{\vee}$, $F_{< \ ,n>}$ for the formulas.

i $\quad <F_{<e,n>}, n+1, <e^n, e>, e, \ldots, e; e>:$

$$F_{<e,n>}(f_{n,i}, a_1, \ldots a_n) = f_{n,i}(a_1, \ldots a_n);$$

ii $\quad <F_{<t,n>}, n+1, <e^n, t>, e, \ldots, e; t>:$

$$F_{<t,n>}(p_n, i, a_1, \ldots a_n) = p_{n,i}(a_1, \ldots a_n);$$

iii $\quad <F_{\neg}, 1, t; t>: F_{\neg}(\phi) = \neg\phi;$

iv $\quad <F_{\vee}, 2, t, t; t>: F_{\vee}(\phi_1, \phi_2) = \phi_1 \vee \phi_2$

v $\quad F_{<\exists,n>}, 1, t; t> F_{<\exists,n>}(\phi) = \exists v_n \phi .$

The syntax of the first order classical logic is the word algebra generated by the above operations from $<X_\delta>_{\delta \in T_1}$.

To define the semantics of the language first we define the possible denotations. Let E be an arbitrary set. Then for E the sets of possible denotations are the followings according to the types:

$D_{e,E} = E$, that is E is the set of the entities;

$D_{t,E} = \{TRUE, FALSE\}$,

$D_{<e^n,e>} = E^{E^n}$, that is the n-ary functions defined on E;

$D_{<e^n,t>} = D_{t,E}^{E^n}$, that is the n-ary relations defined on E.

Now for a given E the set of the possible meanings of an τ-type expression is

$$D_{\tau,E}^{(I \times J_1 \times J_2)}$$

where I is the set of possible worlds, J_1 is the set of input assignments and J_2 is the set of assignments for variables. So the carrier of the "meaning algebras" is

$$\langle D_{\tau,E}^{(I \times J_1 \times J_2)} \rangle_{\tau \in T_1}.$$

The operations on it, the semantical operations are defined in the followings. The semantical operator corresponding to the syntactical operator F_τ is denoted by G_τ. Since its arity, the types of the arguments and the type of the value are the same as at the corresponding syntactical one, only its value is needed to define. Since this value is a function from $I \times J_1 \times J_2$ to the set of the corresponding denotations, it will be given by its value at an arbitrary reference point $r \in I \times J_1 \times J_2$.

i Let $h \in M_{<e^n,e>,E}$ and $a_1, \ldots a_n \in M_{e,E}$. Then $G_{<e,n>}(h, a_1, \ldots a_n)(r) = h(r)(a_1(r), \ldots a_n(r))$;

ii Let $h \in M_{<e^n,t>,E}$ and $a_1, \ldots a_n \in M_{e,E}$. Then $G_{<t,n>}(h, a_1, \ldots a_n)(r) = h(r)(a_1(r), \ldots a_n(r))$;

Let $q_1, q_2 \in M_{t,E}$, and $r = \langle i, j_1, j_2 \rangle$.

iii $G_\neg(q_1)(r) =$ TRUE iff $q_1(r) =$ FALSE;

iv $G_{\vee}(q_1,q_2)(r) = \text{TRUE}$ iff $q_1(r) = \text{TRUE}$ or $q_2(r) = \text{TRUE}$;

v $G_{<\exists,n>}(q_1)(i,j_1,j_2) = \text{TRUE}$ iff there is a $j_2 \in J_2$ such, that for every $m \neq n$ $j_2(v_m) = j_2(v_n)$ and $q_1(i,j_1,j_2) = \text{TRUE}$.

Now a model of the first order classical language is the pair of an algebra

$$<<M_{\tau,E}>_{\tau\in T_1}, G_{\gamma}>_{\gamma\in\{<e,n>,<t,n>, \neg, \vee, <\exists,n>:n<\omega\}}$$

and a function f from $<X_{\tau}>_{\tau \in T_1}$ to $<M_{\tau,E}>_{\tau \in T_1}$ iff f meets the following requirements:

i if $v_n \in Var$, then $f(v_n)(i,j_1,j_2) = j_2(v_n)$;

ii for every r $f(\uparrow)(r) = \text{TRUE}$;

iii if ρ is an element of X_{τ}, where τ is $<e^n, e>$ or $<e^n,t>$, or $\rho \in Con$, then the value of $f(\rho)(r)$ is a constant element of the corresponding $D_{\tau,E}$.

The programming language

We study an "algol-like" language with assignment, *if ... then ... else* and *while* structures. We use the same primitive symbols as in the previous section plus some new ones:

$Id = \{x_n : 0<n<\omega\}$ - identifier symbols;

$Con' = \{c'_n : 0<n<\omega\}$ - constants symbols of the programming language;

$\leftarrow$, *if*, *fi*, *then*, *else*, *while*, *do*, *od* - new special symbols.

The types of the programming language are the followings:

$$T_p = \{st, \langle st, st\rangle, p, \langle p,e\rangle, \langle p,t\rangle\},$$

where the symbols stand for

st	- type of states;
$\langle st, st\rangle$	- type of statements;
p	- type of programs;
$\langle p,e\rangle$	- type of expressions;
$\langle p,t\rangle$	- type of conditions.

The set of basic expressions:

$X_{\langle p,e\rangle} = Id \cup Con'$

$X_{st} = X_{\langle st,\ st\rangle} = X_p = X_{\langle p,t\rangle} = \emptyset.$

The following syntactical operations will be defined: $F_{\langle x,n\rangle}$ for forming expressions, $F_{\langle c,n\rangle}$, $F_{\neg}$, $F_{\vee}$ for conditions, $F_{\langle\leftarrow,n\rangle}$ for creating assignment statements, and F_s, F_o, F_{IF}, F_{WHILE} for building programs. We consider *if* ... *then* ... *else* and *while* structures as program structures, not as statements, to avoid the notion of compound statements.

i $\langle F_{\langle x,n\rangle}, n+1, \langle e^n, e\rangle, \langle p,e\rangle, \ldots \langle p,e\rangle; \langle p,e\rangle\rangle$:

$$F_{\langle x,n\rangle}(f_{n,i}, a_1, \ldots a_n) = f_{n,i}(a_1, \ldots a_n);$$

ii $\langle F_{\langle c,n\rangle}, n+1, \langle e^n, t\rangle, \langle p,e\rangle, \ldots \langle p,e\rangle; \langle p,t\rangle\rangle$:

$$F_{\langle c,n\rangle}(p_{n,i}, a_1, \ldots a_n) = p_{n,i}(a_1, \ldots a_n);$$

iii $\langle F_{\neg}, 1, \langle p,t\rangle; \langle p,t\rangle\rangle : F_{\neg}(\phi) = \phi;$

iv $\langle F_{\vee}, 2, \langle p,t\rangle,\langle p,t\rangle;\langle p,t\rangle\rangle$: $F_{\vee}(\phi_1,\phi_2) = \phi_1 \vee \phi_2$.

v $\langle F_{\langle\leftarrow,n\rangle}, 1,\langle p,e\rangle;\langle st, st\rangle\rangle$: $F_{\langle\leftarrow,n\rangle}(a) = x_n\leftarrow a$;

vi $\langle F_s, 1, \langle st,st\rangle; p\rangle$: $F_s(a) = a$;

vii $\langle F_o, 2, p, p; p\rangle$: $F_o(p_1, p_2) = p_1p_2$;

viii $\langle F_{IF}, 3,\langle p,t\rangle, p, p; p\rangle$:

$F_{IF}(\phi, p_1,p_2) =$ *if* ϕ *then* p_1 *else* p_2 *fi*

ix $F_{WHILE}, 2,\langle p,t\rangle, p; p\rangle$:

$F_{WHILE}(\phi, p_1) =$ *while* ϕ *do* p_1 *od.*

Before defining the semantics of the language one important decision has to be made: whether the meaning of a programs is its possible executions or the mapping defined by them. The first is chosen, so the denotation of a program will be infinite series of states. A state is infinite series consisting of:

- the number of statements in the program;
- which statement will be executed next (its serial number);
- the actual values of the identifiens.

Denotations of a program will be called traces (Fig. 2.)

We introduce a notation: if p is a trace, p_i will denote the i-th state of p, and $p_{i,j}$ will denote the j-th component of p_1. Note that this notation is of the metalanguage, it is not part of the object language.

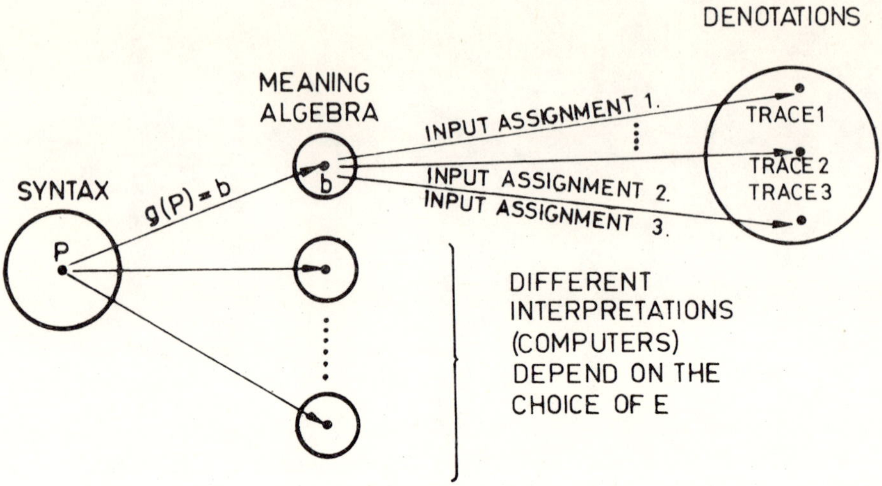

Figure 2.

The possible denotations:

i $D_{st,E} = \omega \times \omega \times E^{\omega}$, where according to the intuitive introduction the first ω is for the number of the statements, the second one is for the serial number of the next statement and E for the actual values of the indentifiers;

ii $D_{<st,st>,E} = D_{st,E}{}^{D_{st,E}}$, that is statements are state-to-state functions;

iii $D_{p,E} = D_{st,E}{}^{I}$, remember that I is isomorph to ω;

iv $D_{<p,e>,E} = E^{D_{p,e}}$

v $D_{<p,t>,E} = \{TRUE, FALSE\}^{D_{p,e}}$

The underlying idea of the two last definitions is that the value of an expression (condition) depends on the program it occurs in.

The semantic operations:

i Let $h \in M_{<e^n, e>, E}$ and $a_1, \ldots a_n \in M_{<p,e>,E}$. $G_{<x,n>}(h, a_1, \ldots a_n)(r)$ is a function from the traces to the entities. Let $t \in D_{p,E}$, then:

$G_{<x,n>}(h, a_1, \ldots a_n)(r)(t) =$

$= h(r)(a_1(r)(t), \ldots a_n(r)(t))$.

ii Let $h \in M_{<e^n, e>, E}$, $a_1, \ldots a_n \in M_{<p,e>,E}$ and $t \in D_{p,E}$. Then:

$G_{<c,n>}(h, a_1, \ldots a_n)(r)(t) = h(r)(a_1(r)(t) \ldots a_n(r)(t))$.

Let $q_1, q_2 \in M_{<p,t>,E}$, $t \in D_{p,E}$. Then

iii $G_{\neg}, (q_1)(r)(t) =$ TRUE iff $q_1\ (r)(t) =$ FALSE;

iv $G_{\vee}, (q_1, q_2)(r)(t) =$ TRUE iff $q_1(r)(t) =$ TRUE or $q_2(r)(t) =$ TRUE

v Let $a \in M_{p,e}$, then $G_{<\leftarrow,n>}(a)(i, j_1, j_2)$ is a function from states to states. Let

$s = <m, k, e_1, \ldots e_{n-1}, e_n, e_{n+1}, \ldots> \in \omega x \omega x\ E^{\omega}$ be

an arbitrary state. Let j_1' be such an input assignment, that for all ℓ $j_1'(x_\ell) = e_\ell$ and t be an arbitrary trace:

$G_{<\leftarrow,n>}(a)(i, j_1, j_2)(s) =$

$= <m, k+1, e_1, \ldots e_{n-1}, a(i_o, j_1', j_2)(t), e_{n+1}, \ldots>$.

vi Let $a \in M_{<st,st>,E}$. Then $G_s(a)(i, j_1, j_2) = t$ is a trace, where $t_o = <1, 1, j_1(v_1), \ldots j_1(v_n), \ldots>$ and for all $0<k$ $t_k = a\ (i, j_1, j_2)\ (t_o)$.

The next operations create programs from programs. To make their definition clearer, we introduce some auxiliary notions. Let $t \in D_{p,E}$ and $a \in E$, then for a trace t $t[k/a]$ will denote such a trace, that for all i $t[k/a]_{i,k} = a$ and for all $j \neq k$ $t[k/a]_{i,j} = t_{i,j}$ that is $[k/a]$ means a rewriting of the k-th element of all states in the trace.

Let us define a relation α and a partial function β on traces. Let t be an arbitrary trace belonging to a program p, then $\alpha(t)$ is true iff there is a state in t (t_i) so, that $t_{i,o}<t_{i,1}$, that is all the statements are executed in the program p. So $\alpha(t)$ is true iff the corresponding execution of program p terminates. Let partial function β be defined on t iff $\alpha(t)$ is true. Then let the value of $\beta(t)$ be the index (i) of the first such state where $t_{i,o}<t_{i,1}$.

In the following $p_1, p_2 \in M_{p,F}$ and $p_1(r)_{o,o} + p_2(r)_{o,o} = m$, that is m is the number of statements in the $F_\tau(p_1,p_2)$ program, where $\tau \in \{o, IF\}$.

The semantic operation G_o makes the programs appended by F_o run each after other.

vii Let us denote the trace $G_o(p_1,p_2)(i,j_1,j_2)$ by τ.

- If $\alpha(p_1(i,j_1,j_2))$ is true, then let $\beta(p_1(i,j_1,j_2)$ be u, and for all $0 \le k \le u$ $\tau_k = p_1(i,j_1,j_2)[0/m]_k$.

  Let j_1' be such an input assignment, that $j_1'(x_n) = \tau_{u,n+1}$. (So the input assignment j_1' will be the same, as the output of p_1.) Then for all $0<k$ $\tau_{u+k} = p_2(i,j_1', j_2)[0/m, 1/(p_1 \ (i,j_1,j_2)_{o,o} + p_2(i,j_1', j_2)_{k,1})]_k$

τ_{u+k} is a state in the corresponding trace of p_2, where the number of statements, and their serial numbers have to be changed for the ones in program p_1p_2.

- If $\alpha(p_1(i,j_1,j_2))$ is false, then $\tau = p_1(i,j_1,j_2)$.

The following rules will be defined in a similar way.

viii Let $a \in M_{<p,t>,E}$, $p_1,p_2 \in M_{p,E}$.

$$G_{IF}(a,p_1,p_2)(r) = \textit{if } a(r)(p_1(r)) \text{ is true, then } p_1(r)[O/m] \;\; \textit{else } p_2(r)[O/m,\ 1/p_1(r)_{o,o} + p_2(r)_{k,1}]$$

ix Let $a \in M_{p,t,E}$. Let us denote the trace $G_{WHILE}(a,p_1)(i,j_1,j_2)$ by τ.

- If $a(p_1(i_o,j_1,j_2))$ is false, then $\tau_o = <p_1(i,j_1,j_2)_{o,o}, 1, j_1(v_1), \ldots j_n(v_n), \ldots>$ and for all $o<k$ $\tau_k = <p_1(i,j_1,j_2)_{o,o}, p_1(i,j_1,j_2)_{o,o} + 1, j_1(v_1), \ldots j_n(v_n), \ldots>$.
- If $a(p_1(i_o,j_1,j_2))$ is true, and $\alpha(p_1(i,j_1,j_2))$ is false, then $\tau = p_1(i,j_1,j_2)$.
- Otherwise let $\beta(p_1\ i,j_1,j_2))$ be u.

  For all $O<k<u$ $\tau_k = p_1(i,j_1,j_2)_k$, and $\tau_u = <p_1(i,j_1,j_2)_{u,o}, 1, p_1(i,j_1,j_2)_{u,2}, \ldots \ldots p_1(i,j_1,j_2)_{u,n}, \ldots>$.

  Let j_1' be such an input assignment, that

$j_1'(v_n) = p_1(i,j_1,j_2)_{u,n+1}$. For all $0<k$

$\tau_{u+k} = (G_{WHILE}(a,p_1(i,j_1', j_2))_k$.

We have defined all the semantic operations. To complete the semantics of the programming language the function f from the basic expressions has to be defined. Let t be an arbitrary trace, then:

i for every $x_n \in Id$

- $f(x_n)(i_o,j_1,j_2)(t) = j_1(x_n)$;

- for any $i \neq i_o$ $f(x_n)(i,j_1,j_2)(t) = t_{i,n+1}$;

ii for every $c_n' \in Con'$ $f(c_n')(r)(t) = f(c_n)(r)$, where $c_n \in Con$.

The semantics of the programming language is quite complicated, but one can easily see, that it covers our intuition about programs. The cause of the difficulties is that semantics has to be built up without any direct reference to syntax.

Utterances about programs

The language we want to define will be the union of the first order logic and the programming language being completed with some expressions about programs. A new type is introduced: ω, $X_\omega = \{0, 1, 2, \ldots\}$. $D_{\omega,E}$ will be the set of natural numbers, and the f function between X_ω and $D_{\omega,E}$ will be the obvious one. We suppose that successor (+ 1) is defined, but do not detail it. The new syntactical operations are the followings:

i <F_L, 1,p,ω>: $F_L(p) = \ell(p)$, intuitively the length of the program (number of statements);

ii <F_{IND}, 3,p,ω, t; t>: $F_{IND}(p, n, \phi) = (p, n, \phi)$, intuitively it says, that ϕ is an inductive assertion [3] at the n-th statement of p;

iii <F_{INT}, 3, p, ω, t; t>: $F_{INT}(p, n, \phi) = [p, n, \phi]$, intuitively it says, that ϕ is an intermittent assertion [4] at the n-th statement of p.

ℓ,] and [are new primitive symbols.

The corresponding semantical operations:

i Let $p \in M_{p,E}$, then $G_L(p)(r) = p(r)_{o,o}$.

ii Let $p \in M_{p,E}$, $j \in \omega$, $\phi \in M_{t,E}$, then $G_{IND}(p,j,\phi)(i,j_1,j_2)$ = TRUE iff *for every* $i \in I$ $p(i,j_1,j_2)_{i,1} = j$ implies that $\phi(i,j_1,j_2')$ is true, where $j_2'(v_n) = p(i,j_1,j_2)_{i,n+1}$. Note that $p(i,j_1,j_2)_{i,n+1}$ is the value of the n-th identifier at the i-th state.

iii Let p,j be the same as above. $G_{INT}(p,j,\phi)$ (i,j_1,j_2) = TRUE IFF *there is an* $i \in I$ so that $p(i,j_1,j_2)_{i,1} = j$ and $\phi(i,j_1,j_2')$ is ture, where j_2' is the same as before.

Now our language can express the most important notions concerning correctness of programs:

$(p, \ell(p)+ 1,\phi)$- p is partial correct with respect to ϕ,
$[p, \ell(p)+ 1,\phi]$- p is totally correct with respect to ϕ,
$[p\ \ \ell(p)+ 1,\uparrow]$- p terminates.

CONCLUDING REMARKS

Our aim has been to show how Universal Grammar can be used for defining semantics of computer language. While doing it we had to make some decisions, which are independent of Universal Grammar. Such decision is to formalise operational semantics instead of denotational ones. The other can be done in the same manner. Similarly, it was an arbitrary choice to define I as an ω-type set. We could demand only that I should be a model of arithmetic e.g. of Peano axioms. In that case a family of denotation depends not only on E, but on E and I. For the discussion of these two possible theories see [5].

The definition of more complicated programming languages can be done in the same way. We want to call the reader's attention to non-deterministic and parallel languages. The main difference from the presented case would be in the structure of time scale. In the first case I could be a certain partial ordered set, in the second a family of ω-type sets, one for every processor.

We want to emphasize that Universal Grammar does not give new ideas for the semantics of programming languages, but helps to formalise the existing ones in a strict, correct way. We claim that the existing logical systems for programming like algerithmic logic, dynamic logic can be presented in this frame.

REFERENCES

[1] Montague, R., Universal Grammar, *Theoria*, 36, 1970, pp. 373-398.

[2] Goguen, J.A., T atcher, J.W., Wagner, E.G, Wright, J.B.: Initial algebra semantics and continuous algebras, *Journal of ACM*, 24, 1977, pp. 68-95.

[3] Floyd, R.W.: Assigning Meanings to Programs in J.T. Schwartz (ed.), *Mathematical Aspects of Computer Science*, Proceeding of Symposium on Applied Mathematics, vol. 19, American Mathematical Society, Providence, R.I., 1967, pp. 19-32.

[4] Manna, Z., Waldinger, R.: Is "sometime" sometime be better than "always". Intermittent assertions in proving program correctness, *Comm. of ACM*, vol.21, 2, 1978, pp. 159-172.

[5] Gergely, T., Úry, L.: Time models for programming logics, this volume.

Z. Márkusz
Computer and Automation Institute,
Hungarian Academy of Sciences,
Budapest, XI., Kende u. 13/17.
Hungary

M. Szőts
Research Institute for Applied Computer Science,
H-1536, Budapest, POB 227.
Hungary

COLLOQUIA MATHEMATICA SOCIETATIS JÁNOS BOLYAI
26. Mathematical Logic in Computer Science
Salgótarján (HUNGARY), 1978.

COMPLETIONS, FACTORIZATIONS AND COLIMITS FOR ω-POSETS[1]

J. Meseguer

1. *INTRODUCTION*

The aim of the present paper is to study properties of the category *Pos*(ω) of ω-complete posets and ω-continuous maps. One reason for doing so is that many of these properties can be lifted to, or at least form the basis for (Cf. Meseguer [21], [22], further details to appear elsewhere) similar properties for varieties of continuous algebras, which are closely related to the mathematical semantics of program schemes (see Scott [26], Nivat [25], Courcelle and Nivat [7], ADJ [2], and references there). Related work on complete posets includes Scott [27], Markowsky [18], [19] Markowsky and Rosen [20], Bloom [5], ADJ [4], Courcelle and Raoult[8], Lehmann and Smith [14] and Lehmann [13]. Several results in Lehmann [13] in particular on the topic of factorizations are closely related to results in this paper.

Pos(ω) is perhaps the simplest category in which semantics can be considered. We have favored it over

---

[1] This research was supported in part by the National Science Foundation, Grant No. MCS72-03633.

Pos(Δ): posets complete for arbitrary chains and continuous maps. However, except for local presentability (see Section 4), a property which has to do with the (infinitary) first-order logical character of ω-completeness, as opposed to the second-order logical character of Δ-completeness pointed out by Courcelle [6], there is a very strong parallel, which normally involves minor modifications. More generally, we conjecture (Cf. Meseguer [23],[24]) that a good number of similar results will stand in adequate form for "Z"-posets, a concept under development - Cf. ADJ [4], Courcelle and Raoult [8], Lehmann [13] - which is going to provide a flexible choice of adequate categories for semantics.

In Section 2 we introduce the three categories *Pos*, ω*Pos* and *Pos*(ω) which we will be using, review completions, i.e. left adjoints for inclusions, and also look at function spaces, pointing out that *Pos* and *Pos* (ω) are cartesian closed.

Section 3 treats factorizations and colimits. The topic of factorizations in a category is developed here from scratch, and a general construction of coequalizers is given. Several factorization systems for monotone and continuous maps are studied in detail. In addition, an account is given of a class of continous maps - the nice ones - which do not originate a factorization system, but have other very useful properties, also in connection with colimits. They have been studied by Markowsky [19] and Courcelle and Raoult [8].

Section 4 deals with additional properties which are of interest in connection with algebras. We look at filtered colimits and generators, and deduce the local presentability of *Pos* (ω), a categorical property which is inherited by the categories of algebras - Cf.

Meseguer [22]. The behavior of finite products with respect to epis is finally considered.

I would like to express my gratitude to Jean Bénabou, Bruno Courcelle, Yves Dyers, Joe Goguen, Klaus Indermark and Ignacio Jané for fruitful conversations on the subject of this paper, to Daniel Lehmann for his help on sending comments on previous versions of these results, and to Jim Thatcher for a stimulating correspondence. The March Foundation and the Spanish Ministry of Education and Science, through successive post doctoral research felloships have made possible my work on these topics.

Some final words on categorical matters: the amount of Category Theory which is used is moderate. However, familiarity with the following concepts (as exposed for instance in Mac Lane [16]) is assumed: limits and colimits - and in particular, products and coproducts, equalizers and coequalizers, pullbacks, kernel-pairs, and intersections - and also adjoint functors and their formulations in terms of universal maps (see for instance Mac Lane [16], Ch. IV, Th. 2). Categories are denoted by: *Pos*, *Pos*(ω), *B*,...etc. Objects are denoted A, B, C $\in$ *B*, and *B*(A,B) stands for the set of morphisms from A to B. Composition of morphisms follows the traditional convention for functions.

2. *COMPLETIONS AND FUNCTION SPACES*

In this section we introduce the basic categories and consider completions and cartesian closedness i.e., the lambda conversion property for maps.

We will use the following three categories: *Pos*, with objects posets, (A, $\leq$ A), and morphisms monotone maps; ωPos, also with posets as objects, but will morphisms ω-*continuous* maps, i.e., maps f: A → B such that if $\{a_n\}_{n\in\omega}$ is a countable chain in (A, ≤ A), which has a supremum or limit $a = \bigsqcup_n a_n$, the $\{fa_n\}_{n\in\omega}$ has limit fa. Such an f is in particular monotonic, which gives an inclusion functor $\omega Pos \hookrightarrow Pos$. The third category, denoted $Pos(\omega)$, has as objects ω-*complete* posets, i.e. posets in which any countable chain has a limit. The morphisms are the ω-continuous maps. So we have a full inclusion functor I: $Pos(\omega) \hookrightarrow \omega Pos$ and hence, also an inclusion J: $Pos(\omega) \hookrightarrow Pos$.

It is well known - cf. Markowsky [18] - that in the definitions of ωPos and $Pos(\omega)$, we can replace countable chains by countable filters, i.e., subsets F ⊂ A such that are countable, and if a, a' ∈ F there exists a" ∈ F with a" ≥ a, a" ≥ a'. In the same vein, it is equivalent to define ω-continuous maps as those preserving limits of countable chains or as those preserving limits of countable filters. From now on, when referring to countable chains, countable filters, ω-complete posets or ω-continuous maps, we will frequently simply say chains, filters, complete posets and continuous maps.

2.1. PROPOSITION

(a) The inclusion functor J: $Pos(\omega) \to Pos$ has a left adjoint.

(b) The inclusion functor I: $Pos(\omega) \to \omega Pos$ has a left adjoint.

Proof. (a) Given a poset (A, ≤ A), take the set $F(A)$ of countable filters F ⊂ A. There is a preorder of domination between filters, namely: $F \sqsupseteq F'$ if $\forall x' \in F' \; \exists x \in F$

s.t. $x \geq x'$. Call $R_{\sqsupseteq}$ to the equivalence relation defined by $\sqsupseteq$: F $R_{\sqsupseteq}$ F' iff F $\sqsupseteq$ F' and F' $\sqsupseteq$ F. The set $\hat{A} = \mathcal{F}(A)/_{R_{\sqsupseteq}}$ is a poset in the obvious way, and it is ω-complete, because if $\{[Fn]\}$ is a chain, then $F = \bigcup_n Fn$ is also a countable filter, and [F] is the limit of the chain. There is a monotone map ηA: $(A, \leq A) \to (\hat{A}, \leq \hat{A})$, defined by: $a \mapsto [\{a\}]$. One sees immediately that $\eta A(a) \geq \eta A(a') \Rightarrow a \geq a'$ (ηA is what we will call a full map) and in particular ηA is injective. The above data defines a left adjoint $\wedge$: *Pos* $\to$ *Pos*(ω) for J, because if f: A $\to$ B is a monotone map from a poset A to B $\in$ *Pos*(ω), then there exists a unique $\overline{f} : \hat{A} \to B$, continuous and extending f : $\overline{f}\eta A = f$. The key point is the fact that any [F] is the limit in $\hat{A}$ of its singletons $[\{x\}]$, $x \in F$, and hence $\overline{f}$ is forced to be: $\overline{f}[F] = \bigsqcup fF$.

The proof of part (b) - and a different proof for part (a)- will follow from Prop. (3.18) in the next section (see Remark (3.19)).

□

2.2. Remark

The above proposition is well-known. For posets complete respect to arbitrary chains (a) and (b) were proved in Markowsky [18], [19] and Markowsky and Rosen [20]; and Bloom [5] proved (a) for the ω-case. (a) and (b) are particular instances of a general completion theorem which has been given by Courcelle and Raoult [8]. See also ADJ [4] and Lehmann [13].

2.3. Example

Consider the set $A = (\omega \times \omega) \cup \{\infty\}$, with ∞ as top element and with $(n,m) \geq (n',m) \iff n \geq n'$ and $m \geq m'$

$n, n', m, m' \in \omega$. It is easy to see that $\hat{A}$ is isomorphic to A together with the following limit points between ∞ and any element in $\omega \times \omega$: $[\{(k,n)\}n\in\omega]$ and $[\{(n,k)\}n\in\omega]$ for any $k \in \omega$. The two chains obtained letting k vary in ω tend to $[\{(n,n)\}n\in\omega] = [\omega \times \omega]$.

2.4. Remark

It is natural to ask for the existence of a third adjoint, namely one for the inclusion $\omega Pos \hookrightarrow Pos$. In Section 3 we will see a number of anomalies which make ωPos of little use, and in particular a proof that such an adjoint does not exist.

An important property of *Pos* and *Pos*(ω) is that they are *cartesian closed* categories, i.e., the homsets themselves are objects of the category, and a lambda conversion between maps $\overline{g}: A \times B \to C$, $g: B \to [A,C]$ exists. What one has to prove is the existence of a right adjoint $[A,-]$ for the endofunctor $[A\times-]$.

2.5. PROPOSITION

Both Pos and Pos(ω) *are cartesian closed categories.*

Proof. Given two maps $f, g: (A, \leq A) \to (B, \leq B)$, define $f \geq g$ iff $\forall a \in A$, $fa \geq ga$. This makes *Pos*(A,B) a poset and, for $A, B \in$ *Pos*(ω), *Pos*(ω) (A,B) a complete poset. Denote in both cases by $[A,B]$ the homset together with its order structure. Letting B vary, we get an endofunctor $[A,-]$. The argument goes in a parallel way for the monotone case. Let $A \in$ *Pos*(ω). The evaluation map $ev_C: A \times [A,C] \to C: (a,f) \mapsto f(a)$, is defined for any C

$\in$ *Pos*(ω) and is continuous. It provides the counit of the adjunction $(A\times -) \dashv [A,-]$, because given f: $A \times B \to C$ continuous, the only possible choice for a map $\overline{g}$: $B \to [A,C]$ such that $\overline{g} = ev_C(1_A \times \overline{g})$ is : $\overline{g}$: $b \to f(-,b)$, which is continuous.

□

2.6. Remark

It is easy to see that $f \geq g$ implies $\hat{f} \geq \hat{g}$. One then says that the adjunction $\wedge \dashv J$ is a *Pos*- adjunction.

2.7. Remark

An important connection between completions and function spaces, which gives conditions for $[\hat{A},\hat{B}]$ to be of the form $\hat{H}$, can be found in Markowsky and Rosen [20].

3. *FACTORIZATIONS AND COLIMITS*

In this section, different classes of monomorphisms, epimorphisms, and of epi-mono factorizations in *Pos*(ω) are investigated. These factorizations are related to colimits. A particular class of epis - which we call nice epis - is of special importance. Their universal property was studied in Markowsky [19], who constructed coequalizers through them, Courcelle and Raoult [8] have studied them for continuous algebras, and have put in evidence their close connection with preorders. We give also an account of these developments, and make some related remarks.

We recall that a morphism $f\colon A \to B$ in a category $\mathcal{B}$ is said to be a *monomorphism*, if for any pair $g, h\colon C \to A$, $fg = fh$ implies $g = h$. Dually, f is an epimorphism if $gf = hf \Rightarrow g = h$, for any pair $g,h\colon B \to C$. In the category of sets, monomorphisms coincide with injections and epimorphisms with surjections. What can be said for *Pos*, ω*Pos*, and *Pos*(ω)? Certainly if f is injective, it has to be a monomorphism, and if it is surjective it will be an epimorphism. On the other hand, if $f\colon A \to B$ is a monomorphism, for each $a, a' \in A$ we have maps $a, a'\colon 1 \to A$, which are trivially monotone and continuous, for the unique poset structure on the one point set 1. Hence, $fa = fa' \Rightarrow a = a'$. In *Pos* we will see f iff f surjective. However, this is not true, neither in ω*Pos* nor in *Pos*(ω).

In a category, the class M(A) of all monomorphisms with a same given codomain A is preordered by: $g \sqsupseteq f$ iff $\exists h$ s.t. $f = gh$. Take $P(A) = M(A)_{/R_{\sqsupseteq}}$, which is an ordered class. It is immediately seen that in P(A), $[f] = [g]$ iff h is an isomorphism. Each equivalence class is called a *subobject*. The category is said *wellpowered* if, for each object A, P(A) is not a class, but a set. In many cases one chooses canonical representatives for P(A): in the category of sets, the subsets of A are such. Dually, two epis with the same domain A are preordered by: $g \sqsupseteq f$ iff $\exists h$ s.t. $f = hg$, and from the class of epis E(A) one defines $Q(A) = E(A)_{/R_{\sqsupseteq}}$, class of *quotients* of A. The category is *cowellpowered* if Q(A) is always a set. We can safely leave it for the reader to check:

3.1. LEMMA

Pos, ωPos and Pos(ω) are wellpowered. □

Our three categories have products, as is well known: one takes the cartesian product, with the componentwise order. The equalizers in *Pos* and *Pos*(ω) are constructed in the same way: given f,g: A → B, one takes $I = \{a \in A \mid fa = ga\}$, $\leq_I = \leq_A \cap I^2$, and the inclusion map I → A. Hence *Pos* and *Pos*(ω) are *complete* in the sense that any small diagrams has a limit (Cf. Mac Lane [16], V, 2, Th. 1). However:

3.2. LEMMA

ωPos does not have equalizers.

Proof. Take $A = \omega \cup \{a,b,c\}$ with the ordinal order in ω, and also: $a \geq n$, $b \geq n$, $c \geq n$, $n \in \omega$. Take $f = 1_A$, and g: A → A with $g|\omega = 1_\omega$, ga = a, gb = c, gc = b. As equalizers are monomorphisms and in ω monomorphisms coincide with injective maps, if I is an equalizer of f an g we may assume $I \subset \omega \cup \{a\}$. Take now $B = \omega \cup \{a', a''\}$, with same order on ω, plus: $a' \geq n$, $a'' \geq n$, $\forall n \in \omega$, and define h: B → A by hn = n $\forall n \in \omega$, h(a') = h a" = a, which is trivially continuous. h equalizers f and g, so if I were an equalizer, we would get a map $\bar{h}$: B → I, forcing $I = \omega \cup \{a\}$, with ordinal order on ω, plus $a \geq n$ $\forall n \in \omega$, but then the inclusion I → A fails to be continuous. □

We will now introduce the concept of epi-mono factorizations and discuss some of their basic properties. They do for general categories what the set -

theoretic image for sets, and the homomorphism theorem for universal algebras. In general, there are not one, but several possible epi-mono factorization systems for the same category. This does not happen for sets or for modules, but happens for rings and topological spaces, and also for posets and for complete posets. It is best to take an axiomatic approach. The following comes from Isbell [12], improving Mac Lane [15]:

3.3. Definition

Let *B* be a category, E a class of epis and M a class of monos, in *B*, such that

(i) both E and M are closed under composition: $f, g \in E \Rightarrow gf \in E$, and the same for M,

(ii) every isomorphism is both in E and in M,

(iii) every $f: A \to B$ in admits a factorization $f = me$ with $e \in E$, $m \in M$, which is unique in the sense that if $f = m' e'$, with $e' \in E$, $m' \in M$, there exists an isomorphism h making

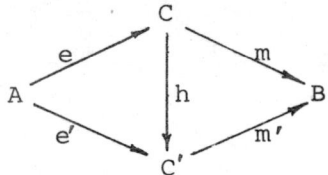

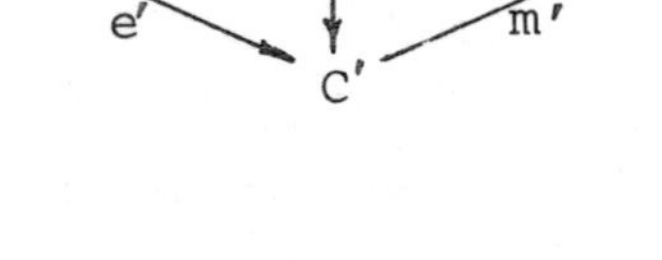

commutative.

Under these conditions we say that *B* is an *E - M category*, or that *B* has an *E - M factorization system.*

(Note that, putting (ii) and (iii) together, f is in $E \cap M$ iff f is iso).

3.4. *PROPOSITION*

If B is an E - M category, the following holds:

(i) *(Duality principle) The opposite category, B^{op}, has a factorization system with epis=M and monos=E.*

(ii) *(Diagonal lemma) given a commutative square*

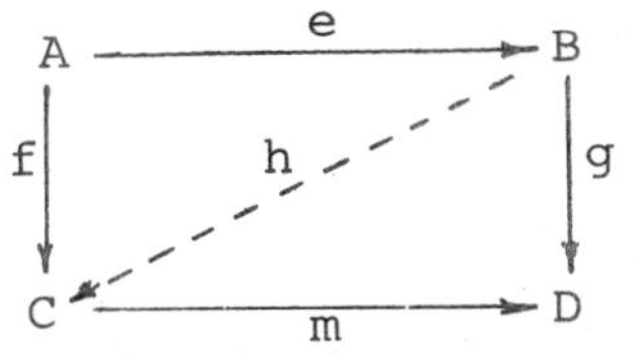

with $e \in E$ and $m \in M$, there exists a unique h commuting the diagram.

(iii) *(E determines M and viceversa)*

$E = \{f \text{ in } B \mid f = mh \text{ and } m \in M \Rightarrow m \text{ iso}\}$

$M = \{f \text{ in } B \mid f = he \text{ and } e \in E \Rightarrow e \text{ iso}\}$

(iv) *(a) any equalizer is in M; (b) $fg \in M$ $g \in M$;*

(c) M is closed under the formation of products, pullbacks, and intersection of subobjects in M.

Proof. See Herrlich and Strecker [11], Ch. IX. Note that - using (i) - (iv) can be dualized. □

Though many categories have epi-mono factorization systems, some have not. An example at hand is:

3.5. *LEMMA*

In ωPos there are maps f which cannot be factorized in any way as an epi followed by a mono.

Proof. Take the map H: B → A defined in (3.2), and note that neither h, nor the restricted maps h': B → A - {b}, h": B → A - {c}, are epimorphisms. □

From (3.4) - (iii), one sees that taking M the class of all monomorphisms, the only *candidate* E for an E - mono factorization system is the class of all epis e such that e = mh, and m mono imply m iso. Such epis are called *extremal epimorphisms*. Dually, one obtain in the same way the class of *external monomorphisms*. Under fairly weak assumptions they give rise to factorization systems in a category:

3.6. *PROPOSITION*

Every well-powered category B with intersections, finite products and equalizers has extremal epi-mono and epi-extremal mono factorization systems.

Proof. See Herrlich and Strecker [11], 34.1 and 34.5. To obtain the factorization, what one does in both cases, is to take the intersection of all the subobjects (resp. extremal subobjects) of the codomain through which the given map factors. □

3.7. *Remark*

In some categories, maps can always be factored as a coequalizer followed by a mono-resp. an epi followed by an equalizer. When this happens, the conditions in

(3.3) follow from the coequalizer-resp. equalizer-property and hence extremal epis coincide with co-equalizers resp. extremal monos with equalizers - and give a factorization system (see Herrlich and Strecker [11], 33.4 or Manes [17], Ch. II, Sec. 1).

Given an E - M category, the subobjects of A which are in M define a subclass MP(A)⊂PA, and the like for E - quotients: EQ(A)⊂Q(A). We speak then of *M - wellpowered* - resp. *E - cowellpowered* - if these classes are sets. For *B* a M - wellpowered category with inter-sections, we know (3.4) - (iii) - (c) - that MP(A) is a complete lattice. For E - quotients, in addition to the dual result, one has:

3.8. LEMMA

Let B be an E - cowellpowered E - M category which has products. Then EQ(A) is a complete lattice for each A ∈ B.

Proof. EQ A has a bottom element, namely the map q in the factorization $A \xrightarrow{q} B \xrightarrow{m} 1$ of the unique map to the terminal object (use the diagonal lemma). For a non-void family $\{q_i : A \to B_i\}i \in I$ in EQ(A), its sup is provided by the E - M factorization of the map $\langle q_i \rangle_{i \in I}$: $A \to \prod_i B_i$, induced by the q_i into the product (use again the diagonal lemma). □

3.9. COROLLARY

A category B as in (3.8), has coequalizers.

Proof. Given f,g: A → B, take q = sup {h∈EQ(B) | hf=hg}. From the construction in (3.8) follows qf = qg.

Considering the E - M factorization of an h with $hf = hg$, one gets the desired unique map $\overline{h}$, with $h = \overline{h}e$. □

What can we say about factorization systems in *Pos* and in *Pos*(ω)? To begin with, (3.6) warrants the existence of extremal epi-mono and epi-extremal mono factorizations in both, but we will examine this result in more detail.

3.10. LEMMA

Pos(ω) *is not a coequalizer-mono category.*

Proof. Consider the following two subsets of $\mathbb{R}^2$ in the figure:

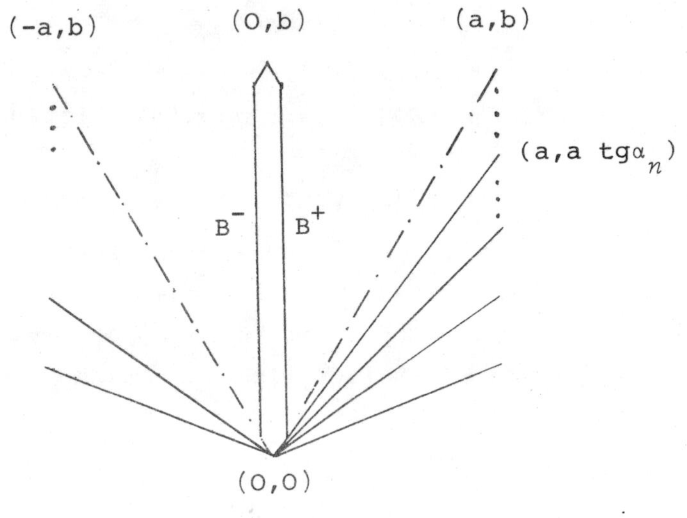

B is the pair of curved segments B^+, B^-, joining (0, 0) and (0, b). The set A is the union of the closed segments joining (0, 0) with points of the form (a, a $tg\alpha_n$), or (-a, a $tg\alpha_n$), for $\{\alpha_n\}$ a strictly increasing sequence of angles $0 < \alpha_n < \alpha_{n+1} < \text{arc tng}\,\frac{b}{a}$.

The order on A and B is as follows: Except for the shared points (0,0) and (0,b), points in different segments are not related. For (x, y), (x', y') in the same segment, (x, y) ≥ (x', y') iff y ≥ y'). Define q: A → B, sending each point (x, y) in A to the corresponding (x', y') in B^+ (if x > 0) or in B^- (if x < 0), and sending (0, 0) to itself. q is continuous and - not having any complete subobject of B as a factor - an extremal epi q': A → qA: a ↦ qa, where qA = B - {(0, b)} has the same order as B, is continuous, but $\tilde{q}A$ is not complete. Take $\tilde{q}A$ equal to qA (with same order) together with two additional points P_1, P_2, one on top of each curved segment, and call $\tilde{\eta}$: qA ↪ $\tilde{q}A$ to the continuous inclusion. Now, if (π_1, π_2) is the kernel-pair of q, $\tilde{\eta} \circ q'$ equalizes π_1, π_2, but there is no way to get a map q" with $\tilde{\eta} \circ q = q'' \circ q$. Hence q is not a coequalizer. Actually $\tilde{\eta} \circ q'$ is the coequalizer. □

A type of monomorphisms which has already appeared is the full one. f: A → B, monotonic, or continuous is said *full* iff fa ≥ fa' ⇒ a ≥ a' ∀a, a' ∈ A.

3.11. LEMMA

In Pos, m is an equalizer iff it is full.

Proof. From the construction of equalizers given before, it follows that they are full. Conversely, any full m: I → A is an equalizer, namely for the two maps h : A → $A \sqcup_I A$: a ↦ if a ∈ mI then a else (a,i), i = 1,2, into the "amalgamated union" $A \sqcup_I A$, which is the cokernel-pair of m, and is the set (A-mI)X{1} ∪ mI ∪ (A-mI)X{2}, with order defined in the obvious way. □

One sees at once that *Pos* has an epi-equalizer factorization, by taking the full order in the set-theoretic image. So epis in *Pos* are surjections and *Pos* is cowellpowered. Actually, $Q(A, \leq_A)$ is known to be antiisomorphic to the complete lattice of all preorders on A which contain $\leq_A$, the join of quotients corresponding to intersection of preorders. To a quotient $[q : A \to B]$ we associate $\sqsupseteq q = (q^2)^{-1} \leq_B$, and to $\sqsubseteq \supset \leq_A$ the quotient $[q_{\sqsupseteq} : A \to A/R_{\sqsupseteq}]$. (3.9) then says that the coequalizer of $f, g : A \to B$ is $q_{\sqsupseteq f, g}$. for $\sqsupseteq f, g$ the intersection of all preorders containing the pairs (fa, ga), (ga, fa), $\forall a \in A$. External epis coincide in fact with coequalizers. $m: A \to B$ is full iff $(m^2)^{-1}(\leq_B) = \leq_A$. Hence for monotone $f: A \to B$ with $(f^2)^{-1}(\leq_B) \supset \sqsubseteq$, one gets a unique $\bar{f}: A/R_{\sqsupseteq} \to B$ with $\bar{f}\, q_{\sqsupseteq} = f$.

In $Pos(\omega)$, we can define for any subset $A' \subset A$ of a complete poset $A \in Pos(\omega)$ its *dense closure* $\overline{A'}$ as the intersection off all full subobjects of a which contain A'. Alternatively, $\overline{A'}$ can be described by the following generation process: $A'_0 = A'$; $A'_{\alpha+1} = A'_\alpha \cup \{a \in A \mid \exists \{a_n\} \subset A'_\alpha$, chain in A, s.t. $a = \sqcup a_n$ in A$\}$; for β a limit ordinal, $A'_\beta = \bigcup_{\alpha<\beta} A'_\alpha$. The process stops for $\alpha = \omega_1$ (we consider countable chains, and ω_1 is regular) and gives $\overline{A'}$. Then we can define a map $f: A \to B$ to be *dense* if it is continuous and $B = \overline{fA}$. Any dense map f is an epimorphism, because if $gf = hf$, and $I \subset B$ is the equalizer of g, h, then $fA \subset I$, and hence, as equalizers are full, $B = \overline{fA} \subset \overline{I} = I$, which forces $g = h$. It is also easy to check, using the definition of the dense closure and the fact that it is a closure operator, that the composition of two dense maps is again a dense map.

3.12. *LEMMA*

Pos(ω) *has a dense-full mono factorization system.*

Proof. The classes of dense maps and of full monos are both closed under composition and contain the isomorphisms. Suppose now that $A \xrightarrow{e} C \xrightarrow{m} B$, $A \xrightarrow{e'} C' \xrightarrow{m'} B$ are two different dense-full mono factorizations of a map f. Without loss of generality we may assume that m and m' are inclusions, which forces $eA = e'A = fA$. As C and C' are full and contain fA, the dense closure of fA in C and in C' is just the one in B, hence $C = \overline{fA} = C'$, which proves the uniqueness. □

3.13. *COROLLARY*

(i) Pos(ω) *is dense-cowellpowered; (ii) Pos*(ω) *is cocomplete.*

Proof. (i) For $|fA| \geq \omega$, one has $|\widehat{fA}| = \bigcup_{\beta<\omega_1} |fA_\beta| \leq$ $\leq \bigcup_{\beta<\omega_1} 2^{|fA|} \leq 2^{|A|} . \omega_1 = 2^{|A|}$ because by transfinite induction on $\beta<\omega_1$ one prove $|fA_\beta| \leq 2^{|fA|}$. As the quotients are determined by the map and the order in the codomain, $|Q(A)| \leq \bigcup_{\beta\leq 2^{|A|}} \beta^{|A|} . 2^{\beta} \leq \bigcup_{\beta\leq 2^{|A|}} 2^{|A|} . 2^{2^{|A|}} =$ $= 2^{|A|} . 2^{2^{|A|}} = 2^{2^{|A|}}$, if $|A| \geq \omega$.

(ii) As coproducts in *Pos*, ω*Pos* and *Pos*(ω) are the disjoint union of the sets with order the disjoint union of the orders, one has, by (3.9), *Pos*(ω) cocomplete. □

In *Pos*(ω), every equalizer is a full monomorphism. However,

3.14. LEMMA

In Pos(ω), not every full mono is an equalizer. Pos(ω) is not an epi-equalizer category.

Proof. See (3.26).

3.15. CONJECTURE[2]

In Pos(ω) epimorphisms coincide with dense maps and extremal monomorphisms coincide with full monos.

We have mentioned in Section 2 that the full inclusion functor I : *Pos*(ω) → *Pos* has a left adjoint: what is called a *reflection* functor. Making *Pos*(ω) *full reflective* in ω*Pos* (denote it ~: ω*Pos* → *Pos*(ω), with unit $\tilde{\eta}X : X \to \tilde{X}$). One might consider it to be of potential interest, because it is well-known that, for a full reflective subcategory, a number of constructions - in particular colimits - can be calculated in the - usually simpler - bigger category and then brought back into the subcategory by the reflection functor (in Sheaf Theory, for instance, this is called the "sheafification" process). However *Pos* happens to be of little help because

3.16. LEMMA

ωPos lacks colimits and coequalizers.

Proof. Consider ω + 1 = ω∪{ω}, with its order as ordinal. All the finite subsets of ω + 1, with the full

2 Note added in proof: This conjecture has been shown false in a recent paper of Lehmann and Pasztor: Lehmann, D., Pasztor, A., "On a conjecture of Meseguer", Technion, Computer Science Tech. Rep. # 170. February 1980.

order inherited from $\omega + 1$, and with inclusion maps between them, give rise to a filtered (two subsets can be merged in their union) diagram in *Pos*, and one sees easily that $\omega + 1$, with its original order, is the colimit of that diagram. Now, the functors $\wedge$ and $\sim$ leave unfixed - up to isomorphism if wanted - finite posets, and in particular the above diagram. If C were a colimit of it in ω*Pos*, $\tilde{C}$ should be its colimit in *Pos*(ω) but, as left adjoints preserve colimits, we know that

$\omega \hat{+} 1 = \omega \cup \{\infty\} \cup \{\omega\}$, $\omega > \infty > n$, $\forall_n \in \omega$ is its colimit in *Pos*(ω), and as $\tilde{\eta}C$ is full mono (see (3.19)) our hipothetical C has to be either $\omega + 1$ or $\omega \hat{+} 1$, with full order. If it is $\omega + 1$, $\tilde{\eta}C$ is not continuous; if it is $\omega \hat{+} 1$, there is no way to obtain an induced continuous map $C \to B$, $B = \omega + 1 \cup \{a\}$ with same order on $\omega + 1$ plus $a \geq n$ $\forall n \in \omega$, for the inclusion of the finite subsets of $\omega + 1$ in B. This proves also that ω*Pos* lacks coequalizers, because it has coproducts. □

3.17. COROLLARY

The inclusion functor J : ωPos $\hookrightarrow$ Pos has no left adjoint.

Proof. If it had, it should preserve colimits, and the above diagram would have one in ω*Pos*. □

We consider now a very important class of epis in *Pos*(ω), what we call nice epis, and which have been investigated by Markowsky [19] and by Courcelle and Raoult [8]. We represent concepts and results in these two papers and make some related remarks.

Given a continuous map $f : A \to B$, A, B not necessarily complete, the preorder $\sqsupseteq_f = (f^2)^{-1} \leq B$ has the property:

(C) if $\{a_n\} \subset A$ is a chain, and $a = \sqcup a_n$, then $a' \sqsupseteq_f a_n \ \forall n \in \omega \Rightarrow a' \sqsupseteq_f a$. Call any preorder $\sqsubseteq \subset \leq_A$ with the property (C) a *continuous preorder* on $(A, \leq_A)$. If $\sqsupseteq$ is continuous, then $q_{\sqsupseteq} : A \to A/R_{\sqsupseteq}$ is continuous. The intersection of an arbitrary family of continuous preorders on A is a continuous preorder, and A x A is also a continuous preorder. Thus they form a complete lattice $CP(A, \leq_A)$. As in the *Pos* case, there is a connection between continuous preorders and quotients in *Pos*(ω), but in this case it is more involved. q and $\tilde{\eta} \cdot q'$ in (3.10) are two different quotients which originate the same preorder. Modifying the example in (3.10) in allowing three radiations of segments instead of two and three curved segments for B, one can find two quotients q', q", (both with two limit points on top) which have the same continuous preorder, but neither $[q'] \geq [q'']$ nor $[q''] \geq [q']$.

3.18. PROPOSITION

Let $f : A \to B$ monotonic. There is a complete poset $\tilde{B}_f$ (called the completion of B along f) and a monotone map $\tilde{\eta} f : B \to \tilde{B}_f$ such that $\tilde{\eta}_f . f$ is continuous, and given C, complete, $g : A \to C$, continuous, and $g' : \tilde{B} \to C$ monotonic, with $g' f = g$ there is a unique continuous map $\bar{g} : \tilde{B}_f \to C$ with $\bar{g} \circ \tilde{\eta}_f = g'$

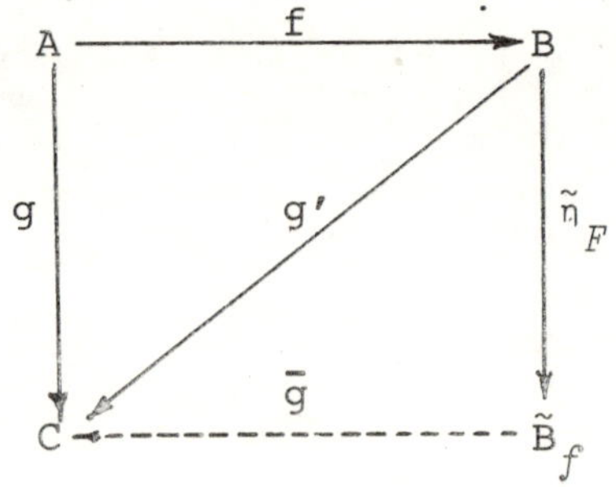

Proof. This is Theorem 2.4 in Markowsky [19]. We will sketch here the main steps. Let $\{\sqsupseteq_i\}i\in I$, be the family of all preorders $\sqsubseteq_i \supset \leq_B$ such that $(f^2)^{-1} \sqsubseteq_i$ is continuous. Then $(f^2)^{-1}(\bigcap_{i\in I}\{\sqsubseteq_i\}) = \bigcap_i\{(f^2)^{-1} \sqsubseteq_i\}$ is also continuous. Let $\tilde{\sqsupseteq} = \bigcap_{i\in I} \sqsupseteq_i$, and $B_f := B/_{R\tilde{\sqsupseteq}}$. Then $q_{\tilde{\sqsupseteq}}\cdot f: A \to B_f$ is continuous. Call an ideal H in B_f f-closed if for a chain $\{a_n\}\subset A$, if $\{q_{\tilde{\sqsupseteq}}fa_n\}\subset H$ and $a = \bigsqcup_n a_n$, then $q_{\tilde{\sqsupseteq}}fa\in H$. As b is f-closed, and the intersection of f-closed ideals is an f-closed ideal, they form a complete lattice $Id_f(B_f)$, under inclusion. The map $\mu: B_f \to Id_f(B_f) : x \mapsto \{y\in B_f | y\leq x\}$ is full and monotonic and the composition $\mu\cdot q_{\tilde{\sqsupseteq}}\cdot f$ is continuous. Let $\tilde{B}_f$ be the smallest (ω-) complete full subobject of $Id_f(B_f)$ which contains B_f, and $\tilde{\eta}_f$ the map $\tilde{\eta}_f : B \xrightarrow{q_{\tilde{\sqsupseteq}}} B_f \hookrightarrow \tilde{B}_f$. Then $\tilde{\eta}_f \circ f$ is continuous because $\mu q_{\tilde{\sqsupseteq}} f$ is. Given C (ω-) complete, $g : A \to C$ (ω-) continuous and $g' : B \to C$ monotonic with $g'f = g$, there is a monotone induced map $g'' : \tilde{B}_f \to C$, and then the unique continuous map $\bar{g}$ from $\tilde{B}_f$ is given by the formula: $\bar{g}H = \bigsqcup g''H$, $H\in\tilde{B}_f$.

□

3.19. Remark

Note that the two left adjoints $\wedge$ and $\sim$ may be obtained as particular instances of the above construction: $\tilde{A} = \tilde{A}_{1_A}$, and $\hat{A} = \tilde{A}j_A$, for $j_A : (A,\Delta_A) \to (A,\leq_A) : a \mapsto a$, Δ_A the diagonal relation. It also follows that the unit $\tilde{\eta}A = \tilde{\eta}_{1_A} : A \to \tilde{A}$, for $\sim$, is full injective.

3.20. Definition

A map $f : A \to B$ in $Pos(\omega)$ is called a *nice epi* if, being $A \overset{e}{\to} fA \overset{m}{\to} B$ the surjective-full factorization of f in *Pos*, the unique continuous map $\overline{f} : f\tilde{A}_e \to B$ with $\overline{f}\tilde{\eta}_e = m$ is an isomorphism. Equivalently, f is nice epi iff it is of the form $\tilde{\eta}_f \circ f: A \to \tilde{B}$ for $f : A \to B_f$ surjective monotonic, up to a commuting isomorphism.

3.21. PROPOSITION

$f : A \to B$ in $Pos(\omega)$ is nice epi if and only if given $g : A \to C$ in $Pos(\omega)$ with $(g^2)^{-1} \leq_C \supset (f^2)^{-1} \leq B$, there exists a unique $\overline{g} : B \to C$, continuous, with $\overline{g}f = g$.

Proof. If f is nice, the unique g" : fA → C with eg" = g, for f = me the surjective-full factorization, extends to a unique $\overline{g}$ as desired. On the other hand, as $\tilde{B}_f$ is determined up to isomorphism by its universal property, the condition in the proposition just says that B is $f\tilde{A}_e$, so f is nice. □

3.22. Remark

The above proposition says that $f : A \to B$ is nice if and only if it is the sup in $Q(A,\leq)$ of all dense quotients $[q_i : A \to B_i]$ such that $(q_i^2)^{-1} \leq_{B_i} = (f^2)^{-1} \leq B$. As a consequence, if e = hg and e is nice epi, then h is nice epi. It is also clear that there is an antiisomorphism of complete lattices between the set $NQ(A,\leq_A)$ of nice quotients and the set $CP(A,\leq_A)$ of continuous preorders of A.

3.23. Remark

Given a pair $f, g : A \to B$ in $Pos(\omega)$, their coequalizer $q : B \to C$ can be characterized as the nice epi corresponding to the smallest continuous preorder on B which contains the pairs (fa, ga), (ga, fa) for each a $\in$ A.

3.24. Remark

Given a diagram $A : I \to Pos(\omega)$, its colimit is the complete poset $\tilde{B}_f$ in the diagram below, where B is the colimit in *Pos* of $J \circ A$ and $\rho : \coprod_i Ai \to B$, the unique monotone map induced from the coproduct of the Ai - note that the coproduct is the same in *Pos* and $Pos(\omega)$ - and where $\{fi : Ai \to C\}_{i \in I}$ is a cone in $Pos(\omega)$.

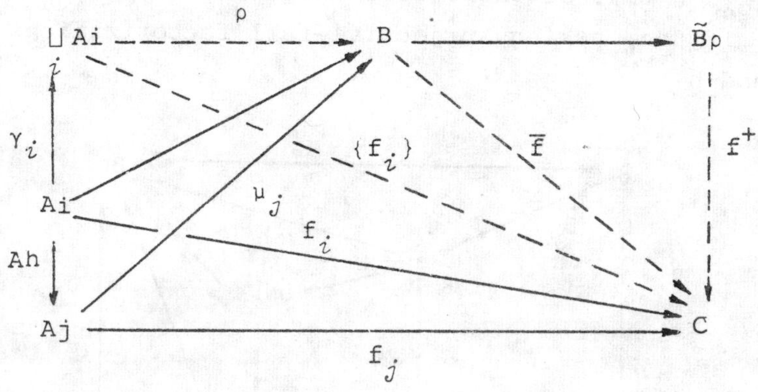

Given a dense map $f : A \to B$ in $Pos(\omega)$, we get a best factorization $f = g_{01} \cdot h_1$ through a nice epi namely for g_{01} the one corresponding to $(f^2)^{-1} \leq B$ $\cdot$ h_1 is again dense and we can do the same: $h_1 = g_{12} \cdot h_2$. There is a natural question: how far is an arbitrary epi from being nice? In other words: how many steps are we to take in the way just hinted, before we stop? In the context of classes of interpretations - Cf. Courcelle and Nivat [7] - if $Q : T_\Sigma \to T$, or $Q' : CT_\Sigma \to T$ is the quotient algebraic theory defined by the class - Cf. Meseguer [21] - (Now Q and Q' are not continuous maps, but countable families of continuous maps) the question for Q is: how far a class of interpretations from being algebraic?, and for Q : how far is it from being "CT_Σ - describable"? The answer - if we stick to the ω-complete case - is given by the generation process of $\overline{fA}$, because fA_1 is contained in the set-theoretic image of h_1, and fA_2 in the one of h_2. The next steps are to define $f_1 = g_{01}$, $f_2 = g_{12}f_1$, so that $f = h_1 f_1 = h_2 f_2 = \dots$, and then for an ordinal α, $f_{\alpha+1} = g_{\alpha\ \alpha+1} \circ f_\alpha$, ($g_{\alpha+1}$ the optimal nice factor of h_α) and for α a limit ordinal we take f_α the inf in Q(A) of all the $f_\beta, \beta<\alpha$, which has to be $[f_\alpha] \geq [f]$. So there is a unique h_α with $h_\alpha f_\alpha = f$. Now, if $\beta \leq \alpha$, taking surjective-full factorizations of h_β and h_α

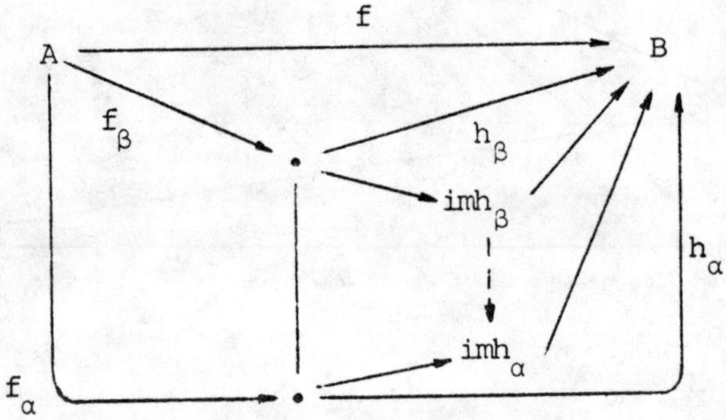

we get - by the diagonal lemma - $\mathrm{im}\ h_\alpha \subset \mathrm{im}\ h_\alpha\ \forall \beta < \alpha$. Hence, by induction, $\mathrm{imh}_\alpha \supset fa_\alpha$, and we make sure that the process stops for $\alpha = \omega_1$, i.e. $f = \inf \{f_\beta \mid 1 \leq \beta < \omega_1\}$

3.25. Remark

As the Lemma (3.10) and the above discussion show, nice epis do not give rise to an epi-mono factorization system. It is also rather obvious from the above that they are not closed under composition: in the context of classes of interpretations, this is the well-known fact that $CT_{\Sigma} \to T$ nice does not imply $T_{\Sigma} \to CT_{\Sigma} \to T$ nice, i.e. algebraic. They play a different, and in a sense complementary, role than factorization systems: they are like generalized coequalizers - the continuous preorder playing the part of the kernel-pair, and provide the simplest approximation (simplest because it can be described exclusively in terms of a continuous preorder on the domain) of dense maps. The following diagram summarizes the situation:

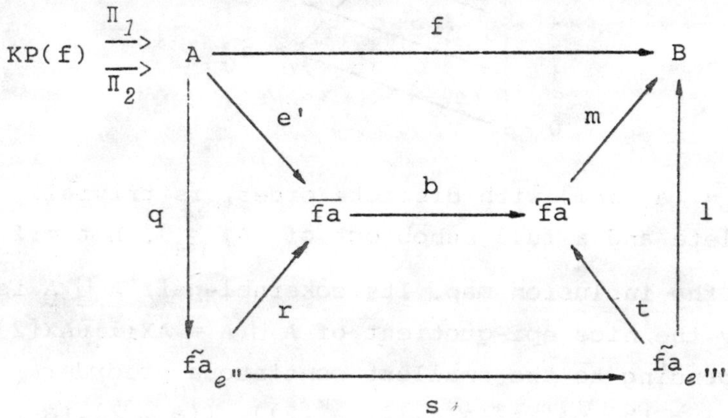

Where e = be' gives me = f dense-full factorization, m' = mb gives m'e' = f extremal epi-mono factorization, and if $A \xrightarrow{e''} fA \xrightarrow{m''} B$ is the coequalizer-mono factorization of f in *Pos* the q is the coequalizer of the kernel--pair of f, KP(f), sq the smallest nice epi through which f factors (for $A \xrightarrow{e'''} fA \xrightarrow{m'''} B$ the surjective-full factorization of f in *Pos*), r is extremal epi, s nice, t dense and b dense and mono, i.e. a bimorphism.

3.26. Proof of 3.14

It is well known and easy to check that, in a category with pushouts, any equalizer map $m\colon I \to A$ is the equalizer of the two cokernel-pair maps $h_1, h_2\colon A \to A \sqcup_I A$ associated to m. Take now $A = \omega \cup \{a_n \mid n \in \omega\} \cup \{\infty\}$, with the ordering described in the figure

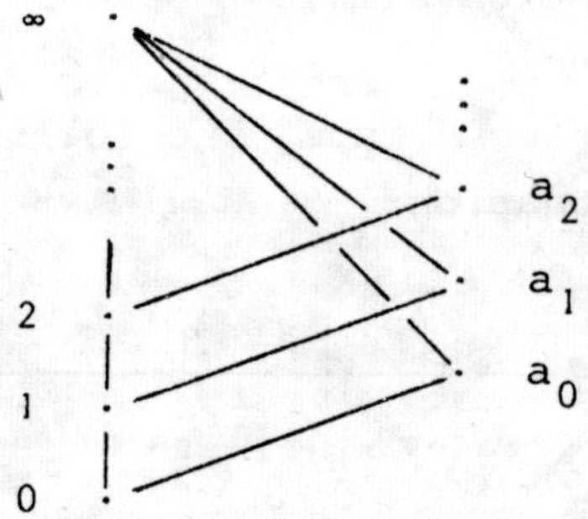

then $I = \{a_n \mid n \in \omega\}$ with discrete order, is trivially ω-complete and a full subobject of $(A, \leq_A)$. Let $m\colon I \to A$ denote the inclusion map. Its cokernel-pair $A \sqcup_I A$ is exactly the nice epi-quotient of $A \sqcup A = A \times \{1\} \cup A \times \{2\}$, corresponding to the smallest continuous preorder containing the pairs $((a_n, 1), (a_n, 2))$, $((a_n, 2), (a_n, 1))$, $n \in \omega$. But then, by transitivity, $\sqsupseteq$ has to contain the pairs $((\infty, 1), (n, 2))$, $((\infty, 2), (n, 1))$, $n \in \omega$. Then, by continuity, $\sqsupseteq$ has also to contain the pairs $((\infty, 1), (\infty, 2))$, $((\infty, 2), (\infty, 1))$, which shows that ∞ belongs to the equalizer of the cokernel-pair maps on m (It is not hard to see that the equalizer is actually $I \cup \{\infty\}$) so m cannot be an equalizer.

To see that $Pos(\omega)$ does not have an epi-equalizer factorization system assume it had, and consider such a factorization for $m: I \to A$ above, say $I \overset{e}{\to} J \overset{j}{\to} A$, where we may assume that j is an inclusion. As e is epi, j equalizes the cokernel-pair of m, hence $J \subset I \cup \{\infty\}$. As J contains I and is equalizer, it has to be indeed $J = I \cup \{\infty\}$. Hence e is the inclusion $I \hookrightarrow I \cup \{\infty\}$, which is not epi, because it equalizes the two different maps $g_i: I\{\infty\} \to \Pi : x \mapsto$ if $x = \infty$ then i, else $\bot$; $i=1,2$, where $\Pi = \{1,2,\bot\}$, with order $1 \geq \bot$, $2 \geq \bot$. □

4. *FURTHER RESULTS*

This section contains results which we have found useful for establishing certain properties of varieties of continuous algebras, which essentially depend on the behavior of the base category $Pos(\omega)$.

We start by looking at filtered colimits in *Pos* and $Pos(\omega)$ and mention the related generation process for extremal epi-mono factorizations. A little more work shows that *Pos* and $Pos(\omega)$ are locally presentable, a property inherited by the categories of algebras, and which states that these categories are "essentially algebraic", i.e., equivalent to categories of limit--preserving functors $A : T \to$ *Set*, for T a small category. We also study how finite products behave with respect to epis.

A category I is said to be *filtered* if: (i) for each i, $j \in I$ there is a $k \in I$ and maps $i \to k$, $j \to k$; (ii) for each pair of maps $f,g: i \to j$ there is a map $h: j \to k$ with $hf = hg$. For α a regular cardinal, a category I is said α-*filtered* if: (i) for each family of objects $(i_\gamma)\gamma\in N$, $|N|<\alpha$, there is an object $k \in I$ and maps f_γ :

$i_\gamma \to k$, $\gamma \in N$; (ii) for each family of maps $(f_\gamma :$ $i \to j)$ $\gamma \in N$, $|N| < \alpha$, there is a map $g : j \to k$ with $gf_\gamma =$ $= gf_{\gamma'}$, $\forall \gamma, \gamma' \in N$. Hence I is filtered iff it is ω-filtered.

An α-*filtered colimit*, (B, hi), in a category B is a colimit for an α-filtered diagram, i.e., a functor $D : I \to B$ in which I is small and α-filtered. We say that $U : B \to C$ *preserves* α-filtered colimits, if (UB, Uhi) is colimit of UD, for any (B, hi) and D as above.

4.1. PROPOSITION

The forgetful functor $U : Pos \to Set$ preserves filtered colimits. The inclusion functor $J : Pos(\omega) \to Pos$ preserves ω_1-filtered colimits.

Proof. For the first part: let $((Ai, \leq i))i \in I$ be a filtered diagram in *Pos*. We will be done if we show that its colimit, say $((A, \leq), hi)$ has uderlying set, the filtered colimit of the sets, and order the filtered colimit of the orders. More precisely: for $f : Ai \to Aj$ in the diagram, we get

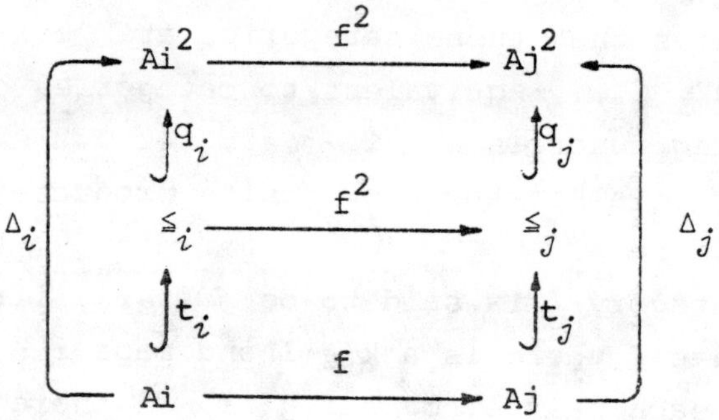

i.e., three filtered diagrams of sets: (Ai) $i \in I$, $(\leq i)$ $i \in I$, (Ai^2) $i \in I$, each included in the next. If A is

the colimit of (Ai) in the category of sets, we have $A^2 = \operatorname{colim}_i A_i^2$, $\Delta = \operatorname{colim}_i \Delta i : A \to A^2$, by the fact (see for instance Mac Lane [16], Ch. IX, Sec. 2, Th. 1) that finite limits commute with filtered colimits in the category of sets. Also for $\leq = \operatorname{colim} \leq_i$ (we have not yet proved that $\leq$ is an order!), we have $q = \operatorname{colim}_i q_i : \leq \to A^2$ a monomorphism (that we can choose to be an inclusion), because filtered colimits commute in particular with pullbacks, and a map $m : A \to B$ is a monomorphism iff

$$\begin{array}{ccc} A & \xrightarrow{=} & A \\ \| \downarrow & & \downarrow m \\ A & \xrightarrow{m} & B \end{array}$$

is a pullback. Hence $\leq$ is reflexive (contains the diagonal). To see that it is also antisymmetric and transitive, one has only to use the construction of A as a quotient set of the disjoint union of the Ai, pick representatives, and use filteredness and the fact that the $\leq i$ have these two properties.

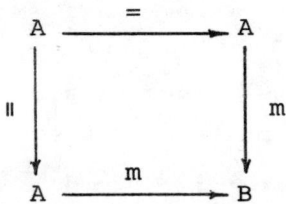

Now, $\leq$ a colimit of the $\leq_i$ says exactly that the maps $h_i : A_i \leq A$ are monotonic, and that the unique induced map $\overline{f} : A \to B$ for a cone $(f_i : A_i \to B$ is monotonic if the f_i are. This completes the first part of the proof. For the second one, as ω_1- filtered $\Rightarrow$ filtered, we will be done if we prove that for $((A_i, \leq_i))$ $i \in I$ a ω_1-filtered diagram in $Pos(\omega)$, the above-described cone $((A, \leq), h\ )$ is in $Pos(\omega)$ and works there as a colimit. But a chain $\{a_n\}$ in A has to come from a chain in some of the A_i, because: suppose $a_n' \in A_n$ are representatives of the a . Then, by ω_1-filteredness we can find a chain of representatives $\{a''_n\}$ in an object A" of the diagram, as the one shown below

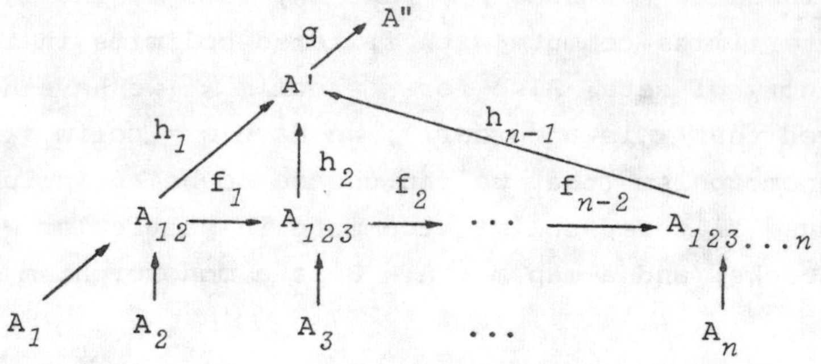

Where in $A_{12\ ...\ n}$, the images of $a_1, \ldots, a_n$ form a finite chain, and g simultaneously equalizes h_1, h_2f_1, $\ldots$, $h_{n1}\ f_{n-2} \ldots f_2f_1, \ldots$. Let $a'' = \sqcup a''_n$, and a its image in A, then $a = \sqcup a_n$, because $a \geq a_n\ \forall n \in \omega$, and $a > a' \leq a_n\ \forall n \in \omega$ would force by filteredness some connecting morphism f: A" → A to be not continuous. Hence A is complete and the h_i continuous. The fact that the induced monotone map $\bar{g} : A \to B$ for a cone $(g_i : A_i \to B)$ in *Pos*(ω) is in fact continuous follows easily from the above. □

4.2. Remark

The extremal epi-mono factorization of a map $f : A \to B$ in *Pos*(ω) can be obtained by a generation process slightly more complicated than the one for the dense-full factorization. Now we also have to take the orders into account and what we do is to calculate a filtered colimit in *Pos*: let $(fA, \leq_{fA})$ be the set-theoretic image with the extremal epi order in *Pos*, and define $(\overline{fA}_0, \leq_{\overline{fA}_0}) = (fA, \leq_{fA})$, $\overline{fA}_{\alpha+1} = \overline{fA}_\alpha \cup \{b \in B \mid \exists \{a_n\}$ chain in $\leq_{\overline{fA}}$ s.t. $b = \sqcup a_n$ in $B\}$, $\leq_{\overline{fA}_{\alpha+1}}$ the smallest order containing $\leq_{fA_\alpha}$ and the pairs $c \geq b$, $b \geq a_n$ for each $c \geq a_n \forall n \in \omega$, $\{a_n\}$ chain in $\leq_{\overline{fA}_\alpha}$, and $b = \sqcup a_n$ in B. Hence $(\overline{fA}_\alpha, \leq_{\overline{fA}_\alpha}) \hookrightarrow (\overline{fA}_{\alpha+1}, \leq_{\overline{fA}_{\alpha+1}})$. For α a limit ordinal define $(\overline{fA}_\alpha, \leq_{\overline{fA}_\alpha}) = \operatorname{colim}_{\beta<\alpha} (\overline{fA}_\beta, \leq_{\overline{fA}_\beta}) = (\cup fA_\beta, \cup_{\beta<\alpha} \leq_{\overline{fA}_\beta})$. As the chains are countable and ω_1 is regular, the process stops for $\alpha = \omega_1$, and gives $(\overline{fA}, \leq_{\overline{fA}}) = (\cup_{\beta<\omega_1} \overline{fA}_\beta, \cup_{\beta<\omega_1} \leq_{\overline{fA}})$, extremal epi image.

An object G in a category B is said to be a *strong generator* if $f : A \to B$ is an isomorphism in B iff f mono and $B(G,f) : B(G,A) \to B(G,B): h \mapsto fh$, surjective.

4.3. LEMMA

2 with its order as ordinal is a strong generator in Pos and Pos(ω).

Proof. The functor $Pos(2,-)$ is naturally equivalent to the functor Ord: $Pos \to Set$ sending each $(A, \leq_A)$ to $\leq_A$, and each monotone $f : A \to B$ to $f^2 : \leq_A \to \leq_B$. But f is an isomorphism in Pos iff $f^2 : \leq_A \to \leq_B$ is bijective, iff $f^2 : \leq_A \to \leq_B$ surjective and $f : A \to B$ mono. The $Pos(\omega)$ case is the same, because $Pos(\omega)(2,A) = Pos(2,A)$ $\forall A \in Pos(\omega)$, and in $Pos(\omega)$, $f : A \to B$ continues to be iso iff $f^2 : \leq_A \to \leq_B$ bijective. □

4.4. *Remark*

Lehmann and Smith [14] have shown that 2 is also a *cogenerator* for *Pos* and *Pos*(ω), i.e. : $f \neq g : A \to B$

$\Rightarrow \exists h : B \to 2$ with $hf \neq hg$, which yields another proof of cocompleteness and cowellpoweredness. Lehmann [13] has pointed out that 2 is actually a cogenerator for categories of "Z" - complete posets and "Z" - continuous maps in the sense of ADJ [4]. Then, through a version of the adjoint functor theorem in Herrlich and Strecker [11], he gets a proof of existence of a left adjoint for the inclusion in *Pos*.

4.5. PROPOSITION

Pos is locally ω-presentable.

Pos(ω) is locally ω_1-presentable (the definition of category locally α-presentable, is due to Gabriel and Ulmer [9], and is the one implicit in the proof, except that, in general, one may need not one, but a set of generators.

Proof. (i) an easy argument using equalizers shows that 2 strong generator is equivalent to 2 "echte generator" in the sense of Gabriel and Ulmer

(ii) *Pos* and *Pos*(ω) are cocomplete

(iii) the functor *Pos*(2, -): *Pos* → *Set* preserves filtered colimits, as follows from the proof in (4.1). It follows also from there that *Pos*(ω) (2, -): *Pos*(ω) → *Set* preserves ω_1-filtered colimits, because *Pos*(ω) (2, -) = *Pos*(2, -) ∘ J, for J: *Pos*(ω) → *Pos* the inclusion functor. □

We finally consider the behavior of finite products with respect to epis.

4.6. *PROPOSITION*

Given $f : A \to B$, $f' : A' \to B$ *in* $Pos(\omega)$, *then*

(*i*) *if* f *and* f' *are extremal epis,* $f \times f'$ *is extremal epi*

(*ii*) *if* f *and* f' *are dense,* $f \times f'$ *is dense.*

Proof. (i) As extremal epis is $Pos(\omega)$ are closed under composition, it will be enough to show that $f \times C : A \times C \to B \times C$ is extremal epi if f is, for any $C \in Pos(\omega)$. Look now to the generation process (4.2) and note that (we omit to write the orders for convenience) (a) $\overline{fA \times C} = fA \times C$; (b) if $\overline{fA \times C}_\alpha = \overline{fA}_\alpha \times C$, then $\overline{fA \times C}_{\alpha+1} = \overline{fA \times C}_\alpha \cup \{(b,c) \in B \times C \mid \exists\{(b, c)\}$ chain in $\leq \overline{fA \times C}_\alpha$ s.t. $(b,c) = \bigsqcup_n (b_n, c_n)\} = \overline{fA}_\alpha \times C$;

(c) for α a limit ordinal with $fA \times C_\beta = \overline{fA}_\beta \times C \quad \forall\beta<\alpha$, $\overline{fA}_\alpha \times C = (\operatorname{colim}_{\beta<\alpha} \overline{fA}_\beta) \times C = \operatorname{colim}_{\beta<\alpha} \overline{fA}_\beta \times C = \overline{fA \times C}_\alpha$, because as $(- \times C) : Pos \to Pos$ is left adjoint, it preserves colimits. Hence $B \times C = \overline{fA}_{\omega 1} \times C = \overline{fA \times C}_{\omega 1}$, as desired.

The proof of (ii) is completely similar to (i), and is left to the reader. □

4.7. *COROLLARY*

If $A_i \xrightarrow{e_i} B_i \xrightarrow{m_i} C_i$, i = 1, ..., n, are extremal epi-mono (respectively dense-full mono) factorizations of maps in $Pos(\omega)$, then

$$A_1 \times \ldots \times A_n \xrightarrow{e_1 \times \ldots \times e_n} B_1 \times \ldots \times B_n \xrightarrow{m_1 \times \ldots \times m_n} C_1 \times \ldots \times C_n$$

is extremal epi-mono (resp. dense-full mono) factorization of their product.

Proof. Apply proposition above and (3.4)-(iv)-(c). □

REFERENCES

[1] ADJ (Goguen, J.A., Thatcher, J.W., Wagner, E.G., Wright, J.B.): Some fundamentals of order - algebraic semantics, in Mazurkiewicz, A. (ed.), *Mathematical Foundations of Computer Science*. Lecture Notes in Computer Science Vol. 45, Springer-Verlag, 1976, pp. 153-168.

[2] ADJ: Rational algebraic theories and fixed-point solutions (unpublished manuscript - summarized in *Proceedings of 17th IEEE Symposium of Foundations of Computer Science*, 1976, pp. 147-158).

[3] ADJ: Initial algebra semantics and continuous algebras, *Journal of ACM*, Vol. 24, (1977), pp. 69-95.

[4] ADJ (Thatcher, J.M., Wagner, E.G., Wright, J.B.): A uniform approach to inductive posets and inductive closure, in Gruska, J. (ed.). *Mathematical Foundations of Computer Science*, Lecture Notes in Computer Science, Vol. 53, Springer-Verlag, 1977, pp. 192-212.

[5] Bloom, S.L.: Varieties of ordered algebras. *Journal of Computer and System Science*, Vol. 13, (1976), pp. 200-212.

[6] Courcelle, B.: On the definition of classes of interpretations, in Salomaa, A., Steinby, M. (Eds.), *Automata, Languages and Programming*, Lecture Notes in Computer Science, Vol. 52, Springer Verlag, 1977.

[7] Courcelle B., Nivat, M.: Algebraic families of interpretations, *Proceedings of 17th IEEE Symposium of Foundations of Computer Science*, 1976, pp. 137-146.

[8] Courcelle, B., Raoult, J.C.: Completions of ordered magmas, 1977 (manuscript).

[9] Gabriel, P., Ulmer, F.: *Lokal präesentierbare Kategorien*, Lecture Notes in Mathematics, Vol. 221., Springer-Verlag, 1971.

[10] Guessarian, I.: Schémas de programme récursifs polyadiques: Equivalence sématique et classes d'interprétations, thése d'Etat, Université Paris VII, 1975.

[11] Herrlich, H., Strecker, G.E.: *Category Theory*, Allyn and Bacon Inc., Boston, 1973.

[12] Isabell, J.R.: Algebras of uniformly continuous functions, *Annals of Mathematics*, Vol. 68. (1958), pp. 96-125.

[13] Lehmann, D.: On the algebra of order, *Proceedings of 19th IEEE Symposium of Foundations of Computer Science*, Ann. Arbor, 1978, pp. 214-220.

[14] Lehmann, D., Smith, M.B.: Data Types, University of Warwick, Theory of Computation Report, No.19. 1977.

[15] Mac Lane, S.: Duality for groups, *Bulletin of AMS*, Vol.56. (1950), pp. 485-516.

[16] Mac Lane, S.: *Categories for the Working Mathematician*, Springer-Verlag, New York, 1971.

[17] Manes, E.G.: *Algebraic Theories*, Springer-Verlag, New York, 1976.

[18] Markowsky, G.: Chain-complete posets and directed sets with applications, *Algebra Universalis*, Vol. 6.(1976), pp. 53-68.

[19] Markowsky, G.: Categories of chain-complete posets. *Theoretical Computer Science*, Vol. 4. (1977), pp. 125-135.

[20] Markowsky, G., Rosen, B.K.: Bases for chain-complete posets. *IBM Journal of Research and Development*, Vol. 20. (1976), pp. 138-147.

[21] Meseguer, J.: On order-complete universal algebra and enriched functional semantics, in Karpinski, M. (ed.) *Fundamentals of Computation Theory*, Lecture Notes in Computer Science, Vol. 56. Springer Verlag, 1977, pp. 234-301.

[22] Meseguer, J.: A Birkhoff variety theorem for continuous ordered algebras. *Notices AMS*, Vol. 25. (1978), No.5. A-482-483.

[23] Meseguer, J.: Ideal monads and Z-posets, *Notices AMS*, Vol. 25. (1978), No.6. A-579-580.

[24] Meseguer, J.: Order completion monads (manuscript).

[25] Nivat, M.: On the interpretation of polyadic recursive schemes, *Symposia Mathematica XV*, Instituto Nationale di Alta Matematica Italy, 1975, pp. 225-284.

[26] Scott, D.: The lattice of flow diagrams, in Engeler, E. (ed.) *Symposium on Semantics of Algorithmic Languages*, Lecture Notes in Mathematics, Vol. 188. Springer-Verlag, 1971, pp. 311-366.

[27] Scott, D.: Continuous lattices, in *Toposes, Algebraic Geometry and Logic*, Lecture Notes in Mathematics, Vol. 274, Springer-Verlag, 1972, pp. 97-136.

[28] Tiuryn, J.: Fixed-points and algebras with infinitely long expressions, Warsaw University, Institute of Mathematics, 1977, (manuscript).

J. Meseguer
Computer Science Dept. UCLA
Los Angeles, CA 90024, USA
and
Mathematics Department
University of California
Berkely, CA 94720, USA

COLLOQUIA MATHEMATICA SOCIETATIS JÁNOS BOLYAI
26. Mathematical Logic in Computer Science,
Salgótarján (HUNGARY), 1978.

ALGORITHMIC LOGIC WITH NONDETERMINISTIC PROGRAMS

G. Mirkowska

INTRODUCTION

Up to present the logic of programs, developed in Poland science 1968 [13], aimed to study the properties of deterministic programs and deterministic program-connectives: begin... end if... then...else, while...do. From 1974 we began to study also the phenomena connected with procedures, their recursitivity and ways of transmitting parameters to procedures. In 1975 a definition of a tree of possible computation-histories for parallel program appeared [cf.7.]. It made use of nondeterministic choice. Yet, another sign of growing interest in nondeterminism appeared in a paper by Radziszowski [cf.9.].

In 1976 V.R.Pratt initiated dynamic logic (DL) which may be also called logic of nondeterministic programs. The language of DL makes use of the program connective or - of nondeterministic choice. The motivations for nondeterminism may be found in [2, 3]. Here we wish to add two further arguments: parallelism and the lower cost of the nondeterministic definitions of relations in comparison to deterministic ones.

Example

```
cobegin  s || if  α  then K else M coend
begin    s : if  α  then K else M or
             if  α  then cobegin  s || K coend else
             cobegin  s || M coend  or
             cobegin  s || x:=α coend; if x then K
             else M     end
```

where x is a propositional variable which does not appear in considered programs.

Looking on the above example we see that in some cases parallel programs can be considered as equivalent to nondeterministic ones.

To see the second sort of motivations let us observe that a relation frequently may be defined both in deterministic and nondeterministic ways.

Example

For example we shall study the data structure of trees, i.e. two sorted (Atoms and Trees) algebraic system of binary trees with information in leaves only and with operations: left, right, cons and relations: atom, empty. Let us define the relation: is an atom *a* a member of the tree *t*?

```
D  member(a,t) ≡ [begin t':= t; bool := false:
                 while ¬ bool ∧ ¬ empty(t')do
                 begin if atom(t')∧a=t' then
                 bool :=true
                 else if(atom(left(t'))∧a = left(t')then
                 bool := true else if empty(left(t'))then
                       t' := right(t')else
                       t':=cons(left(left(t')),
                           cons(right(left(t')),
                           right(t'))
                end]end(bool∧¬ empty(t')).
```

ND member(a,t) $\underset{df}{\equiv}$ ∇[begin t':=t;while$\neg$atom(t')$\wedge\neg$empty(t') do
t':=left(t') or t':=right(t')
end] (a = t' $\wedge\neg$ empty(t')).

The cost of the first definition is proportional to the number n of nodes in a tree t, when the cost of nondeterministic definition is proportional to log n.

This is why we decided to study nondeterministic program.

Now let us present the language of algorithmic logic. Denote by W the set of all valuations in some relational structure and by B_o two-element Boolean algebra.

Every formula in classical logic may be interpreted as a mapping that to every valuation assigns a Boolean value.

$$W \xrightarrow[\alpha]{} B_o$$

In deterministic algorithmic logic we consider formulas of the form $K\alpha$ as a superposition of two mappings indicated by K and α.

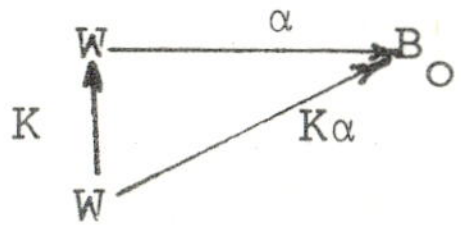

If the result of computation of program K is undefined we put false as a value of the formula $K\alpha$.

In the constrast with deterministic case, in nondeterministic one an initial configuration <valuation of variable, program> may have at a fixed realization many different configurations, i.e. computation histories. Below, we shall give the definition of acceptable computation. Given initial configuration determines a set of acceptable computations. So in nondeterministic algorithmic logic we are forced to consider two kinds of formulas denoted by $\nabla K\alpha$ and $\Delta K\alpha$:

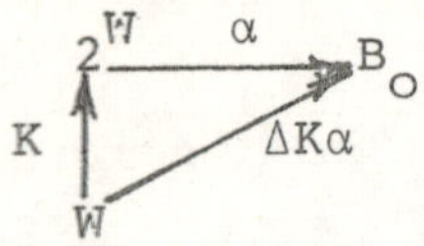

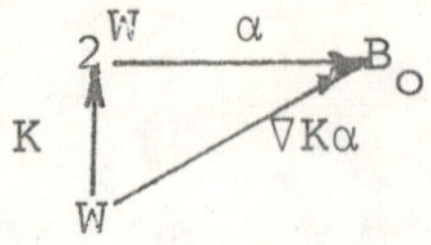

The meaning of these formulas is as follows

$R \models \Delta K\alpha[v]$ iff all computations of the program K in the realization R at the valuation v are finite and all results satisfy α.

$R \models \nabla K\alpha[v]$ iff there exists a finite computation of the program K in the realization R at the valuation v such that its result satisfy the formula α.

These formulas are corresponding to $[K]^{+}\alpha$ and $\langle K\rangle\alpha$ from the augmented Dynamic Logic DL^{+} [cf.3].

The notations of acceptable computation and of graph of possible computations enable us to observe many different semantical phenomena:

- a computation of the program K may be finite or not
- graph of the possible computations for a given initial configuration $\langle v,K\rangle$ may be finite or not,
- a program K may be weakly correct w.r.t. conditions α and β,
- a program K may be "strongly" correct, etc.

It turns out that all the semantical properties of programs are expressible by means of algorithmic formulas containing subformulas of the form ΔKα and ∇Mα , e.g. the formula $(\alpha \Rightarrow \nabla K\beta)$ expressed the strong correctness of K with respect to α and β.

The paper contains a system of logical axioms and inference rules and we prove its completeness. It means that both semantical and syntactical (axiomatized) consequence operations coincide.

The system proposed by us differs from the axiomatization of dynamic logic since it does not assume all first order formulas valid in standard model of arithmetic. Theories based on our axiomatization need not be extensions of arithmetics. We consider this fact important since many interesting algorithmic theories may be investigated of data structures which are not arithmetical, [cf. 11, 12].

1. Language and its realization

Let A be an alphabet in which we have symbols for individual variables, relations, functions, logical connectives and program connectives. The set of proper expressions of the language is divided into three parts: terms, formulas and programs.

Denote by T_o and F_o the sets of terms and open classical formulas and by S the set of all expressions of the form:

$[X_1/\tau_1 \ldots x_n/\tau_n]$ (called substitutions) where τ_i are terms, $i \leq n$, and $x_1, \ldots, x_n$ are different individual variables

The set of programs is the least set FS containing all substitutions $s \in S$ and such that if K, M are programs and α is an open formula, $\alpha \in F_o$, then "begin K; M end", "[K or M]","if α then K else M", "while α do K" are in FS.

The programs without symbol "or" will be called deterministic and the others nondeterministic.

The set of formulas FSF is the least set such that all open classical formulas belong to FSF and if α, β are formulas in FSF and K is a deterministic program, M is an arbitrary program then all formulas of the following from are in FSF:

$$\alpha \;,\; (\alpha \cup \beta) \;,\; (\alpha \cap \beta), (\alpha \Rightarrow \beta) \;,\; \alpha \;,\; \Delta M\alpha \;,\; \nabla M\alpha .$$

By the language of the nondeterministic algorithmic logic we shall understand system $\langle A, T_o, FS, FSF \rangle$.

By a realization R of the language L in the nonempty set J and Boolean algebra $B_o = \langle \{true, false\}, \neg, \cup, \cap, \Rightarrow \rangle$ we understant mapping which to every n-argument functor φ assigns an n-argument function φ_R in the set J and to every m-argument predicate ρ - the characteristic function ρ_R of m-arguments relation in J.

We can extend the realization to every proper expression of the language. Let v be any valuation in J, i.e. a function from the set of all variables V into the set J. For every formula α, term τ and program K, τ_R is a function from the set of all valuations J^V into the set J, α_R is a function from the set of all valuations J^V into the set B_o and K_R is a binary relation in J^V. The strict definitions are by induction with respect to the length of expressions and for classical formulas and terms are as usual.

The notion of realization of the program K is defined by means of the notion of computation. Let us assume the following definitions.

By a configuration in the realization R we shall mean any orderred pair $\langle v, K_1, \ldots, K_n \rangle$ where v is a valuation in R and $K_1, \ldots, K_n$ are programs.

$\underset{R}{\to}$ is the least binary relation in the set of all configurations such that

$\langle v, [x_1/\tau_1, \ldots, x_n/\tau_n]; rest \rangle \underset{R}{\to} \langle v', rest \rangle$

where v' is a valuation such that $v'(x_i) = \tau_{iR}(v)$, $i = 1, \ldots, n$ and $v'(z) = v(z)$ for $z \notin \{x_1, \ldots, x_n\}$.

$\langle v, [K \text{ or } M]; rest \rangle \underset{R}{\to} \langle v, K; rest \rangle$

$\langle v, [K \text{ or } M]; rest \rangle \underset{R}{\to} \langle v, M; rest \rangle$

$\langle v, \text{begin } K;\ M \text{ end}, rest \rangle \underset{R}{\to} \langle v, K; M; rest \rangle$

$\langle v, \text{if } \alpha \text{ then } K \text{ else } M, rest \rangle \underset{R}{\to} \langle v, K; rest \rangle$ iff $R \models \alpha[v]$

$\langle v, \text{if } \alpha \text{ then } K \text{ else } M, rest \rangle \underset{R}{\to} \langle v, M; rest \rangle$ iff $R \models \neg \alpha[v]$

$\langle v, \text{while } \alpha \text{ do } K; \text{ rest}\rangle \underset{R}{\to} \langle v, \text{rest}\rangle$ iff $R \models \alpha[v]$

$\langle v, \text{while } \alpha \text{ do } K; \text{ rest}\rangle \underset{R}{\to} \langle v, K; \text{while } \alpha \text{ do } K, \text{rest}\rangle$

iff $R \models \neg\alpha[v]$

The sequence $\{c_i\}_{i \in I}$ of configurations is called a computation of the program K at the valuation v iff for every $i \in I$

$$c_i \underset{R}{\to} c_{i+1} \quad \text{and} \quad c_o = \langle v, K\rangle.$$

If the computation is a finite sequence $c_1, \ldots, c_n$ and the last configuration is equal to $\langle v', \emptyset\rangle$ then the valuation v' is called the result of the computation. So, K_R is a binary relation such that $(v, v') \in K_R$ iff there exists a finite computation of the program K in the realization R at the initial valuation v such that v' is its result.

The set of all results of the program K at the valuation v in the realization R will be denoted by $K_R(v)$. Now we may continue the definition of the mapping α_R.

$$(K\alpha)_R(v) = \begin{cases} \alpha_R(v') & \text{iff } K_R\, v \neq \emptyset \text{ and } v' \in K_R(v) \\ \text{false} & \text{in the opposite case} \end{cases}$$

$(\Delta M\alpha)_R(v) = \text{true}$ iff all computations of the program M at the valuation v in the realization R are finite and for all v', if $v' \in M_R(v)$ then $\alpha_R(v') = \text{true}$

$(\nabla M\alpha)_R(v) = \text{true}$ iff M has a finite computation at the valuation v in the realization R and there exists $v' \in M_R(v)$ that $\alpha_R(v') = \text{true}$

In the above definitions K denotes a deterministic program and M an arbitrary one.

In the last place let us see the following sketch of a tree of possible computations.

Example

Let us consider program

```
begin
    [while β do s  or  if α then [s′ or s″] else K]
end
```

in the fixed realization R, where s, s′, s″ are substitutions and K is a program. Assume that R ⊨ α[v] and β is valid in R.

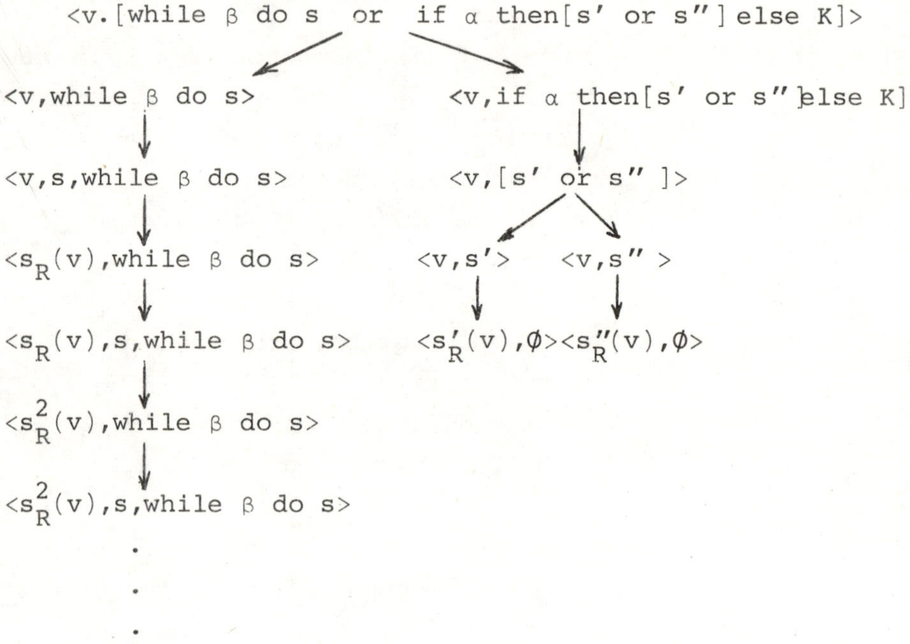

All branches of this tree are computations of considered program: there are two possible finite computations and one infinite.

2. *Algorithmic properties*

The following list contains chosen semantical properties and formulas of algorithmic language with nondeterministic programs describing them.

a/ All computations of the program K in the realization R at the initial valuation v are finite - R⊨ΔKtrue[v] b/ There exists finite computation of the program K in the realization R at the valuation v - $R \models \Delta K true[v]$

c/ There exists infinite computation of the program K in the realization R at the valuation v -

$R \models \neg \Delta K true[v]$

From now on our notation will be less formal

d/ All computations of the program K are infinite -

$\neg(\nabla K true)$

e/ If all computations of the program K are finite, then all results satisfy the formula α - $(\Delta K true \Rightarrow \Delta K\alpha)$

f/ If there exists finite computation of the program K, then there exist results that satisfy the formula α -

$(\nabla K true \Rightarrow \nabla K\alpha)$

g/ If the data have the property α then there exists a result that has the property β - $(\alpha \Rightarrow \nabla K\beta)$

h/ If the data have the property α then all results of the program K have the property β - $(\alpha \Rightarrow \Delta K\beta)$

i/ If the data have the property α and all computations of the program K are finite then all results have the property β - $((\alpha \cap \Delta K true) \Rightarrow \Delta K\beta)$

j/ If the data have the property α and there exists a finite computation of the program K, then there exists a result that has the property β - $((\alpha \cap \nabla K true) \Rightarrow \nabla K\beta$

k/ All results of the program K have the property α -

$\neg(\nabla K \neg\alpha)$

l/ Every computation of the program [while α do K] includes at least i+1 iterations of the program K -

$([\text{if } \alpha \text{ then } K]^{i}\alpha)$

m/ Every computation of the program [while α do K] includes at most i iterations of the program K -

$([\text{if } \alpha \text{ then } K]^{i}\neg\alpha)$

3. Logic

In this section we introduce the deductive system that enables us to make the notion of tautology precise and allow us to investigate when a formula can be recognized as a logical consequence of a certain set of axioms.

The set of axioms contains all axiom of classical propositional calculus [10], algorithmic axioms [1,4] and all formulas of the following schemata, where α, β, γ are formulas, s is a substitution. K,M are programs and the symbol $\boxtimes$ denote ∇ or Δ or empty.

$\nabla K(\alpha \cup \beta) \equiv \nabla K\alpha \cup \nabla K\beta$

$\Delta K(\alpha \cap \beta) \equiv \Delta K\alpha \cap \Delta K\beta$

$\boxtimes[\text{begin } K;M \text{ end}]\alpha \equiv (K(M\alpha))$

$\boxtimes[\text{if } \gamma \text{ then } K \text{ else } M]\alpha \equiv [\gamma \cap \boxtimes K\alpha] \cup [\gamma \cap \boxtimes M\alpha)$

$\boxtimes[\text{while } \gamma \text{ do } K]\alpha \equiv (\gamma \cap \alpha) \cup (\gamma \cap \boxtimes K \boxtimes[\text{while } \gamma \text{ do } K]\alpha$

$\nabla[K \text{ or } M]\alpha \equiv (\nabla K\alpha \cup \nabla M\alpha)$

$\Delta[K \text{ or } M]\alpha \equiv (\Delta K\alpha \cap \Delta M\alpha)$

$s\exists x\alpha(x) \equiv \exists y(s[x/y]\alpha(x))$ where y is an individual variable not occuring in $s\alpha$.

We admit the following inference rules:

$$\frac{\alpha, (\alpha \Rightarrow \beta)}{\beta} \qquad \frac{\alpha \Rightarrow \beta}{(\boxtimes K\alpha \Rightarrow \boxtimes K\beta)}$$

$$\frac{\{(s\boxtimes[\text{if } \gamma \text{ then } K]^{i}(\alpha \cap \neg\gamma) \Rightarrow \beta)]_{i\in\omega}}{(s\boxtimes[\text{while } \gamma \text{ do } K]\alpha \Rightarrow \beta)}$$

$$\frac{(s[x/y]\alpha(x) \Rightarrow \beta)}{(s\exists x\alpha(x) \Rightarrow \beta)} \qquad \frac{(\beta \Rightarrow s[x/y]\alpha(x))}{(\beta \Rightarrow s\forall x\alpha(x))}$$

$$\frac{(s\exists x\alpha(x) \Rightarrow \beta)}{(s[x/y]\alpha(x) \Rightarrow \beta)} \qquad \frac{(\beta \Rightarrow s\forall x\alpha(x))}{(\beta \Rightarrow s[x/y]\alpha(x))}$$

In rules of introducting quanifiers we assume that $s\beta$ contains no occurence of x and α contains no quantifiers binding x.

Observe that one of the inference rules is of infinite character. The formula $(\mathbb{X}[\text{while } \gamma \text{ do } K]\alpha \Rightarrow \beta)$ is the consequence of the infinite set of premises. It arise from [4] that such a rule cannot be omitted.

Now let us consider the connections between the notions of validity and of theorem.

Let T = <L,C,A> be a consistent algorithmic theory, where L is an algorithmic language (see §1) C - is a consequence operation defined by axioms and rules given above and A is a set of formulas (specific axioms of this theory).

The following theorem holds:

COMPLETENESS THEOREM:

For every consistent theory T *the following conditions are equivalent:*

(i) α *is a theorem of* T.

(ii) α *is valid in every model of* T. □

The proof makes use of algebraic method of Rasiowa--Sikorski given in [10].

This theorem assures us that the consequence operation C given in this section is a good formalization of the semantic consequence. For a given set of formulas A, a formula α belongs to C(A) if and only if it is a semantic consequence of the set A i.e. if it is valid in every model of A .

The next theorem give the connections between the consistency of a theory and the existence of model.

MODEL EXISTENCE THEOREM:

An algorithmic theory posseses a model iff it is consistent. □

As a simple consequence of these two theorems we have the downward Skolem-Löwenheim theorem

THEOREM:

If an algorithmic theory T has an infinite model then it has an enumerable model. □

The proofs of all the above theorems may be found in [5,6].

OPEN PROBLEMS

At the end of our considerations let us quote some problems connected with nondeterministic programs.

1. Are formulas $\Delta K\alpha$ and $\nabla K\alpha$ mutually expressible?

2. Let r be a relation in a fixed algebraic system . The relation $r(x_1,\ldots,x_n)$ is definable by a formula iff for all $a_1,\ldots,a_n$ from .

$$r(a_1,\ldots,a_n) \quad \text{iff} \quad \models \alpha(x_1,\ldots,x_n)[a_1,\ldots,a_n]$$

The relation r is weakly nondeterministically (strongly nondeterministically or deterministically) programmable by a program K iff it is definable by a formula $\nabla K\alpha$ ($\Delta K\alpha$ or $K\alpha$).

Let D . WND . SND denote the sets of all deterministically (weakly nondeterministically or strongly nondeterministically) programmable relation in $\mathfrak{A}$. Let H denotes the class of all relations that are programmable by a program K such that the formula ΔKtrue is valid in $\mathfrak{A}$.

The questions arise:

A/ D = WND ∪ SND?

B/ WND H = SND?

3. Let r be a WND relation defined by $\nabla K\alpha$ with the cost $O(T(n))$[ ] and let ΔKtrue be valid in $\mathfrak{A}$. Does there exists a formula $\Delta K\alpha$ defining the relation r with the same cost $O(T(n))$?

REFERENCES

[1] Banachowski,L., Investigations of properties of programs by means of the extended algorithmic logic. *Fundamenta Informaticae*, vol. 1. 1977, 93-119, 169-194.

[2] Harel,D., Meyer,A., Pratt,V., Computability and Completeness in logic of programs. MIT Cambridge. Mess. May 1977.

[3] Harel,D., Pratt,V., Nondeterminism in logic of programs. MIT. Cambridge. Mass. 1977.

[4] Mirkowska,G., On formalized systems of algorithmic logic. *Bull. Acad. Polon. Ser. Math.* vol. 19. 1971. 241-428.

[5] Mirkowska,G., Model existance theorem in algorithmic logic with nondeterministic programs, to appear in Fundamenta Informaticae.

[6] Mirkowska,G., Algorithmic logic in with nondeterministic programs, to appear in Fundamenta Informaticae

[7] Muldner,T., On properties of certain synchronizing tool for parallel computations in M. Karpinski (ed.) *Fundamentals of Computation Theory*. Lecture Notes in Computer Science vol. 56. Springer-Verlag, 1977. 459-465.

[8] Pratt,V., Semantical considerations on Floyd-Hoare logic MIT. Cambridge. 1977.

[9] Radziszowski,S., Programmability and P = NP conjecture. in M. Karpinski (ed.) *Fundamental of Computer Theory*. Lecture Notes in Computer Science vol. 56., Springer-Verlag 1977.

[10] Rasiowa,H., Sikorski,R., Mathematics of metamatematics. PWN. Warsaw, 1963.

[11] Salwicki,A., Applied algorithmic logic, in.J. Gruska (ed.) *Mathematical Foundations of Computer Science*, Lecture Notes in Computer Science, vol.53. Springer-Verlag, 1977.

[12] Salwicki,A., On algorithmic theory of stacks, to appear in Fundamenta Informaticae.

[13] Salwicki,A., Formalized algorithmic languages. *Bull. Acad. Polon. Sci. Ser. Math.* vol. 18. 1970. 227-232.

G.Mirkowska
Warszawa,
Kochanowskiego 28/9
Poland

COLLOQUIA MATHEMATICA SOCIETATIS JÁNOS BOLYAI
26. Mathematical Logic in Computer Science
Salgótartján (HUNGARY), 1978.

CONNECTIONS BETWEEN CYLINDRIC ALGEBRAS AND INITIAL ALGEBRA SEMANTICS OF CF LANGUAGES

I. Németi

This paper is a continuation of [13] which can be found in this volume. We shall use the notations listed at the beginning of [13] without introducing them. Here we elaborate some details as it was promised in [13]. Computer science motivations of this paper are in [13] and also in the note on applications to the Burstall-Goguen approach to stepwise refinement of program specifications at the end of this paper. This paper contains results on cylindric algebras (CA-s) and on the connection between model theory and CA-s which can be read independently of [13]. The fundamental connection between model theory and cylindric algebras is established in Proposition 1 (ii).

The content of the present section can be interpreted in two ways: One is "algebraization of first order Modal Logic S5". The other is: semantic algebraization of *theories* of $L_1 = \langle S_1, M_1, k_1 \rangle$ defined in Example 1 of [13]. By a semantic algebraization of a theory $Th \subseteq S_1$ we understand an algebraization of a set $K \subseteq M_1$ of its models such that Th is the set of all formulas valid in K; i.e.: $Th=\{\varphi \in S_1 : K \models \varphi\}$ where $K \models \varphi$ iff $(\forall \mathfrak{A} \in K)\ \mathfrak{A} \models \varphi$. For more about this second interpretation see the end of this paper. In § 3.B. of [13] an algorithm was defined "to turn $\langle M,k \rangle$ into a class $\mathcal{M}$ of algebras". Below we shall execute this algorithm for a concrete choice M_2 of M. This M_2 can be viewed from two aspects: *Aspect* (i): M_2 is the class of all Kripke models of the first order Modal Logic S5. *Aspect* (ii): Define the language $L_{12}=\langle S_1,M_2,\models \rangle$ as follows: M_2 consists of all subsets of M_1 and $\models$ is the usual validity relation between sets of models and formulas, see above. This language L_{12} is equivalent to L_1, but L_{12} has the advantage that every theory can be represented by a single element of M_2. Hence M_2 is the class of all models of an improved version L_{12} of Classical First order Language. Below the algorithm given in § 3.B. of [13] is executed for $\langle M_2,\models \rangle$. The reader is asked to consult Example 1 and § 3.B of [13] before reading on.

By a *Kripke model* we understand a *set* $K \subseteq M_1$ of classical first order models of the same similarity type. E.g.: $K=\{\mathfrak{A},\mathfrak{B}\}$ is a Kripke model if the similarity types of the classical models $\mathfrak{A},\mathfrak{B}\in M_1$ coincide. M_2 denotes the class of all Kripke models as defined above. More precisely, $K\in M_2$ iff K is a set $\{\mathfrak{A}_i : i\in I\}$ of classical first order models of the same similarity type with disjoint universes i.e. $A_i \cap A_j = 0$ if $i,j\in I$ and $i\neq j$. This disjointness condition is not essential but will simplify proofs to come. By this the class M_2 is defined.

Now we define the *meaning function* $k_{12} : S_1\times M_2 \to$ "Sets" analogously to part (ii) of Example 1 of [13]. The reader is asked to recall the definition of $k_1 : S_1\times M_1 \to$ "Sets" from [13] Example 1.(ii). Let $K\in M_2$ and $\varphi\in S_1$. Then we define $k_{12}(\varphi,K) \stackrel{d}{=} \cup\{k_1(\varphi,\mathfrak{A}) : \mathfrak{A}\in K\}$. For example, $k_{12}((x_0=x_0),\{\mathfrak{A},\mathfrak{B}\}) = {}^{\omega}A\cup{}^{\omega}B$.

By this the language $L_{12}=\langle S_1,M_2,k_{12} \rangle$ is defined. The algorithm of [13]§3.B can be applied to L_{12}

and it associates a class M_2 of meaning algebras to $\langle M_2, k_{12} \rangle$ i.e. to the semantics of L_{12}. Consider e.g. the Kripke model $\{ \mathfrak{A}, \mathfrak{B} \} \in M_2$. The associated meaning algebra in M_2 will be a Boolean algebra of unit ${}^{\omega}A \cup {}^{\omega}B$ enriched with some additional operations. For more detail see Proof of Prop.1(ii) below. Following the tradition of [1], [2] etc. we shall denote the class M_2 of meaning algebras by Lrg. First we shall define the class Lrg and afterward in Proposition 1 we shall prove that it indeed corresponds to the language L_{12} i.e. to M_2 in the sense of [13] § 3.B..

In Def.1 below we shall recall the definition of the class Lrg of algebras from Andréka [1], [2]. Throughout this paper α is an arbitrary set, but for some readers it may be convenient to think that α is an ordinal. Recall from the last paragraph of [13] the definition of g_1, g_1^{α}, Crs_{α} and Gs_{α}. g_1^{α} was a similarity type with operations $\{\cdot, -, c_i, d_{ij} : i,j \in \alpha\}$. $Alg(g_1^{\alpha})$ was the class of algebras of similarity type g_1^{α} and $Gs_{\alpha} \subseteq Crs_{\alpha} \subseteq Alg(g_1^{\alpha})$ were defined in detail and illustrated by figures. Recall that if $\underset{\sim}{A} \in Alg(g_1^{\alpha})$ then we use the notation $\underset{\sim}{A} = \langle A, \cdot, -, c_i, d_{ij} \rangle_{i,j \in \alpha}$ and $1^A = -(d_{00} \cdot -d_{00})$ if $\alpha \neq 0$. 1^A is said to be the unit of $\underset{\sim}{A}$.

DEFINITION 1 (i-finiteness and $Lrg_{\alpha} \subseteq Gs_{\alpha}$)

(i) Let $\underset{\sim}{A} \in Crs_{\alpha}$. Let $x \in A$. Then x is said to be *i-finite* in A iff there is a finite $H \subseteq \alpha$ such that $x \supseteq \{f \in \cup A : H \upharpoonright f \subseteq g$ for some $g \in x\}$. A is i-finite iff every $x \in A$ is i-finite in A.

(ii) $Crs_{\alpha}^{if} \overset{d}{=} \{\underset{\sim}{A} \in Crs_{\alpha} : \underset{\sim}{A}$ is i-finite$\}$.
$Lrg_{\alpha} \overset{d}{=} Gs_{\alpha} \cap Crs_{\alpha}^{if}$. $Lrg \overset{d}{=} Lrg_{\omega}$.

(iii) $Cs_{\alpha} \overset{d}{=} \{\underset{\sim}{A} \in Gs_{\alpha} : \cup A = {}^{\alpha}U$ for some set $U\}$.

(iv) $Lr_{\alpha} \overset{d}{=} Cs_{\alpha} \cap Lrg_{\alpha}$ and $Lr \overset{d}{=} Lr_{\omega}$. □

Remark: The abbreviation Lr stands for "*L*ocally finite *r*egular cylindric set algebra." Lrg stands for "*L*ocally finite *r*egular generalized cylindric set algebra". The adjectives "locally i-finite" and "locally finite regular" are synonyms, see also the remarks about Lr at the end of § 3 of [13]. The abbreviation i-finite stands for "independently finite". The origin of this name is that if X is i-finite then there is a *finite* $H \subseteq \omega$ such that, for every $f \in {}^{\omega}U$, the truth of the statement "$f \in X$" is independent of the behaviour of f outside of H. For more on Def.1. above see [4] p.18 and [5] p.10 under the name "i-finite"-ness. □

(i) of Proposition 1 below is taken from Andréka [1], [2] which have not yet been translated into English. Recall from [13]§ 3.B the meaning algebra $\underset{\sim}{K}$ associated to $K \in M_2$.

PROPOSITION 1

(i) i-finite elements generate i-finite ones in every $\underset{\sim}{A} \in Gs_{\omega}$.

(ii) There exists a pair p,q of functions $q : M_2 \twoheadrightarrow Lrg$ and $p : Lrg \rightarrowtail M_2$ *both definable in set theory* with a single absolute formula such that 1-3 below hold.

1. p is a right inverse of q, i.e. $q \circ p$ is the identity function on Lrg.
2. Let $K \in M_2$. Then its image $p(q(K))$ is first order definitionally equivalent to K in the sense of [16] p.56. Therefore $q(K_1) = q(K_2)$ implies that K_1 and K_2 are definitionally equivalent.
3. $q(K) = \underset{\sim}{K}$ for every $K \in M_2$. Therefore Lrg is the class of meaning algebras associated to L_{12} in [13] § 3.B.

Proof:

Before going into details, we make some considerations on the notions involved. We shall use the list of notations of [13] without warning or any mention. E.g. $Sb\ V = \{X : X \subseteq V\}$ for any class V.

DEFINITION 1.1

(i) Let $\underset{\sim}{A}$ be a Crs_α with unit V. Let $x \in A$ and $H \subseteq \alpha$. Then x is *H-independent* (in V) iff $(\forall f \in V)(\forall g \in x)[H \upharpoonleft f \subseteq g \Rightarrow f \in x]$.

(ii) Let $\underset{\sim}{A} \in Alg(g_1^\alpha)$ and $x \in A$. Then $\Delta^A(x) \stackrel{d}{=} \Delta x = \{i \in \alpha : c_i^A x \neq x\}$.

(iii) Let $V \subseteq {}^\alpha Sets$ and $i,j \in \alpha$. Then $c_i^V \stackrel{d}{=} \langle \{f \in V : (\exists q \in x)\alpha \sim \{i\} \upharpoonleft f \subseteq q\} : x \subseteq V \rangle$ and $D_{ij}^V \stackrel{d}{=} \{f \in V : f_i = f_j\}$ and $_V\!\sim \stackrel{d}{=} \langle V \sim x : x \subseteq V \rangle$.

Note that $c_i^V : Sb\ V \to Sb\ V$ and that $\langle Sb\ V, \cap, {}_V\!\sim, c_i^V, D_{ij}^V \rangle_{i,j \in \alpha}$ is the full Crs_α with unit V if V is a set.

End of Definition 1.1

LEMMA 1.1 Let $|\alpha| \geq \omega$. Let $\underset{\sim}{A}$ be a Crs_α with unit V. Let $x \in A$. Then a.-d. below hold.

a. x is i-finite iff [x is H-independent for some *finite* $H \subseteq \alpha$].

b. Let $\underset{\sim}{A} \in Cs_\alpha$. Then x is i-finite iff [x is Δx-independent and $|\Delta x| < \omega$].

c. Let $\underset{\sim}{A} \in Gs_\alpha$. Then x is i-finite iff [x is $(1 \cup \Delta x)$-independent and $|\Delta x| < \omega$].

d. Statement b. above is not true for Gs_α and c. is not true for Crs_α.

The proof is immediate by checking the definitions.

End of Lemma 1.1

REMARK 1.1 If Δx is finite then by Proposition 1(ii) and by Lemma 1.1 c. we know that $(1\cup\Delta x)$ -independence of x means that x is some meaning in some model of L_{12}. What is the situation if Δx is *infinite?* Let $\mathfrak{A} = \langle \omega,\leq \rangle$. Let some ordinal $\alpha>\omega$ be fixed and let $\{x_i : i\in\alpha\}$ be the set of variables of an *infinitary* language of $\mathfrak{A}$. Let φ be the infinitary formula $\exists x_\omega (\bigwedge_{i<\omega} x_i\leq x_\omega)$. The free variables of φ are $\{x_i : i<\omega\}$. The meaning $k(\varphi,\mathfrak{A})$ of φ in $\mathfrak{A}$ is then $k(\varphi,\mathfrak{A}) = \{q\in{}^\alpha\omega : (\exists n\in\omega)(\forall i<\omega)q(i)\leq n\}$. Let $Y \stackrel{d}{=} k(\varphi,\mathfrak{A})$. The meaning algebra $\underset{\sim}{\mathfrak{A}}$ corresponding to $\mathfrak{A}$ and to the infinitary language (of α variables) will consist of subsets of $V={}^\alpha\omega$, i.e. it will be a Gs_α instead of a Gs_ω. Clearly $Y \subseteq V$ and Y *is a meaning*. Despite of this, Y is *not* $(1\cup\Delta Y)$ -independent! Namely $\Delta Y=0$ since $C_i^V Y = Y$ for every $i<\alpha$ and Y is not 1-independent. Hence in the case of infinitary logic $(1\cup\Delta x)$ - independence has no logical or model theoretical meaning. However, the following generalization of i-finiteness *has* some.

DEFINITION 1.2 Let $x \subseteq V \subseteq {}^\alpha$Sets. Then x is *independently-small* (i-small) in V iff $(\exists H \subseteq \alpha)[|H|<|\alpha|$ and x is H-independent$]$.

End of Definition 1.2.

Now in an α-ary language we allow conjuctions that are shorter than $|\alpha|$ and relation symbols with arities smaller than $|\alpha|$. (I.e. an α-ary similarity type is a function $t\in{}^\Sigma|\alpha|$. These α-ary languages are usually denoted by $L_{|\alpha|,\omega}$.) Then, analogously to Proposition 1 (ii), the meaning algebras of the α-ary analogues of L_{12} are exactly the ($|\alpha|$-complete) independently-small Gs_α -s. Proposition 1(i) also holds: i-small elements generate i-small ones in every Gs_α (the present proof works to show this). *End of Remark 1.1*

Now we return to prove Proposition 1.

Proof of (i): (i) says that i-finite elements generate i-finite ones in every $\underset{\sim}{A} \in Gs_\omega$. Let $\underset{\sim}{A}$ be a Gs_ω . Let $R \subseteq A$ be the set of all i-finite elements of $\underset{\sim}{A}$. We shall show that R is a subalgebra. Hence there is a greatest i-finite subalgebra $\underset{\sim}{R} \subseteq \underset{\sim}{A}$ in every Gs_ω $\underset{\sim}{A}$.

REMARK: The assumption that the unit of $\underset{\sim}{A}$ is a Gs_ω -unit will be used only at one place below, namely when checking that the set $R \subseteq A$ of all i-finite elements of $\underset{\sim}{A}$ is closed under the operations C_i^V . Hence: for *every* Crs_ω $\underset{\sim}{B}$ with unit V we know that in its reduct $\langle B, \cap, {}_V\sim, D_{ij}^V \rangle_{i,j\in\omega}$ the i-finite elements generate i-finite ones. *End of Remark.*

Let $\underset{\sim}{A} \in Gs_\omega$ with unit $\cup\{{}^\omega W_i : i \in I\} \overset{d}{=} V$ where of course $W_i \cap W_j = 0$ if $i \neq j$. Recall from Lemma 1.1 a. that for every $x \subseteq V$ we have that x is i-finite iff there exists some finite $H \subseteq \omega$ such that x is H-independent. Now let $x \subseteq V$, $y \subseteq V$, and let $H, G \subseteq \omega$ be such that $|H| < \omega$, $|G| < \omega$, x is H-independent, and y is G-independent.

Consider first ${}_V\sim X$.:

For every $z \subseteq V$ we have that z is H-independent iff $(\forall q, k \in V)[H \upharpoonleft k = H \upharpoonleft q \Rightarrow (q \in z \leftrightarrow k \in z)]$. Since the statements $(q \in z \leftrightarrow k \in z)$ and $(q \in ({}_V\sim z) \leftrightarrow k \in ({}_V\sim z))$ are equivalent, we have that ${}_V\sim x$ is H-independent, i.e. *i-finite.*

Consider now $x \cap y$.:

We prove that $x \cap y$ is $H \cup G$ -independent. Let $f \in V$ and let $(H \cup G) \upharpoonleft f \subseteq g \in x \cap y$. Since x is H-independent and y is G-independent, we have $f \in x \cap y$. Thus $x \cap y$ *is i-finite* since $H \cup G$ is finite.

Consider now $C_i^V x$.:

We may suppose that $H \neq 0$. Suppose that $h \in V$, $H \upharpoonleft h \subseteq f$, and $f \in C_i^V x$. We want to prove that $h \in C_i^V x$. This will imply that $C_i^V x$ is i-finite, see Definition 1.1 and Lemma 1.1 a. above. $f \in C_i^V x$ means that $f \in V$ and $(\omega \sim \{i\}) \upharpoonright f \subseteq z$ for some $z \in x$. We define $h' \in {}^{\omega}U$ to be the function

$$h'(j) \overset{d}{=} \begin{cases} h(j) & \text{if} \quad j \neq i \\ z(i) & \text{if} \quad j = i \end{cases} .$$

Now $(\omega \sim \{i\}) \upharpoonright h \subseteq h'$ and $h \in V$. Therefore to prove $h \in C_i^V x$, it is enough to prove $h' \in x$. Now, $H \upharpoonright h' \subseteq z \in x$ and $(\omega \sim \{i\}) \upharpoonright f \subseteq z \in x$. By i-finiteness of x, we have $h' \in x$ provided that $h' \in V$. Remains to show $h' \in V$. $H \upharpoonright h = H \upharpoonright f$, $\{h,f\} \subseteq V$, and $H \neq 0$ imply that there is a unique $j \in I$ such that $\{f,h\} \subseteq {}^{\omega}W_j$. Therefore $(\omega \sim \{i\}) \upharpoonright f \subseteq z \in V$ implies that $z \in {}^{\omega}W_j$, i.e. that $z(i) \in W_j$. Since $h \in {}^{\omega}W_j$ and $z(i) \in W_j$, by the definition of h', also $h' \in {}^{\omega}W_j \subseteq V$. We have seen that $h' \in V$, and therefore $h \in C_i^V x$. This means that $C_i^V x$ *is i-finite.*

Notice that in this step we made use of the hypothesis that the unit of $\underset{\sim}{A}$ is $\cup\{{}^{\omega}W_i : i \in I\}$ where $W_i \cap W_j = 0$ if $i \neq j$. This disjointness condition is essential: see Remark following the proof.

For every $i,j \in \omega$, $D_{ij}^V = \{f \in V : f(i) = f(j)\}$ is obviously i-finite.

This proves that i-finite elements generate i-finite ones in every $\underset{\sim}{A} \in Gs_{\omega}$. *QED(i).*

Proof of (ii): (Cf.III.6. and V.5. of [3]). We shall define a pair of functions $q : M_2 \to Lrg$ and $p : Lrg \to M_2$ and then we shall prove that p is a right inverse of q. Then we shall show that for every $K \in M_2$, the model $p(q(K))$ is first order definitionally equivalent to K.

Let $t : R \to \omega$ *be a similarity type* where R is a set of relation symbols. Let M_t denote the class of all models of similarity type t . The definition of M_2 was the following:

$M_2 \stackrel{d}{=} \{ K \in Sets : (\exists \text{ similarity type } t) \ K \subseteq M_t$ and $(\forall \mathfrak{A}, \mathfrak{B} \in K) A \cap B = 0 \}$. Roughly:

$M_2 = \cup \{ Sb(M_t) : t \in {}^R\omega$ and $R \in Sets \}$.

First we define $q : M_2 \twoheadrightarrow Lrg$. Let $K \in M_2$ be arbitrary. Let $\underset{\sim}{C}$ be the full Gs_ω with unit $V = \cup \{ {}^\omega A : \mathfrak{A} \in K \}$. Let $K \subset M_t$ where $t : R \to \omega$ is a similarity type. For every relation symbol $r \in R$ let $\mathfrak{A}(r)$ denote the meaning of r in the model $\mathfrak{A} \in M_t$. I.e. $\mathfrak{A}(r) \subseteq {}^{t(r)}A$ and $\mathfrak{A} = \langle A, \mathfrak{A}(r) \rangle_{r \in R}$. We define a map $g : R \to C = Sb\ V$ as follows: For every $r \in R$ let

$$g(r) \stackrel{d}{=} \cup \{ \{ s \in {}^\omega A : t(r) \restriction s \in \mathfrak{A}(r) \} : \mathfrak{A} \in K \} =$$
$$= \{ s \in V : t(r) \restriction s \in \cup \{ \mathfrak{A}(r) : \mathfrak{A} \in K \} \}.$$

Clearly $\{ g(r) : r \in R \}$ is a subset of C consisting of i-finite elements only. (The set H corresponding to the element $g(r) \in C$ is $t(r)$.) Let $\underset{\sim}{B} \subseteq \underset{\sim}{C}$ be the subalgebra of $\underset{\sim}{C}$ *generated* by $\{ g(r) : r \in R \}$. *By (i)* we have $\underset{\sim}{B} \in Lrg$. Now, we define

$$q(K) \stackrel{d}{=} \underset{\sim}{B} .$$

This q is certainly a function $q : M_2 \to Lrg$ everywhere defined on M_2 . To see that it is onto, we shall construct its inverse p. Let $\underset{\sim}{B} \in Lrg$ have unit $V = \cup \{ {}^\omega W_i : i \in I \}$ such that $W_i \cap W_j = 0$ whenever $i \neq j$. (Recall that every element of Lrg does have such a unit.) We have to define $p(\underset{\sim}{B}) \in M_2$. Let $h : \underset{\sim}{B} \to \omega$ be such that for every $b \in B$ the set $h(b) \subseteq \omega$ "makes b look i -finite". Since $\underset{\sim}{B} \in Lrg$, every element of B is i-finite, and thus to every $b \in B$ there is such a finite $h(b) \subset \omega$ and by basic properties of ω we can suppose $h(b) \in \omega$ as well. We may also suppose that $h(b) \in \omega$ is the *smallest* element of the set $\{ n \in (\omega \sim 1) : n$ is "a good H for b "$\}$. This makes h unique to every $\underset{\sim}{B}$. *Observe* that $h : B \to \omega$ is a similarity type. The set $p(\underset{\sim}{B})$ of models will be of type h, i.e. $p(\underset{\sim}{B}) \subseteq M_h$. For every $i \in I$ we shall define a model $\mathfrak{A}_i = \langle A_i, \mathfrak{A}_i(b)\rangle_{b\in B}$ of similarity type h. Let $i\in I$. $A_i \overset{d}{=} W_i$. For every $b\in B$ we define the relation $\mathfrak{A}_i(b)$, the meaning of the relation symbol b in $\mathfrak{A}_i$, as

$$\mathfrak{A}_i(b) \overset{d}{=} \{h(b) \upharpoonright s : s \in ({}^{\omega}W_i \cap b)\}.$$

Clearly $\mathfrak{A}_i(b)$ is a $h(b)$-ary relation over A_i as required. This completes the definition of $\mathfrak{A}_i \in M_h$. Now we define $p(\underset{\sim}{B}) \overset{d}{=} \{\mathfrak{A}_i : i\in I\}$. Clearly, $p(\underset{\sim}{B}) \in M_2$. Thus we have a function $p : Lrg \to M_2$. It is easy to check that $q(p(\underset{\sim}{B}))=\underset{\sim}{B}$ for every $\underset{\sim}{B}\in Lrg$. I.e. $q\circ p$="Identity on Lrg". Thus p is one-one and q is onto.

Let $K\in M_2$. We have to check that $p(q(K))$ is definitionally equivalent to K. Notation: $q(K)=\underset{\sim}{B}\in Lrg$ and $p(q(K))$ is of similarity type $h : B \to \omega$. By the definition of M_2 there is a similarity type $t : R \to \omega$ such that $K \subseteq M_t$. Recall that $q(K)=\underset{\sim}{B}$ was defined by fixing some generators $\{g(r) : r\in R\} \subseteq B$ of $\underset{\sim}{B}$. I.e. we have a function $g : R \to B$. Now every relation symbol $r\in R$ of K corresponds to the relation symbol $g(r)\in B$ of $p(q(K))$. It is easy to check that the relation denoted by r in K is exactly the same as that denoted by $g(r)$ in $p(q(K))$. Of course, first one has to check that the set of universes $\{A_i : i\in I\}$ of K coincides with the set of universes $\{W_i : i\in I\}$ of $p(q(K))$. But this is obvious by the definition of p and q. Let $b\in B\sim\{g(r) : r\in R\}$. Clearly the relation denoted by b in $p(q(K))$ is definable by a first order formula φ of type h such that all the relation symbols occurring in φ are in $\{g(r) : r\in R\}$. See Defs. 11.5-11.7 in [20]. This completes the proof of definitional equivalence.

Now we show that q and p are *definable in set theory*. I.e. there exist two formulas $\gamma(x,y)$ and $\pi(y,x)$ of the language of say ZFC such that, whenever $\underset{\sim}{B}\in Lrg$

and $K\in M_2$ then $\pi(\underset{\sim}{B}, K)$ is true iff $p(\underset{\sim}{B})=K$. Similarly for for γ and q . This can be done by inspecting the above definitions of p and q . We omit it for its being long though very easy. It is easy to see that the formulas $\gamma(xy)$ and $\pi(yx)$ of ZFC are *absolute* in the usual set theoretic sense.

It remains to show that $q(K)$ is the meaning algebra $\underset{\sim}{K}$ for every $K\in M_2$ in the sense of [13]. Consider the grammar G_1 given in §4 of [13]. Clearly G_1 generates the syntax S_1 of L_{12} . We shall show that *G_1 is adequate to L_{12}* in the sense of Def.3 of [13]. Recall that $\underset{\sim}{S}_1$ is the syntax algebra obtained from G_1 . The statement that " G_1 is adequate to $L_{12}=\langle S_1, M_1, k_{12}\rangle$ " means that the function $k_{12}(-, K)$ is a homomorphism on $\underset{\sim}{S}_1$, for every $K\in M_2$.

Let $K\in M_2$ be arbitrary. We have to show that there is an algebra $\underset{\sim}{C}$ such that $k_{12}(-, K) : \underset{\sim}{S}_1 \to \underset{\sim}{C}$ is a homomorphism. Let $\underset{\sim}{C}$ be the full Gs_ω with unit $V = \cup\{{}^\omega A : \mathfrak{A}\in K\}$. Note that $\underset{\sim}{C}$ was defined the same way. Clearly

$k_{12}(-, K) : S_1 \to C$ since

$$k_{12}(\varphi, K)=\cup\{k_1(\varphi, \mathfrak{A}) : \mathfrak{A}\in K\} \subseteq \cup\{{}^\omega A : \mathfrak{A}\in K\}=V ,$$

for every $\varphi\in S_1$. Remains to check the conditions for homomorphism. Recall that $\underset{\sim}{S}_1=\langle S_1, \langle F_i\rangle_{i\in T}\rangle$, where $T=\{\neg, \wedge, \exists_i, =_{ij} : i,j\in\omega\}$. (See the Remark in [13] §4 too.) Let $\varphi,\psi\in S_1$ be arbitrary. We have to show that the following equations hold:

$$k_{12}(F_\neg(\varphi), K) = {}_V\!\sim\!(k_{12}(\varphi, K)) \text{ i.e. } k_{12}(\neg\varphi, K) = V\sim k_{12}(\varphi,K)$$
$$k_{12}((\varphi\wedge\psi), K) = k_{12}(\varphi, K) \cap k_{12}(\psi, K) ,$$
$$k_{12}(\exists x_i\varphi, K) = C_i^V\, k_{12}(\varphi, K) \quad , \text{ and}$$
$$k_{12}(x_i=x_j, K) = D_{ij}^V .$$

To see these is a straighforward job by the definition of the usual satisfaction relation " $\mathfrak{A}\models\varphi[s]$" . As an example, we show $k_{12}(\exists x_i\varphi, K)=C_i^V\, k_{12}(\varphi, K)$.

$k_{12}(\exists x_i\varphi, K)=\cup\{k_1(\exists x_i\varphi, \mathfrak{A}) : \mathfrak{A}\in K\}$, by the definition of k_{12} . $k_1(\exists x_i\varphi, \mathfrak{A})=\{s\in {}^{\omega}A : \mathfrak{A}\models\exists x_i\varphi[s]\}=$
$=\{s\in {}^{\omega}A : (\exists z\in {}^{\omega}A)[\mathfrak{A}\models\varphi[z] \text{ and } (\omega\sim\{i\})\restriction s\subseteq z]\}=$
$=\{s\in {}^{\omega}A : (\exists z\in k_1(\varphi, \mathfrak{A}))(\omega\sim\{i\})\restriction s\subseteq z\}=$
$= C_i^{{}^{\omega}A}(k_1(\varphi, \mathfrak{A}))=C_i^V(k_1(\varphi, \mathfrak{A}))$, by the disjointness condition in the definition of M_2 .
Now $k_{12}(\exists x_i\varphi, K)=\cup\{C_i^V(k_1(\varphi, \mathfrak{A})) : \mathfrak{A}\in K\}=$
$=C_i^V(\cup\{k_1(\varphi, \mathfrak{A}) : \mathfrak{A}\in K\})=C_i^V(k_{12}(\varphi, K))$, by the definition of C_i^V . By these we have seen that the *First Criterion of Adequateness* is satisfied by G_1 and L_{12} . We have also proved that $k_{12}(-, K) : \underset{\sim}{S}_1\to \underset{\sim}{C}\in Gs_\omega$. Recall that [13] §3.B defines the meaning algebra $\underset{\sim}{K}$ associated to K to be the image of $\underset{\sim}{S}_1$ by the homomorphism $k_{12}(-, K)$. This image is, obviously, the subalgebra of $\underset{\sim}{C}$ with universe $B=\{k_{12}(\varphi, K) : \varphi\in S_1\}$. Hence $\underset{\sim}{B}\subseteq\underset{\sim}{C}$ and $\underset{\sim}{B}=\underset{\sim}{K}$. Clearly, B is generated in $\underset{\sim}{C}$ by $\{k_{12}(r(x_1\ldots x_n), K) : \langle r,n\rangle\in t\}$ where $t : R\to\omega$ is the similarity type of the elements of K , i.e. $K\subseteq M_t$. Recall that $q(K)$ was defined to be the subalgebra of $\underset{\sim}{C}$ generated by the set $\{g(r) : r\in R\}$, where $g : R\to C$ was defined in the proof of (ii), too. Remains to check that $g(r)=k_{12}(r(x_1\ldots x_n), K)$, for every $r\in R$. This is immediate by the definition of $g : R\to C$.
QED Proposition 1.

Remark: Proposition 1 (i) is not true for Gws_ω as defined in Henkin-Monk-Tarski [17]. There exists an i-finite x such that $C_i x$ is not i-finite. However, Proposition 1 (i) is true for *normal* Gws_ω as defined in [17]. □

PROPOSITION 2 **HSP** Lrg = **HSP** Lr = **I** Gs_ω .

Proof: By Corollary 3.41.(a) of [4] on p.25 there, **HSP** Lr = **HSP** Cs_ω . In that paper Cs_α was denoted by Ha_α.

A purely algebraic direct proof of **HSP** $Cs_\alpha = \mathbf{I}\,Gs_\alpha$ for $|\alpha|>1$ is in [17]. Observing $Lr \subseteq Lrg \subseteq Gs_\omega$ completes the proof. □

DEFINITION 2 Let α and U be arbitrary sets and $p\in{}^\alpha U$. Then the *weak Cartesian space* ${}^\alpha U^{(p)}$ with base U and dimension α is defined to be

${}^\alpha U^{(p)} \overset{d}{=} \{q\in{}^\alpha U : |\{i\in\alpha : q_i\neq p_i\}|<\omega\}$.

$Ws_\alpha \overset{d}{=} \{\underset{\sim}{A}\in Crs_\alpha : \cup A={}^\alpha U^{(p)}$ for some U and $p\in{}^\alpha U\}$. □

If $0<|\alpha|<\omega$ then $Ws_\alpha = Cs_\alpha$, for Cs_α see [13] §9.

By Proposition 1 the class of meaning algebras of the language L_{12} is Lrg. The variety generated by Lrg is $\mathbf{I}\,Gs_\omega$ by Proposition 2. The class of meaning algebras of L_1 is Lr, and the variety generated by Lr is again $\mathbf{I}\,Gs_\omega$. According to the general plan in [13] (see [13]§5), the next step to do is the algebraic investigation of the variety $\mathbf{I}\,Gs_\omega$ generated by the meaning algebras. We shall do a little of this here.

To investigate $\mathbf{I}\,Gs_\omega$, Tarski introduced a larger variety $CA_\omega \supseteq Gs_\omega$ which is simpler in certain aspects. We shall also use Ws_ω in our investigations. Some facts that make Ws_ω interesting here are: $Lr \subseteq \mathbf{I}\,Ws_\omega \subseteq \mathbf{I}\,Cs_\omega$, $Lf_\omega \cap \mathbf{I}\,Ws_\omega = \mathbf{I}\,Lr$, and **SP** $Ws_\omega = \mathbf{I}\,Gs_\omega$. See [17].

Recall from [16] that an algebra is said to be *simple* if it has no proper congruences.

PROPOSITION 3 Let $\alpha\neq 0$. Then there exists a simple finitely generated Ws_α which is not generated by a single element. I.e. $(\exists\underset{\sim}{A}\in Ws_\alpha)(\exists X \subseteq A)[|X|<\omega$ and X generates $\underset{\sim}{A}$ but $(\forall y\in A)(y$ does not generate $\underset{\sim}{A})$ and $\underset{\sim}{A}$ is simple].

Outline of Proof: Let $\alpha\neq 0$. Q is the set of rational numbers. $V \overset{d}{=} \{q \in {}^{\alpha}Q : |\{i \in \alpha : q_i \neq 0\}| < \omega\}$.

$Y \overset{d}{=} \{q \in V : 0 = \sum_{i\in\alpha} q_i\}$ and $Z \overset{d}{=} \{q \in V : 1 = \sum_{i\in\alpha} q_i\}$.

Let $\underset{\sim}{C}$ be the full Crs_α with unit V, that is $C = Sb\, V$ etc. Let $\underset{\sim}{B}$ be the subalgebra of $\underset{\sim}{C}$ generated by $\{Y,Z\}$. Clearly $\underset{\sim}{B}$ is a finitely generated Ws_α. To show that $\underset{\sim}{B}$ is not generated by a single element, we have to check that there is no single element in B which would generate both Y and Z. This is proved in Andréka-Németi [10] (by elimination of cylindrifications) where it is also proved that $\underset{\sim}{B}$ is simple. □

The variety $CA_\alpha \subseteq Alg(g_1^\alpha)$ is defined in [16] for all α, see also [13] §4, §9. Note that $Alg(g_1^\alpha)$ is the class of all algebras similar to CA_α-s.

DEFINITION 3 Let $\beta \supseteq \alpha$ be two arbitrary sets. Let $\underset{\sim}{A} = \langle A, \cdot, -, c_i, d_{ij} \rangle_{i,j\in\beta} \in Alg(g_1^\beta)$. Then we define $\underset{\sim}{Rd}_\alpha \underset{\sim}{A} \overset{d}{=} \langle A, \cdot, -, c_i, d_{ij} \rangle_{i,j\in\alpha}$. Note that $\underset{\sim}{Rd}_\alpha \underset{\sim}{A} \in Alg(g_1^\alpha)$.
Let $K \subseteq Alg(g_1^\beta)$. Then we define

(i) $\mathbf{Rd}_\alpha K \overset{d}{=} \{\underset{\sim}{Rd}_\alpha \underset{\sim}{A} : \underset{\sim}{A} \in K\}$.

(ii) $\mathbf{Nr}_\alpha K \overset{d}{=} \{\underset{\sim}{B} \in Alg(g_1^\alpha) : (\exists \underset{\sim}{A} \in Alg(g_1^\beta))[\underset{\sim}{B} \subseteq \underset{\sim}{Rd}_\alpha \underset{\sim}{A}$ and $B = \{a \in A : \Delta^A(a) \subseteq \alpha\}]\}$,

where the function $\Delta^A : A \to Sb\,\beta$ for any $\underset{\sim}{A} \in Alg(g_1^\beta)$ was introduced in Definition 1.1. □

Remark: $\mathbf{Rd}_\alpha\, Alg(g_1^\beta) = Alg(g_1^\alpha)$. Let $\underset{\sim}{A}$ be a Gs_2 with unit V. Then
$\mathbf{Nr}_1 \{\underset{\sim}{A}\} = \{\langle \{x \in A : c_1^V x = x\}, \cap, {}_V\sim, c_0^V, D_{00}^V \rangle\}$ and
$\mathbf{Nr}_0 \{\underset{\sim}{A}\} = \{\langle \{x \in A : c_1^V x = c_0^V x = x\}, \cap, {}_V\sim \rangle\}$. Let $\underset{\sim}{A} \in Alg(g_1^\beta)$ and $\underset{\sim}{B} \in Alg(g_1^\alpha)$. Then $\underset{\sim}{B} \in \mathbf{Nr}_\alpha \{\underset{\sim}{A}\}$ iff $\underset{\sim}{B} \subseteq \underset{\sim}{Rd}_\alpha \underset{\sim}{A}$ and

$B = \{a\in A : (\forall i\in\beta\sim\alpha) c_i a=a \text{ in } \underset{\sim}{A}\}$. □

The operator **Nr** is interesting for us here because **I** Gs_α = **SNr**$_\alpha$ $CA_{\alpha+\omega}$ by [16].

PROPOSITION 4 Let $\beta>\alpha>1$. Then

$(\forall \underset{\sim}{A}\in CA_\beta)|\mathbf{Nr}_\alpha\{\underset{\sim}{A}\}| = 1$ but $(\exists\underset{\sim}{A}\in Crs_\beta)\ \mathbf{Nr}_\alpha\{\underset{\sim}{A}\} = 0$.

Outline of proof: If $\underset{\sim}{A}\in CA_\beta$ then the unique element of $\mathbf{Nr}_\alpha\{\underset{\sim}{A}\}$ is said to be *the neat reduct* of $\underset{\sim}{A}$ in [16]. Let $V = \{\langle 0,0\rangle, \langle 1,0\rangle, \langle 1,1\rangle\}$ and let $\underset{\sim}{A}$ be the full Crs_2 with unit V. If $\underset{\sim}{B}\in\mathbf{Nr}_1\{\underset{\sim}{A}\}$ then $\{\langle 0,0\rangle\}\in B$ since $c_1^V\{\langle 0,0\rangle\} = \{\langle 0,0\rangle\}$. But then $c_0^V\{\langle 0,0\rangle\} =$ $= \{\langle 0,0\rangle, \langle 1,0\rangle\}\in B$. But $c_1^V c_0^V\{\langle 0,0\rangle\} \neq c_0^V\{\langle 0,0\rangle\}$. This is a contradiction. Thus $\mathbf{Nr}_1\{\underset{\sim}{A}\}=0$. The same argument works for arbitrary $\beta>\alpha>1$. □

For any $K \subseteq Alg(t)$ the class $Mod\ Th\ K \subseteq Alg(t)$ is defined to be the first order axiomatizable hull of K. **Uf** K and **Up** K were defined in [16]. By the Keisler--Shelah theorem we have Mod Th K = **Uf Up** K.

PROPOSIITION 5 Let $\beta>\alpha>1$. Then (i)-(ii) below hold.

(i) $Ws_\alpha \nsubseteq Mod\ Th\ \mathbf{Nr}_\alpha CA_\beta = \mathbf{Up}\ \mathbf{Nr}_\alpha CA_\beta$, and
$\mathbf{Uf}\ \mathbf{Nr}_2 CA_\beta \nsubseteq \mathbf{Nr}_2 CA_\beta$.

(ii) Let $K \subseteq CA_\beta$ be such that $Ws_\beta \subseteq K$. Then
$\mathbf{S}\ \mathbf{Nr}_\alpha K \nsubseteq Mod\ Th\ \mathbf{Nr}_\alpha K$.

(iii) $\mathbf{I}\ Crs_\alpha = \mathbf{Nr}_\alpha\ \mathbf{I}\ Crs_\beta$.

Outline of Proof: First we define a formula in the language of CA_α-s. The term $\tau(x)$ is defined to be $[c_0(d_{01}\cdot c_1 x)\cdot c_1(d_{01}\cdot c_0 x)]$. At(x) is the formula $\forall y[y\leq x \rightarrow (y=0 \vee y=x)]$. Let φ be the formula $(\forall z[At(z)\rightarrow z=\tau(\tau(z))] \rightarrow \forall x\exists y\forall z[At(z)\rightarrow(z\leq x\leftrightarrow\tau(z)\leq y)]$. By 1.5.16 of [16] $\mathbf{Nr}_\alpha CA_\beta \models \varphi$ (if $\beta>\alpha>1$). We shall show $Ws_\alpha \not\models \varphi$. First we construct an $\underset{\sim}{A} \in Ws_\alpha$. Let $\underset{\sim}{C}$ and V be as in the proof of Proposition 3. $X \overset{d}{=} \{q \in V : q_0+1 = \sum_{0<i<\alpha} q_i\}$. Let $\underset{\sim}{A}$ be the subalgebra of $\underset{\sim}{C}$ generated by $\{X\} \cup \{\{q\} : q \in V\}$. Clearly $\underset{\sim}{A} \in Ws_\alpha$. To prove that $\underset{\sim}{A} \not\models \varphi$, one observes $\underset{\sim}{A} \models \forall z[At(z) \rightarrow z=\tau(\tau(z))]$ and then one shows that to our X there is no $y \in A$ of the required kind. This was proved in Andréka-Németi [10] by the method of eliminating cylindrifications. In [17] $Ws_\alpha \subseteq \mathbf{S}\,\mathbf{Nr}_\alpha\, Ws_\beta \subseteq CA_\alpha$ is proved. $\mathbf{Up}\,\mathbf{Nr}_\alpha\,\mathbf{S}\,\mathbf{Up}\, K = \mathbf{Nr}_\alpha\,\mathbf{S}\,\mathbf{Up}\, K$ for every $K \subseteq CA_\beta$ is proved in Andréka-Németi [10]. We omit the proof of $\mathbf{Uf}\,\mathbf{Nr}_2\, CA_\beta \not\subseteq \mathbf{Nr}_2\, CA_\beta$ because it is long. *QED of (i) and (ii)* Let $\underset{\sim}{A}$ be a Crs_α with unit V. Let $q : \beta \sim \alpha \rightarrow \beta$. Then there is a Crs_β $\underset{\sim}{B}$ with unit $\{p \cup q : p \in V\}$ such that $\underset{\sim}{A} \in \mathbf{Nr}_\alpha\{\underset{\sim}{B}\}$. □

PROBLEM 1 Let $\beta>\alpha>1$. Is $\mathbf{Uf}\,\mathbf{Nr}_\alpha\, CA_\beta \not\subseteq \mathbf{Nr}_\alpha\, CA_\beta$ true?

PROPOSITION 6

(i) Let $\beta \geq \alpha$ and $K \subseteq CA_\beta$. Then
$\mathbf{H}\,\mathbf{Nr}_\alpha\,\mathbf{HS}\, K = \mathbf{Nr}_\alpha\,\mathbf{HS}\, K$, $\mathbf{Up}\,\mathbf{Nr}_\alpha\,\mathbf{S}\,\mathbf{Up}\, K = \mathbf{Nr}_\alpha\,\mathbf{S}\,\mathbf{Up}\, K$
$\mathbf{P}\,\mathbf{Nr}_\alpha\, K = \mathbf{Nr}_\alpha\,\mathbf{P}\, K$.

(ii) There are $\beta>\alpha$ and $K \subseteq Gs_\beta$ such that
$\mathbf{H}\,\mathbf{Nr}_\alpha\, K \not\subseteq \mathbf{Nr}_\alpha\,\mathbf{H}\, K$, $\mathbf{H}\,\mathbf{Nr}_\alpha\,\mathbf{S}\, K \not\supseteq \mathbf{Nr}_\alpha\,\mathbf{H}\, K$,
$\mathbf{Up}\,\mathbf{Nr}_\alpha\,\mathbf{S}\, K \not\supseteq \mathbf{Nr}_\alpha\,\mathbf{Up}\, K$, $\mathbf{Up}\,\mathbf{Nr}_\alpha\, K \not\subseteq \mathbf{Nr}_\alpha\,\mathbf{Up}\,\mathbf{S}\, K$.

Outline of Proof: (i) goes by carefully checking the definitions, see Andréka-Németi [10]. To see (ii), observe that $\mathbf{H}\,\mathbf{Nr}_0\, Ws_\omega \not\supseteq \mathbf{Nr}_0\,\mathbf{H}\, Ws_\omega$, $\mathbf{Up}\,\mathbf{Nr}_0\, Ws_\omega \not\supseteq \mathbf{Nr}_0\,\mathbf{Up}\, Ws_\omega$. Let $0<\alpha<\omega\leq\beta$. Let $K \overset{d}{=} \{\underset{\sim}{A} \in Ws_\beta : |\{x \in A : \Delta x \subseteq \alpha\}|<\omega\}$. Then $K = \mathbf{S}\, K$ and $\mathbf{Up}\,\mathbf{Nr}_\alpha\, K \not\subseteq \mathbf{Nr}_\alpha\,\mathbf{Up}\,\mathbf{S}\, K$ because $(\exists \underset{\sim}{A} \in \mathbf{Up}\,\mathbf{Nr}_\alpha\, K)[|Zd\, \underset{\sim}{A}|=2$ and $|A|\geq\omega]$, and by $\beta\geq\alpha$ there is no such member of $\mathbf{Nr}_\alpha$ $\mathbf{Up}$ $\mathbf{S}$ K . Our counterexample for $\mathbf{H}$ $\mathbf{Nr}_\alpha$ K $\not\subseteq$ $\mathbf{Nr}_\alpha$ $\mathbf{H}$ K is too complicated to describe here. □

COROLLARY 7 Let $0<\alpha<\beta$. Then

$\mathbf{Nr}_\alpha$ $\mathbf{I}\,Gs_\beta = \mathbf{HUpP}$ $\mathbf{Nr}_\alpha$ Gs_β and $\mathbf{Nr}_\alpha$ $CA_\beta = \mathbf{HUpP}$ $\mathbf{Nr}_\alpha$ CA_β . □

More universal algebraic results on $\mathbf{Nr}$ are in [7] which are closer to the spirit of [13] than the present CA-oriented results. Propositions 3-6 above solve Problems 2.3 and 2.11 of [16]. About Problem 2.2 of [16] we note that $\mathbf{Uf}$ $Cr_\alpha \neq \mathbf{S}$ Cr_α if $\alpha\geq 3$. Proposition 8 below improves 2.6.9 of [16].

PROPOSITION 8 $\mathbf{Rd}_\alpha$ was introduced in Definition 3.

(i) Let $\beta\geq\alpha\geq\omega$. Then
$CA_\alpha = \mathbf{Uf}$ $\mathbf{Rd}_\alpha$ CA_β and $\mathbf{I}\,Gs_\alpha = \mathbf{Uf}$ $\mathbf{Rd}_\alpha$ Gs_β .

(ii) Let $\beta\geq\alpha>1$. Then $\mathbf{I}\,Crs_\alpha = \mathbf{Rd}_\alpha$ $\mathbf{I}\,Crs_\beta$ and
$Ws_\alpha \subseteq \mathbf{SRd}_\alpha$ $\mathbf{I}\,Ws_\beta$ but $Ws_\omega \not\subseteq \mathbf{Uf}$ $\mathbf{Rd}_\omega$ $Ws_{\omega+2}$.

Outline of Poof: (i) is a corollary of Proposition 9 and Theorem 10 below. To prove (ii), let $\underset{\sim}{A}$ be a Crs_α with unit V . Let $q : \beta\sim\alpha \to \{V\}$. Let $f = \langle \{p\cup q : p\in X\} : X\in A \rangle$. Then there is a Crs_β $\underset{\sim}{G}$ with unit $W \overset{d}{=} f(V)$ such that $f : \underset{\sim}{A} \rightarrowtail\!\!\!\twoheadrightarrow \langle G,\cap,{}_W\!\sim ,C^W_i,D^W_{ij}\rangle_{i,j\in\alpha}$ is an isomorphism. This proves $Crs_\alpha \subseteq \mathbf{Rd}_\alpha$ $\mathbf{I}\,Crs_\beta$. The proofs of $Ws_\alpha \subseteq \mathbf{S}$ $\mathbf{Rd}_\alpha$ $\mathbf{I}\,Ws_\beta$ and $Gs_\alpha \subseteq \mathbf{S}$ $\mathbf{Rd}_\alpha$ $\mathbf{I}\,Gs_\beta$ are similar. To see $\mathbf{I}\,Crs_\alpha \supseteq \mathbf{Rd}_\alpha$ Crs_β let $\underset{\sim}{A}\in Crs_\beta$ be of unit V . For every $X\in A$ we define $f(X) \overset{d}{=}$
$\overset{d}{=} \{\langle\langle q_i,(\beta\sim\alpha)\upharpoonright q \rangle : i\in\alpha \rangle : q\in X\}$. Let $W \overset{d}{=} f(V)$. Now
$f : \langle A,\cap,{}_V\!\sim',C^V_i,D^V_{ij} \rangle_{i,j\in\alpha} \rightarrowtail \langle Sb\,W,\cap,{}_W\!\sim ,C^W_i,D^W_{ij} \rangle_{i,j\in\alpha} \in Crs_\alpha$

is a one-one homomorphism if $\alpha>1$. This proves $\mathbf{Rd}_\alpha \mathrm{Crs}_\beta \subseteq \mathbf{I}\,\mathrm{Crs}_\alpha$. It is easy to see that if $\underset{\sim}{A}\in \mathrm{Gs}_\beta$ then W is a Gs_α unit proving $\mathbf{Rd}_\alpha \mathrm{Gs}_\beta \subseteq \mathbf{I}\,\mathrm{Gs}_\alpha$. To see $\mathrm{Ws}_\omega \not\subseteq \mathbf{Uf}\ \mathbf{Rd}_\omega \mathrm{Ws}_{\omega+2}$, let $\underset{\sim}{A}\in \mathrm{Ws}_\omega$ and $\underset{\sim}{B} \in \mathbf{Rd}_\omega\{\underset{\sim}{C}\}$, $\underset{\sim}{C}\in \mathrm{Ws}_{\omega+2}$, and assume $\underset{\sim}{A} \models (d_{01}\neq 1 = d_{01}+d_{12}+d_{02})$. If $\underset{\sim}{B} \in \mathbf{Up}\,\{\underset{\sim}{A}\}$ then $\omega>|\mathrm{Zd}\ \underset{\sim}{B}|>3>|\mathrm{Zd}\ \underset{\sim}{A}|$. Now, $\underset{\sim}{A} \notin \mathbf{Uf}\ \mathbf{Rd}_\omega \mathrm{Ws}_{\omega+2}$ follows from observing that $(\forall \underset{\sim}{N}\in \mathbf{Up}\{\underset{\sim}{A}\})\ [\,|\mathrm{Zd}\ \underset{\sim}{N}|$ is either 2 or is 2^ω, hence $\underset{\sim}{N} \neq \underset{\sim}{B}]$. □

DEFINITION 4 ([6]) For any class Ax of formulas of similarity type t we define $\mathrm{Mod}\ Ax \overset{d}{=} \{\underset{\sim}{A}\in \mathrm{Alg}(t) : \underset{\sim}{A} \models Ax\}$. A *type scheme* is a triple $t = \langle T,\delta,\nu \rangle$ such that $\delta,\nu\in {}^T\omega$. For a type scheme $t = \langle T,\delta,\nu \rangle$ and for any set α we define

$$t_\alpha \overset{d}{=} \{\langle f_{i_1\ldots i_n},\ \nu(f)\rangle : f\in T,\ \ n=\delta(f)\ \text{and}\ i_1,\ldots,i_n\in\alpha\}.$$

Note that t_α is a similarity type. Let Eq be a set of equations of type t_α and let $\xi : \alpha \to \beta$. Then $\xi\mathrm{Eq}$ is a set of equations of type t_β obtained from Eq by replacing every occurrence of $f_{i_1\ldots i_n}$ by $f_{\xi(i_1)\ldots \xi(i_n)}$ for all $f\in T$ and $n=\delta(f)$.

Let $(\forall\alpha\in \mathrm{Ord})\ K_\alpha \subseteq \mathrm{Alg}(t_\alpha)$. Then $\langle K_\alpha : \alpha\in \mathrm{Ord}\rangle$ is said to be a *system of varieties definable by schemes* of equations iff there is a set Eq of equations of type t_ω such that $(\forall\alpha\in \mathrm{Ord})\ K_\alpha = \mathrm{Mod} \cup\{\xi\mathrm{Eq} : \xi\in {}^\omega\alpha \text{ is one-one}\}$. □

Systems of varieties definable by schemes were shown to be rather important for the theme of [13] and also for universal algebraic logic in [6] and [2]. The notion of systems of varieties definable by schemes of equations in the sense of Def.4 above was introduced first in Monk [19] last 8 lines of page 339.

PROPOSITION 9 $\langle CA_\alpha \rangle_{\alpha\in Ord}$, $\langle \mathsf{I}Gs_\alpha \rangle_{\alpha\in Ord}$, $\langle \mathsf{I}Crs_\alpha \rangle_{\alpha\in Ord}$, $\langle CA_\alpha \cap \mathsf{I}Crs_\alpha \rangle_{\alpha\in Ord}$ are systems of varieties definable by schemes of equations.

Outline of proof: By using the results proved so far it is enough to prove that $\mathsf{I}Crs_\alpha$ is a variety. This was proved by ultrapower construction in Németi [21], using the following idea: Let $\underset{\sim}{A}$ be a Crs_α with unit V. Then $V \subseteq {}^\alpha U$ for some set U. All the elements of A are α -ary relations on U and V is the greatest among them. Consider the (possibly infinitary) relational structure $\langle U, R : R\in A \rangle$. We can consider the ultrapowers of this structure but if $\alpha\geq\omega$ then we have to modify the definition. A partial Łos lemma is needed in the sense of Chap IV. of [4] pp.36-39. (See the proof of Thm 6.5 in [4].)

□

Let t be a type-scheme as in Def.4 above. Let $\beta \supseteq \alpha$ be two sets. Then $t_\alpha \subseteq t_\beta$ by definition. Let $K \subseteq \subseteq Alg(t_\beta)$. Then $\mathbf{Rd}_\alpha K \subseteq Alg(t_\alpha)$ is defined to be the class of t_α -type reducts of members of K in the usual sense, see Def.3 above.

THEOREM 10 Let $\langle V_\alpha \rangle_{\alpha\in Ord}$ be a system of varieties definable by schemes of equations. Let $\beta\geq\alpha\geq\omega$. Then

$$V_\alpha = \mathbf{Uf}\ \mathbf{Rd}_\alpha\ V_\beta .$$

Proof: Let $\langle V_\alpha \rangle_{\alpha\in Ord}$ be defined by the scheme Eq of equations. Recall that Eq is of type t_ω. For every $\alpha\in Ord$ let $Eq^\alpha \overset{d}{=} \cup\{\xi Eq : \xi\in {}^\alpha\omega \text{ is one-one}\}$. Then $V_\alpha = Mod\ Eq^\alpha$. Let $\omega\leq\alpha\leq\beta$. Then $Eq^\alpha \subseteq Eq^\beta$ and therefore $\mathbf{Rd}_\alpha V_\beta \models Eq^\alpha$. Hence $V_\alpha \supseteq \mathbf{Rd}_\alpha V_\beta$. $\mathbf{Uf}\ V_\alpha = V_\alpha$ since V_α is a variety and $\mathbf{Uf}\ K = \{\underset{\sim}{A}\in Alg(t_\alpha) : \text{some ultrapower of } \underset{\sim}{A} \text{ is in } K\}$, for any $K \subseteq Alg(t_\alpha)$.

Therefore $V_\alpha = \mathbf{Uf}\ V_\alpha \supseteq \mathbf{Uf}\ \mathbf{Rd}_\alpha\ V_\beta$. To prove $V_\alpha \subseteq \mathbf{Uf}\ \mathbf{Rd}_\alpha\ V_\beta$, first we show $V_\alpha \models \mathrm{Th}(\mathbf{Rd}_\alpha\ V_\beta)$. Let φ be any first order formula of similarity type t_α such that $\mathbf{Rd}_\alpha\ V_\beta \models \varphi$. Then $V_\beta \models \varphi$, i.e. $Eq^\beta \models \varphi$ by $V_\beta = \mathrm{Mod}\ Eq^\beta$. Then $E \models \varphi$ for some finite $E \subseteq Eq^\beta$, by the compactness theorem. Let H be the smallest set such that $E \cup \{\varphi\}$ are formulas of type t_H . Let $\xi \in {}^\beta\beta$ be a permutation of β such that ξ is identity on $H \cap \alpha$ and ξ maps $H \sim \alpha$ into α . Such a ξ exists by $|H| < \omega \leq \alpha$. Then $\xi E \models \varphi$ since φ is a formula of type $t_{H\cap\alpha}$. Now by the definitions of H , Eq^β , Eq^α the facts $\xi : H \rightarrowtail \alpha$ and $E \subseteq Eq^\beta$ imply $\xi E \subseteq Eq^\alpha$. Now $V_\alpha \models \models \xi E \models \varphi$ shows $V_\alpha \models \varphi$. We have seen $V_\alpha \models \mathrm{Th}(\mathbf{Rd}_\alpha\ V_\beta)$. Then $V_\alpha = \mathrm{Mod}\ \mathrm{Th}\ V_\alpha \subseteq \mathrm{Mod}\ \mathrm{Th}\ \mathbf{Rd}_\alpha\ V_\beta = \mathbf{Uf}\ \mathbf{Up}\ \mathbf{Rd}_\alpha\ V_\beta = \mathbf{Uf}\ \mathbf{Rd}_\alpha\ V_\beta$ because $V_\beta = \mathbf{Up}\ V_\beta$ and it is known from universal algebra that $\mathbf{Up}\ \mathbf{Rd}_\alpha\ K = \mathbf{Rd}_\alpha\ \mathbf{Up}\ K$ for any $K \subseteq \mathrm{Alg}(t_\beta)$, see Thm 0.5.13 in [16]. □

In Corollary 11 below we use the notion $\mathbf{Rd}^\xi\ K$ for $K \subseteq \mathrm{Alg}(t_\beta)$ and $\xi : \alpha \rightarrowtail \beta$ introduced in [16] and [24] without recalling it. Here $\mathbf{Rd}^\xi\ K \subseteq \mathrm{Alg}(t_\alpha)$ if $\xi : \alpha \rightarrowtail \beta$.

COROLLARY 11 Let $\langle K_\alpha \rangle_{\alpha\in Ord}$ be such that $(\forall\alpha\in Ord)$ $[K_\alpha \subseteq \mathrm{Alg}(t_\alpha)$ is a variety$]$. Then (i) and (ii) below are equivalent.

(i) $\langle K_\alpha \rangle_{\alpha\in Ord}$ is definable by a scheme of equations.

(ii) For every infinite α,β and one-one $\xi : \alpha \rightarrowtail \beta$ we have $K_\alpha = \mathbf{Uf}\ \mathbf{Rd}^\xi\ K_\beta$. □

For more results on systems of varieties see [24], [6],[2]. Note that **Uf** cannot be omitted from Cor.11 above since e.g. $(\forall\alpha\in Ord)Gs_\alpha \neq \mathbf{Rd}_\alpha\ Gs_{\alpha+1}$, and

$CA_\alpha \neq \mathbf{Rd}_\alpha\ CA_{\alpha+1}$ if $\alpha>1$.
The above results improve Thm 2.6.9 of [6].

More representation theory

$LF_\alpha \stackrel{d}{=} \{\underset{\sim}{A}\in Alg(g_1^\alpha) : (\forall x\in A)\,|\Delta x|<\omega\}$.
$DC_\alpha \stackrel{d}{=} \{\underset{\sim}{A}\in Alg(g_1^\alpha) : (\forall x\in A)\,|\alpha\sim\Delta x|<\omega\}$.

An abstract characterization of $\mathbf{HSP}\ Lrg_\alpha$ is $\mathbf{HSP}\ Lrg_\alpha = \mathbf{I}\ Gs_\alpha = \mathbf{S}\ \mathbf{Nr}_\alpha\ CA_{\alpha+\omega}$. Thm.12 below is an abstract characterization of the class Lrg_α of meaning algebras of the language L_{12} .

THEOREM 12 Let $|\alpha|\geq\omega$. Then $\mathbf{I}\,Lrg_\alpha = CA_\alpha \cap LF_\alpha$.
The proof is not hard if we use [5] or [4]. □

Thm.13 below says that CA_α is only one among infinitely many systems of varieties which share all the nice representation properties of CA_α w.r.t. Gs_α and Lrg_α and which are definable by *finite* schemes just as CA_α is. Moreover, CA_α is neither the largest nor the smallest among these equivalently nice varieties.

THEOREM 13 Let Z be the set of integers. There is a system $\langle V_z : z\in Z \rangle$ of systems $V_z = \langle V_{z\alpha} \rangle_{\alpha\in Ord}$ of varieties definable by *finite* schemes of equations such that (i) and (ii) below hold.

(i) $(\forall m,z\in Z)(\forall\alpha\in Ord)([\alpha\geq\omega \text{ and } m<z] \Rightarrow V_{m\alpha} \subset V_{z\alpha})$ and $V_0 = \langle CA_\alpha \rangle_{\alpha\in Ord}$. That is for all $\alpha\geq\omega$.

$$.. \subset V_{-2\alpha} \subset V_{-1\alpha} \subset CA_\alpha \subset V_{1\alpha} \subset V_{2\alpha} \subset V_{3\alpha} \subset .. .$$

(ii) Let $\alpha\geq\omega$ and $z\in Z$. Then $\mathbf{I}\,Lrg_\alpha = V_{z\alpha} \cap LF_\alpha$, $\mathbf{I}\,Gs_\alpha = \mathbf{HSP}\ (V_{z\alpha} \cap DC_\alpha) = \mathbf{S}\ \mathbf{Nr}_\alpha\ V_{z\alpha+\omega}$.

Proof: Let $V_0 \overset{d}{=} \langle CA_\alpha \rangle_{\alpha\in Ord}$. V_0 satisfies (ii) by Thm.12, Prop.2. and by Cor.3.14 of [4] p.25 which says $\mathbf{HSP}\, Lr_\alpha = \mathbf{HSP}\, (CA_\alpha \cap DC_\alpha) = \mathbf{S}\, \mathbf{Nr}_\alpha\, CA_{\alpha+\omega}$. Let $n\in\omega$ and assume that $V_{-n} = \langle V_{-n\alpha} \rangle_{\alpha\in Ord}$ has already been defined by a finite scheme E_{-n} of equations satisfying (i)-(ii). Then $Gs_\omega \subseteq V_{-n\omega}$. By [19], $\mathbf{I}\, Gs_\omega$ is not definable by any finite scheme of equations. Hence there is an equation e such that $Gs_\omega \models e$ and $E_{-n} \not\models e$. We define $E_{-n-1} \overset{d}{=} \{e\} \cup E_{-n}$. Let $V_{-n-1} = \langle V_{-n-1\alpha} \rangle_{\alpha\in Ord}$ be the system of varieties defined by E_{-n-1}. Then $Gs_\alpha \subseteq V_{-n-1\alpha} \subset V_{-n\alpha}$ and therefore V_{-n-1} satisfies (ii) since V_{-n} does so. By recursion we have defined $\langle V_z : z\in Z, z\leq 0 \rangle$ satisfying (i) and (ii).

Let Eq be the set of equations C_1-C_7 formulated for $\alpha=6$ on p.162 of [16]. Let $Eq' \overset{d}{=} Eq \cup \{B_0,\dots,B_2\}$ where B_i, $i\leq 3$ are defined on p.161 of [16]. Then Eq' is a finite set of equations of type g_1^6. Let $n\in\omega$. Let $B_3^n \overset{d}{=} \{c_1\dots c_n x + -c_1\dots c_n x = 1,\ c_1\dots c_n x \cdot -c_1\dots c_n x = 0\}$. Let $E_n \overset{d}{=} Eq' \cup B_3^n$. Let $V_n = \langle V_{n\alpha} \rangle_{\alpha\in Ord}$ be the system defined by the finite scheme E_n of equations. Then $V_0 = \langle CA_\alpha \rangle_{\alpha\in Ord}$ by Def.1.1.1 of [16]. Let $0\leq n<m\in\omega\leq\alpha$. Then it is easy to see that there is $\underset{\sim}{A}\in V_{m\alpha}$ and $x\in A$ such that $\alpha\sim\Delta x=\{1,\dots,n\}$ and $x+-x\neq 1$. Then $\underset{\sim}{A}\notin V_{n\alpha}$ since $\underset{\sim}{A} \not\models B_3^n$ by $x=c_1\dots c_n x$. (Actually, the Boolean reduct $\langle A,\cdot,+,- \rangle$ of $\underset{\sim}{A}$ is not a Boolean algebra but only a distributive lattice.) Thus $V_{n\alpha} \neq V_{m\alpha}$. Since $E_n \models E_m$ we have $V_{n\alpha} \subset V_{m\alpha}$. We proved that $\langle V_z : z\in Z\rangle$ satisfies (i). Let $n\in\omega\leq\alpha$. Let $\underset{\sim}{A} \in V_{n\alpha} \cap DC_\alpha$ and $x\in A$. Then there are $i_1,\dots,i_n \in \alpha\sim\Delta x$ such that $|\{i_1,\dots,i_n\}|=n$. Since $\underset{\sim}{A} \models \xi E_n$ for any $\xi : \omega \rightarrowtail \alpha$ and $c_{i_1}\dots c_{i_n} x=x$ we have $x+-x=1$ and $x\cdot -x=0$ in $\underset{\sim}{A}$. Then $\underset{\sim}{A} \models B_3^0$ and thus $\underset{\sim}{A} \in CA_\alpha \cap DC_\alpha$. We have seen $V_{n\alpha}\cap DC_\alpha \subseteq CA_\alpha \cap DC_\alpha$. By $CA_\alpha \subseteq V_{n\alpha}$ and $LF_\alpha \subseteq DC_\alpha$ this implies $V_{n\alpha}\cap DC_\alpha =$

$= CA_\alpha \cap DC_\alpha$ and $V_{n\alpha} \cap LF_\alpha = CA_\alpha \cap LF_\alpha$. The proof of $\mathbf{Nr}_\alpha V_{n\alpha+\omega} = \mathbf{Nr}_\alpha CA_{\alpha+\omega}$ is completely analogous to the above DC_α -argument. Therefore V_n satisfies (ii), since $V_0 = \langle CA_\alpha \rangle_{\alpha\in Ord}$ satisfies (ii). □

PROBLEM 2 Find a meaningful aspect in which CA_α is special or extremal among the finite-scheme definable varieties satisfying (ii) of Thm.13.

DEFINITION 5 $Lf_\alpha \stackrel{d}{=} CA_\alpha \cap LF_\alpha$ and $Dc_\alpha \stackrel{d}{=} CA_\alpha \cap DC_\alpha$. Let $\underset{\sim}{A} \in Alg(g_1^\alpha)$. Then $Zd\ \underset{\sim}{A} \stackrel{d}{=} \{x\in A : \Delta^A(x)=0\}$. □

Let $n\in\omega$. We define $\sigma_n \stackrel{d}{=} c_0 \ldots c_{n-1}\Pi\{-d_{ij} : i<j<n\}$. $\tau_n \stackrel{d}{=} \sigma_n \cdot -\sigma_{n+1}$.
$Ax \stackrel{d}{=} \{(c_0 - d_{01} = 1 \vee x=0 \vee x=1),\ (\tau_n = 0 \vee \tau_n = 1) : n\in\omega\}$.

THEOREM 14 Let $\alpha \geq \omega$. Let E be the set of all equations valid in Lf_α. Then

(i) $\mathbf{SUp}\ Lr_\alpha = Mod(Ax \cup E) = \mathbf{SUp}\ Cs_\alpha$.

(ii) $\mathbf{SUp}\ Cs_\alpha = \mathbf{HSUp}\ Cs_\alpha = \mathbf{SUp}\ (Cs_\alpha \cap Lf_\alpha) =$
$=\{\underset{\sim}{A} \in \mathbf{I}\,Gs_\alpha : (\exists \underset{\sim}{B} \subseteq \underset{\sim}{A})|Zd\ \underset{\sim}{B}| \leq 2$ and $(d_{01}^A = 1 \Rightarrow |A| \leq 2)\}$.

Outline of proof: (ii) Let $K \stackrel{d}{=} \{\underset{\sim}{A} \in \mathbf{I}\,Gs_\alpha : (\exists \underset{\sim}{B} \subseteq \underset{\sim}{A})$ $|Zd\ \underset{\sim}{B}| \leq 2$ and $(d_{01}^A = 1 \Rightarrow |A| \leq 2)\}$. For every $n\in\omega$ let $\varphi_n \stackrel{d}{=} (\sigma_n = 1 \wedge \sigma_{n+1} = 0)$. Let ${}_\infty K \stackrel{d}{=} \{\underset{\sim}{A} \in K : (\forall n\in\omega)\sigma_n = 1\}$ and ${}_\infty Cs_\alpha \stackrel{d}{=} Cs_\alpha \cap {}_\infty K$. By [16] 2.4.61-63, it is not hard to see that $K = Mod\ Ax \cap \mathbf{I}\,Gs_\alpha = {}_\infty K \cup \bigcup\{Mod\{\varphi_n\} \cap \mathbf{I}\,Gs_\alpha : n\in\omega\}$. Then $\mathbf{HSUp}\ Cs_\alpha \subseteq K$ since $Cs_\alpha \subseteq K = \mathbf{HSUp}\ K$ by $Cs_\alpha \models$ $\models Ax$ and $\mathbf{I}\,Gs_\alpha = \mathbf{HSP}\ Gs_\alpha$. Let $\underset{\sim}{A} \in K$. If $\underset{\sim}{A} \in {}_\infty K$ then $\underset{\sim}{A} \in \mathbf{SUp}\ (Cs_\alpha \cap Lf_\alpha)$ since ${}_\infty K = \mathbf{HSP}\ {}_\infty K = \mathbf{I}\ {}_\infty Cs_\alpha$ by [17] and $\mathbf{I}\ {}_\infty Cs_\alpha = \mathbf{SUp}\ ({}_\infty Cs_\alpha \cap Lf_\alpha)$ can be proved by a proof similar to that of $\mathbf{I}\,Gs_\alpha = \mathbf{SUp}\ Lf_\alpha$ (using the fact that

$\langle \mathbf{I}\,{}_\infty Cs_\alpha \rangle_{\alpha\in Ord}$ is a system definable by schemes of equations. Suppose $\underset{\sim}{A} \in K \sim {}_\infty K$. Then $\underset{\sim}{A} \in Mod\{\varphi_n\} \cap \mathbf{I}\,Gs_\alpha$ for some $n\in\omega$. $\mathbf{I}\,Gs_\alpha = \mathbf{SUp}\ Lf_\alpha$ by [16] and by the properties of $\mathbf{Up}$ then $\underset{\sim}{A} \in \mathbf{SUp}\ (Mod\{\varphi_n\} \cap Lf_\alpha) \subseteq \mathbf{SUp}\ (Mod\ Ax \cap Lf_\alpha)$. It is proved in Andréka-Németi [9] that the class of all directed unions of $Cs_\alpha \cap Lf_\alpha$-s is just $K \cap Lf_\alpha = Mod\ Ax \cap Lf_\alpha$. Therefore $Mod\ Ax \cap Lf_\alpha \subseteq \mathbf{SUp}(Cs_\alpha \cap Lf_\alpha)$. Now $\mathbf{SUp}\ \mathbf{SUp} = \mathbf{SUp}$ finishes the proof of $K \subseteq \mathbf{SUp}(Cs_\alpha \cap Lf_\alpha)$. By $\mathbf{I}\,Gs_\alpha = \mathbf{HSP}\ Lf_\alpha$ we have $Mod\ Ax \cap \mathbf{I}\,Gs_\alpha = Mod\ (Ax \cup E)$. This proves (ii). We omit the proof of (i). It can be found in Németi [22]. □

Different notions of regularity and i-finiteness

Now we return to the investigation of the different notions of regularity and i-finiteness. See also Def.1, Def.1.1, Lemma 1.1 and Remark 1.1 in this paper.

DEFINITION 6 Let α and $x \subseteq V \subseteq {}^\alpha U$ be arbitrary sets. We shall use H-independence introduced in Def.1.1.

$\Delta^V(x) \overset{d}{=} \{i\in\alpha : c_i^V x \neq x\}$.

x is defined to be *regular* in V iff
x is $\Delta^V(x)\cup\{0\}$-independent in V.

x is defined to be *cregular* in V iff
$(\forall i\in\alpha)x$ is $\Delta^V(x)\cup\{i\}$-independent in V.

x is defined to be *oregular* in V iff
x is $\Delta^V(x)$-independent in V.

Let $\underset{\sim}{A}\in Crs_\alpha$. Then $\underset{\sim}{A}$ is said to be regular (cregular, oregular) if $(\forall x\in A)x$ is regular(cregular, oregular).

Let $K \subseteq Crs_\alpha$. Then

$K^{reg} \overset{d}{=} \{\underset{\sim}{A}\in K : \underset{\sim}{A}$ is regular$\}$,

$K^{creg} \overset{d}{=} \{\underset{\sim}{A} \in K : \underset{\sim}{A}$ is cregular$\}$,

$K^{oreg} \overset{d}{=} \{\underset{\sim}{A} \in K : \underset{\sim}{A}$ is oregular$\}$. $\square$

The notion of cregularity seems to be more natural than that of regularity.

About the relationship of the notions introduced in Def.6 see Figure 1. In the figure, if K is in a double box like $\boxed{\boxed{K}}$ then $K = \mathbf{HSP}\ K$. If K is in a plain box like $\boxed{K}$ then $K = \mathbf{SPUp}\ K \neq \mathbf{H}\ K$. All the inclusions are proper, the inclusions not indicated do not hold, i.e. $Cs_\alpha \not\subseteq \mathbf{I}\,Crs_\alpha^{oreg}$. $\overset{?}{---}$ means that it is an open problem whether or not the inclusion holds. Thick arrows like $K \overset{\mathbf{SP}}{\longrightarrow} L$ indicate that the $\mathbf{SP}$ -closure of K is L, i.e. $\mathbf{SP}\ K = L$. Part of the statements indicated on the figure will be proved (or sketched) below. The remaining statements (e.g. $\mathbf{I}\,Gs_\alpha = \mathbf{SP}\ Ws_\alpha$ and $Ws_\alpha \subseteq \mathbf{I}\,Cs_\alpha^{reg}$) are proved in [17]. The direct proof of $\mathbf{SP}\ Ws_\alpha = \mathbf{I}\,Gs_\alpha$ uses Prop. 23 quoted below from [17].

PROPOSITION 15

(i) Let $|\alpha| \geq 4$. Then

$$\mathbf{I}\,Crs_\alpha \supset \mathbf{I}\,Crs_\alpha^{reg} \supset \mathbf{I}\,Crs_\alpha^{creg} \supset \mathbf{I}\,Crs_\alpha^{oreg} ,$$

$$\mathbf{I}\,Gs_\alpha = \mathbf{I}\,Gs_\alpha^{reg} = \mathbf{I}\,Gs_\alpha^{creg} \supset \mathbf{I}\,Gs_\alpha^{oreg} ,$$

$$Gs_\alpha \supset Gs_\alpha^{reg} = Gs_\alpha^{creg} \supset Gs_\alpha^{oreg} ,$$

$$Cs_\alpha \supset Cs_\alpha^{reg} = Cs_\alpha^{creg} = Cs_\alpha^{oreg} ,$$

$$Ws_\alpha = Ws_\alpha^{reg} = Ws_\alpha^{creg} = Ws_\alpha^{oreg} .$$

(ii) $Crs_\alpha^{oreg} = \{\underset{\sim}{A} \in Crs_\alpha^{creg} : |Zd\ \underset{\sim}{A}| \leq 2\}$.

Outline of proof: $(\forall \underset{\sim}{A} \in Crs_\alpha^{oreg})\,|Zd\ \underset{\sim}{A}| \leq 2$ is obvious. Let $\underset{\sim}{A} \in Crs_\alpha^{creg}$ and $|Zd\ \underset{\sim}{A}| \leq 2$. Then every element of $Zd\ \underset{\sim}{A}$ is oregular. Let $z \in A \sim Zd\ \underset{\sim}{A}$, $i \in \Delta z$. By cregularity z is $\Delta z \cup \{i\} = \Delta z$ - independent, hence oregular. *(i)*:

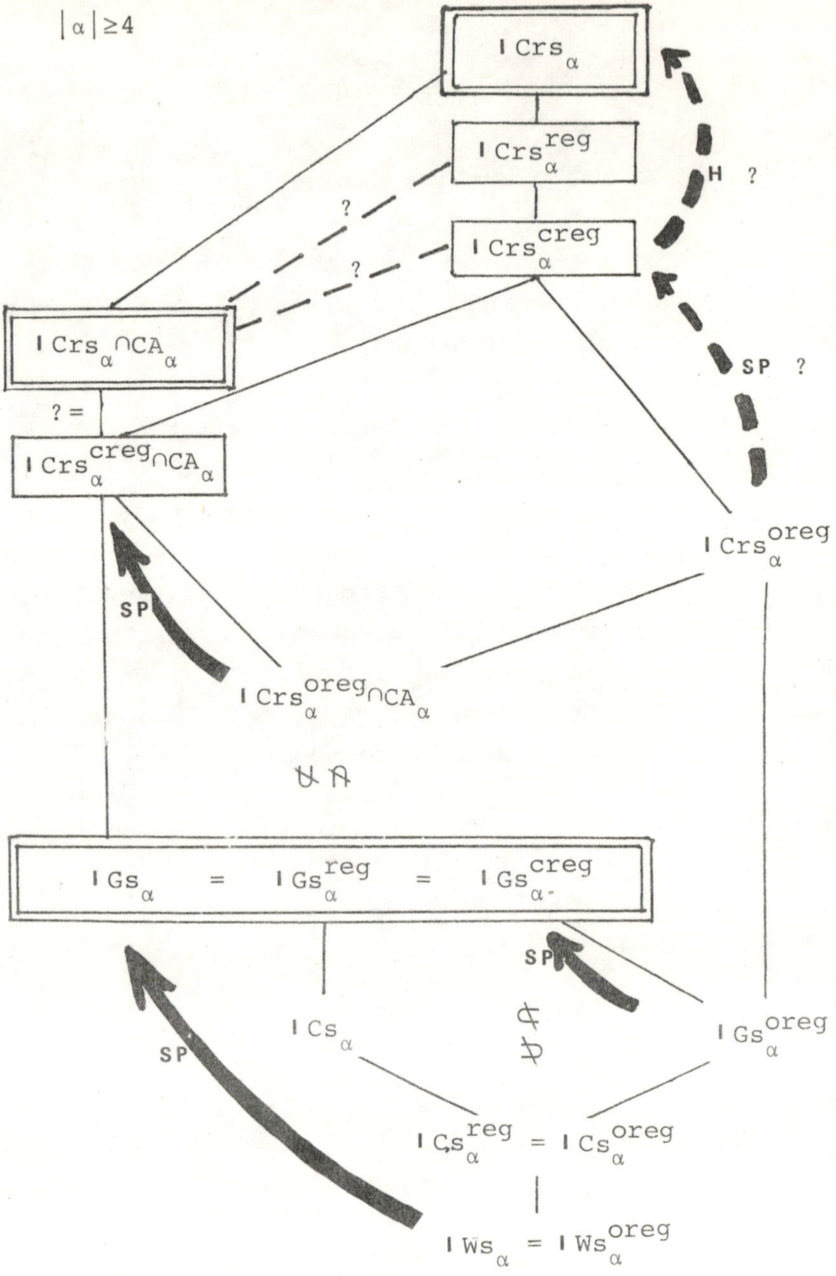
|α|≥4
I Crs_α
I Crs_α^reg
I Crs_α^creg
H ?
?
?
I Crs_α∩CA_α
SP ?
? =
I Crs_α^creg∩CA_α
I Crs_α^oreg
SP
I Crs_α^oreg∩CA_α
I Gs_α = I Gs_α^reg = I Gs_α^creg
SP
SP
I Cs_α
I Gs_α^oreg
I Cs_α^reg = I Cs_α^oreg
I Ws_α = I Ws_α^oreg

Figure 1

A proof of $\mathbf{I}\,Gs_\alpha = \mathbf{I}\,Gs_\alpha^{reg}$ can be found in [17]. The other positive statements are easy by checking the definitions. $Gs_\alpha^{creg} \not\subseteq \mathbf{I}\,Crs_\alpha^{oreg}$ by the last figure in [13] which is a Gs_2^{creg} and has more than 2 zerodimensional elements, hence $\notin \mathbf{I}\,Crs_2^{oreg}$ by (ii). To prove $Crs_\alpha^{reg} \not\subseteq \mathbf{I}\,Crs_\alpha^{creg}$ let $V =$

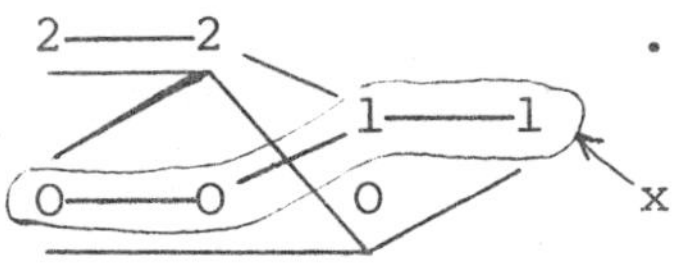

. Then $V \subseteq {}^\alpha 3$ for $\alpha=4$ and x is regular but not cregular in V. Let $\underset{\sim}{A}$ be the full Crs_4 with unit V. Then one can check $\underset{\sim}{A} \in Crs_4^{reg}$. Let φ be the quasiequation

$[c_0c_1x=x\cdot d_{23} \to (d_{23}\cdot c_2c_0c_1c_2x)\leq x]$. It is not hard to see $Crs_4^{creg} \models \varphi$. But $\underset{\sim}{A} \not\models \varphi$ and hence $\underset{\sim}{A} \notin \mathbf{I}\,Crs_4^{creg}$. To prove $Crs_\alpha \not\subseteq \mathbf{I}\,Crs_\alpha^{reg}$ we use the same construction as above with the indices $0\in\alpha$ and $3\in\alpha$ interchanged both in $\underset{\sim}{A}$ and in φ. Then we obtain a quasiequation φ' valid in Crs_4^{reg} and an $\underset{\sim}{A}' \in Crs_4$ such that $\underset{\sim}{A}' \not\models \varphi'$. The proof for $|\alpha|\geq 4$ is practically the same. $Gs_\alpha \neq$ $\neq Gs_\alpha^{reg}$ and $Cs_\alpha \neq Cs_\alpha^{reg}$ are easy to see. □

PROBLEM 3 Is $\mathbf{I}\,Crs_\alpha \cap CA_\alpha = \mathbf{I}\,Crs_\alpha^{reg} \cap CA_\alpha$ or $\mathbf{I}\,Crs_\alpha^{reg} \cap CA_\alpha = \mathbf{I}\,Crs_\alpha^{creg} \cap CA_\alpha$ true?

PROPOSITION 16

(i) $Lrg_\alpha = Gs_\alpha^{reg} \cap LF_\alpha \neq Gs_\alpha^{oreg} \cap LF_\alpha$ if $\alpha\neq 0$ and

$Lr_\alpha = Cs_\alpha^{reg} \cap LF_\alpha$.

(ii) $Gs_\alpha^{oreg} \cap LF_\alpha \subseteq \mathbf{I}\,Cs_\alpha$ but if $\alpha\geq\omega$ then

$Gs_\alpha^{oreg} \not\subseteq \mathbf{I}\,Cs_\alpha$.

Outline of proof: (i) The positive statements are easy by checking the definitions and by Lemma 1.1. $Lrg_\alpha \neq$ $\neq Gs_\alpha^{oreg} \cap LF_\alpha$ by Prop.15(ii) since an Lrg_α may have more than 2 zerodimensional elements. *(ii)* To see

$Gs_\alpha^{oreg} \cap LF_\alpha \subseteq \mathsf{I}Cs_\alpha$, let $\underset{\sim}{A} \in Gs_\alpha^{oreg} \cap LF_\alpha$. Then $\underset{\sim}{A}$ is simple by $|Zd\ \underset{\sim}{A}| \le 2$ and therefore $\underset{\sim}{A} \in \mathsf{I}Cs_\alpha$ by Prop.17 below. Let $\alpha \ge \omega$, $V \overset{d}{=} {}^\alpha\{0,1\} \cup {}^\alpha\{2,3\}$ and $X \overset{d}{=} \{{}^\alpha\{i\} : : i<4\}$. Let $\underset{\sim}{A}$ be the Crs_α with unit V an generated by the set X of elements. Then $\underset{\sim}{A} \in Gs_\alpha^{oreg}$ follows from results in [9]. Since $\{y \in A : (\forall i,j \in \alpha) y < d_{ij}$ and y is an atom of $\underset{\sim}{A}\}$ has more than 3 elements, $\underset{\sim}{A} \notin \mathsf{I}Cs_\alpha$. The positive statement is not hard. □

In the following, we shall need the notion of relativization. It is a tool which can be used to show $K = \mathbf{P}\ K$ for various $K \subseteq Crs_\alpha$.

DEFINITION 7 Let W be a set. Then

$rl_W \overset{d}{=} \langle x \cap W : x \in Sets \rangle$. $rl_W^A \overset{d}{=} A \upharpoonright rl_W$.

We shall often omit the superscript A . Let $\underset{\sim}{A} \in Crs_\alpha$.

Then $Rl_W A \overset{d}{=} \{x \cap W : x \in A\} = Rng\ rl_W^A$ and

$\underset{\sim}{Rl}_W A \overset{d}{=} \langle Rl_W A , \cap , {}_W\!\sim , C_i^W , D_{ij}^W \rangle_{i,j \in \alpha}$.

Notation: $rl^A(W) \overset{d}{=} rl(W) \overset{d}{=} rl_W^A$ and $\underset{\sim}{Rl}(W)A \overset{d}{=} \underset{\sim}{Rl}_W A$.

Note that $\underset{\sim}{Rl}_W A$ is always a partial algebra but sometime $\underset{\sim}{Rl}_W A$ is *not an algebra* since C_i^W may be partial on $Rl_W A$. I.e. sometime $\underset{\sim}{Rl}_W A \notin Alg(g_1^\alpha)$ because it may happen that $Rng(Rl_W A \upharpoonright C_i^W) \nsubseteq Rl_W A$ for some $i \in \alpha$. □

Notation: $h : \underset{\sim}{A} \to \underset{\sim}{B}$ means that h is a homomorphism between the algebras $\underset{\sim}{A}$ and $\underset{\sim}{B}$. $h : \underset{\sim}{A} \twoheadrightarrow \underset{\sim}{B}$ means that h is a *surjective* homomorphism, i.e. $Rng\ h = B$ and $h : \underset{\sim}{A} \rightarrowtail \underset{\sim}{B}$ means that h is an *injective* homomorphism, i.e. $(\forall a,b \in A)[a \ne b \Rightarrow h(a) \ne h(b)]$.

PROPOSITION 17 Let $\underset{\sim}{A} \in Crs_\alpha$ with unit V. Let $Z \subseteq V$.

(i) If $\Delta^V(Z)=0$ then $rl_Z : \underset{\sim}{A} \twoheadrightarrow \underset{\sim}{Rl}_Z A \in Crs_\alpha$.

I.e. $\underset{\sim}{Rl}_Z A$ is a Crs_α and rl_Z is a homomorphism from $\underset{\sim}{A}$ onto $\underset{\sim}{Rl}_Z A$.

(ii) Assume $Z \in A$. Then $\Delta^V(Z)=0$ iff

$rl_Z : \underset{\sim}{A} \rightarrow \underset{\sim}{B} \in Alg(g_1^\alpha)$ for some $\underset{\sim}{B}$ iff

$rl_Z : \underset{\sim}{A} \twoheadrightarrow \underset{\sim}{Rl}_Z A \in Crs_\alpha$.

Proof: Immediate by Thm.1 and Prop.2(ii)2 of [23]. □

The following Prop.18 implies that the conditions (i) rl_W is a homomorphism, (ii) rl_W is a homomorphism into a Crs_α and (iii) $\underset{\sim}{Rl}_W A \in Crs_\alpha$ are independent to the extent that (i) implies neither (ii) nor (iii) and (i)& (iii) does not imply (ii).

Notation: Let $\underset{\sim}{A} \in Crs_\alpha$. Then $base(A) \stackrel{d}{=} \cup\{Rng\ f : f \in \cup A\}$.

PROPOSITION 18 Let $|\alpha| \geq 2$.

(i) There are an $\underset{\sim}{A} \in Lr_\alpha$ and $U \subseteq base(A)$ such that $rl({}^\alpha U)$ is an isomorphism, i.e.
$rl({}^\alpha U) : \underset{\sim}{A} \rightarrowtail \underset{\sim}{B} \in Alg(g_1^\alpha)$ for some $\underset{\sim}{B}$, but
$\underset{\sim}{Rl}({}^\alpha U)A \notin Alg(g_1^\alpha)$ and
$rl({}^\alpha U) : \underset{\sim}{A} \rightarrow \underset{\sim}{B}$ implies $\underset{\sim}{B} \notin Crs_\alpha$.

(ii) There are $\underset{\sim}{A} \in Crs_\alpha$ and $U \subseteq base(A)$ such that $\underset{\sim}{Rl}({}^\alpha U)A \in Crs_\alpha$ and $rl({}^\alpha U)$ is an isomorphism but $rl({}^\alpha U)$ is not a homomorphism between $\underset{\sim}{A}$ and $\underset{\sim}{Rl}({}^\alpha U)A$. Moreover, $rl({}^\alpha U) : \underset{\sim}{A} \rightarrowtail \underset{\sim}{B} \in Alg(g_1^\alpha)$ for some $\underset{\sim}{B}$ but $rl({}^\alpha U) : \underset{\sim}{A} \rightarrow \underset{\sim}{B}$ implies $\underset{\sim}{B} \notin Crs_\alpha$.

Proof: Let $|\alpha| \geq 2$. *(i)* Z denotes the set of integers. $x \stackrel{d}{=} \{q \in {}^\alpha Z : q_1 = q_0 + 1\}$. $V \stackrel{d}{=} {}^\alpha Z$. Let $\underset{\sim}{B}$ be the full Cs_α

with unit V. Let $\underset{\sim}{A}$ be the subalgebra of $\underset{\sim}{B}$ generated by the element $x \in B$. Then $\underset{\sim}{A} \in Lr_\alpha$ by Prop.1 (i). Let $U \stackrel{d}{=} \omega$. Then $U \subseteq Z$. Let $W \stackrel{d}{=} {}^\alpha U$. Let $h \stackrel{d}{=} rl_W^A$.

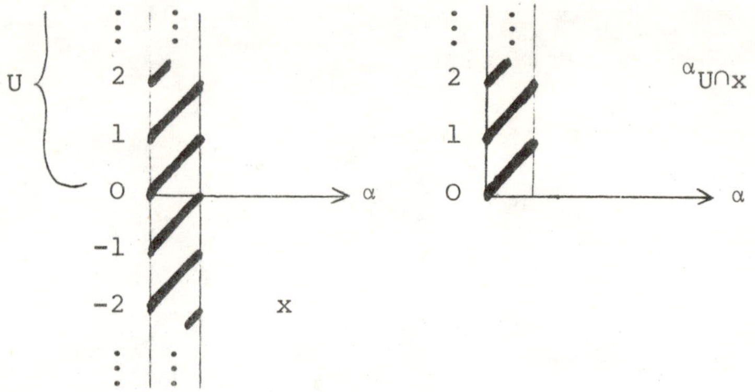

We shall prove that $h : \underset{\sim}{A} \rightarrowtail \underset{\sim}{D}$ for some algebra $\underset{\sim}{D}$ and that $(\forall \underset{\sim}{D} \in Alg(g_1^\alpha))[h : \underset{\sim}{A} \to \underset{\sim}{D} \Rightarrow \underset{\sim}{D} \notin Crs_\alpha]$ and $\underset{\sim}{Rl}_W A \notin Crs_\alpha$.

CLAIM 1 $\quad (\forall a \in A)[a \neq 0 \Rightarrow h(a) \neq 0]$.

Proof: Let $y \in A$, $y \neq 0$. Let $q \in y$. Then $(\exists p \in y)(\forall i \in \alpha \sim \Delta y)$ $p_i > 0$ by Lemma 1.1.b since y is i-finite. $(\exists b \in Z)$ $(\forall i \in \Delta y) p_i > b$ since $|\Delta y| < \omega$. There is $z \in Z$, $z > 0$ such that $b+z > 0$. Let $f \stackrel{d}{=} \langle w+z : w \in Z \rangle$. Let $\tilde{f} \stackrel{d}{=} \langle \{f \circ q : q \in a\} : a \in A \rangle$. Then $V \cap \tilde{f} y \neq 0$ by $f \circ p \in {}^\alpha\omega$. By [17], $\tilde{f}$ is an automorphism of $\underset{\sim}{B}$. Then $\tilde{f}$ is identity on A, since $\tilde{f}x = x$ by $(\forall q \in {}^\alpha Z)[q_1 = q_0 + 1 \iff (f \circ q)_1 = (f \circ q)_0 + 1]$. Thus $V \cap \tilde{f} y = V \cap y \neq 0$. *QED of Claim 1.*

CLAIM 2 $\quad \underset{\sim}{Rl}_V A \notin Alg(g_1^\alpha)$.

Proof: Let $y \stackrel{d}{=} V \sim C_1^V(x \cap V)$. Then $Y = \{q \in {}^\alpha\omega : q_1 = 0\}$. Assume $(\exists y \in A) Y = y \cap V$. Let $\tilde{f}$ be as in the proof of Claim 1. Recall that $f(w) = w+z$ where $0 < z \in \omega$. Then $\langle z : i \in \alpha \rangle \in Y$, since $\langle 0 : i \in \alpha \rangle \in Y$ and $(\forall q \in Y) f \circ q \in Y$ by $\tilde{f}y = y$ and $\tilde{f}V \subseteq V$. A contradiction. Therefore $Y \notin Rl_V A$, and thus $C_1^V(x \cap V) \notin Rl_V A$. *QED of Claim 2.*
By Claim 1 and by Boolean algebra theory we have

$h : \underset{\sim}{A} \rightarrowtail \underset{\sim}{D}$ for some algebra $\underset{\sim}{D}$. By Claim 2 and by the definitions of $Rl_V A$, rl_V it is clear that $(\forall \underset{\sim}{D} \in Alg(g_1^{\alpha}))[h : \underset{\sim}{A} \to \underset{\sim}{D} \Rightarrow \underset{\sim}{D} \notin Crs_{\alpha}]$.

(ii) Let $i,j \in \alpha$. Then $\langle i,j,..\rangle$ denotes the element $p \in {}^{\alpha}\omega$ for which $p(0)=i$, $p(1)=j$ and $(\forall k \in \alpha)[k>1 \Rightarrow p(k)=0]$. Let $R \stackrel{d}{=} \{\langle 1,0,..\rangle , \langle 4,3,..\rangle\}$ and $S \stackrel{d}{=} \{\langle 1,2,..\rangle, \langle 4,5..\rangle\}$. $W \stackrel{d}{=} R \cup S$. Let $\underset{\sim}{A}$ be the Crs_{α} with unit W and generated by $\{R,S\}$. Then $A = \{0,R,S,W\}$. Let $U \stackrel{d}{=}$ $\stackrel{d}{=} \{1,2,3,4\}$. Let $V \stackrel{d}{=} {}^{\alpha}U$. Then $Rl_V A = \{0,\{\langle 4,3,..\rangle\}, \{\langle 1,2,..\rangle\}, W \cap V\}$. Clearly, rl_V is an isomorphism and $\underset{\sim}{Rl}_V A \in Crs_{\alpha}$. But rl_V is not a homomorphism between $\underset{\sim}{A}$ and $\underset{\sim}{Rl}_V A$ because $C_1^W R = W$ while $C_1^{(W \cap V)}(R \cap V) = R \cap V \neq$ $\neq W \cap V$. From the figure below it can be seen that actually, there is no homomorphism between $\underset{\sim}{A}$ and $\underset{\sim}{Rl}_V A$.

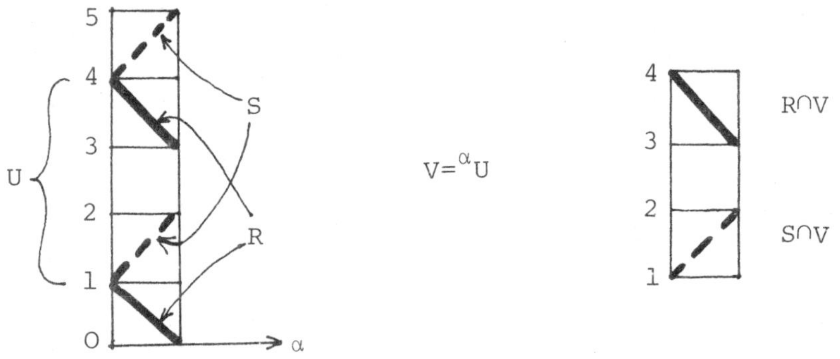

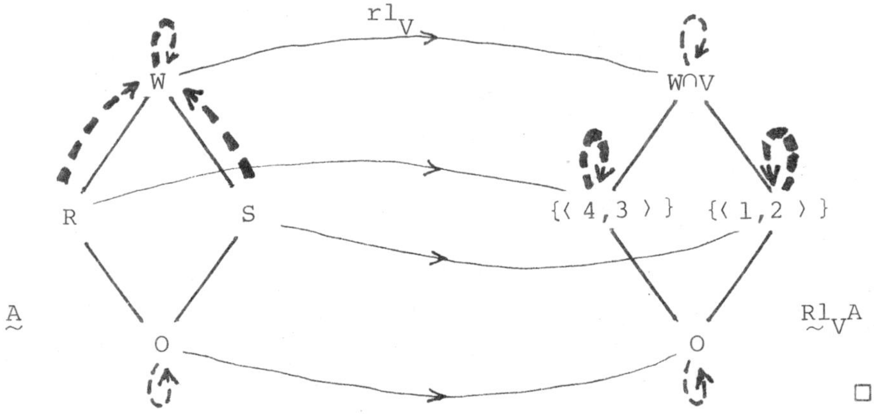

□

Prop.18(i) above says that there are $\underset{\sim}{A} \in Lr_\alpha$ and $U \subseteq base(A)$ such that $rl^A({}^\alpha U)$ is an isomorphism on $\underset{\sim}{A}$ but $\underset{\sim}{Rl}({}^\alpha U)A \notin Crs_\alpha$, and for every algebra $\underset{\sim}{B}$ if $rl({}^\alpha U) : \underset{\sim}{A} \to \underset{\sim}{B}$ then $\underset{\sim}{B} \notin Crs_\alpha$.

PROBLEM 4 Are there an $\underset{\sim}{A} \in Lr_\alpha$ and $U \subseteq base(A)$ such that $\underset{\sim}{Rl}({}^\alpha U)A \in Alg(g_1^\alpha)$ and $rl({}^\alpha U)$ is an isomorphism but $rl({}^\alpha U) : \underset{\sim}{A} \to \underset{\sim}{B}$ implies $\underset{\sim}{B} \notin Crs_\alpha$?

THEOREM 19 Let $|\alpha| \geq 2$. Then

(i) $\mathbf{I}\, Crs_\alpha^{creg} = \mathbf{SPUp}\ Crs_\alpha^{creg}$ and $\mathbf{I}\, Crs_\alpha^{reg} = \mathbf{SPUp}\ Crs_\alpha^{reg}$, i.e. $\mathbf{I}\, Crs_\alpha^{creg}$ and $\mathbf{I}\, Crs_\alpha^{reg}$ are quasiequational classes.

(ii) Let $|\alpha| \geq 3$. Then $\mathbf{H}\ Crs_\alpha^{oreg} \not\subseteq \mathbf{I}\, Crs_\alpha^{reg}$, hence $\mathbf{I}\, Crs_\alpha^{creg} \neq \mathbf{H}\ Crs_\alpha^{creg}$, $\mathbf{I}\, Crs_\alpha^{reg} \neq \mathbf{H}\ Crs_\alpha^{reg}$ and $\mathbf{I}\, Crs_\alpha^{oreg} \neq \mathbf{H}\ Crs_\alpha^{oreg}$.

Outline of proof: To prove (i), let $|\alpha| > 1$. By Cor.9 of [23] the category theoretical products exist in Crs_α, because of the followings. Let $\langle V_i : i \in I \rangle$ be such that $(\forall i,j \in I)[V_i \subseteq {}^\alpha Sets$ and $(i \neq j \Rightarrow V_i \cap V_j = 0)]$. Let $\langle \underset{\sim}{A}_i : i \in I \rangle \in {}^I Crs_\alpha$ be such that $(\forall i \in I) A_i = Sb\ V_i$. Let $W = \cup\{V_i : i \in I\}$. There is $\underset{\sim}{B} \in Crs_\alpha$ such that $B = Sb\ W$. By Prop.17 $rl(V_i) : \underset{\sim}{B} \twoheadrightarrow \underset{\sim}{A}_i$ for all $i \in I$ since $(\forall i \in I)\ \Delta^W(V_i) = 0$. By Cor.9 of [23] $\underset{\sim}{B}$ is a category theoretic product of $\langle \underset{\sim}{A}_i : i \in I \rangle$ with projections $\langle rl^B(V_i) : i \in I \rangle$. By [23] this category theoretic product coincides with the universal algebraic direct product up to isomorphisms. Clearly, if $x \in B$ and $(\forall i \in I) rl(V_i)(x)$ is cregular or regular in V_i then so is x in W. (This is not true for oregular, however.) Thus $\mathbf{SP}\ Crs_\alpha^{creg} = \mathbf{I}\, Crs_\alpha^{creg}$ and $\mathbf{SP}\ Crs_\alpha^{reg} = \mathbf{I}\, Crs_\alpha^{reg}$. It remains to see that Crs_α^{creg} and Crs_α^{reg} are closed under **Up** .

DEFINITION 19.1 Let $\underset{\sim}{A}\in Crs_\alpha$, $V \overset{d}{=} \cup A$, $U \overset{d}{=} base(A)$. Let $x\in A$. Then

x is defined to be *lregular* iff

$(\forall q\in V)(\forall f \in x\cap{}^\alpha U^{(q)})[\{0\}\cup\Delta x)\upharpoonright f \subseteq q \Rightarrow q\in x]$.

x is defined to be *lcregular* iff

$(\forall q\in V)(\forall f \in x\cap{}^\alpha U^{(q)})(\forall i\in\alpha)[(\{i\}\cup\Delta x)\upharpoonright f \subseteq q \Rightarrow q\in x]$.

Let $K \subseteq Crs_\alpha$. Then

$K^{lreg} \overset{d}{=} \{\underset{\sim}{A}\in K : (\forall x\in A)x$ is lregular in A$\}$

$K^{lcreg} \overset{d}{=} \{\underset{\sim}{A}\in K : (\forall x\in A)x$ is lcregular in A$\}$.

End of Definition 19.1

$\mathbf{I}\, Crs_\alpha^{lreg} = \mathbf{Up}\ Crs_\alpha^{lreg}$ and $\mathbf{I}\, Crs_\alpha^{lcreg} = \mathbf{Up}\ Crs_\alpha^{lcreg}$ are proved in [21] by using a result from [17]. $Crs_\alpha^{creg} \subseteq$ $\subseteq Crs_\alpha^{lcreg}$ and $Crs_\alpha^{reg} \subseteq Crs_\alpha^{lreg}$ are obvious. Let $\underset{\sim}{A}\in Crs_\alpha^{lreg}$. Let $V \overset{d}{=} \cup A$ and $q\in V$. Then $\Delta^V(V\cap{}^\alpha Sets^{(q)})=0$. Then by applying Cor.9 of [23] twice one can prove that $\underset{\sim}{A} \simeq \underset{\sim}{B} \in Crs_\alpha^{lcreg}$ for some $\underset{\sim}{B}$ having the following property: Let $W \overset{d}{=} \cup B$ and $Y \overset{d}{=} base(B)$. Then $(\forall p,q\in W)$ $[W\cap{}^\alpha Y^{(p)} \neq W\cap{}^\alpha Y^{(q)} \Rightarrow W\cap{}^\alpha Y^{(p)}\cap{}^\alpha Y^{(q)} = 0]$. Then $\underset{\sim}{B} \in$ $\in Crs_\alpha^{creg}$. Hence $Crs_\alpha^{lcreg} \subseteq \mathbf{I}\, Crs_\alpha^{creg}$. The same idea works to show $Crs_\alpha^{lreg} \subseteq \mathbf{I}\, Crs_\alpha^{creg}$. (Warning: There are $\underset{\sim}{A}\in Crs_\alpha^{lreg}$, $q\in V \overset{d}{=} \cup A$ such that $\underset{\sim}{Rl}_Z A \notin Crs_\alpha^{lreg}$ for $Z \overset{d}{=} V\cap{}^\alpha Sets^{(q)}$. But this does not affect the validity of the above claimed construction of $\underset{\sim}{B}\in Crs_\alpha^{lreg}$.)

By this, (i) is proved.

(ii) Let $\alpha>3$ be an ordinal and Z be the set of integers.

$T \overset{d}{=} \{q\in{}^\alpha Z : (\forall i\in\alpha\sim3)q_i=0\}$. Then $T \subseteq {}^\alpha Z$.

$Y \overset{d}{=} \{q\in T : 3\upharpoonright q\in\{\langle 0,n,n+1\rangle,\langle 0,n,n-1\rangle : n\in Z\}\}$.

$W \stackrel{d}{=} \{q \in T : 3 \upharpoonright q \in \{\langle 2n,2n,2n+1\rangle, \langle 2n,2n-1,2n+1\rangle, \langle 2n,2n-1,2n\rangle : n \in Z\}\}$,
$X \stackrel{d}{=} \{q \in Y : q_1 = 2n \text{ for some } n \in \omega\}$,
$V \stackrel{d}{=} Y \cup W$.
Let $\underset{\sim}{A}$ be the Crs_α with unit V and generated by the element X. First we show that $\underset{\sim}{A}$ is oregular. One can prove (*) $(\forall y \in A)[y = c_1 c_2 y \Rightarrow y \cap Y \in \{0,Y\}]$. Then (*) implies $|Zd\, \underset{\sim}{A}| = 2$.
Define $H(i,z) \stackrel{d}{=} \{q \in Y : q_i = z\}$ for all $i<3$ and $z \in Z$.

CLAIM Let $\{i,j,k\}=3$ and $z \in Z$. Let $f,g \in H(i,z)$, $\langle i,z\rangle \neq \langle 0,0\rangle$. Then $f \in c_j^V c_k^V c_j^V c_k^V \{g\}$.

The claim can be proved by distinguishing 6 cases of $\langle i,z\rangle$ as follows: $\langle i,\text{even}\rangle, \langle i,\text{odd}\rangle$ for $i<3$. *QED of Claim*

Let $g \in y \in A$ and $f \in V \sim y$. Then $\Delta y \neq 0$ by (*). By definition of V we have $\Delta y \subseteq 3$. Assume $\Delta y \upharpoonright f \subseteq g$. Let $H \stackrel{d}{=} \{i \in \alpha : f_i \neq g_i\}$. Then $H \subseteq 3 \sim \Delta y$. Clearly, $|H|>1$. Then $\Delta y = \{i\}$ for some $i<3$. Then $f,g \in H(i,z)$ for $z \stackrel{d}{=} g_i$. Let $\{j,k\} = 3 \sim \{i\}$. Then $c_j c_k y = y$. If $\langle i,z\rangle = \langle 0,0\rangle$ then by (*), otherwise by the Claim then $f \in y$, a contradiction! Then y is oregular. Thus $\underset{\sim}{A} \in Crs_\alpha^{oreg}$. Let $I \stackrel{d}{=} \{x \in A : |x| < \omega\}$. Then I is an ideal of $\underset{\sim}{A}$. Let $\varphi \stackrel{d}{=} (x \cdot d_{03} = c_1 c_2 x \rightarrow d_{03} \cdot c_0 c_2 c_1 c_0 x \leq x)$. It can be seen that $\underset{\sim}{A}/I \not\models \varphi$ while $Crs_\alpha^{reg} \models \varphi$. □

By Prop.15(i) and by Thm.19(i) we have **SP** $Crs_\alpha^{oreg} \subseteq$ $\subseteq$ **I** Crs_α^{creg} .

PROBLEM 5 Let $|\alpha| \geq 2$. Is then **I** Crs_α^{creg} = **SP** Crs_α^{oreg} true?

To motivate this problem, see Thm.20 below and Prop. 15(ii). Note that by [23], every subdirectly irreducible

Crs_α^{creg} is oregular. Thm.21 below lists some cases when the answer to Problem 5 is positive.

Let $V = \cup\{{}^\alpha U_i : i\in I\}$ be such that $(\forall i,j\in I)[i\neq j \Rightarrow \Rightarrow U_i \cap U_j = 0]$. Then $Subu(V) \stackrel{d}{=} \{{}^\alpha U_i : i\in I\}$. Let $\underset{\sim}{A}\in Gs_\alpha$. Then $Subu(A) \stackrel{d}{=} Subu(\cup A)$.

Clearly, $\langle rl_Y : Y\in Subu(A)\rangle$ is a subdirect decomposition of $\underset{\sim}{A}\in Gs_\alpha$ into Cs_α -s, i.e. $\underset{\sim}{Rl}_Y A\in Cs_\alpha$ for all $Y\in Subu(A)$. For a detailed proof of this fact see [17]. This decomposition is called the *natural subdirect decomposition* of a Gs_α into Cs_α -s.

THEOREM 20 Let $\alpha\geq\omega$. Then there is $\underset{\sim}{A}\in Gs_\alpha^{oreg}$ such that $(\forall Y\in Subu(A))\ \underset{\sim}{Rl}_Y A \notin \mathbf{I}\, Cs_\alpha^{reg}$. As a contrast, $(\forall \underset{\sim}{A}\in Lrg_\alpha)(\forall Y\in Subu(A))\ \underset{\sim}{Rl}_Y A \in Cs_\alpha^{oreg}$.

The proof can be found in Németi [22]. □

Thm.20 above says that the natural subdirect decomposition of a Gs_α into Cs_α -s can destroy regularity in every component while it cannot destroy i-finiteness.

$$B_\alpha \stackrel{d}{=} \{\underset{\sim}{A}\in Alg(g_1^\alpha) : (\forall m\in\omega)(\forall i \in {}^m\alpha)(\forall x\in A)(\exists n\in\omega)$$
$$(c_{i_0}\dots c_{i_{m-1}})^n x = (c_{i_0}\dots c_{i_{m-1}})^{n+1} x\},$$

where $(c_{i_0}\dots c_{i_m})^0 x \stackrel{d}{=} x$

and $(c_{i_0}\dots c_{i_m})^{n+1} x \stackrel{d}{=} c_{i_0}\dots c_{i_m}(c_{i_0}\dots c_{i_m})^n x$.

Note that $CA_\alpha \subseteq B_\alpha$.

THEOREM 21 Let $|\alpha|\geq 2$.

(i) Let $K = \mathbf{H}\, K \subseteq B_\alpha$. Then

$\mathbf{I}(Crs_\alpha^{creg} \cap K) \subseteq \mathbf{SP}(Crs_\alpha^{oreg} \cap K)$.

If $K = \mathbf{HSP}\, K \subseteq B_\alpha$ then

$\mathbf{I}(Crs_\alpha^{creg} \cap K) = \mathbf{SP}(Crs_\alpha^{oreg} \cap K)$.

(ii) $\mathbf{I}(Crs_\alpha^{creg} \cap CA_\alpha) = \mathbf{SP}(Crs_\alpha^{oreg} \cap CA_\alpha)$ and

$\mathbf{I}\, Gs_\alpha^{creg} = \mathbf{SP}\, Gs_\alpha^{oreg}$.

(iii) $\mathbf{I}\, Crs_\alpha^{reg} \cap B_\alpha \neq \mathbf{I}\, Crs_\alpha^{creg} \cap B_\alpha$, for $|\alpha| \geq 4$.

Proof: (i) Let $|\alpha| \geq 2$. Let $K = \mathbf{H}\, K \subseteq B_\alpha$. Let $\underset{\sim}{A} \in Crs_\alpha^{creg} \cap K$ be of unit V . Let Z be the set of atoms of $Zd\, Sb\, V$. Let $W \in Z$. By Prop.17 we have $rl_W : \underset{\sim}{A} \twoheadrightarrow \underset{\sim}{Rl}_W A \in Crs_\alpha$. Assume that $\underset{\sim}{Rl}_W A \notin Crs_\alpha^{oreg}$. Then there are $x \in A$, $f \in W \sim x$, $g \in x \cap W$ such that $\Delta^W(x \cap W) \upharpoonright f \subseteq g$. Let $H \overset{d}{=} \{i \in \alpha : f_i \neq g_i\}$. Then $|H| < \omega$ since W is an atom. Let $m \in \omega$ and $i \in {}^m\alpha$ be such that $H = \{i_1, \ldots, i_m\}$. Let $c_H^V \overset{d}{=} c_{i_1}^V \ldots c_{i_m}^V$. By $\underset{\sim}{A} \in B_\alpha$ there is $n \in \omega$ such that $(c_H^V)^n(-x) = (c_H^V)^{n+1}(-x)$. Let $y \overset{d}{=} x \sim [(c_H^V)^n(-x)]$. Then $\Delta^V(y) \cap H = 0$ by the choice of n. $c_H^W(W \sim x) = W \sim x$ by $\Delta^W(x \cap W) \cap H = 0$. Then $rl_W(y) = x \cap W$ since rl_W is a homomorphism. $|Zd\, \underset{\sim}{Rl}_W A| \leq 2$ since W is an atom of $Zd\, \underset{\sim}{A}$, hence there is $i \in \Delta^W(x \cap W) \subseteq \Delta^V(y)$. By cregularity of $\underset{\sim}{A}$, y is $\Delta y \cup \{i\}$ -independent, contradicting $H \cap \Delta y = 0$, $g \in y$, $f \notin y$. This proves $\underset{\sim}{Rl}_W A \in Crs_\alpha^{oreg}$. $\underset{\sim}{Rl}_W A \in K$ since rl_W is a homomorphism and $K = \mathbf{H}\, K$. By Cor.9 of [23] $\langle rl_W : W \in Z \rangle$ is a subdirect decomposition of $\underset{\sim}{A}$ i.e. $\underset{\sim}{A} \in \mathbf{SP}\{\underset{\sim}{Rl}_W A : W \in Z\}$. This proves $Crs_\alpha^{creg} \cap K \subseteq \mathbf{SP}(Crs_\alpha^{oreg} \cap K)$.

Suppose further $K = \mathbf{SP}\, K$. Then $\mathbf{SP}(Crs_\alpha^{oreg} \cap K) \subseteq \mathbf{I}(Crs_\alpha^{creg} \cap K)$ by $\mathbf{SP}\, Crs_\alpha^{oreg} \subseteq \mathbf{I}\, Crs_\alpha^{creg}$. *(ii)* follows from (i) and $CA_\alpha \subseteq B_\alpha$, $\mathbf{I}\, Gs_\alpha = \mathbf{HSP}\, Gs_\alpha \subseteq CA_\alpha$.

(iii) follows from the proof of Prop.15. □

Def.8 and Prop.23 below are tools to construct homomorphisms of Crs_α -s explicitly. They can be used to obtain direct algebraic proofs for $\mathbf{H}\, K = K$ for various $K \subseteq Crs_\alpha$, see Thm. 24 below.

DEFINITION 8 Let α be any set and $\underset{\sim}{A} \in Crs_\alpha$. Let I be a set and F be a filter on I . Let ${}^I U/F$ denote the usual reduced power of U modulo F, see [16]. Let $c : \alpha \times {}^I U/F \to {}^I U$ be arbitrary. We say that c is a *choice function* if $(\forall j\in\alpha)(\forall u \in {}^I U/F)\ c(j,u) \in u$.
We define the function $ud^A_{cF} \overset{d}{=} ud_{cF} : A \to Sb\,({}^I U/F)$ as follows:

$$ud^A_{cF}(x) \overset{d}{=} \{q\in {}^\alpha({}^I U/F) \ : \ \{i\in I : \langle c(j,q_j)_i : j\in\alpha \rangle \in x\}\in F \},$$

for every $x\in A$. See Figure 2. □

PROPOSITION 22 Let $q : M_2 \twoheadrightarrow Lrg$ be as in Prop.1. Let $\mathfrak{A}$ be a model, i.e. $\mathfrak{A}\in M_1$. Let F be an ultrafilter on some set I . Let $c : \alpha \times {}^I U/F \to {}^I U$ be any choice function. Then

$$ud_{cF} \ : \ q(\{\mathfrak{A}\}) \rightarrowtail\!\!\!\twoheadrightarrow q(\{{}^I\mathfrak{A}/F\}) ,$$

i.e. ud_{cF} is an isomorphism between $q(\{\mathfrak{A}\})$ and $q(\{{}^I\mathfrak{A}/F\})$.

The proof can be reconstructed from the proof of Prop.1. □

Prop.23 below is quoted from [17].

PROPOSITION 23 Let α and $\underset{\sim}{A}\in Crs_\alpha$ be arbitrary. Let F be an ultrafilter on some set I . Let $U \overset{d}{=}$ $\overset{d}{=}$ base(A). Let $c : \alpha \times {}^I U/F \to {}^I U$ be any choice function.

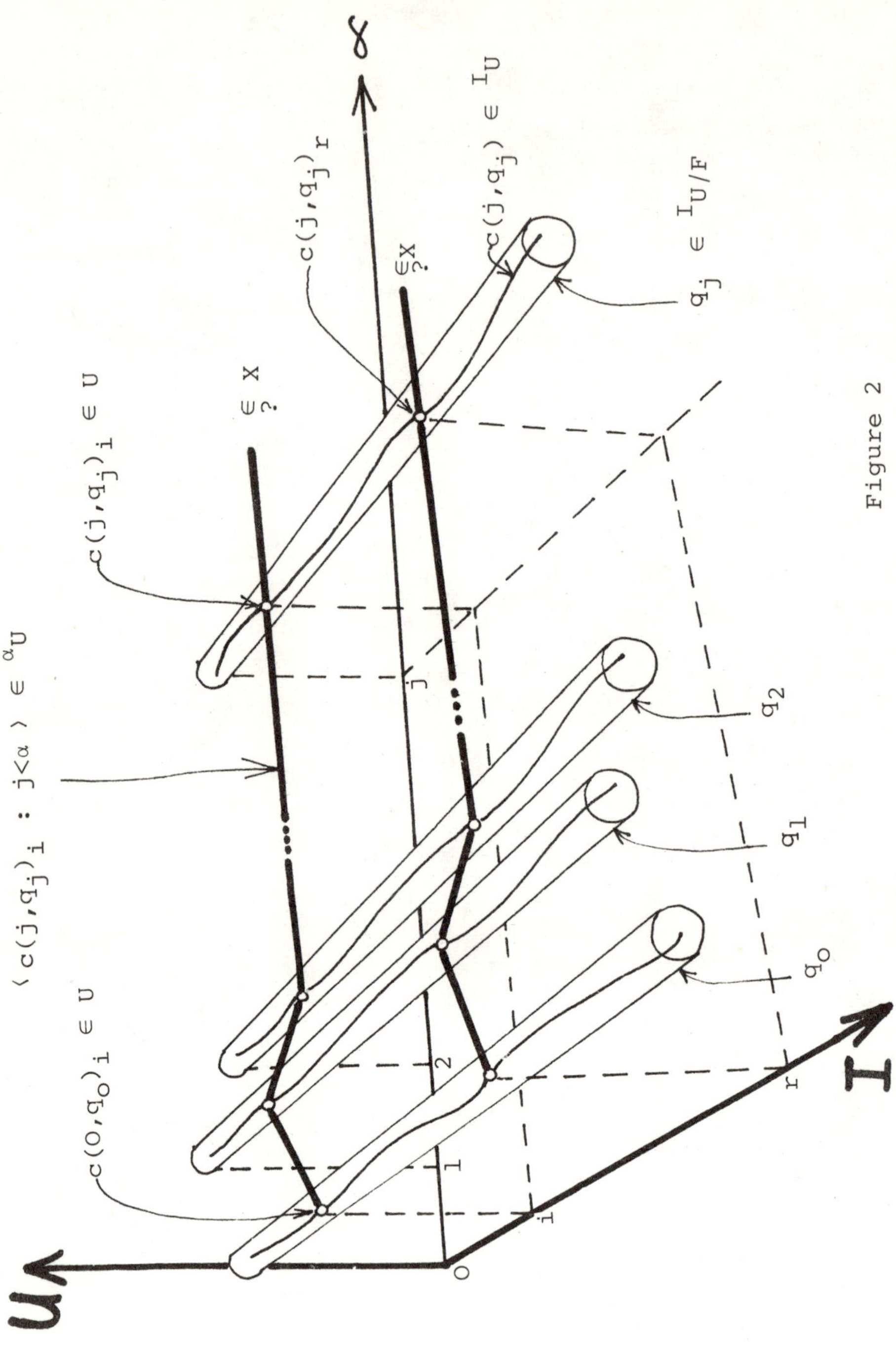

⟨ c(j,q_j)_i : j<α ⟩ ∈ ^αU
c(O,q_o)_i ∈ U
c(j,q_j)_i ∈ U
c(j,q_j)_r
c(j,q_j) ∈ I_U
q_j ∈ I_U/F
∈ X ?
∈X ?
q_o
q_1
q_2
U
I
O
i
r
1
2
j

Figure 2

Then $ud^{A}_{cF} : \underset{\sim}{A} \to \underset{\sim}{B} \in Crs_\alpha$ for some $\underset{\sim}{B}$.

That is, ud_{cF} is a homomorphism on $\underset{\sim}{A}$, moreover the homomorphic image of $\underset{\sim}{A}$ is again a Crs_α.

The proof is in [17]. □

Prop.23 above is a partial Łos lemma in algebraic form.

PROBLEM 6 Let $|\alpha| \geq \omega$ and $\underset{\sim}{A}, \underset{\sim}{B} \in Cs_\alpha$. Assume $\underset{\sim}{A} \cong \underset{\sim}{B}$. Are there ultrafilters F,D choice functions c,d and $\underset{\sim}{A}^{+}, \underset{\sim}{B}^{+} \in Cs_\alpha$ such that $ud^{A}_{cF} : \underset{\sim}{A} \rightarrowtail\!\!\!\twoheadrightarrow \underset{\sim}{A}^{+}$ and $ud^{B}_{dD} : \underset{\sim}{B} \rightarrowtail\!\!\!\twoheadrightarrow \underset{\sim}{B}^{+}$ and for some $f : base(A^{+}) \rightarrowtail\!\!\!\twoheadrightarrow base(B^{+})$ there holds $\tilde{f} \overset{d}{=} \langle \{f \circ q : q \in x\} : x \in A \rangle : \underset{\sim}{A}^{+} \rightarrowtail\!\!\!\twoheadrightarrow \underset{\sim}{B}^{+}$?

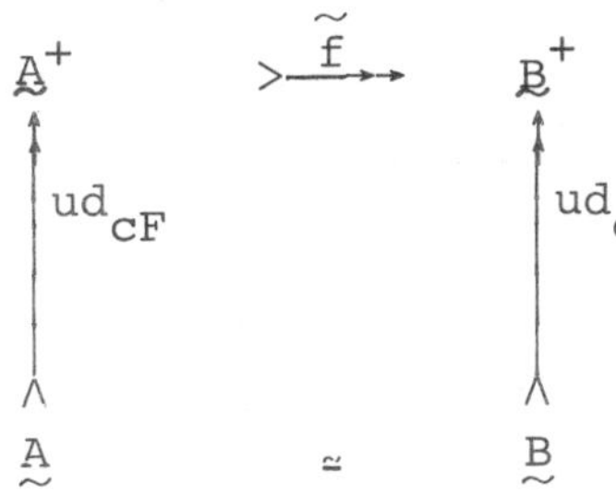

A positive answer to Problem 6 would be a generalization of the Keiser-Shelah ultraproduct theorem.

It was proved in [17] that $\mathbf{I}\, Gs_\alpha \subset \mathbf{I}\,(CA_\alpha \cap Crs_\alpha) \subset CA_\alpha$. Below we prove that $\mathbf{I}\, Crs_\alpha$ is a variety. The method of the proof below can be used to prove directly that $\mathbf{I}\, Gs_\alpha$ and $\mathbf{I}\, Gs^{reg}_\alpha$ are varieties for $\alpha \geq 2$.

THEOREM 24 Let $|\alpha| \geq 2$. Then $\mathbf{HSP}\, Crs_\alpha = \mathbf{I}\, Crs_\alpha$, i.e. $\mathbf{I}\, Crs_\alpha$ and $CA_\alpha \cap \mathbf{I}\, Crs_\alpha$ are varieties.

Outline of proof: The proof of $\mathbf{SP}\, Crs_\alpha = \mathbf{I}\, Crs_\alpha$ goes like the proof of $\mathbf{SP}\, Crs^{reg}_\alpha = \mathbf{I}\, Crs^{reg}_\alpha$, see the proof of Thm.19. To prove $\mathbf{H}\, Crs_\alpha \subseteq \mathbf{I}\, Crs_\alpha$, let $\underset{\sim}{A} \in Crs_\alpha$ and let R be a congruence on $\underset{\sim}{A}$. It is enough to prove that $\underset{\sim}{A}/R \in \mathbf{SP}\ Crs_\alpha$. Let $I \overset{d}{=} 0/R$ i.e. $I = \{x \in A : \langle 0, x\rangle \in R\}$. It is enough to prove that $(\forall x \in A)[x \neq 0 \Rightarrow (\exists h : \underset{\sim}{A} \to \underset{\sim}{B} \in Crs_\alpha)[h(x) \neq 0$ and $(\forall y \in I) h(y) = 0]$. Let $x \in A$, $x \neq 0$. Then one can find an ultrafilter F on I and a function c such that $W \overset{d}{=} ud_{cF}(x) \sim \cup\{ud_{cF}(y) : y \in I\} \neq 0$. By Prop.23 $ud_{cF} : \underset{\sim}{A} \to \underset{\sim}{B} \in Crs_\alpha$ for some full $Crs_\alpha\ \underset{\sim}{B}$. Let $q \in W$. One can prove that $Z \overset{d}{=} \cap\{X \in Zd\ \underset{\sim}{B} : q \in X\} \in Zd\ \underset{\sim}{B}$ and that $(\forall y \in I) ud_{cF}(y) \cap Z = 0$. Then by Prop.17 $rl_Z^A : \underset{\sim}{B} \to \underset{\sim}{Rl}_Z B \in Crs_\alpha$. Then $h \overset{d}{=} rl_Z \circ ud_{cF}$ is the desired homomorphism. □

PROBLEM 7 Is $\mathbf{H}\ Crs_\alpha^{reg} = \mathbf{I}\ Crs_\alpha$ true?

About Problems 1-2 of [16] we can only prove the following.

PROPOSITION 25 Let $\alpha \supseteq 3$. Let $Cr_\alpha \subseteq Alg(g_1^\alpha)$ be as defined in [16] p.261.

Then $\mathbf{S}\ Cr_\alpha \not\subseteq Mod\ Th\ Cr_\alpha$ and $Crs_\alpha \not\subseteq Mod\ Th\ Cr_\alpha$.

There is a Π_2-formula φ such that $Cr_\alpha \models \varphi$ but $Crs_\alpha \not\models \varphi$.

Outline of proof: Let $\alpha \supseteq 3$. Let φ be the following formula $\forall x \forall z \exists y[(x+z \leq d_{02} \wedge c_1 x \not\leq z) \to (x \leq y = c_0 c_1 y \not\leq z)]$. Clearly φ is a Π_2 first order formula of similarity type g_1^3. Hence φ is also of type g_1^α. $Cr_\alpha \models \varphi$ can be proved by computing formulas in $\underset{\sim}{Rl}_b A$ for any $\underset{\sim}{A} \in CA_\alpha$ and $b \in A$ starting with the assumption $x, z \in A$ and $b \cdot x + b \cdot z \leq b \cdot d_{02}$ and $b \cdot c_1 x \not\leq b \cdot z$ and then choosing $y \overset{d}{=} b \cdot c_0 c_1 x$. Then one can construct a $Crs_\alpha\ \underset{\sim}{B}$ such that $\underset{\sim}{B} \not\models \varphi$. One can choose a $Ws_\alpha\ \underset{\sim}{D}$ with infinite base and $V \subseteq \cup D$ such that $\underset{\sim}{B} \subseteq \underset{\sim}{Rl}_V D$. It is crucial that $\underset{\sim}{B}$ should contain an infinite zigzag that is $\underset{\sim}{B} \notin B_\alpha$ where B_α was defined above Thm.21. The construction is basically the same as the Henkin-Resek construction quoted on p.26 of [16]. □

Category theoretic connections

Next we look into category theoretic properties of the category Lf_α . This is interesting e.g. because the category TH of all first order theories and interpretations is isomorphic to the category Lf_ω in the sense of Def.14.1 in [18] p.86.

In Prop.26 below we shall use the notions strongly algebroidal category, strongly small object and to have enough projectives which can be found in e.g. [11],[25], [12],[14]. The quoted papers deal with category theoretic model theory. All the results of the quoted papers apply to Lf_α , e.g. characterizations of classes $K \subseteq Lf_\alpha$ such that K is closed under Lf -ultraproducts (see [11]) and subalgebras etc.. In 22E of the book [18] on p.155 what we call here strongly small object is called strongly finitary object and is defined there. Strongly algebroidal category is practically the same as locally ω -presentable category in the sense of Gabriel-Ulmer.

PROPOSITION 26 Let $|\alpha|\geq\omega$. Then the category Lf_α is strongly algebroidal and has enough projectives. Lf_α is a complete and cocomplete category in the sense of [18] p.156 Def.23.1. Specially, category theoretic ultraproducts exist in Lf_α in the sense of [11]. Lf_α is locally ω-presentable in the sense of Gabriel-Ulmer.

Outline of proof: Let $|\alpha|\geq\omega$. Let $\beta\in Ord$ and $t : \beta \to$ $\to Sb\,\alpha$. Then the CA_α freely generated by β elements under defining relations t was defined in [16] p.348 and denoted by $\underset{\sim}{Fr}^{(t)}_\beta CA_\alpha$. *Notation:* Let H be any class. Then $Sb_\omega H \overset{d}{=} \{x\in Sets : |x|<\omega$ and $x \subseteq H\}$. Let

$\underset{\sim}{A} \in Lf_\alpha$. Then $\underset{\sim}{A}$ is strongly small in Lf_α iff $\underset{\sim}{A} \cong (\underset{\sim}{Fr}_n^{(t)} CA_\alpha)/R$ for some $n \in \omega$, $t : n \to Sb_\omega \alpha$ and R is a *finitely generated* congruence on $\underset{\sim}{Fr}_n^{(t)} CA_\alpha$. Note that the above $\underset{\sim}{Fr}_n^{(t)} CA_\alpha$ is not finitely presented in CA_α hence it is not small in CA_α but it is such in Lf_α. Observing that every $\underset{\sim}{B} \in Lf_\alpha$ is a directed colimit of strongly small ones proves that Lf_α is strongly algebroidal. Let $\beta \in Ord$ and $t : \beta \to Sb_\omega \alpha$. Then $\underset{\sim}{Fr}_\beta^{(t)} CA_\alpha$ is a projective object of Lf_α. Then $Lf_\alpha = \mathbf{H}\{\underset{\sim}{Fr}_\beta^{(t)} CA_\alpha : \beta \in Ord$ and $t \in {}^\beta Sb_\omega \alpha\}$ proves that Lf_α has enough projectives. We note that there are other projectives too. To see completeness and cocompleteness of Lf_α observe that Lf_α is a mono-coreflective subcategory of CA_α, which by the duals of 36.12-36.18 of [18] pp.278-280 implies the desired property. □

PROPOSITION 27 Let $\alpha \geq \omega$. Then Lrg_α and Lf_α are isomorphic as categories in the sense of Def.14.1 in [18] p.86.

The proof of isomorphism from equivalence goes by the Cantor-Bernstein argument. □

Burstall-Goguen stepwise refinement of program specifications programme via Cylindric algebras

Burstall-Goguen [15] explains why it is important for Computer Science to correlate algebras to theories. In short: For structured programming it is useful to investigate the class of algebras corresponding to theories (or to investigate the category of theories).

Let us see how to correlate an algebra to an arbitrary theory of classical first order logic. In what follows, by a theory we always mean a classical first order theory. A theory can always be represented by a *set* K of models (with mutually disjoint universes). This follows from Gödel's completeness theorem. The proof of Prop.(ii) shows how to correlate an algebra $\underset{\sim}{K}\in$Lrg to a set K of models. Prop.1(ii) proved that this correlation is invertible (and definable). Now, to a theory Th correlate a set K of models such that Th is the set of all formulas valid in K, and consider the algebra $\underset{\sim}{K}\in$Lrg. We shall say that $\underset{\sim}{K}$ is *the* algebra corresponding to Th. Then by Prop.1(ii) and by the above observation we have a one-one correspondence between theories and algebras in Lrg. By Prop.s 12,26 Lrg is a complete and cocomplete category in the sense of [18] and therefore Prop.s 1,12, 26 may serve as a foundation for realizing the Burstall-Goguen programme for *all* first order theories.

REFERENCES

[1] Andréka,H.: Algebraic logic. Dissertation, Eötvös L. Univ., Budapest, 1972.

[2] Andréka,H.: Universal algebraic logic. Dissertation, Hung.Acad.Sci., Budapest, 1975.

[3] Andréka,H. Gergely,T. Németi,I.: Purely algebraical construction of first order logics. Central Res.Inst. Phys.H.A.S., Preprint No.KFKI-73-71, 1973.

[4] Andréka,H. Gergely,T. Németi,I.: On universal algebraic construction of logics. *Studia Logica* vol.36, No.1-2, 1977, pp.9-47.

[5] Andréka,H. Németi,I.: Simple purely algebraic proof of completeness of logics. *Algebra Universalis* vol. 5, 1975, pp.8-15.

[6] Andréka,H. Németi,I.: Varieties definable by schemes of equations. *Algebra Universalis* vol.11, No.1.

[7] Andréka,H. Németi,I.: Neat reduct of varieties.

Studia Sci.Math.Hung., to appear.

[8] Andréka,H. Németi,I.: Dimension complemented and locally finite dimensional cylindric algebras are elementarily equivalent. *Algebra Universalis*, to appear.

[9] Andréka,H. Németi,I.: Some constructions of cylindric set algebras. Manuscript, Budapest, 1979.

[10] Andréka,H. Németi,I.: Solution of Problems 2.3 and 2.11 of Henkin-Monk-Tarski 71. Math.Inst.H.A.S., Preprint No.14/1979, 1979.

[11] Andréka,H. Németi,I.: Formulas und ultraproducts in categories. *Beiträge zur Algebra und Geometrie* vol. 8, 1979, pp.133-151.

[12] Andréka,H. Németi,I.: Los lemma holds in every category. *Studia Sci.Math.Hung.*, to appear.

[13] Andréka,H. Sain,I.: Connections between algebraic logic and initial algebra semantics of CF languages, in this volume.

[14] Banaschewski,B. Herrlich,H.: Subcategories defined by implications. *Houston J.Math.* vol.2, No.2, 1976, pp.149-171.

[15] Burstall,R.M. Goguen,J.A.: Putting theories together to make specifications. In: *Proc. 5th Int. Joint. Conf. Artificial Intelligence*, Cambridge Mass., pp.1045-1058.

[16] Henkin,L. Monk,J.D. Tarski,A.: *Cylindric algebras, Part I.* North-Holland, 1971.

[17] Henkin,L. Monk,J.D. Tarski,A.: Cylindric set algebras and related structures. Springer Lecture Notes in Math. Series, to appear.

[18] Herrlich,H. Strecker,G.E.: *Category theory.* Allyn and Bacon Inc., Boston, 1973.

[19] Monk,J.D.: Nonfinitizability of classes of representable cylindric algebras. *Journal of Symbolic Logic*

vol.34, No.3, 1969. pp.331-343.

[20] Monk,J.D.: *Mathematical logic.* Springer Verlag,1976.

[21] Németi,I.: The class of cylindric-relativized set algebras is a variety. Math.Inst.H.A.S., Preprint, 1979.

[22] Németi,I.: **SUp** Lr = **SUp** Cs. Math.Inst.H.A.S., Preprint.

[23] Németi,I.: Some constructions of cylindric algebra theory applied to dynamic algebras of programs. *CL&CL*, Budapest, vol.14.

[24] Németi,I. Andréka,H.: On universal algebraic logic and cylindric algebras. *Bull.Section of Logic,* Wroclaw. vol.7, No.4, 1978.pp.152-158.

[25] Németi,I. Sain,I.: Cone injective subcategories and Birkhoff-type theorems. In: Csákány,B. Fried,E. Schmidt,E.T.(eds.) *Universal Algebra.* North-Holland, to appear.

Németi,I.
Mathematical Institute of the
Hungarian Academy of Sciences
Budapest, Reáltanoda u.13-15.
H-1053 Hungary

COLLOQUIA MATHEMATICA SOCIETATIS JÁNOS BOLYAI
26. Mathematical Logic in Computer Science
Salgótarján (HUNGARY), 1978

FUNCTIONAL INTERPRETATION OF λ-TERMS

A. Obtułowicz and A. Wiweger

We shall outline certain approach to the problems of interpretation λ-terms of the type free λ-calculi. The main feature of our approach is that it can be easily generalized to obtain interpretation of λ-terms in certesian closed categories.

We shall use λ-terms in slightly modified form, using different symbols for free and bound variables. Let $(x_1, x_2, \ldots)$ and $(\xi_1, \xi_2, \ldots)$ be two infinite sequences with disjoint sets of values such that if $i \neq j$, then $x_i \neq x_j$ and $\xi_i \neq \xi_j$ (x_i is a *free variable* and ξ_i is a *bound variable*). The set Exp of λ-terms is defined by induction as follows: 1/ $x_i \in$ Exp for any i; 2/ if $M, N \in$ Exp, then $(MN) \in$ Exp; 3/ if $M \in$ Exp and ξ_j does not occur in M, then $\lambda\xi_j \cdot (x_i/\xi_j)M \in$ Exp, where $(x_i/\xi_j)M$ is the result of substituting ξ_j for x_i in M.

We define the *rank of λ-term* M to be 0 if there is no free variable occuring in M, and $\max\{i : x_i$ occures in M$\}$ otherwise.

A *labelled λ-term* is an ordered pair (M,n), where M is λ-term and n is non-negative integer greater or equal to the rank of M. The set of all labelled λ-terms is equal to the set Exp* defined by induction as follows: 1/ $(x_i, n) \in \text{Exp}^*$ for any $1 \leq i \leq n$; 2/ if $(M,n),(N,n) \in \text{Exp}^*$, then $((MN),n) \in \text{Exp}^*$ $(n \geq 0)$; 3/ if $(M,n+1) \in \text{Exp}^*$ and ξ_j does not occur in M, then $(\lambda\xi_j \cdot (x_{n+1}/\xi_j)M, n) \in \text{Exp}^*$ $(n \geq 0)$.

Let us consider a pair $(\mu : A \to C, \nu : C \to A)$ of functions, where $C \subset A^A$ and $\mu \circ \nu = id_C$. We adopt the following notation: $A^0 = \{0\}$; $A^1 = A$; $A^{n+1} = A^n \times A$ $(n \geq 1)$; p_i^n will denote the ith projection from A^n into A, that is, for each $(a_1,\dots,a_n) \in A^n$ we have $p_i^n(a_1,\dots,a_n) = a_i$ $(1 \leq i \leq n)$; $<f_1,f_2>$ will denote the function from A^n into A^2 defined for functions f_1, f_2 from A^n into A in the following way: $<f_1,f_2>(x) = (f_1(x),f_2(x))$ for each $x \in A^n$ $(n \geq 0)$; $\lambda_A[g]$ will denote the function from A^n into A^A defined for any $g : A^{n+1} \to A$ as follows: $\lambda_A[g](x)(a) = g(x,a)$ for any $x \in A^n$ and $a \in A$ $(n > 0)$, and $\lambda_A[g](0) = g$ in the case $n=0$; $\varepsilon_\nu^\mu : A^2 \to A$ is the function with values given by $\varepsilon_\nu^\mu(a_1,a_2) = \mu(a_1)(a_2)$ for all $a_1,a_2 \in A$. We say that the composition $f \circ g$ of functions f,g is defined iff the set of values of g is contained in the domain of f. Now we shall define a family $(\Lambda^n(\mu,\nu) : n \geq 0)$ of sets by induction: $E_0^0 = \{\emptyset\}$, $E_0^1 = \{id_A\}$, $E_0^2 = \{p_1^2, p_2^2, \varepsilon_\nu^\mu\}$, and for $n > 2$ $E_0^n = \{p_i^n : 1 \leq i \leq n\}$; for $n \geq 0$ and $k \geq 0$ $E_{k+1}^n = E_k^n \cup \{\varepsilon_\nu^\mu \circ <f_1,f_2> : f_1,f_2 \in E_k^n\} \cup \{g : \exists f \in E_k^{n+1}$ .the composition of $\lambda_A[f],\nu$ is defined and $g = \nu \circ \lambda_A[f].\}$; finally $\Lambda^n(\mu,\nu) = \bigcup_{k>0} E_k^n$. We shall call the pair (μ,ν) a *regular pair* if for each $n \geq 1$ and each $f \in \Lambda^n(\mu,\nu)$, the composition of $\lambda_A[f]$, ν is defined. The examples of regular pairs are the homeomorphism $\varphi : D_\infty \to [D_\infty \to D_\infty]$ and its converse, and fun: $P_\omega \to [P_\omega \to P_\omega]$, graph: $[P_\omega \to P_\omega] \to P_\omega$, where D_∞, P_ω are Scott's models of the type free λ-calculi, see [1]. Since Exp^* is inductively defined, for any regular pair (μ,ν) there is the unique function, called the *functional interpretation of λ-terms* $I^{\mu}_{\nu}:\mathrm{Exp}^* \to \bigcup_{n>0} \Lambda^n(\mu,\nu)$ such that $I^{\mu}_{\nu}(x_i,n)=p^n_i$ $(1 \leq i \leq n)$,

$$I^{\mu}_{\nu}((MN),n)=\varepsilon^{\mu}_{\nu}\circ\langle I^{\mu}_{\nu}(M,n),I^{\mu}_{\nu}(N,n)\rangle (n\geq 0),$$

$$I^{\mu}_{\nu}(\lambda\xi_j\cdot(x_{n+1}/\xi_j)M,n)=\nu\circ\lambda_A[I^{\mu}_{\nu}(M,n+1)] \quad (n \geq 0).$$

We note that a pair (μ,ν) is regular iff this pair gives interpretation of λ-terms in the style of [2].

We use the following generalization of the notion of an abstract algebra. A (partial) *hyperoperation* on a set A is a (partial) function

$$\omega:(A^{(A^p)})^n \to A^{(A^q)}$$

where p,n,q are non-negative integers. A (partial) *hyperalgebra* is a pair $=(A,\omega)$, where $\omega=(\omega_t : t\in T)$ is a family (indexed by some set T) of (partial) hyperoperations on A. A type of (partial) hyperalgebras is defined in an obvious way. To introduce the concept of a homomorphism of hyperalgebras we use the following notation: for any $f:X\to Y$ we define $f^Z:X^Z\to Y^Z$ and $Z^f:Z^Y\to Z^X$ as the functions given by $f^Z(r:Z\to X)=f\circ r$, $Z^f(s:Y\to Z)=s\circ f$, respectively. If $\mathcal{A}$ and $\mathcal{B}$ are hyperalgebras of the same type (with $\omega^{\mathcal{A}}_t$ corresponding to $\omega^{\mathcal{B}}_t$), then a *hyperhomomorphism* from $\mathcal{A}$ to $\mathcal{B}$ is any function $h:A\to B$ such that for any t in T the following diagral is commutative (where p,n,q depend on t):

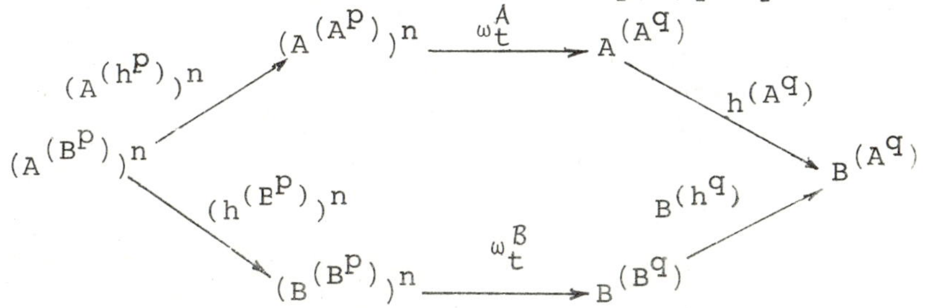

In a similar way one may define hyperhomomorphisms of partial hyperalgebras.

The construction of sets $\Lambda^n(\mu,\nu)$ gives rise to a partial hyperalgebra with the underlying set A and partial hyperoperations

$$ap_n : (A^{(A^n)})^2 \to A^{(A^n)}, \quad ap_n(f_1,f_2)=\varepsilon_\nu^\mu o\langle f_1,f_2\rangle,$$

$$ab_n : A^{(A^{n+1})} \to A^{(A^n)}, \quad ab_n(f)=\nu o\lambda_A[f],$$

$$c_i^n : (A^A)^o \to A^{(A^n)}, \quad c_i^n(o)=p_i^n.$$

A regular pair (μ,ν) gives rise to a hyperalgebra with hyperoperations

$$\mu:(A^{(A^o)})^1 \to A^A,\ c_{(M,n)}:(A^A)^o \to A^{(A^n)},\ (M,n)\in Exp^*,$$

where $c_{(M,n)}(o)=I_\nu^\mu(M,n)$. (One may also consider $\nu:C\to A$ as a partial hyperoperation from A^A into $A^{(A^o)}$.)

REFERENCES

[1] H.P.Barendregt, The type free lambda-calculus in J.Barwise (ed),*Handbook of Mathematical Logic* North-Holland, Amsterdam, 1977.

[2] C.P.Wadsworth, The relation between computational and denotational properties for Scott's D_∞--models of the lambda-calculus, *SIAM Journal of Computation*, 5(3), 1976, pp. 488-521.

A. Obtulowicz, A. Wiweger

Instytut Matematyczny P.A.N.,
ul. Sniadeckich 8.
Skrytka Pocztowa 137,
00950 Warszawa
Poland

COLLOQUIA MATHEMATICA SOCIETATIS JÁNOS BOLYAI
26. Mathematical Logic in Computer Science,
Salgótarján (HUNGARY), 1978.

THE SEMANTICS OF PARALLELISM AND CO-ROUTINING IN LOGIC PROGRAMMING

L.M. Pereira and L.F. Monteiro

ABSTRACT

We begin with an introduction to a simple but powerful logic programming language called Prolog, in order to provide a rigorous context of presentation of ideas and results.

Next we present definitions of sequential, parallel and co-routined executions of programs, in strictly logic programming terms, and go on to define in logic a parallel interpreter for logic programs, obtained by a simple program transformation from a purely sequential interpreter. We then show how similar transformations may be directly applied to programs to obtain transforms that achieve parallelism or co-routining without re-course to special interpreters. Afterwards, we apply our results to data base lookup and to problems arising from the use of negation as nonderivability, and suggest the basis of a rudimentary control language for logic programs.

We conclude by examining the features of logic which make logic programming specially suitable for parallel and co-routined modes of processing. In the final sections we furnish rigorous proofs of our results, to suplement the informal and intuitive basis used to motivate and derive them.

INTRODUCTION TO THE PROLOG LANGUAGE

Prolog is a simple but powerful programming language founded on symbolic logic, developed at the University of Marseille, as a practical tool for "logic programming" ([3], [6], [7], [8], [12]). A major attraction of the language, from a user's point of view, is ease of programming. Clear, readable, concise programs can be written quickly with minimum error. Recently, an efficient compiler and an interpreter were implemented on the DECsystem-10 ([9], [13]).

Like Lisp, Prolog is an interactive language designed primarily for symbolic data processing. Both are founded on formal mathematical systems - Lisp on the lambda calculus, Prolog on a powerful subset of calassical logic [11] Pure Lisp in fact can be viewed as a specialization of Prolog [14].

Introductory syntax

Here is a Prolog program, consisting of two clauses, for relating a tree with the list of its leaves (or frontier):

```
leaves (t(void, N, void), [N,..Z] - Z).

leaves (t(Stl, N, Str), L - Z): - leaves (Stl,
      L - X), leaves (Str, X - Z).
```

In general, a Prolog *program* consists of a set of *procedures*, where each procedure comprises a number of *clauses*. The procedure name is called a *predicate* ("leaves" above), and has an *arity* which is the number of its arguments (2 above). A clause begins with a *head* or procedure entry point, and continues with a *body*. If the body is not empty it is separated from the head by ":-" (2nd clause above). Every clause terminates with a ".". The head displays a possible form of the arguments to the procedure's predicate. The body consists of a number (possibly zero) of *goals* or procedure calls, which impose conditions for the head to be true. If the body is empty we speak of a *unit* clause (1st clause above).

In general, all Prolog objects are *terms*. A clause is a term, a predicate or a goal with their respective arguments, and the arguments themselves are terms. A term is either a *variable* (distinguished by an initial capital letter), an *atom* ("void" above), or a *compound term*. A compound term comprises a *functor* ("leaves" or "t" above) of some arity N $\geq$ 1, and a sequence of N terms as its arguments ("t(void, N, void)" above). An atom is treated as a functor of arity Ø. A term of the form [H,..T] stands for the list. (H, T), whose head is H and tail is T. The empty list is denote by [], and a list with exactly two elements by [A, B].

The term [N,..Z] - Z, where "-" is a binary functor written in (optional) infix notation, stands for a difference list [2]. A difference list L - Z stands for the list which concatenated with list Z produces list L. Difference lists provide a convenient way for appending the lists they denote. Above, the (difference) list L - Z is conveniently split into the (difference) lists L - X and X - Z.

The second clause above is just infix notation for the term

```
:- (leaves (t(stl, N, Str), L - Z),
      "," (leaves  Stl, L - X), leaves (Str,Z)))
```

where":-"and "," are binary functors. ":-" takes as arguments the head and the body of the clause, and "," the goals. This term stands for a clause because it figures in the set of clauses for a procedure. It is distinguished by a final".".

Apart syntax conventions, the names and arities of terms (and their number) are arbitrary, except for a pre-defined set of procedures which are built into the implementation of the language, and which achieve input, output, arithmetic, etc.

Semantics

Prolog differs from most programming languages in that there are two quite distinct ways to understand its semantics. The *procedural* or operational semantics is the more conventional, and describes as usual the sequence of states passed through when executing a program. In addition a Prolog program can be understood as a set of descriptive statements (one for each clause) about the state space of a problem. The *declarative* or denotational semantics, which Prolog inherits from logic, provides a formal base for such an understanding. Informally, one interprets terms as shorthand for natural language phrases by applying a uniform translation of each functor. e.g.:-

void = "the empty tree"

t(Stl, N, Str) = "the binary tree with root N, left subtree Stl and right subtree Str" leaves (T, L - Z) = "the leaves of tree T are the elements of L - Z"

L - Z = "the list of elements of L after Z is tail subtracted from it"

A clause "P̲: - Q̲, R̲, S̲." where P̲, Q̲, R̲ and S̲ are metavariables standing for terms, is interpreted as

"P̲ if Q̲ and R̲ and S̲"

A clause "P̲." is interpreted as "P̲ is true".

Each variable in a clause should be interpreted as some arbitrary object (i.e. variables are universally quantified). The type of the object will be appropriate to the functor(s) where it figures by using terms consistently throughout the program.

The declarative semantics then simply defines (recursively) the set of terms which are asserted to be true according to a program. A term is *true* if it is the head of some clause instance and each of the goals (if any) of that clause instance is true, where an *instance* of a clause (or term) is obtained by substituting, for each of zero or more of its variables, some term for all occurences of the variable.

Thus the only true instance of the goal: -

leaves (t(t(void, a, void), b, Str), [A, c] - []).

is

leaves (t(t(void, a, void), b, t(void, c, void)), [a, c] - []).

It is the declarative aspect of Prolog which is responsible for promoting clear, rapid, accurate programming. It allows a program to be broken down into small, independently meaningful units (clauses), and it allows some understanding of a program without looking into the details of how it is executed.

Procedural semantics

It is the procedural semantics that describes the way a goal is executed. The objective of execution is to produce true instances of the goal. It then becomes important to know that the ordering of clauses in a program, and of goals within a clause, which are irrelevant as far as the declarative semantics is concerned, constitute crucial *control information* for the procedural semantics.

To *execute* a goal, the system searches for the first clause whose head *matches* or *unifies* with the goal. The unification process [10] finds the most general common instance of the two terms, which is unique if it exists. If a match is found, the matching clause instance is then *activated* by executing in turn, from left to right, each of the goals of its body (if any). If at any time the system fails to find a match for a goal it backtracks, i.e. it rejects the most recently activated clause, undoing any substitutions made by the match with the head of the clause. Next it reconsiders the original goal which activated the rejected clause, and tries to find a subsequent clause which also matches the goal. Execution terminates if no goals remain to be executed (the system has then found a true instance of the original goal). Backtracking may then be provoked to find other true instances of the goal. Execution fails when no true instances of the original goal are found, and terminates if it cannot find any more true instances. Termination however cannot be guaranteed, even if there are no more true instances (eg. if there are infinite branches).

Note that the execution just defined is a left to right depth-first process. Note also that because unification always provides the most general common instance between a goal and a matching clause, all the most general true instances of a goal can potentially be found (i.e. aside termination issues).

Let us now briefly look at how the goal: - leaves (t(t(void, a, void), b, Str), [A, c] - []) is actually executed. The goal only matches the second clause for "leaves". The body of the matching clause instance is: - leaves (t(void, a, void), [A, c] - X), leaves (Str, X - []). The result of executing the first of these two goals against the only clause it matches is to instantiate A to a and X to c. The second goal also matches the first clause, thereby instantiating Str to t(void, c, void) since X is already instantiated to c.

If the second goal were matched with the second clause, an attempt would be made to generate an infinite tree, which need not concern us now, but showing just how crucial can a convenient ordering of clauses be.

The connection between the execution mechanism defined above and logical derivability, and also the proof of completeness of the search space can be found in [5].

Basically, each execution step is justified by Robinson's Resolution Principle [10]. This principle subsumes in a single inference rule the classical rules of "modus ponens" and "generalization" in formulations of first order predicate calculus. For example, from: -

```
p(X): - q(a, X), r(X).
```

and

```
q (Y, f(Y, Z)): - s (a, Z).
```

it allows to conclude

p(f(a, Z)) : s (a, Z), r (f(a, Z)).

by "execution" of q(a, X).

Note that computer-wise it is advantageous to have a single inference rule uniformly applicable.

Besides the ordering of clauses and the sequencing of goals within clauses Prolog provides just one other essential mechanism for specifying control information. This is the "cut" symbol, written "!". It is inserted in a program just like a goal, but it is not be regarded as part of the logic of the program and should be ignored as far as the declarative semantics is concerned.

The effect of the "cut" is as follows: when first encountered, as a goal, "cut" succeeds immediately. If backtracking should later return to the "cut", the effect is to fail the goal which caused the clause containing the "cut" to be activated. In other words, the "cut" operation *commits* the system to all choices made since execution of the goal activating the clause begun. I.e. other alternatives for that goal are not considered, as well as for all goals occuring in the matching clause before the "cut". By means of a "cut" one can ensure that some goals, once partly executed by a clause up to a "cut", either must continue that partial execution or fail. The "cut" renders deterministic the whole partial execution made by the activated clause up to it.

Example of the effect a "cut" in the flux of control, when goal F fails: -

```
P: -   A,  B,   C.

B: -   D,  E,   !,    F.
```

where A, B, C, D, E, F, and P are metavariables standing for predicate instances.

Backtracking returns to goal A immediately before B, the goal that activated the clause with the "cut". If there was no A in P control would return to the goal calling P.

Outstanding features of Prolog

To end this introduction to Prolog, let us briefly review the combination of features which make Prolog a powerful but simple to use programming language.

(1) A declarative semantics inherited from Logic in addition to the usual procedural semantics.

(2) Identity of form of program and data - clauses can be employed for expressing data, and can be manipulated as terms by interpreters written in Prolog.

(3) The input and output arguments of a procedure do not have to be distinguished in advance, but may vary from one call to another. Procedures can be multi-purpose. The procedure for "leaves" may be given completely or incompletely specified trees or lists of nodes in any combination.

(4) Procedures may have multiple outputs as well as multiple inputs.

(5) Procedures may generate, through backtracking, a sequence of alternative results. This amounts to a high level form of iteration.

(6) Terms provide general record structures with any number of record types may be used, and there are no type restrictions on the fields of a record.

(7) Pattern matching replaces the use of selector and constructor functions for operating on structured data.

(8) Incomplete data structures may be returned (i.e. containing free variables) which may later be filled in by other procedures.

(9) All communication between co-routined or concurrent procedures is ensured by the variables through unification. No explicit interfacing is needed.

(10) Prolog dispenses with *go to*, *do*, *for* and *while* loops, *assignment*, and *references* (pointers).

(11) The procedural semantics of a syntactically correct program is totally defined. It is impossible for an error condition to arise or for an undefined operation to be performed. This totally defined semantics ensures that programming errors do not result in bizarre program behaviour or incomprehensible error messages.

(12) No part of the program is concerned with the details of the underlying machine or implementation.

SEMANTICS OF PARALLELISM AND CO-ROUTINING IN LOGIC PROGRAMS

We begin this section by supplying definitions of sequential, parallel and coroutined execution of logic programs. Next, we describe a sequential interpreter in clausal form, and from it derive an interpreter which achieves parallel execution of two goals, whilst preserving both declarative and procedural semantics. This parallel interpreter, obtained by a simple transformation of the clausal program for the sequential interpreter, is then generalized to execute n goals in parallel. This same transformation is then shown to apply directly to any program for obtaining parallel execution of a number of goals using just the sequential interpreter. Finally, another program transformation is provided which illustrates co-routining, and examination is made of the features of logic programming which make logic a unique language for parallel and co-routined processing.

Definitions of sequential, co-routined and parallel executions

We take unification to be, from the logic programmer's point of view, a single event. When a goal unifies with the head of a clause, all unification of arguments is considered simultaneous. It is not relevant for the programmer to know how the executor of the program performs unification, or to specify how it should be performed. Thus we take the match of a goal with the head of a clause as the single elementary unanalysed event in logic programs.

We say two goals are executed *sequentially* when execution of one goal begins only after execution of the other is completed.

We say two goals are executed *in parallel*, or concurrently, when execution of one goal is interleaved with the execution of the other, such that at most only a single elementary event takes place in one execution before another elementary event takes place in the other. I.e. an execution waits for the other no more than the completion of a match with the head of a clause.

We say two goals are executed in *co-routining* fashion in all other cases.

The above definitions have the virtue of providing rigorous, implementation and hardware independent notions of parallelism and co-routining which are meaningful from the logic programming point of view. Indeed, sequencing of goal calls is what traditional search strategies are all about.

Recall the program for "leaves", displayed in the introduction to Prolog: -

```
leaves (t(void, N, void), [N,.. Z] - Z).

leaves (t(Stl, N, Str), L - Z): - leaves
      (Stl, L - X), leaves (Str, X - Z).
```

The procedure: -

```
same-leaves (T1, T2): - leaves (T1, L - []),
      leaves  T2, L - []).
```

will execute the two calls to leaves sequentially to find whether T1 and T2 have the same list L of leaves. Sequential execution in this case can be largely inefficient. For suppose the two trees differ in the first leaf. This will only be discovered after the first tree is already totally processed. Parallel execution of the two goals however would provide the best search strategy for arbitrary trees assuming each tree is going to be searched in a depth-first fashion as stated by the "leaves" procedure.

In general, parallelism of n goals has the further advantage that it always fails as soon as the goal which fails soonest does. For example, one or both of the trees might be ill-formed.

A sequential interpreter for logic programs

Imagine the sequential interpreter we are about to define reads in the clauses for "leaves" and re-writes them as: -

```
clause (( leaves (t(void, N, void), [N,..Z] - Z),
         C : - C )).
clause (( leaves (t(Stl,N,Str),L - Z), C: - leaves
         (Stl,L - X), leaves (Str, X - Z), C )).
```

I.e. each clause is re-written as the single argument of a unit clause for predicate "clause", after a variable C is conjuncted to both its consequent and its antecedent. Introduction of this variable on both sides of the implication does not modify the semantics of the program as long as C is guaranteed to be bound, during execution, to some predicate instance or conjunction of predicate instances. Nor does it go outside fist order logic, on the same condition. The extra parentheses around the single argument of "clause" are just syntax due to the infix notation used for ":-". The original clauses for "leaves" in an equivalent form, are thus supplied as data to the interpreter, by means of predicate "clause".

The clauses for the interpreter which accepts and executes this data are: -

```
i(succeed).
i(P) : - clause ((P: - C)), i(C).
```

where a call to the original program such as

    leaves (T, L - [])

is replaced by an equivalent call to the interpreter of the form

    i((leaves (T, L - []), succeed))

The effect of "i" is to accept a conjunction of goals terminated with "succeed" and to execute it as follows. It considers the first goal in the conjunct, looks for the first clause that matches the conjunction of that goal with the remaining sequence of goals, and accepts from the clause a new conjunction of goals to be executed. The new conjunction of goals is just the old one, where the first goal of the conjunct has been replaced by the goals in the body of the original clause. Eventually, execution terminates when the goal "succeed" is reached.

Thus, execution of a goal by "i" exactly mimicks the procedural semantics defined for Prolog. The difference lies in that "i" explicitly carries along the conjunction of outstanding goals awaiting execution. This requires the introduction of a variable C, on both sides of the implication in a clause, for receiving and passing along the conjunction of goals still awaiting execution (sometimes referred to as the "continuation"). "succeed" is then needed to express the empty continuation.

Proof of the equivalence between the declarative and procedural semantics of the original clauses and the program made up of the clauses for "i" and the clauses for "clause" is given in a later section.

Next, we show how to obtain a parallel interpreter "ip" from the above sequential one.

Define the new predicate "ip" as: -

```
ip(P1, P2): - i(P1), i(P2).
```

Now symbolically evaluate the calls to "i", by replacing them with the body of the clauses for "i" which they match - a process also known as "unfolding" [1].

Four clauses obtain: -

```
ip(succeed, succeed).

ip(succeed, P2): - clause ((P2: - C2)), i(C2).

ip(P1, succeed): - clause ((P1: - C1)), i(C1).

ip(P1, P2): - clause ((P1: - C1)), clause ((P2: -
                    C2)), i(C1), i(C2).
```

Finally, replace calls to "i" by calls to "ip" - a process known as "folding" [1] - using the facts

i(P) = ip(P, succeed) = ip(succeed, P) where "succeed" is the recursion base argument of "i", and the definition of "ip" to obtain: -

```
ip(succeed, succeed).

ip(succeed, P2): - clause ((P2: - C2)),
                   ip(succeed, C2).

ip(P1, succeed): - clause ((P1: - C1)),
                   ip(C1, succeed).

ip(P1, P2): - clause ((P1: - C1)), clause ((P2: -
                   C2)), ip(C1, C2).
```

The symmetry of "ip" on its arguments allows a simplification. The result is: -

ip(succeed, succeed).

ip(succeed, P2): - ip(P2, succeed).

ip(P1, P2): - clause ((P1: - C1)), ip(P2, C1).

Symbolical evaluation of these clauses plus symmetry reproduce the previous ones.

Again, formal proof of the procedural equivalence between the call i(P) and the corresponding calls ip(P, succeed) and ip(succeed, P) is left to another section as well as the declarative equivalence between the conjunct i(P1),i(p2) and ip(P1,P2). The proof is more general though. Let P and Q be metavariables standing for two predicates. Define a new predicate P & Q with the single clause.

P & Q: - P, Q.

where the arguments of P & Q are obtained by adjoining the arguments of P and of Q. We prove that the declarative semantics of P & Q is equivalent to that of P and Q, where the clauses for P & Q result from a simple transformation of the clauses for P and Q. Furthermore, the declarative and procedural semantics of P & Q preserves the original semantics of P or of Q. The transformation in question is just a simple one-step unfolding produced by symbolic evaluation of the defining clause, followed by an appropriate folding step.

A program transformation for direct parallelism

Take the clauses for "i". Evaluate "clause" with respect to the "leaves" procedure. Call "1" the resulting procedure: -

```
l(succeed).

l((leaves(t(void,N, void), [N,..Z] - Z),C)):-l(C).

l((leaves(t(Stl,N,Str),L - Z), C)): - l((leaves
        (Stl,L - X),leaves (Str,X - Z), C)).
```

It directly mimicks the behaviour of "i" over "leaves". The only argument to this procedure specifies a sequence of disjoint subtrees of the original tree. The effect of the procedure is to flatten the original tree into terminal trees, by successively flattening its subtrees. Every time a terminal tree is found its only leaf is inserted into the list of leaves (cf. [1]). Similarly, evaluation in "ip" of "clause" with respect to "leaves", produces a procedure "lp" which directly mimicks the behaviour of "ip" over "leaves": -

```
lp(succeed, succeed).

lp(succeed, P): - lp(P, succeed).

lp((leaves(t(void, N, void), [N,..Z] - Z), C),
          P): - lp(P, C).

lp((leaves(t(Stl,N,Str),L - Z),C), P): - lp(P,
          (leaves(Stl,L - X), leaves(Str,X-Z),C)).
```

Of course, the transformation t relating "l" to "lp" is the same that relates "i" to "ip". In other words, let i[l] informally denote both the declarative and denotational semantics of program l when interpreted by interpreter i. We have: -

$$i[l] \subset i\,[t(l)] \equiv t(i)\,[l]$$

that is $i\,[l] \subset i\,[t(l)]$ and $i\,[l] \subset t(i)[l]$. The transformation t preserves both semantics.

Interpretation of non-transformed clauses. Bootstrapping

Both interpreters "i" and "ip" may access non-transformed clauses. To do so the following clause must be provided: -

```
clause ((P, C: - C)): - P.
```

which directly executes P whilst leaving the continuation C untouched. The only requirement, as before, is that P, during execution, be instantiated to some term instance, so not to remain as a free variable.

An instance of this clause, viz: -

```
clause (( clause((P: - C)), Cl: - Cl)): -
                clause ((P: - C)).
```

allows any of the interpreters to interpret the other or itself (bootstrapping).

A parallel interpreter of n goals

The interpreter "ip" can be readily generalized from two to n goals: -

```
ipn([] - []).

ipn([succeed,..PS] - Z): - ipn(PS - Z).

ipn([P,..PS] - [C,..Z]): - clause ((P: - C)),
        ipn(PS - Z).
```

This interpreter takes as argument a difference list of goal conjuncts. It processes the first goal of the first conjunct in the list for just one elementary execution step, inserts its continuation on the back of the list of conjuncts, and continues processing on the next conjunct in the list. If one conjunct succeeds, it continues processing the remaining ones, until eventually the list of conjuncts becomes empty.

A co-routining transformation

Reconsider the original "leaves" program. Define the new predicate

```
co-leaves (T1, L1, T2, L2):- leaves (T1, L1),
        leaves (T2, L2).
```

Now unfold it by symbolically evaluating the two calls to "leaves": -

```
co-leaves (t(void, N1, void), [N1,..Z1] - Z1, t(void,
        N2, void), [N2,..Z2] - Z2).

co-leaves (t(void, N1, void), [N1,..Z1] - Z1, t(Stl2,
N2, Str2), L2 - Z2): -
     leaves (Stl2, L2 - X2), leaves (Str2, X2 - Z2).

co-leaves (t(Stl1, N1, Str1), L1 - Z1, t(void,[N2,
void), [N2,..Z2] - Z2): -
     leaves (Stl, L1 - X1), leaves (Str1, X1 - Z1).

co-leaves (t(Stl1, N1, Str1), L1 - Z1, t(Strl2, N2,
Str2), L2 - Z2): -
     leaves (Stl1, L1 - X1), leaves (Str1, X1 - Z1),
     leaves (Stl2, L2 - X2), leaves (Str2, X2 - Z2).
```

Folding is now accomplished by substituting calls to "co-leaves" for pairs of calls to "leaves". If there were single calls to "leaves" these would be replaced by a call to "co-leaves" where the missing arguments would be those of the unit clause of "leaves". In general, they can be any arguments known to make true the predicate in question.

A number of sets of pairings are possible. Only one such set though preserves the procedural semantics of "leaves", in the sense that a call leaves (T, L - []) is procedurally equivalent to the call co-leaves (T, L - [], T, L - []) : -

```
co-leaves (t(void, N1, void), [N1,..Z1] - Z1, t(void,
N2, void),[N2,..Z2] - Z2).

co-leaves (t(void, N1, void), [N1,..Z1] - Z1, t(Stl2,
N2, Str2), L2 - Z2): -
        co-leaves (Stl2, L2 - X2, Str2, X2 - Z2).

co-leaves (t(Stl1, N1, Str1), L1 - Z1, t(void, N2, void),
[N2,..Z2] - Z2): -
        co-leaves (Stl1, L1 - X1, Str1, X1 - Z1).

co-leaves (t(Stl1, N1, Str1), L1 - Z1, t(Stl2, N2,
Str2), L2 - Z2): -
        co-leaves (Stl1, L1 - X1, Stl2, L2 - X2),
                co-leaves (Str1, X1 - Z1, Str2, X2 - Z2).
```

The first and fourth clauses cover the cases where both trees find a leaf and where both trees are further decomposed into subtrees, respectively. The second and third clauses are responsible for proper co-routining, i.e. they cover the cases where one tree finds a leaf but processing on that ree is suspended until the other tree produces some more leaves.

Other choices of pairings of calls to "leaves" produce different search behaviours, making interesting the corresponding transformations. We refrain here from the details.

Parallel and co-routining processing features of logic programming

The potential of logic programming for parallel and co-routining processing is described in [6], [7]. Parallel or co-routined execution of goals is advocated to be indicated by the user in some sort of control language or, alternatively, an especially smart interpreter could have the initiative of recognizing its usefulness in some parts of a program. In our opinion a control language, including specification of parallelism or co-routining, is needed anyway for allowing the user to freely specify the control he thinks best.

The importance and utility of such a control language has been argued in [4]. It should provide the programmer with the ability to specify appropriate sequencing of goal calls without affecting the meaning of programs, influencing only their efficiency and/or their manner of finding solutions for goals.

Four features of logic stand out, in parallel co-routined processing of logic programs.

First, sequencing of goals is arbitrary from the declarative semantics point of view, and thus meaning is not altered by the particular type of execution chosen.

Second, logic programs may be used as data for other logic programs, thus allowing clauses (which are terms) to be submitted as data do particular interpreters, written in logic. The clauses for the interpreters provide thus a semantics of parallelism.

Third, in logic instantanenous communication between predicates is ensured by the logic variable. Thus parallel or co-routined executions do not require any explicit interfacing for data flow. Unification does it all. Moreover, partly specified data structures in the form of terms containing variables may be constructed by one process and further completed by another through instantiation of those variables. Execution control can be effected either by a special interpreter, or by the clauses themselves, as we have shown. Fourth it is meaningful to define sequential, parallel and co-routined executions in pure logic programming terms, without recourse to implementation or machine dependent concepts and capabilities.

Next we show how parallel processing might be indicated in Prolog programs, by means of a simple syntax. This syntax simply specifies the arguments to a particular interpreter. In the following section we shall exhibit an interpreter which performs parallel execution of all the goals in a clause. Likewise, we could show an interpreter for executing several clauses at once for the same predicate, which combined with the previous one would give the ability, to perform breadth-first executions. We believe that composition of calls to special interpreters forms the basis of a simple control language.

In the two examples shown next we also display a mechanism for directing an execution to wait for another, concurrently executed with it, until some condition is met.

Consider the following definition of grandparent: -

grandparent (X, Z): - parent (X, Y), parent (Y, Z).

which acesses data base of unit clauses for "parent" indexed on both arguments. According to whether only X or Z are already instantiated in a call to "grandparent", it becomes convenient to execute one of its two goals first. This is so because the indexing of the clauses gives direct access to the relevant unit clauses if one of the arguments in the call is already instantiated. Thus, in the case where X or Z alone are instantiated, one would like one of the goals to await execution until execution of the other goal instantiates Y. This can be done using parallel processing. Rewrite "grandparent" as follows: -

```
grandparent (X, Z): - or ((atom (X), atom (Z))||wait
     or (atom (X), atom (Y)), parent (X, Y)|wait  or
     (atom (Z), atom (Y)), parent (Y, Z)||.

grandparent (X, Z): - and (var(X), var(Z)), parent (X,
     Y), parent (Y, Z).

or (P, Q): - P,!..

or (P, Q): - Q.

wait (P): - P,!.

wait (P): - wait (P).

and (P, Q): - P, Q.
```

where atom (A) is an implementation defined predicate which tests whether A is instantiated to an atom, and var (V) is also an implementation defined predicate that tests whether V is a variable which is not yet bound.

The effect of "wait" is to postpone its execution until the condition expressed in the argument is true.

The vertical bars indicate which groups of goal sequences are to be executed in parallel, where each sequence by itself is sequentially executed.

The first of these clauses would then be rewritten by the Prolog interpreter as: -

```
grandparent (X, Z): - or (atom (X), atom (Z)),
     ipn([(wait or (atom (X), atom (Y)),
     parent (X, Y), succeed), (wait or (atom (Z),
     atom (Y)), parent (Y, Z), succeed),..W] - W).
```

Another example of the use of "wait" combined with parallel execution refers to the use of negation as non-provability. This type of negation is accomplished in Prolog by the clauses: -

```
not (P): - P, !, fail.

not (P).
```

where "fail" always fails. A problem with this definition is that it does not in general respect the semantics of "not" in the case where P contains some unbound variables. For example, different solutions are found for

```
r(X): - not (p(X)), q(X).
```

and

```
r(X): - q(X), not (p(X)).
```

given

```
p(a).

q(b).
```

Now, for reasons of efficiency one might want the "not" to be executed first, unless X is not instantiated. This and similar problems can be solved by the use of "wait" in conjunction with parallelism. The above clause is then written: -

```
r(X): -|| wait (notvar (X)), not (p(X))| q(X)||.
```

where "notvar" is an implementation defined predicate which tests that X is not an unbound variable. This way, execution of not (P(X)) is postponed until X is bound, while execution of Q(X) goes on, eventually binding X.

Other interpreters

The next interpreter provides parallel execution of all the goals in a clause. I.e. it performs a breadth-first execution of the *and* nodes of a derivation tree.

```
ipb (succeed - succeed).

ipb((P, L) - (C, Z)): - clause ((P: - C)),
                        ipb (L - Z).
```

where parallel execution of two goals, P and Q say, is achieved by the call

```
ipb((P, Q, Z) - Z)
```

What "ipb" does it to insert each goal in a clause at the end of the sequence of goals to be executed in parallel, where each goal gives rise to a new separate execution.

A final interpreter is shown for completeness. It is just the sequential interpreter we started out from, re-written to account for the use of the "cut" in Prolog programs. It is meant only for Prolog knowledgeable readers.

Take a clause of the form

```
P: - Q, R, ! , S, !, T.
```

Imagine the interpreter re-writes it as

```
P: - ! ((Q, R, succeed)), ! ((S, succeed)), T.
```

The clauses for the interpreter which appropriatelly deals with such clauses are

```
i (succeed).

i (P): - clause ((P: - C)), (ifcut (C, Cl, C2),
               i(Cl), !, i(C2); ifnocut (C),i(C)).
```

```
i((!(C1), C2)): - i(C1), !, i(C2).
ifcut ((!(C1), C2), C1, C2).
ifnocut ((!(C1), C2)): - !, fail.
ifnocut (C).
```

THE SEMANTICS OF THE INTERPRETERS

The semantics of Prolog Programs

In this section we formalize the behaviour of Prolog programs interpreted by the interpreters presented above, thus complementing the intuitive descriptions made at the appropriate places. Our formalization is what might be called the procedural (operational) semantics of the interpreters, because it describes the computation sequences obtained when the interpreters are supplied with data.

We begin by formalizing Prolog programs themselves and their semantics. This is easy, if we consider only Prolog programs consisting exclusively of strictly logical features, which we shall do henceforth. Such programs will be called "declarations", following [12].

It is advantageous to depart here a little from the notations employed in the earlier part of the paper. This will be done as needed. For the time being, let us write unit clauses in the form "A: - " instead of the form "A.". We define now a *declaration* to be a (finite) set of clauses of the form

$$A: - A_1, \ldots, A_n$$

where $A, A_1, \ldots, A_n$ are (positive) literals and $n \geq 0$. We shall consider the literals inside each clause ordered from left to right and the clauses inside each declarations also ordered, from top to bottom as they are written.

The inputs to declarations are *goal statements*, which are clauses of the form

$$:- A_1,\ldots, A_n$$

where $A_1,\ldots, A_n$ are positive literals, called *goals* of the goal statement, and $n \geq 0$. The special case in which $n = 0$ is the *null clause*, usually denoted $\square$. As for any other clause, we consider the goals inside each goal statement as ordered.

Let $\underline{D}$ be a declaration. We say goal statement G *derives directly* a goal statement G′ iff

$$G \text{ is } :- A_1,\ldots, A_n \quad \text{with} \quad n > 0$$

and

$$G' \text{ is } :- B_1\theta,\ldots, B_m\theta, A_2\theta,\ldots, A_n\theta,$$

where there is a clause $B:-B_1,\ldots, B_m$ in $\underline{D}$ and θ is the most general unifier of A_1 and B. (Notice that if $n = 1$ and $m = 0$ then G′ is the null clause.) In this definition, the clause $B:- B_1,\ldots, B_m$ is assumed to have no variable in common with G; if necessary, the variables occurring in the clause may be renamed.

The goal statement G is said to *derive* a goal statement G′ iff there is a seguence of goal statements $G_o, G_1,\ldots, G_n$ $(n \geq 0)$, called a *derivation* of G′ from G, such that $G_o = G$, $G_n = G'$ and G_{i-1} derives directly G_i for $i = 1,\ldots, n$. A refutation of G is a derivation of the null clause from G. It is well known [5] that $\underline{D} \cup \{G\}$ is unsatisfiable iff there exists a refutation of G.

We now define a *derivation tree* for every declaration $\underline{D}$ and goal statement. G, as a tree containing all possible derivations from G. The root of the tree is labelled with G. If some node is labelled with a goal statement G_1, then there is an arc from this node to a node labelled with a goal statement G_2 iff G_1 derives directly, G_2. (When determining the direct descendents of a node labelled with G_1, a new node is created for every G_2 such that G_1 derives directly G_2, even if there is already a node labelled with G_2. Bearing this in mind, we shall from now on confuse the name of a node with its label, in order to simplify the exposition.) It is clear that every path starting at the root is a derivation from $\underline{G}$ and conversely. Thus $\underline{D} \cup \{G\}$ is unsatisfiable iff there is a finite branch of the derivation tree of G whose leaf is the null clause.

We order the direct descendents of each node of a derivation tree as follows: if G_2 and G_3 are direct descendents of a common node G_1, we say G_2 is *generated before* G_3 iff the clause used to derive G_2 from G_1 occurs first in $\underline{D}$ than the clause used to derive G_3 from G_1. With this definition we are apt to order the derivations from a goal statement G. Given two distint derivations

(D) $\quad G = G_o, G_1, \ldots, G_n$

(D') $\quad G = G'_o, G'_1, \ldots, G'_m$

we say D is *generated before* D' iff

- either $n < m$ and $G_i = G'_i$ for $i = 0, \ldots, n$;
- or, for some index $i \leq \min\{n, m\}$, $G_i \neq G'_i$, and if k is the least such index then G_k is generated before G'_k.

Further, we say D *precedes immediately* D' iff for no derivation D" is D generated before D" and D" generated before D'.

Let us say a derivation D is *final* iff it is a refutation or no other derivation is generated after it.

By the *computation* of the declaration D̲ with input goal statement G we mean the sequence of derivations beginning with the derivation consisting of a G alone and such that, for every derivation D in the sequence:

- either D is final, in which case it is the last derivation in the sequence;
- or precedes immediately the derivation that follows it in the sequence.

A computation will be said to be terminating if it is finite, otherwise *non-terminating*. A successfull computation is a terminating computation whose last derivation is a refutation.

We thus see that the computation of D̲ with input G is the top-down depth-first search of the first final derivation in the derivation tree of G. Notice that it may well happen that $\underline{D} \cup \{G\}$ is unsatisfiable, yet the computation of D̲ with input G is non-terminating. This is so iff there exists an infinite branch of the derivation tree of G which is generated before any finite branch whose leaf is the null clause.

This concludes our description of the procedural semantics of Prolog programs.

The sequential interpreter

Recall that the sequential interpreter consists of the two following clauses:

$$i(\sigma): -$$

$$i(P): - c((P:-C)),\ i(C)$$

where we are using for short 'σ' and 'c' instead of 'succeed' and 'clause' respectively.

Given a declaration $\underline{D}$ we define the declaration $i(\underline{D})$ to consist of the two above clauses plus a unit clause

$$c((A,\ C: -\ A_1, \ldots,\ A_n,\ C)): -$$

for each clause $A: -\ A_1, \ldots,\ A_n$ in $\underline{D}$. We assume that the order of these clauses in $i(\underline{D})$ is the same as the order of the respective original clauses in $\underline{D}$. We assume further that the predicate symbols 'i' and 'c' as well as the constant 'σ' are not among the predicate symbols and constants occurring in $\underline{D}$.

If G is an input goal statement to $\underline{D}$, say

$$: -\ A_1, \ldots,\ A_n$$

we denote the term $A_1, \ldots,\ A_n,\ \sigma$ by G^σ (if n = o, so that G is □, we have $G^\sigma = \sigma$). This way,

$$: -\ i\ (G^\sigma)$$

is the input goal statement $-\ i((A_1, \ldots,\ A_n,\ \sigma))$ to $i(\underline{D})$. (Note that in the term $A_1, \ldots,\ A_n,\ \sigma$ the comma is just a binary functor used in infix notation, with association on the right implicitly assumed. The extra-parenthesis above specify that $A_1, \ldots,\ A_n,\ \sigma$ is a single term and not a sequence of arguments.)

We want to relate the computation of $\underline{D}$ with input G with the computation of $i(\underline{D})$ with input $:- i(G^\sigma)$. For the rest of this section we shall keeps this notation fixed.

Let D be a derivation $G = G_o, G_1, \ldots, G_k$ from G in $\underline{D}$. Denote by D_{i1} the following sequence of goal statements:

$$:- i\,(G_o^\sigma)$$

$$:- c((G_o^\sigma \;:- C)),\; i(C)$$

$$:- i(G_1^\sigma)$$

$$\vdots$$

$$:- i(G_k^\sigma\;).$$

Denote further by D_{i2} the previous sequence D_{i1} followed by the null clause, if G_k is the null clause, or by $:- c((G_k^\sigma \;:- C));\; i(C)$ otherwise

LEMMA 1.

(a) D_{i1} and D_{i2} are derivation from $:- i(G^\sigma)$ in $i(D)$, and every such derivation has one of these two forms.

(b) D_{i1} precedes immediately D_{i2}.

(c) If D' is another derivation from G in $\underline{D}$ such that D precedes immediately D', then D_{i2} precedes immediatelly D'_{i1}.

Proof

(a) We shall only prove that D_{i1} is a derivation $:- i(G^\sigma)$ in $i(\underline{D})$. It is clearly enough to prove that if G_j

derives directly G_{j+1} in $\underline{D}$ then :- $i(G_j^\sigma)$ derives directly :- $c((G_j^\sigma$:- $C))$, $i(C)$ and this last goal statement derives directly :- $i(G^\sigma{}_{j+1})$ in $i(\underline{D})$. Let G_j and G_{j+1} be respectively

$$:- A_1, \ldots, A_n \quad (n < 0)$$

and

$$:- B_1 \ , \ldots, B_m \ , A_2 \ , \ldots, A_n$$

where there is a clause B:- $B_1, \ldots, B_m$ in $\underline{D}$ such that is the most general unifier of A_1 and B. Then :- $i(G_j^\sigma)$ is

$$:- i((A_1, \ldots, A_n, \sigma))$$

which derives directly

$$:- c((A_1, \ldots, A_n, \sigma :- C)), i(C)$$

that is

$$:- c((G_j^\sigma \ :- C)), i(C)$$

by means of the clause $i(P)$:- $c((P$:- $C))$, $i(C)$ of $i(D)$. Now the clause.

$$c((B, C' :- B_1, \ldots, B_m, C')):-$$

is in $i(\underline{D})$ and $c((A_1, \ldots, A_n, \sigma$:- $C))$ is unifiable with it, with most general unifier the substitution σ' given by

$$\theta' = \theta \ \{C/(B_1 \ \theta, \ldots, B_m \ \theta, C')\} \ \{C'/(A_2 \ \theta, \ldots, A_n \ \theta, \sigma)\}.$$

Therefore $:- c((G_j^\sigma \; :- C)), i(C)$ derives directly

$$:- i((B_1\,\theta, \ldots, B_m\,\theta, A_2\,\theta, \ldots, A_n\,\theta, \sigma))$$

which is $:- i(G_{j+1}^\sigma)$.

(b) Clear.

(c) It is clear that if D is generated before D' then D_{i2} is generated before D'_{i1}. The statement of (c) follows immediately from this. □

THEOREM 2.

If $D^o, D^1, \ldots, D^n, \ldots$ is the computation of $\underline{D}$ with G then

$$D^o{}_{i1}, D^o{}_{i2}, D^1{}_{i1}, D^1{}_{i2}, \ldots, D^n{}_{i1}, D^n{}_{i2}, \ldots$$

is the computation of $i(\underline{D})$ with input $:- i(G^\sigma)$.

Proof

Use the lemma and induction on k to prove that the (2k)-th (resp. (2k+1)-th) derivation in the computation of $i(\underline{D})$ with input $:- i(G^\sigma)$ is D_{i1}^k (resp. D_{i2}^k). □

We may obtain another description of the "behaviour" of $i(\underline{D})$ with input $:- i(G^\sigma)$ if we restrict our attention to goal statements of the form $:- i(G'^\sigma)$. By an *i - derivation* from $:- i(G^\sigma)$ in $i(\underline{D})$ we mean the sequence of goal statements of the form $:- i(G'^\sigma)$ obtained from some derivation from $:- i(G^\sigma)$ in i(D). By the lemma, every i - derivation is of the form $:- i(G_o^\sigma)$, $:- i(G_1^\sigma), \ldots, :- i(G_k^\sigma)$ for some derivation $G = G_o, G_1, \ldots, G_k$ from G. If we denote this last derivation from G in $\underline{D}$ by D, the resulting i - derivation will be denoted i(D). By the *i - computation* of i($\underline{D}$) with input :- i(G^{σ}) we mean the sequence of i - derivations, without repetition, obtained from the computation of i($\underline{D}$) with input :- i(G^{σ}). From the theorem it follows that if D^{o}, D^{1},..., D^{n},... is the computation of $\underline{D}$ with input G then i(D^{o}), i(D^{1}),..., i(D^{n}),... is the i - computation of i($\underline{D}$) with input :- i(G^{σ}).

The parallel interpreter

The parallel interpreter consists of the following clauses

i_p (σ, σ): -

i_p (σ, P): - i_p (P, σ).

i_p (P, Q): - c ((P:- C)), i_p (Q, C).

As for the sequential interpreter, we associate with every declaration $\underline{D}$ a declaration $i_p(\underline{D})$, which is defined in exactly the same way as i($\underline{D}$). It consists of the three above clauses plus a clause

c((A, C:- A_1,..., A_n, C)):-

for every clause A: - A_1,..., A_n in $\underline{D}$.

Let G and H denote respectively the terms $A_1,\ldots,A_n$ and $B_1,\ldots, B_m$ (n, m ≥ 0), where the A's and B's are (positive) literals. Denote further $A_1,\ldots,A_n$, σ and $B_1,\ldots, B_m$, σ by G^{σ} and H^{σ} respectively (this notation departs a little from the one previously used), so that :- G, H and :- $i_p(G^{\sigma}, H^{\sigma})$ are inputs to $\underline{D}$ and $i_p(\underline{D})$ respectively.

It would be interesting to compare the computations of $\underline{D}$ with input :- G, H and of $i_p(\underline{D})$ with input :- $i_p(G^\sigma, H^\sigma)$. (In the first case the terms G and H are computed sequentially, and in the latter they are computed in parallel.) It may happen, however, that the computation of $\underline{D}$ is successfull and the computation of $i_p(\underline{D})$ is non-terminating, and conversely. Consider, for example, the following declaration $\underline{D}$:

q(f,(X), Y):- q(X, Y)
q(X, g(Y)):- q(X, O)
q(O, O):-
m(O, O):-
m(f (X), O):- m(X, O)
m(f (X), g (Y)):- m(f (X), g(Y))

The computation of $\underline{D}$ with input :-q(f(O), Y), m(f (O), Y) is non-terminating and the computation of $i_p(\underline{D})$ with input :- i_p((q(f(O), Y),σ), (m(f(O), Y), σ)) is successfull. This fact makes it difficult to compare the above mentioned computations.

We shall follow another approach to describe the computation of $i_p(\underline{D})$. Reconsider the definition of "direct derivation" presented before. According to this definition, to derive directly a given goal statement from another goal statement we had to match the *first* literal in the last goal statement with the head of some clause, and the first goal statement was obtained by resolution without factoring or merging. The reason for using such a restrictive notion of "direct derivation" is because this is the way Prolog has been implemented. We shall now abandon this restriction and consider the possibility of selecting other literals than the first in goal statements for matching with the heads of clauses. Consider the following method of obtaining a derivation in D with top goal statement :- G, H. Every goal statement occurring in the derivation may be written in the form :- G_k, H_k, where G_k or H_k or both may be empty, so that the derivation itself is of the form :- G_o, H_o; :- G_1, H_1;...; :- G_n, H_n. Here G_o is G and H_o is H. Consider now the goal statement :- G_k, H_k. If G_k and H_k are both empty then :- G_k, H_k is the null clause and consequently is the last goal statement in the derivation. Otherwise let us distinguish two cases:

(1) k is even: if G_k is not empty select the first literal in G_k, else the first literal in H_k.

(2) k is odd: if H_k is not empty select the first literal in H_k, else the first literal in G_k.

In either case, the selected literal is matched with the head of some clause in D (if such a clause exists) and the next goal statement is obtained by resolving the given goal statement and clause without factoring or merging. The resulting goal statement can be written in the form :- G_{k+1}, H_{k+1} where if the literal chosen in :- G_k, H_k belongs to G_k, then :- G_k derives :- G_{k+1} and H_{k+1} is an instance of H_k; else :- H_k derives :- H_{k+1} and G_{k+1} is an instance of G_k.

Such a derivation will be called a *p - derivation* from :- G, H in D, where the 'p' indicates that G and H are executed in parallel (see above for the justification of the last statement). We speak of a *p - refutation* when the last goal statement in a p - derivation is the null clause. We may say of two p - derivations when one of them is generated before the other, and therefrom we define the concepts of "p - derivation tree" and "p-computation".

We now describe the computation of $i_p(\underline{D})$ with input :- $i_p(G^\sigma, H^\sigma)$ by describing its i_p - computation, to be defined below, and comparing it with the p - computation of $\underline{D}$ with input :- G, H. The concept of i_p - computation is analoguous to the concept of i - computation defined in the section on the sequential interpreter. By an i_p - *derivation* of :- $i_p(G^\sigma, H^\sigma)$ in $i_p(\underline{D})$ we mean the sequence of goal statements of the form :- $i_p(-,-)$ obtained from some derivation from :- $i_p(G^\sigma, H^\sigma)$ in $i_p(\underline{D})$. The i_p - *computation* of $i_p(\underline{D})$ with input :- i_p (G^σ, H^σ) is the sequence of i_p - derivations, without repetitions, obtained from the computation of $i_p(\underline{D})$. By a lemma analoguous to lemma 1, we may associate with every p - derivation D from :- G, H a unique i_p - derivation $i_p(D)$ from :- i_p (G^σ, H^σ), and every i_p - derivation may be obtained in this way. Next, by analogy with Theorem 2, we may say that if $D^o, D^1, \ldots, D^n, \ldots$ is the p - computation of D with input :- G, H then i_p (D^o), $i_p(D^1), \ldots, i_p(D^n), \ldots$ is the i_p-computation of $i_p(\underline{D})$ with input :- $i_p(G^\sigma, H^\sigma)$.

This finishes our description of the behaviour of $i_p(\underline{D})$ with input :- $i_p(G^\sigma, H^\sigma)$. From this description we may conclude that the parallel interpreter is an implementation in Prolog of the "parallel strategy" described above for the execution of $\underline{D}$ when the goals in the input goal statements for $\underline{D}$ are separated into two sets G and H. Notice that if G (or H) is empty then the parallel execution of :- G, H coincides with the usual (sequential) execution of :- H (or:-G). We conclude this section by showing that this parallel strategy is complete.

THEOREM 3.

Of $\underline{D} \cup \{:- (G, H)\}$ is unsatisfiable then there is a p - refutation of :- G, H in $\underline{D}$.

Proof

Our proof will be directed towards reducing the problem of finding a p-refutation of :- G, H to the problem of finding a refutation of :- G, H with respect to a convenient selection function defined for goal statements based on $\underline{D}$. We therefore start by defining such a selection function ρ and then use a result by [5] (Theorem 3, p. 26) which assures us of the existence of a refutation of :- G, H with respect to ρ. We conclude the proof by showing that this refutation is also a p-refutation.

We say a goal statement is "based" on $\underline{D}$ iff all the function and predicate symbols occurring in it also occur in $\underline{D}$. By a "selection function" for $\underline{D}$ we mean a function ρ assigning to each goal statement based on $\underline{D}$ (except the null clause) a literal occurring in it. A goal statement G "ρ-derives directly" a goal statement H iff H is obtained from G and a clause in $\underline{D}$ whose head matches ρ(G) by resolution without factoring or merging. Let us define a selection function ρ for all non-null goal statements based on $\underline{D}$. This will be done by constructing a sequence $\underline{G}_o$, $\underline{G}_1, \ldots, \underline{G}_k, \ldots$ of ordered sets of goal statements based on $\underline{D}$, and ρ will be defined on these sets by induction on k. Further, every goal statement in any set $\underline{G}_k$ will have the form :- G , H . $\underline{G}_o$ is to consist of the single goal statement :- G, H, which we may assume to be different from the null clause, and ρ selects the first literal in G if G is not empty, otherwise the first literal in H. Assume that $\underline{G}_k$ and the selection function σ on $\underline{G}_k$ have already been defined. A goal statement :- G", H" belongs to $\underline{G}_{k+1}$ iff some :- G', H' in $\underline{G}_k$ σ-derives directly :- G", H" and :- G", H" does not belong to $\underline{G}_i$ for any $i \leq k$. The goal statements in $\underline{G}_{k+1}$ are ordered in the obvious fashion: for :- G''_1, H''_1 and :- G''_2 in $\underline{G}_{k+1}$, pick the first :- G'_1, H'_1 and :- G_2, in $\underline{G}_k$ which ρ-derive directly :- G''_1, H''_1 and :- G''_2, H''_2 respectively; if :- G_1, H_1 is equal to :- G'_2, H'_2 then :- G''_1, H''_1 occurs in $\underline{G}_{k+1}$ before :- G''_2, H''_2 iff the clause used to generate :- :- G''_1, H''_1 occurs in $\underline{D}$ before the clause used to generate :- G''_2, H''_2; otherwise, :- G''_1, H''_1 occurs before :- G''_2, H''_2 iff :- G'_1, H'_1 occurs before :- G'_2, H'_2. To define ρ on $\underline{G}_{k+1}$ let :- G", H" belong to $\underline{G}_{k+1}$ and :- G', H' be the first goal statement in $\underline{G}_k$ which ρ-derives directly :- G", H". If G" (resp. H") is empty then ρ selects the first literal in H" (resp. G"), unless H" (resp. G") is also empty, in which case :- G", H" is the null clause; otherwise ρ selects the first literal in G" if it selected the first literal in H , else it selects the first literal in H". This completes the definition of ρ for goal statements belonging to the sets $\underline{G}_k$. For any other goal statement based on $\underline{D}$, ρ is defined arbitrarily. Now according to the already mentioned result by [5] it follows that there is a ρ-refutation of :- G, H in $\underline{D}$. This implies that the null clause belongs to some G_k. We may then construct a ρ-refutation of :- G, H consisting of a sequence of k+1 goal statement such that the i-th goal statement belongs to $\underline{G}_{i-1}$. It is clear that this refutation is also a p-refutation, thus finishing the proof of the theorem. □

The n-parallel interpreter

This interpreter is entirely analogous to the previous one, with the only difference that any number of goal statements may be executed in parallel, instead of two; thus the use of lists instead of n-tuples. For this reason we shall attempt no description of the computation of this interpreter, which would be in all relevant ways similar to the description of the computation of the parallel interpreter.

Co-routining

Let there be given a declaration D and suppose p and q are respectively an n-place and an m-place predicate symbols occurring in D. When presented with an input :- $p(t_1, \ldots, t_n)$, $q(t'_1, \ldots, t'_m)$ D will compute the literals $p(t_1, \ldots, t_n)$ and $q(t'_1, \ldots, t'_m)$ sequentially. We shall define a transformation on D such that when the transform of D is presented with a convenient input it can be said to compute the two above literals co-routiningly. This transformation consists in eliminating from D the symbols p and q, in creating a new (n+m) - place predicate symbol p & q, and in substituting the clauses in D where p & q occur by new clauses. The transform of D will be denoted D (p & q).

Let us call a p-*literal* a literal of the form $p(x_1, \ldots, x_n)$ and a p-*clause* a clause whose head is a p-literal. We define similarly q-literal and q-clause. We shall begin by defining D(p & q) only in the case where there exist a unit p-clause and a unit q-clause in D. Later we shall indicate how to define D(p & q) when this requirement is not met.

The declaration $\underline{D}(p \& q)$ is the union of the three following sets of clauses:

(1) The set of all clauses in $\underline{D}$ where neither p nor q occur.

(2) The set of all clauses C obtained by the following procedure applied to clauses C' in $\underline{D}$ containing p or q but which are neither p-clauses nor q-clauses:

(a) if the number of p-literals in the body of C' is different from the number of q-literals, adjoin to the body of C' as many p-literals or q-literals as necessary so as to make their numbers equal; these p-literals (resp. q-literals) may be any instances of heads of unit p-clauses (resp. q-clauses) in $\underline{D}$;

(b) set up a bijective correspondance between the p-literals and the q-literals in the body of the clause obtained in (a) and substitute each pair $(p(x_1,\ldots, x_n), q(y_1,\ldots, y_m))$ in correspondance by the p & q-literal $p \& q(x_1,\ldots, x_n, y_1,\ldots, y_m)$.

(3) The set of all clauses C_{12} obtained by applying the following procedure to a p-clause C'_1 and a q-clause C'_2 in $\underline{D}$:

(a) rename the variables occurring in C'_2 (for instance) so that they are different from those occurring in C'_2 (for instance) so that they are different from those occurring in C'_1;

(b) if the head of C'_1 is $p(x_1,\ldots, x_n)$ and the head of C'_2 is $q(y_1,\ldots, y_m)$, construct a clause C''_{12} whose head is $p \& q(x_1,\ldots, x_n, y_1,\ldots, y_m)$ and whose body is the set of literals which belong to the body of C'_1 or of C'_2;

(c) finally, transform C''_{12} into C_{12} by applying to C''_{12} the procedure applied to C' in (2).

The input to D(p & q) associated with the input :- $p(t_1,\ldots, t_n)$, $q(t'_1,\ldots, t'_m)$ to D is :- p & q $(t_1,\ldots, t_n, t'_1,\ldots, t'_m)$. Let us abbreviate these inputs to :- p, q and :- p & q respectively.

THEOREM 4.

D ∪ {:- (p, q)} is unsatisfiable iff D(p & q) ∪ {:- p & q} is unsatisfiable (i.e. the declarative semantics are equivalent).

Proof

Suppose there is a model of D ∪ {:- (p,q)}. This model is a subset M of the Herbrand base of D ∪ {:-(p, q)}.

Let M(p & q) be the subset of the Herbrand base of D(p & q) ∪ {:- p & q} consisting of the literals in M which are neither p-literals nor q-literals, plus a p & q-literal p & $q(x_1,\ldots, x_n, y_m)$ for every p-literal $p(x_1,\ldots, x_n)$ and every q-literal $q(y_1,\ldots, y_m)$ in M. It is clear that M(p & q) is a model of D(p & q) ∪ {:-p & q}. Conversely, suppose M(p & q) is a model of D(p & q) ∪ {:- p & q}. A model M of D ∪ {:- (p, q)} can be constructed by letting M consist of all the non-p p & q-literals in M(p & q) plus literals $p(x_1,\ldots, x_n)$ and $q(y_1,\ldots, y_m)$ for every p & $q(x_1,\ldots, x_n, y_1,\ldots,y_m)$ in M(p & q). This finishes the proof. □

In the definition of D(p & q) we did not bother to order the clauses inside D(p & q) and the literals inside each clause for two main reasons. In the first place, this was not necessary to state and prove Theorem 4. More importantly, we did not want to impose unnecessary restrictions from the outset on the kinds of co-routining of p and q that could be obtained. But now we want to compare the computation of D with input :- $p(x_1, \ldots, x_n)$ with the computation of D(p&q) with input p & $q(x_1, \ldots, x_n, b_1, \ldots, b_m)$ where $q(b_1, \ldots, b_m)$ is an instance of the head of some unit q-clause. (By symmetry, we also want to compare the computation of D with input :- $q(y_1, \ldots, y_m)$ with the computation of D(p & q) with input :- p & $q(a_1, \ldots, a_n, y_1, \ldots, y_m)$, where $p(a_1, \ldots, a_n)$ is an instance of the head of some unit p-clause, but this comparison is made in a manner entirely analogous to the previous one so we will not mention it any further.) This will dictate the orderings D(p & q) must be supplied with.

We may assume that the clauses in D are ordered thus: all p-clauses come first, then come q-clauses, and finally the remaining clauses. Indeed, as far as any computation of D is concerned, we may permute at will the clauses of D, provided that the clauses in each of the three sets above maintain their relative sequential positions. We may arrange the clauses in D(p & q) so that the p & q-clauses come first, followed by the remaining clauses. The relative positions of the remaining clauses are the same as the relative positions of the respective clauses in D. As to the p & q-clauses, let C_{12} (resp. C_{34}) be a p & q-clause obtained from a p-clause, C'_1 (resp. C'_3) and a q-clause C'_2 (resp. C'_4); then C_{12} occurs before C_{34} iff either C'_1 occurs before C'_3 or $C'_1 = C'_3$ and C'_2 occurs before C'_4.

In order to arrange the literals inside the clauses of D(p & q) we have to alter slightly the definition of

D(p & q). Theorem 4 still remains valid, however. We fix an instance $p(a_1,\ldots, a_n)$ of the head of a unit p-clause, and assume that this unit p-clause occurs in D before any other p-clause with whose head $p(a_1,\ldots, a_n)$ matches. We fix similarly $q(b_1,\ldots, b_m)$ and make the corresponding assumption. Now recall that D(p & q) is the union of three sets of clauses. Accordingly, we order the literals in the clauses of D(p & q) in three steps:

(1') The clauses of this set also belong to D, so the order in which the literals are written is the same as in D.

(2') Here the clause C is now obtained from C' by rewriting each p-literal $p(x_1,\ldots, x_n)$ (resp. each g-literal $q(y_1,\ldots, y_m)$) as $p \,\&\, q(x_1,\ldots, x_n, b_1,\ldots,b_m)$ (resp. $p \,\&\, q\ (a_1,\ldots, a_n, y_1,\ldots, y_m)$), while leaving the ordering of the remaining literals unchanged.

(3') Let C_{12} be the p & q-clause obtained from the p-clause C'_1 and the q-clause C'_2.

a') If neither C_1 nor C_2 are unit clauses, or if they are both unit clauses, then C_{12} is defined as before and the literals are ordered arbitrarily.

b') Otherwise order the literals in the body of C''_{12} as they are ordered in the one of C'_1 and C'_2 which is not a unit clause, and transform C''_{12} into C_{12} by applying to C''_{12} the procedure applied to C' in (2').

For any goal statement G′ based on D let G be the goal statement based on D(p & q) obtained by applying to G' the procedure applied to C' in (2'). We may now state the following theorem, whose (easy) proof we omit.

THEOREM 5.

(a) If $D' = (G'_o, G'_1, \ldots, G'_k)$ is a derivation from $:- p(x_1, \ldots, x_n)$ in $\underline{D}$, then $D = (G_o, G_1, \ldots, G_k)$ is a derivation from $:- p \,\&\, q(x_1, \ldots, x_n, b_1, \ldots, b_m)$ in D (p & q).

(b) If $D'_1, D'_2, \ldots, D'_r, \ldots$ is the computation of $\underline{D}$ with input $:- p(x_1, \ldots, x_n)$, then $D_1, D_2, \ldots, D_r, \ldots$ is the computation of $\underline{D}$(p & q) with input $:- p \,\&\, q(x_1, \ldots, x_n, b_1, \ldots, b_m)$.

Thus the declaration $\underline{D}$(p & q) can be used to simulate the original declaration $\underline{D}$.

We end this section by indicating how $\underline{D}$(p & q) can be defined whenever there do not exist in $\underline{D}$ unit p-clauses or unit q-clauses or both. Suppose there do not exist unit p-clauses but a unit q-clause exists. We assume there is a ground p-literal $p(a_1, \ldots, a_n)$ such that $\underline{D} \cup \{:- p(a_1, \ldots, a_n)\}$ is unsatisfiable. Then we define $\underline{D}'$ to be $\underline{D} \cup \{p(a_1, \ldots, a_n): -\}$, where this new clause is placed before any other clause, and let $\underline{D}$(p & q) to be $\underline{D}'$(p & q). Notice that Theorem 5 is no longer valid as stated. The q-version of it, however, remains valid.

CONCLUSIONS

Logic programming is highly suitable for parallel and co-routined modes of processing. First, it supports a natural self-contained definition of parallelism in terms of a single elementary event - the match of a goal with a clause. Second, it allows freedom in the ordering of goal executions, whilst preserving the declarative semantics of programs. Third, the logical variable, through unification, automatically provides for process interfacing without execution or code overheads. Fourth, because logic clauses are terms they can be given as data to a parallel interpreter also written in logic. This provides a clear semantics of parallelism.

REFERENCES

[1] Burstall, R.M., Darlington, J.: A transformation system for developing recursive program *Journal of ACM*, Vol. 24, (1977), No 1, pp. 44-67.

[2] Clark, K., Tärnlund S. A.: A first order theory of data and programs, *Proceedings of the IFIP Congress' 77*, North-Holland, Amsterdam, 1977.

[3] Colmerauer, A.: Les grammaires de metamorphose, Groupe d'Intelligence Artificielle, Université d'Aix-Marseille II, Marseille, 1975.

[4] Hayes, P.J.: Computation and deduction, *Proceedings of MFCS Conference*, Czechoslovakian Academy of Sciences. 1973.

[5] Hill, R.: LUSH resolution and its completeness, DCL memo 78, Dept of AI, Edinburgh, 1974.

[6] Kowalski, R.: Logic for problem solving, DCL memo 75, Dept. of AI, Edinburgh, 1974.

[7] Kowalski, R.: Predicate Logic as a programming language, *Proceedings of the IFIP Congress' 74*, North-Holland, Amsterdam, 1974, pp. 569-574.

[8] Pereira, L.M.: Prolog - uma linguagem de programacao em lógica Divisao de Informática, Laboratório Nacional de Engenharia Civil Lisbon. 1977.

[9] Pereira, L.M., Pereira, F., Waren D.H.O.: User's guide to DECsystem-10 Prolog. Divisao de Informática, Laboratório Nacional de Engenharia Civil Lisbon. 1978.

[10] Robinson, J.A.: A machine - oriented logic based on the resolution principle, *Journal of ACM*, vol. 12 (1965), pp. 23-24.

[11] Tärnlund, S.-A.: Horn clause computability, *BIT*, vol. 17 (1977), pp. 215-226.

[12] van Emden, M.H.: Programming with resolution logic, Report Cs-75-30, Dept. of Computer Science, University of Waterloo, Canada, 1975.

[13] Warren, D.H.D.: Implementing Prolog - compiling predicate logic programs, Report no 39, Dept. of AI, Edinburgh, 1977.

[14] Warren, D.H.D., Pereira, L.M., Pereira, F.C.N.: Prolog - the language and its implementation compared with Lisp, ACM Symposium on AI and Programming Languages, Rochester, *SIGART Newsletter*, *SIGPLAN Notices*, August 1977.

Pereira, L.M., Monteiro, L.F.

Departamento de Informática
Universidade Nova de Lisboa

QUINTA DO CABEÇO, 1899 LISBOA
PORTUGAL

COLLOQUIA MATHEMATICA SOCIETATIS JÁNOS BOLYAI
26. Mathematical Logic in Computer Science,
Salgótarján (HUNGARY), 1978.

ON THE GENTZEN TYPE PROOF THEORY FOR PROGRAM ANALYSIS

R. Pliuskevicius

1. *Introduction*

As shown by the Bakker in [1] the axiomatic Hoare's system [2] does not give sufficient characteristics of the loop operator: in Hoare's system it is impossible to prove the true equivalence of these programs: "*while* A *do* α" and "*if* A *then* (α ; *while* A *do* α)". That is why it is necessary for program analysis to have axiomatic systems prossessing better deductive possibilities than Hoare's system. Logical-program systems (LP-systems) are suggested as possible axiomatic systems of such a kind. The objects of consideration in these systems are LP-sequents, whose members are the so-called LP-formulas, i.e. the objects, constituted from Hoare's formulas A{α}B by means of second order logic. In LP-formulas the quantifiers are allowed according to both dummy program variables and predicate ones that enables us to apply LP-systems not only to program analysis but also to program synthesis. Though in LP-systems the property of total correctness is not expressed, it is possible to prove the total correctness of programs in these systems too, analogously as in [3] or [4].

This paper deals with some questions of the Gentzen proof theory for LP-systems: we form systems H_1 and H_2 in which LP-structural rules are eliminable, including LP-cut. Some corollaries from the theorem on the elimination of LP-structural rules are presented, namely, consistency of LP-systems H_i /i=1,2/, analogues of the "midsequent" theorem, analogues of Kreig and Harrop's theorems, etc. Some structural transformations of LP-formulas are considered, particular by the generalization of the famous theorem on "the monotone replacement".

Bounded intuitionistic second order logic being a logical basis of the LP-systems under consideration enables us to avoid non-constructive deductive structures.

2. *Initial definitions*

Formulas for second order logic L-formulas and terms are constructed in a usual way (see, e.g. [5]). Basic programs in the consedered version of LP-systems are

1/ dummy program variables (interpreted as indivisible program constructions);

2/ one-place operator, named a test*/by Pratt in [6] and denoted by A? where A is an L-formula;

---

* Tests are analogous to "guards" introduced by Dijkstra in [7]; apart from that test A? corresponds to the notion of the marked assertion A-if in the system VCG [8].

3/ a simple assignment $(x:=t)$. Programs are formed of basic programs with the help of operations $\circ$, $|$, $*$. A degenerate case is admitted when an empty word is a program. $(\alpha \circ \beta)$ stands for successive realization of programs α and β, $(\alpha \mid \beta)$ stands for non-deterministic realization of α or β, α^* denotes a repeated realization of α for a non-deterministic number of times. It is known that tests and non-deterministic operations $|$ and $*$ allow us to express the operators "*while do*" and "*if then else*": *if* A *then* α *else* β $= (A? \circ \alpha)|(\neg A? \circ \beta)$; *while* A *do* $\alpha = (A? \circ \alpha)^* \circ \neg A?$. Primary LP-formulas are expressions of the type $\Theta\{\alpha\}\Delta$ where Θ, Δ are lists*/ of L-formulas /L-lists/, α is a program. The case when α is an empty word will be denoted by $\Theta \supset \Delta$.

Let us define the concept of the LP-formula:

1/ an arbitrary primary one is the LP-formula.

2/ If $\mathfrak{A}$, $\mathfrak{B}$ are LP-formulas, then expressions $(\mathfrak{A} \odot \mathfrak{B})$ /where $\odot \in \{\supset, \&, \vee\}$) and $\neg(\mathfrak{A})$ are LP-formulas.

3/ If $\mathfrak{A}$ is an LP-formula, x is a first or second order variable or a dummy program variable, then expression $\forall x \mathfrak{A}$ and $\exists x \mathfrak{A}$ are LP-formulas.

LP-sequence are expressions of the type $\Gamma \rightarrow \Omega$,

---

*/ Expression $A_1,\ldots,A_n$ $\{\alpha\}$ $B_1,\ldots,B_m$ is interpreted as an assertion stating that if L-formula $\underset{i=1}{\overset{n}{\&}} A_i$ is true before execution of program α and α halts, then L-formula $\bigvee_{i=1}^{n} B_i$ holds afterward.

where Γ is a list /possibly empty/ consisting of LP-formulas, Ω is an arbitrary LP-formula or an empty word. Γ is called antecedent while Ω is succedent of the sequent $\Gamma \rightarrow \Omega$. Let us consider that no variable occurs in the LP-sequent both free and bound simultaneously. Let us call an LP-sequent, the members of which are only primary LP-formulas, a P-sequent.

Consider any part φ of some given LP-formulas; this part may be a LP-formula, logical connective, program operator or any type of variable. Following Herbrand let us define positive and negative occurence of part φ in the LP-formula under consideration. First consider a case when φ is neither a quantifier complex $\exists x$ nor a program operator :=, * or o. Part φ occurs in φ positively. Connectives &, $\vee$, $\forall$, $\exists$ as well as program operators do not change the sign of occurence. If φ occurs in the LP-formula $\mathfrak{A}$ positively /negatively/, then φ occurs in the LP-formula $\neg \mathfrak{A}$ negatively /positively/, i.e. symbol $\neg$ changes its sign of occurence. The sign of occurence in the LP-formula of the type / $\mathfrak{A} \supset \mathfrak{B}$ / is calculated in the same way as that of occurence in the LP-formula /$\neg \mathfrak{A} \vee \mathfrak{B}$ /. Let part φ be a quantifier complex $\exists x$ or a program operator :=, *, o. If the part controlled by φ occurs in the LP-formula positively /negatively/ then φ occurs negatively /or positively, correspondingly/. For instance, in the LP-formula $A\{\alpha^*\}B \supset C\{\beta^*\}D$ occurence of the LP-formula $A\{\alpha^*\}B$ is negative, occurence of the LP-formula $C\{\beta^*\}D$ is positive; left occurence of operator * is positive while the right one is negative. The sign of occurence of part φ in the LP-sequent S is identical with that of part φ in the LP-formula, if the considered occurence of part φ is in the succedent of S; while, it is opposite to the sign of occurence of part φ in the LP-formula if φ occurs in the antecedent of the LP-sequent S.

Let us call a LP-formula $\mathfrak{A}$ well-formed, if

1/ all the dummy program variables are bound by negative occurences of quantifiers;

2/ if x is the left part of assignments in the LP-formula $\mathfrak{A}$, then x is not the eigen-variable of a quantifier complex in the LP-formula. In the sequel only well-formed LP-formulas will be considered.

Let S_1 and S_2 be arbitrary LP-formulas or LP-sequents, I be an arbitrary system, considered in this work. Assume S_1 to be I-provable if S_1 is provable by means of system I; say S_1 to be I-equivalent to S_2 if $S_1 \supset S_2$ and $S_2 \supset S_1$ to be I-provable. The proof in system I is called I-proof.

Traditional concepts of the logical proof theory, namely, ancestral relationships in proof, height of proof, side and principal formulas, an admissible and inversible rule, etc. are determined in a usual way /see, e.g. [5], [10]/.

3. *Systems* H_1 *and* H_2

Let us describe the Gentzen type LP-systems H_1 and H_2. Arbitrary LP-sequents of the type $\Gamma_1, \Sigma, \Gamma_2 \rightarrow \Omega$ are axioms for system H_1, if $\Sigma \rightarrow \Omega$ is a logical sequent provable by second order predicate calculus.

The rules of inference of system H_1 will be divided into three groups. The first group consists of LP--structural rules /weakening, contraction, exchange and cut rules/:

$$\frac{\Gamma_1, \Gamma_2 \to \Omega}{\Gamma_\lambda, \mathfrak{A}, \Gamma_2 \to \Omega} \text{ LPW}; \quad \frac{\Gamma_1, \mathfrak{A}, \Gamma_2, \mathfrak{B}, \Gamma_3 \to \Omega}{\Gamma_1, \mathfrak{B}, \Gamma_2, \mathfrak{A}, \Gamma_3 \to \Omega} \text{ LPE};$$

$$\frac{\Gamma_1, \mathfrak{A}, \Gamma_2, \mathfrak{A}, \Gamma_3 \to \Omega}{\Gamma_1, \mathfrak{A}, \Gamma_2, \Gamma_3 \to \Omega} \text{ LPC}; \quad \frac{\Gamma_1 \to \mathfrak{A} ; \Gamma_2, \mathfrak{A}, \Gamma_3 \to \Omega}{\Gamma_1, \Gamma_2, \Gamma_3 \to \Omega} \text{ LP-cut};$$

$$\frac{\Gamma_1 \to \theta \supset B ; \Gamma_2 \to \theta_1, B, \theta_2 \{\alpha\} \Delta}{\Gamma_1, \Gamma_2 \to \theta_2, \theta, \theta_2 \{\alpha\} \Delta} \text{ C1}; \quad \frac{\Gamma_1 \to \theta \{\alpha\} \Delta_1, B, \Delta_2 ; \Gamma_2 \to B \supset \Delta}{\Gamma_1, \Gamma_2 \to \theta \{\alpha\} \Delta_1, \Delta, \Delta_2} \text{ C2}.$$

The second group contains the rules for program operators:

$$\frac{\Gamma \to \sigma_1 \; ; \; \Gamma \to \sigma_2}{\Gamma \to \sigma} \to := ; \quad \frac{\Gamma_1, \sigma_1', \sigma_2', \Gamma_2 \to \Omega}{\Gamma_1, \sigma, \Gamma_2 \to \Omega} := \to ,$$

where

$$\sigma \leftrightharpoons \theta\{\gamma \circ (x:=t)\} \Delta \quad ;$$
$$\sigma_1 \leftrightharpoons \theta\{\gamma\}[R]^x_t \quad ;$$
$$\sigma_2 \leftrightharpoons R \supset \Delta \;;$$
$$\sigma_1' \leftrightharpoons \theta\{\gamma\}[C^n(x_1,\ldots,x_n)]^x_t \;;$$
$$\sigma_2' \leftrightharpoons C^n(x_1,\ldots,x_n) \supset \Delta$$

where C^n is a n-place /n=0,1,.../ predicate variable, not occuring in the conclusion of the rule of inference $:= \to$.

$$\frac{\Gamma \to \nu_1}{\Gamma \to \nu} \to ? \quad ; \quad \frac{\Gamma_1, \nu_1, \Gamma_2 \to \Omega}{\Gamma_1, \nu, \Gamma_2 \to \Omega} ? \to ,$$

where

$$\nu_1 \rightleftharpoons \theta\{\gamma\}\neg P, \Delta \quad ; \quad \nu \rightleftharpoons \theta\{\gamma_0 P?\}\Delta .$$

$$\frac{\Gamma \to \lambda_1 ; \ \Gamma \to \lambda_2}{\Gamma \to \lambda} \to | \quad ; \quad \frac{\Gamma_1, \lambda_1, \lambda_2, \Gamma_2 \to \Omega}{\Gamma_1, \lambda, \Gamma_2 \to \Omega} | \to ,$$

where $\lambda_1 \rightleftharpoons \theta\{\gamma \circ \alpha\}\Delta$; $\lambda_2 \rightleftharpoons \theta\{\gamma \circ \beta\}\Delta$; $\lambda \rightleftharpoons \theta\{\gamma \circ (\alpha | \beta)\}\Delta$.

$$\frac{\Gamma \to \tau_1 ; \Gamma \to \tau_2 ; \Gamma \to \tau_3}{\Gamma \to \tau} \to * , \quad \frac{\Gamma_1, \tau_1', \tau_2', \tau_3', \Gamma_2 \to \Omega}{\Gamma_1, \tau, \Gamma_2 \to \Omega} * \to$$

where $\tau_1 \rightleftharpoons \theta\{\gamma\}R$; $\tau_2 \rightleftharpoons R\{\alpha\}R$; $\tau_3 \rightleftharpoons R \supset \Delta$; $\tau \rightleftharpoons \theta\{\gamma \circ \alpha^*\}\Delta$; τ_i' /i=1,2,3/ is obtained from τ_i by substituting a given concurence of the L-formula R with the L-formula $C^n (x_1, \ldots, x_n)$, where C^n is a n-place /n=0,1,.../ predicate variable, not occuring in the conclusion of the rule of inference $* \to$.

The third group includes the rules of intuitionistic second order logic without comprehension axioms /see, e.g. [5]/. The shape of these rules is absolutely analogous to that of rules from the intuitionistic sequential system without structural rules of inference /see, e.g. [9]/. For instance the rules for introduction and elimination of the second order universal quantifier are of the following shape:

$$\frac{\Gamma \to [\mathfrak{A}]^{x^n}_{y^n}}{\Gamma \to \forall x^n \mathfrak{A}} \to \forall \quad ; \quad \frac{\Gamma_1, [\mathfrak{A}]^{x^n}_{y^n}, \forall x^n \mathfrak{A}, \Gamma_2 \to \Omega}{\Gamma_1, \forall x^n \mathfrak{A}, \Gamma_2 \to \Omega} \forall \to ,$$

where y^n in $\forall\to$ stands for arbitrary second order n-places variable or the n-places predicate constant; y^n in $\to\forall$ stands for arbitrary second order n-places variable not occuring in the conclusion of $\rightarrow \forall$. Since only well-formed LP-formulas are considered to be the quantifier rules of inference, connecting dummy program variables are the rules of introduction of an existential quantifier and elimination of a universal quantifier. For instance, the rule $\rightarrow \exists$ has the form:

$$\frac{\Gamma \rightarrow [\mathfrak{A}]^x_\alpha}{\Gamma \rightarrow \exists x\ \mathfrak{A}},$$

where x is a dummy program variable, α is an arbitrary program.

System H_2 is obtained from system H_1 by eliminating the given occurence of program γ from the definition of the second group rules of inference and by adding the following rules for the composition operator:

$$\frac{\Gamma \rightarrow \rho_1; \Gamma \rightarrow \rho_2}{\Gamma \rightarrow \rho} \rightarrow\circ \quad ; \quad \frac{\Gamma_1, \rho_1', \rho_2', \Gamma_2 \rightarrow \Omega}{\Gamma_1, \rho, \Gamma_2 \rightarrow \Omega} \circ\rightarrow ,$$

where

$$\rho_1 \leftrightharpoons \theta\{\alpha\}R \ ; \quad \rho_2 \leftrightharpoons R\{\beta\}\Delta \ ; \quad \rho \leftrightharpoons \theta\{\alpha\circ\beta\}\Delta \ ;$$

ρ_i' /i=1,2,/ is obtained from ρ_i by substituting the given occurence of the L-formula R with the L-formula $C^n(x_1,\ldots,x_n)$, where C^n /n=0,1,.../ has the same meaning as in $*\rightarrow$.

System H_i^r /i=1,2/ is obtained from system H_i by eliminating all the rules of the first group.

From the shape of operator rules it is obvious that the new formula in the premises of the rules appears not only for the loop operator /such an effect is observed in the succedent rule in usual Hoare's system/, but also in the premises of assignment rules.

Note that the number of arguments in the new predicate variable C^n in antecedent rules $:=\rightarrow$ and $*\rightarrow$ of system H_i is not fixed in advance and in general it is not sufficient to take a new propositional variable as $C^n(x_1,\ldots,x_n)$. Let us take some examples confirming these assertions. Let I_1 be a system, obtained from system H_1^r by substituting rule $\rightarrow:=$ with the following one:

$$\frac{\Gamma \rightarrow \theta\{\gamma\}[\Delta]_t^x}{\Gamma \rightarrow \theta\{\gamma\circ(x:=t)\}\Delta}.$$

Then in system I_1 LP-sequent $A(t), \neg A(x) \rightarrow M\{x:=t\}N$, where A, M, N are different predicate variables, is not provable. On the other hand this LP-sequent is provable in the system, obtained from system I_1 by adding the structural rule C2.

Let I_2 be a system, obtained from system H_1^r by substituting rule $:=\rightarrow$ with

$$\frac{\Gamma_1,\theta\{\gamma\}[\Delta]_t^x,\Gamma_2 \rightarrow \Omega}{\Gamma_1,\theta\{\gamma\circ(x:=t)\}\Delta,\Gamma_2 \rightarrow \Omega};$$

I_3 be a system, obtained from system I_2 by substituting rule $*\rightarrow$ with the rule, in which a new propositional variable is instead of $C^n(x_1,\ldots,x_n)$. Then in system I_3 LP-sequent

$$A\{(x:=t)^*\}B \rightarrow \exists y^1(y^1(x) \supset y^1(t))$$

is not provable. At the same time this sequent is provable in system I_2 as well as in system H_1^r.

4. *Some structural transformations of LP-formulas*

For illustration we shall give several examples of the application of LP-system for some structural transformations of LP-formulas.

THEOREM 1. /Implicative replacement of programs/. *Let* α, β, γ *be arbitrary programs,* γ' *be a program resulting from the replacement of program* β *instead of some occurence of program* α *in* γ *. Then the following LP-sequent*

$$\forall y^n \mathfrak{B}^n(y^n(x_1,\dots,x_n)\{\alpha\}\mathfrak{B}^n(x_1,\dots,x_n) \supset y^n(x_1,\dots,x_n)\{\beta\}\mathfrak{B}^n(x_1,\dots,x_n)) \to$$
$$M^n(x_1,\dots,x_n)\{\gamma\}N^n(x_1,\dots,x_n) \supset M^n(x_1,\dots,x_n)\{\gamma'\}N^n(x_1,\dots,x_n)$$

is H_i*-provable, where* M^n , N^n *are n-place predicate variables.*

THEOREM 2. /Equivalent replacement of program/. *Let* α, β, γ, γ' *be programs, related to each other as in the above theorem. Then the following LP-sequent*

$$\forall y^n \mathfrak{B}^n(y^n(x_1,\dots,x_n)\{\alpha\}\mathfrak{B}^n(x_1,\dots,x_n) \equiv y^n(x_1,\dots,x_n)\{\beta\}\mathfrak{B}^n(x_1,\dots,x_n)) \to$$
$$M^n(x_1,\dots,x_n)\{\gamma\}N^n(x_1,\dots,x_n) \equiv M^n(x_1,\dots,x_n)\{\gamma'\}N^n(x_1,\dots,x_n)$$

is H_i*-provable.*

THEOREM 3. /Monotone replacement of LP-formulas/. *Let* $\mathfrak{A}$, $\mathfrak{B}$, $\mathfrak{M}$ *be arbitrary LP-formulas and LP-formula* $\mathfrak{M}^*$ *be obtained as a result of the replacement of LP-formula* $\mathfrak{B}$ *instead of some positive /negative/ occurence of LP-formula* $\mathfrak{A}$ *in* . *Then the LP-sequent* $\tilde{\forall}(\mathfrak{A} \supset \mathfrak{B}) \to (\mathfrak{M} \supset \mathfrak{M}^*)$ /$\forall(\mathfrak{A} \supset \mathfrak{B}) \to (\mathfrak{A}^* \supset \mathfrak{M})$, *respectively/ is* H_i*-provable, where* $\forall(\mathfrak{A} \supset \mathfrak{B})$ *denotes the closure of free variables in the LP-formula* $(\mathfrak{A} \supset \mathfrak{B})$.

5. Some properties of systems H_1 *and* H_2

We shall present some accertions specifying certain properties of the introduced LP-systems H_i and H_i^r /here and in the sequel $i=1,2$/.

THEOREM 4. *The following LP-sequents are* H_i^r*-provable:*

a/ $\mathfrak{A} \rightarrow \mathfrak{A}$

b/ $\theta\{\alpha\}\theta$ *under the condition that* $L[\alpha] \cap V[\theta] = \emptyset$, *where* $L[\alpha]$ *is a set of the left part of assignments in* α*;* $V[\theta]$ *is a set of all the first order free variables in list* θ*;*

c/ $\rightarrow\theta\{\alpha\}\Delta_1, T, \Delta_2$, *where* T *is an arbitrary true* L*-formula;* $\rightarrow\theta_1, F, \theta_2\{\alpha\}\Delta$, *where* F *is an arbitrary false L-formula*

THEOREM 5. *Logical and structural transformation rules in* L*-lists of the primary* LP*-formulas are admissible in system* H_i.

Note, that in Hoare's system the admissibility of two transformation rules in L-list,correspoinding,to the rules $\rightarrow\&$ and $V\rightarrow$ /in logical sequential systems/ is proved in [8].

THEOREM 6. *Whatever the* LP*-sequents* S *and* S', *if the* LP*-sequent* S' *is obtained from* S *by means of the structural rule* LPW *or* LPE *then according to the* H_i^r*-proof of the* LP*-sequent* S *it is possible to construct the* H_i^r *-proof of the* LP*-sequent* S', *and the heights of both proofs are equal.*

THEOREM 7. *All the rules of system* H_1 *are inversible, excluding:*

1/ structural rules of LP*-cut and* Cl, C2

2/ the rules of the second group $\rightarrow *$, $*\rightarrow$, $\rightarrow:=$, $:=\rightarrow$

3/ the rules of the third group $\neg\rightarrow$, $\rightarrow\vee$ *and* $\rightarrow\exists$.

THEOREM 8./Elimination of structural rules in system H_i/. *Rules* LPW, LPE, LPC, LP*-cut,* Cl *and* C2 *are admissible in system* H_i^r, *i.e. systems* H_i *and* H_i^r *are equivalent.*

THEOREM 9. *Systems* H_1 *and* H_2 *are equivalent.*

THEOREM 10. *According to any* H_i^r*-proof of the* LP*-sequent* S *it is possible to construct the* H_i^r*-proof of the same* LP*--sequent* S, *in which all the axioms are such that they do not contain program operators.*

6. *Some corollaries of the structural rules elimination theorem*

Let us illustrate some application of theorem 8. H_i-proof will be regarded to be regular /semiregular/, if it does not contain the applications of the first group rules /if only the LPW rule is apllied of all of them, respectively/.

Since there is no possibility to construct a regular H_i-proof of an empty LP-sequent then from Theorem 8 follows:

THEOREM 11. *Systems* H_1 *and* H_2 *are simply consistent.*

Any LP-formula is an immediate subformula of the given LP-formula $\mathfrak{A}$, if it may be a side formula to $\mathfrak{A}$ as principal formula in any rule of H_i^r-proof. Then the subformula of the given LP-formula $\mathfrak{A}$ is $\mathfrak{A}$ itself, its immediate subformulas, the immediate subformulas of each of the latter, etc. From the traditional concept of a subformula it follows, however, that even arbitrary LP-formulas, containing only those predicate symbols which are missing in the LP-formula $\mathfrak{A}$, can be subformulas of the given LP-formula $\mathfrak{A}$. However, the following theorem holds.

THEOREM 12. /Ancestor and subformula property/. *In* H_i^r*-proof of any* LP*-sequent* S*:*

a/ each LP*-formula occurence in* S *is identifiable as an ancestor of a specific* LP*-formula occurence in the* LP*-endsequent; and the former* LP*-formula is a subformula of the latter;*

b/ each occurence of a program operator in S *is identifiable as an ancestor /or ancestral image/ of a specific occurence of the same operator in the* LP*-endsequent.*

THEOREM 13. /Sign property/. *In* H_i^r*-proof of any* LP*-sequent* S *each image of a positive /negative/ occurence of part* φ *in* S *is positive /negative/ in* S_1 *which is higher than* S.

THEOREM 14. /Theorem on P-midsequent/. *According to an arbitrary semiregular* H_i*-proof of* LP*-sequent* S, *not containing either*

a/ negative occurences of quantifiers $\forall$ *and* $\exists$

or

b/ positive occurences of symbols $\circ$, $*$ *and* $:=$, *it is possible to construct a* H_i*-proof* $\vee$ *of sequent* S, *each branch of which contains some* P*-sequents /called the* P*-midsequent/, and a part of the branch from this occurence of the* P*-midsequent to* LP*-endsequent* S, *consists of the* LPW *rules and the rules of the third group.*

THEOREM 15. /Interpolation theorem for LP-formulas/ *Let an* LP*-formula* $(\mathfrak{A} \supset \mathfrak{B})$ *be* H_i*-provable, then there exists such a* L*-formula* C *that*

1/ $(\mathfrak{A} \supset C)$ *and* $(C \supset \mathfrak{B})$ *are* H_i*-provable,*

2/ the list of predicate symbols of LP*-formula coincides with that of* L*-formula* C.

It ought to be noted that the interpolant i.e. L-formula C can be constructively formed from H_i-proof of LP-formula $(\mathfrak{A} \supset \mathfrak{B})$. On the other hand if in the statement of the theorem not L-formula but LP-formula were used for an interpolant, then the theorem would be trivial: LP-formula $\mathfrak{A}$ would be the desired interpolant.

THEOREM 16. /analogue of Harrop's theorem /see [5]//. *Let* Γ *be a finite sequence of* LP*-formulas such that* Γ *does not contain positive /negative/ occurences of symbols* $\vee$ */*$\exists$, $*$ *and* $:=$, *respectively/, then*

1/ *LP-sequent* $\Gamma \to \mathfrak{A} \vee \mathfrak{B}$ *is* H_i*-provable if and only if* $\Gamma \to \mathfrak{A}$ *or* $\Gamma \to \mathfrak{B}$ *are* H_i*-provable;*

2/ $\Gamma \to \exists x\ \mathfrak{A}$ *is* H_i*-provable if and only if for some* φ $\Gamma \to [\mathfrak{A}]^x_\varphi$ *is* H_i*-provable, where*

   a/ φ *is some program, if* x *is a dummy program variable;*

   b/ φ *is some term if* x *is a first order variable,*

   c/ φ *is a* n*-place second order variable or a* n*-place predicate constant, if* x *is a* n*-place second order variable;*

3/ $\Gamma \to \theta\{\alpha\circ\beta^*\}\Delta$ *is* H_i*-provable if and only if for some* L*-formula* R $\Gamma \to \theta\{\alpha\}R$; $\Gamma \to R\{\beta\}R$; $\Gamma \to R \supset \Delta$ *are* H_i*-provable;*

4/ $\Gamma \to \theta\{\alpha\circ(x:=t)\}\Delta$ *is* H_i*-provable if and only if for some* L*-formula* R $\Gamma \to \theta\{\alpha\}[R]^x_t$ *and* $\Gamma \to R \supset \Delta$ *are* H_i*-provable.*

7. *Conclusions and some perspectives*

The mathematical proof theory represents a topical branch of mathematical logic. There are good reasons to belive that the program proof theory will become an independent branch of the programming theory having a direct relation to programming in practice. In this paper some problems of the so-called structural proof theory for operators of simple assignment, tests, non--deterministic choice and iteration were considered, using some extension of Hoare's system [2]. It would be reasonable

1/ to consider analogous questions for richer program constructions, including the procedures and abstract data types, and also for various constructions of parallel programming;

2/ to investigate other aspects of the program proof theory connected with the model theory and the complexity of algorithms;

3/ to develop the program proof theory based on dynamic logic [6] and temporal logic.

References

1 J.W. de Bakker, The fixed point approach in semantics: theory and applications, *Foundations of Computer Science* (J.W.de Bakker,ed.) Mathematical Centre Tracts 63, 1975, 3-53.

2 C.A.R.Hoare An axiomatic basis for computer programming, *Comm.ACM*, 12, 1969, 576-580, 583.

3 D.C.Luckham,Suzuki, Proof termination within a weak logic of programs, *Acta Informatica* 8, 1977, 21-36.

4 D.Harel, A.Pnueli, J.Stavi, A complete axiomatic system for proving deductions about recursive programs, *Proc. 9-th Ann. ACM Symp. on theory of computing*, 1977, 249-260.

5 G.Takeuti, *Proof theory*, North-Holland Publ.Co.N.Y., 1975.

6 S.L.Litvinchouk, V.R.Pratt, A proof-checker for dynamic logic, *Proc. 5-th Intern.Conf. on Artif. Intel.*, 1977, 552-558.

7 E.W.Dijkstra, Guarded commands, non-determinacy and formal derivation of programs. *Comm.ACM*, 18,1975,453-457.

8 S.Igaraschi, R.London, D.C.Luckham, Automatic Program Verification I: A logical basis and its implementation, *Acta Informatica*, 4, 1975, 145-182.

9 R.A.Pliuskevicius, On a variant of the constructive predicate calculus without structural rules of inferences, *Dokl. Akad. Nauk SSSR*, 161, 1965, 292-295.

10 S.C.Kleene, *Mathematical Logic*, John Wiley & Sons, 1967.

R. PLIUSKEVICIUS

Institute of Mathematics and Cybernetics
of the Academy of Sciences of the
Lithuanian SSR
232600 Vilnius, 54 Poželos str.
USSR

COLLOQUIA MATHEMATICA SOCIETATIS JÁNOS BOLYAI
26. Mathematical Logic in Computer Science,
Salgótarján (HUNGARY), 1978.

LOGIC AND COMPLEXITY OF SYNCHRONOUS PARALLEL COMPUTATIONS

S. Radziszowski

ABSTRACT

We investigate a certain model of synchronous parallelism. Syntax, semantics and complexity of programs within it are defined. We consider algorithmic properties of synchronous parallel programs in connection with sequential programs with arrays. The complexity theorem states that the class PP-time (polynomial-time bounded parallel languages) is equal to P-space (languages requiring polynomial amount of memory).

INTRODUCTION

In the recent years many papers appeared investigating different kinds of models for synchronous parallel computations. In general, they are divided into two groups, which are dealing with:

1. Creation of new formal algerbraic models for parallel computations, such as vector machines (Pratt, Stockmeyer [9]), alternating Turing machines (Chandra, Stockmeyer [2]), M-Ram, C-Ram (Simon [6]), conglomerates (Goldschlager [5]) and others.
2. Practical parallel programming in languages not formally defined, with intuitive semantics. Even examples of algorithms from the first group of papers are often written in such languages.

Our goal is to present a certain very natural language with the complete definition of syntax and semantics. On the basis of this language we investigate some algorithmic properties of parallelism, the complexity of programs written in it and we give some examples of algorithms. Such an approach might turn out to be directly applicable to future parallel computers.

The general idea of our model is as follows:

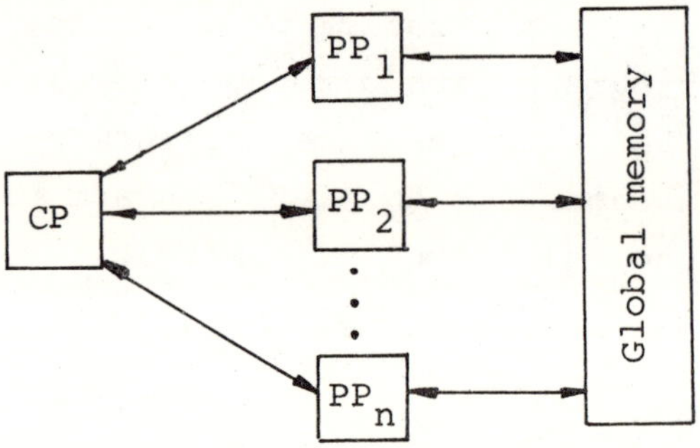

CP - the control processor stores the text of a program and synchronizes the actions of other processors $PP_1, PP_2, \ldots, PP_n, \ldots$

PP_i - $i>0$, there is an unbounded number of processors indexed by natural numbers, acting in parallel, all of them using one global memory M. In the language syntax there does not exist any notion of processor. We have only to specify which instructions should be executed in parallel.

The whole structure acts in sequential and parallel steps. A sequential step is the execution of one standard statement at a given moment. One parallel step has two stages. First the number of active processors and their allocation are computed by the control processor CP. After that every active processor PP_i performs one sequential instruction (which can be a sequential program in general). This model is stronger than SIMDG - single instruction stream, multiple data stream, global memory (Flynn [4]) because in our model different processors PP_i can perform in one step different programs, contrary to SIMDG, where two parallel processors must be executing the same instruction if they are both active in a given time. However, in our model the number of different instructions executed in parallel is syntactically bounded by the text of the program stored in the control processor CP.

By synchronous parallel program we shall intuitively mean the program whose computation is deterministic and all parallel instructions are mutually separated. All processors act in tacts, that means there are syntactically defined points in the program, where all processors are forced to synchronize their actions.

We use only synchronous parallel computations, because we are interested in fast algorithms solving particular problems. Up to now asynchronous model of parallelism is mostly used for on-line computations, operating system and so on. There exist also some numerical asynchronous parallel algorithms (Kung [8]).

Asynchronous parallel programming does not apply to particular problem solving, such as language recognition. It seems that by removing synchronization we canot essentially improve the complexity of algorithm, for the worst case complexity of asynchronous algorithm will always remain not smaller than the complexity of the corresponding synchronized version. At most, an average running time of the program can be decreased.

I. PARALLEL PROGRAMS

Let R be a relational system:

$$R = \langle A, \{f_l\}, \{r_j\} \rangle,$$

where the set of natural numbers forms a subset of A and f_i, r_j are the functors and predicates of R. Let us denote by FS_R the set of sequential programs in R, i.e. containing substitutions and closed under composition (block statement), condition (if statement) and iteration (while statement), (see Banachowski et al [1]).

The set of variables V consists of infinite sets V_b, V_s, V_n, V_i:

$$V = V_b \cup V_s \cup V_n \cup V_i$$

where all of them are pairwise disjoint and

- V_b - boolean variables,
- V_s - standard individual variables valuated into the set A,

V_i - index variables valuated into the set of natural numbers.

Let $\overline{V}_n$ be an infinite set of names, then:

$V_n = \{x(k): x \in \overline{V}_n \text{ and } k \in N\}$ is the set of simple indexed variables.

The terms and the formulas are built by induction on the basis of variables from the set V and functors and predicates from R.

If the valuation $v: V_b \cup V_s \cup V_i \cup V_n \to A$ is given, then we shall use an extended valuation v for complex indexed variables of the form $x(\tau)$, where τ is a term. In this case we define:

$$x(\tau)\,(v) = \begin{cases} x(\tau(v)) & \text{if } \tau(v) \text{ is a natural number} \\ \text{undefined} & \text{otherwise} \end{cases}$$

Let us denote by $\overline{FS}_R$ an extended set of sequential programs, where the set of variables is equal to $\overline{V} = V_b \cup V_s \cup V_i \cup \{\text{simple and complex indexed variables}\}$.

Definition:

Parallel instruction is the program of the form:

cobegin $(I_1 \square \rho_1), \ldots, (I_r \square \rho_r)$ *coend*

where:

1. ρ_j is the relation programmable in R and it is writen in the form $K\alpha$ for some program $K \in \overline{FS}_R$ and an open formula α, for $j=1,\ldots,r$. (cf. [1]).
2. I_j for $j=1,\ldots,r$ is a sequential program from $\overline{FS}_R$.

3. For all $j=1,\ldots,r$ the set of free index variables in I_j and ρ_j is the same.

4. For all $j=1,\ldots,r$ any index variable in I_j can not occur as a left side of substitution. (This restriction is implied by semantics, because ρ_j will assign those variables on which program I_j will be executed in parallel)[1]. □

In some cases, which are not involving any confusion, we will use the simplified notation for parallel instructions, for instance:

a) *cobegin* I□ρ *coend* if r = 1 ,

b) *cobegin* I, J *coend* if I, J does not contain any index variable .

Definition:

The set of *parallel programs* PP is the smallest set satisfying the following conditions:

1. $\overline{FS}_R$ is included in PP;

2. Parallel instruction is a parallel program;

3. PP is closed under composition, condition and iteration on the basis of programs from $\overline{FS}_R$. □

Before the definition of semantics let us give an example of program sorting n different elements given in array B[1:n]. Our example is a somewhat improved

---

1) This condition is to avoid the variable conflict as in

```
begin  k:=2;  i:=2;
    cobegin  X(k):=4;  X(i):=0 coend
end
```

version of the algorithm from Goldschlager [5]. In the formalism of PP-programs we can restrict the range of an index j in *while* statement I2 to the interval $1 \le j \le n/2^k$.

Example 1

B[1],...,B[n] - elements from the ordered set A to be sorted

$R = \langle A \cup N, \le_A, \text{arithmetic in } N \rangle$.

For the sake of convenience we shall use multiindexing of variables, i.e. a (i,j) instead of a (number of the pair (i,j)).

Program K:

```
begin comment perform all comparisons;
 I1: cobegin if B[j]≥B[i] then less(i,j):=1 else
                              less(i,j):=0□
                              1≤i,j≤n coend;
     comment compute the number of inpute less than
             or equal to input B[i];
 I2: K:=1; while k≤log2n do
           begin
            cobegin less(i,j):=less(i,2j-1) +
            less (i,2j) □1≤i≤n∧1≤j≤n/2^k coend;
             k:=k+1
           end;
     comment re-arrange the input numbers;
 I3: cobegin B[less(i,1)]:= B[1]□ 1≤i≤n  coend
end;
```

Our program K sorts elements B[1],...,B[n] in time O(logn), assuming n is a power of 2.

II. SEMANTICS OF PARALLEL PROGRAMS

Let K be a parallel program from PP and v an initial valuation into the set A. We shall define output valuation $v' = K_R(v)$.

1. $K \in \overline{FS}_R$, K is a sequential program, by standard inductive way we put $v' = K_R(v)$. (cf.[1]).
2. K = *cobegin* $(I_1 \Box \rho_1), \ldots, (I_r \Box \rho_r)$ *coend*

   Let S_j be the set of all free variables from ρ_j.

   Denote:

   $$T_j = \{(n_1, \ldots, n_{k_j}) : \rho_j(n_1, \ldots, n_{k_j})(v) = 1\} \text{ for}$$

   j=1,... The set T_j is the set of all sequence of index variables satisfying ρ_j. For each such sequence $n_1, \ldots, n_k$ a separate processor will execute program I_j, assuming it will not lead to the conflict. In order to omit conflicts we have to force the actions of all processors to be independent inside the parallel instruction. Formally, if we define t^j_ξ for $\xi \in T_j$:

   t^j_ξ = {the set of all variables occuring in I_j as a left side of substitution, while the initial valuation for program I_j is given by v changed by ξ on variables from S_j (denote it by v^j_ξ)}

   then the resulting valuation v′ will be defined if:

   a) all sets T_j, $j=1,\ldots,r$, are finite
   b) all sets t^j_ξ, for $j=1,\ldots,r$ and ξ ranging over T_j, are pairwise disjoint

   and v′ is given by:

$$v' = K(v) = w \cup \bigcup_{j,\xi} I_j(v_\xi^j) \qquad (*)$$

where w is equal to the valuation v restricted to the set $V \setminus \bigcup_j S_j \setminus \bigcup_{j,\xi} t_\xi^j$. The formula (*) describes independent actions of $\sum_j |T_j|$ processors, which are allowed to change the valuation in the separate parts of memory t_ξ^j, but they have the possibility to read all variables.

Following conditions a), b) v' is correctly defined. Note, that v' can be undefined on the part of set $\bigcup_j S_j$.

3. if $K = K_1;K_2$ then $K_R(v) = K_{2R}(K_{1R}(v))$

if $K =$ *if* α *then* K_1 *else* K_2 then:

$$K_R(v) = \begin{cases} K_{1R}(v) & \text{if } \alpha_R(v) = 1 \\ K_{2R}(v) & \text{if } \alpha_R(v) = 0 \end{cases}$$

if $K =$ *while* α *do* K_1 then:

$$K_R(v) = \begin{cases} v & \text{if } \alpha_R(v) = 0 \\ K_R(K_{1R}(v)) & \text{if } \alpha_R(v) = 1 \end{cases}$$

Note that the output valuation v' can become undefined in three cases:

i) by the infinite loop in the *while* statement as in sequential programs

ii) by variable conflict as in 2b) of semantics definition, i.e. instruction of the form:

cobegin x:=a; x:=b *coend*

iii) by the requirement of an infinite number of processors (point 2a) of semantics definition).

Let us notice also that the conflict of the form:

cobegin x:=b; y:=x *coend*

is solved in the definition. Following the remark 4b) from the syntax definition appropriate sets will be like this:

$$T_1 = T_2 = \{\emptyset\}, \qquad S_1 = S_2 = \{\emptyset\}$$

$$v^1 = v^2 = v \;, \quad t^1 = \{x\}, \quad t^2 = \{y\}, \quad t^1 \cap t^2 = \emptyset$$

$$v' = (v \text{ restricted to } V - \{x,y\}) \cup [x/b](v^1) \cup [y/x](v^2)$$

The result is equivalent to the following sequential program:

```
begin  yl:=y; xl:=x;
       xl:=b; yl:=x;
       y:=yl; x:=xl  end
```

In a real computer acting in parallel the computation can be performed in the following way:

All sequential statements outside the parallel instructions are executed in a standard way. The general assumption about hardware is that all processors have access to the whole global memory. Each variable can be read many processors at the same moment, but only one processor can change the value of the variable at a given time. Before each parallel instruction the control processor computes the set of all sequences of index values satisfying ρ_j, whose cardinality gives the number of required processors. This can be done in parallel by the operating system. Control processor does not have to check whether the number of processors is finite. CP can simply print out the computated number or inform that its capacity is too small to activate all desired processors.

For every sequence satisfying ρ_j a new processor is activated which executes the program I_j in its local memory. The results are copied into the global memory. This is described in the semantics definition. The variable conflicts as in 2b) can be checked in running time by special marking of changed variables inside parallel instruction. The execution of parallel instruction is terminated when all processors have executed their programs.

III. ALGORITHMIC PROPERTIES OF PARALLEL PROGRAMS

Let $K \in PP$ be a parallel program, v - a valuation of its variables and α a formula. By $K\alpha$ we shall mean the formula with the following definition of valuation:

$$(K\alpha)_R(v) \overset{df}{=} \alpha_R(K_R(v))$$

In the same manner as in sequential case we would like to prove partial and total correctness of a program K in the structure R with respect to the formulas α and β, i.e. to prove the following formulas:

$(K1 \wedge \alpha) \rightarrow K\beta$ - partial correctness

$\alpha \rightarrow K\beta$ - total correctness

In the last two chapters of the paper we have to make some restrictions on the form of the programmable relation ρ_j. If we allow ρ_j to be an arbitrary formula $K\alpha$, then even the problem of finiteness ρ_j could become undecidable. In order to obtain effectiveness and complexity theorem we assume that ρ_j is given by a system of linear inequalities (with respect to index variables), where coefficients are arithmetic terms over standard variables.

In this case there exist fast algorithms checking satisfiability of ρ_j and computing the set of solutions for ρ_j (the special case of linear programming). On the other hand this restriction seems to be reasonable, because the structure of parallelism still remains powerful enough in practice.

THEOREM 1

There exists an effective semantically equivalent translation between the parallel and sequential programs with arrays.

Proof:

$\Leftarrow$ every sequential program is a parallel one. The cell of array can be treated as a single indexed variable. In such a transformed program, that is equivalent to the initial one, there are no simultaneous statements.

$\Rightarrow$ It is enough to give an efficient semantically equivalent transformation of an arbitrary parallel program to the sequential one with arrays. It is obviously sufficient to do this for a parallel instruction. One parallel instruction can be transformed as follows:

1. For every ρ_j, $j=1,\ldots,r$ compute the set T_j of sequences satisfying ρ_j. If T_j is not finite (in the case of system of linear inequalities the problem is decidable) then stop without result.
2. Following the semantics definition check if there exists unavoidable variable conflict (point ii)). If yes then stop with undefined result.

3. Treat indexed variables with the same name as an array. Working on local memory for every sequence from $\bigcup_j T_j$ execute program I_j (sequentially).

4. Copy results into the global memory.

As an immediate consequence we have the following:

COROLLARY

The problem of partial (total) correctness of parallel program is equivalent to the problem of partial (total) correctness of sequential programs with arrays.

The transformation in the proof of the theorem gives the equivalent program since we have followed the definition of semantics. Our program is a sequential one with arrays. For the original study of algorithmic properties of programs with arrays see Dańko [3].

The correctness proofs of parallel programs can be made in particular case much simpler than in general by combining the transformation to the sequential program with arrays and proving its correctness. If one parallel instruction is not very complicated we can treat it as a single instruction and we apply standard methods to the whole program.

For instance, let us prove the total correctness of program K (Example 1) in natural numbers with respect to the formulas:

input formula α: $n \geq 1 \wedge B[i] \in N \wedge (i \neq j \rightarrow B[i] \neq B[j])$
output formula β: $i, j \in N \wedge (i<j \rightarrow B[i]<B[j])$

We would like to prove that:

$$\alpha \rightarrow K\beta$$

is a valid formula in the natural numbers. The program K has the form:

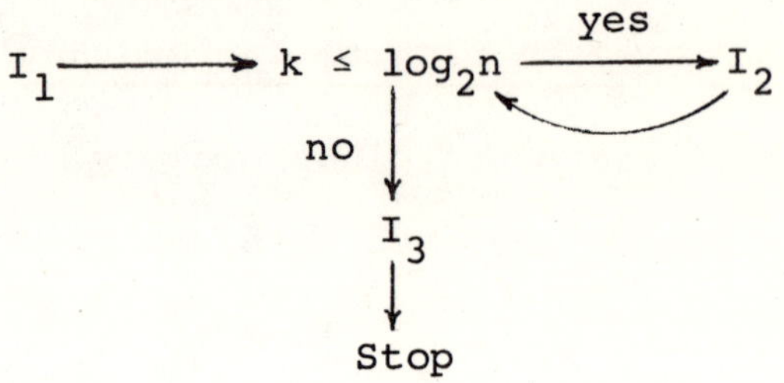

After executing I_1, k is equal to 1 and less(i,j) codes the relation B[i]≤B[j]. The program halts, because every step of loop execution increases k and hence the loop will operate exactly $\log_2 n$ times. Then it is sufficient to show that the following formula γ is an invariant of the loop statement I_2:

$$\gamma = \bigwedge_{1\leq i\leq n} \Big(\sum_{1\leq j\leq n/2^k} \text{less}(i,j) = \text{the number of elements smaller than or equal to } B[i] \text{ in the array} \Big)$$

It can be proved by simple induction. For $k = \log_2 n$ less [i,j] gives the number of elements smaller than B[i] and hence instruction I_3 leads to the correct result.

Such parallel programs can be used as a device for recognizing languages, by setting x_1:=input word at the start of the computation and leading out the result at the distinguished boolean variable.

IV. COMPLEXITY OF PARALLEL COMPUTATIONS

The main theorem of this section is a generalization of results from [2, 9].

The complexity of program from FS_R is measured by the number of instructions performed during the computation (Radziszowski [10]). The length of parallel instruction is the maximum of length of sequential computations together with the cost of relation ρ. (The cost of ρ depends on the way of compilation, but there exist fast algorithms for computing ρ. We do not specify these algorithms, because we are interested in parallel complexity up to a polynomial.)

The length of parallel computation is the sum over all lengths of sequential and parallel instructions executed during computation. The parallel complexity of the language will be the minimum over all complexities of parallel programs recognizing this language. We shall say, that a language L has parallel complexity $T(n)$ if there exists a parallel program accepting L, where all its accepting computations for a data of the length n have length not greater than $T(n)$.

Let PP-time denote the class of languages acceptable by a polynomial parallel program.

Finally assume, that relational system R includes effective arithmetic, that means arithmetic operations are polynomially programmable in R and all functions and predicates of R are polynomially programable on Turing machines. Then we can prove the following:

THEOREM 2

PP-time = P-space

where P-space is the class of languages acceptable by polynomial space bounded Turing machines.

Proof:

The proof consists of two simulations:

1. Every Turing machine, which uses polynomial space can be simulated by parallel program running in polynomial time.

2. Every polynomial time bounded parallel program can be simulated by a sequential one requiring polynomial amount of memory.

Part one

Let M be a q state one-tape deterministic Turing machine using $T(n)$ cells of memory for some polynomial T. Without loss of generality we can assume, that M has a two letter alphabet. Hence for a data of length n there exist at most

$$2^{T(n)} \cdot q \cdot T(n) \leq 2^{p(n)}$$

different configurations of M, where $p(n)$ - is a polynomial. The polynomial time bounded algorithm simulating M can be written as follows:

```
begin
 cobegin c(i):= the i-th configuration of M□1≤i≤2^p(n)
        coend;
 cobegin comment use the next-move function of M;
        k(i):= the index of the next configuration
        after c(i)□1≤i≤2^p(i) coend;
s := 1;
 while s≤p(n) do
   begin s := s+1;
     cobegin k(i) := k(k(i))□1≤i≤2^p(n) coend;
   end;
 if k(1) is an index of accepting configuration
                         then accept else reject;
end.
```

To consider only the computations of the length $2^{p(n)}$ we define the successor of the terminal configuration as the same configuration. Indices k(i) computed after s steps of while loop code the configuration which follows after c(i) by application 2^s times next move function of M to c(i). The length of the computation M is bounded by $2^{p(n)}$, hence after p(n) steps of the loop we can simulate all possible computations of M in the memory bounded by T(n). Obviously the running time of our parallel program is of the range O(p(n)), that means it is bounded by some polynomial.

Part two

To complete the proof of the theorem it is sufficient to construct a polynomial space bounded nondeterministic Turing machine simulating an arbitrary given parallel program K running in polynomial time. Instead of writing a next-move function of the Turing machine, we will construct an algorithm easy by transformable to the formal one.

From the assumption about the relational system R we can rewrite our original program K into the parallel program K' with only boolean operations and 0-1 boolean variables by:

a) replacing every variable by 0-1 array coding the current value of this variable;

b) substituting for occurences of functors, predicates and arithmetic operations suitable polynomial programs;

c) substituting for every formula α occuring in the program K_α a new part of program K_α which carries out the value of α at a special boolean variable b_α.

Resulting parallel program K′ has exactly the same structure of parallelism as K and it remains polynomial (counting binary operations).

We will construct a recursive procedure find (i,t) which returns the value of i-th 0-1 variable after t steps of computation K′, assuming all variables and cells of arrays are numbered by integers. The steps are counted at external level, that means every parallel instruction is treated as one step and all binary instructions outside parallel instructions are counted separately. The idea of the procedure find is taken from [9]. By treating parallel instruction as a single statement our program is sequential. Build up the graph G whose vertices are conditions, substitutions and parallel instructions of K′. The edges are implied by the structure of K′. For the sake of simplicity assume that the terminal instruction is a successor of itself. A nondeterministic algorithm simulating K′ acts as follows:

1. Choose a path $p=p_1 \dots p_{T(n)}$ of the length T(n) in G such that p begins at the start vertex, where T(n) is a polynomial bounding time of K′. (It is done in T(n) nondeterministic steps.)

2. Compute find(0,T(n)) for the chosen path p. If the path p does not code the valid computation of K′ then procedure find will loop infinitely.

3. *if* find(0,T(n)) = 1 *then* accept *else* reject.

The proof will be completed if we construct polynomial space bounded algorithm for function find, since the acceptance of the input word by the program K is equivalent to find(0,T(n)) = 1, where 0 is the index of boolean variable denoting the acceptance of K′.

procedure find(i, t)

1. if t=0 then
   if i is an index of input variable *then return* x_i
   else return 0

2. if p_t is not a parallel instruction at the choosen path p then
   a) if p_t is a substitution which does not change x_i then return find(i,t-1);
   b) if p_t is of the form $x_i := x_j \otimes x_k$ for some boolean operation $\otimes$ then return find(j,t-1) $\otimes$ find(k,t-1);
   c) if p_t is a formula of the form $x_i = \alpha$ for $\alpha \in \{0,1\}$ then compute find(i,t-1) and check if the edge in G we have passed was in agreement with the structure of K'. If not then loop infinitely, otherwise return find(i,t-1);

3. if p_t is a parallel instruction:
   cobegin $I_1 \square \rho_1$, ..., $I_r \square \rho_r$ *coend*
   then
   Find k such that in I_k x_i can be changed. (There exists at most one such integer between 1 and r - it can be found having at the disposal procedure find(j,t-1) for pertinent j.) If such k does not exist then return find(i,t-1).
   Let x_i be the variable occuring as a left side of substitution in I_k at the instruction p_t. Then return findl(i,t,T(n)) where procedure findl is almost exactly the same as find, but constructed for "internal level" of program K', that means for I_k written in details, where each binary instruction is counted as a single instruction. Function findl uses the second parameter t and

the function find when the third parameter decreases to 0. It plays the same role as input word for procedure find.

This completes the description of the algorithm. To observe that it operates in polynomial space note that in stack implementation of find the depth of recursion is bounded by $T^2(n)$ and the memory required for recording path p_t and all parameters at every level of recursion is also bounded by some polynomial.

This proves our theorem.

V. EXAMPLES OF PROGRAMS

Example 2

Boolean matrix multiplication in time O(logn). A is a n×n boolean matrix and n is a power of 2. Then after execution of program M matrix A contains its previous square.

```
M:
cobegin M(i,j,k):=A(i,k)∧A(k,j) □ 1≤i,j,k≤n coend;
s:=1;
while s≤log₂n do
  begin s:=s+1;
    cobegin M(i,j,k):=M(i,j,2k-1)∧M(i,j,2k) □
            1≤i,j≤n∧1≤k≤n/2^(s-1)      coend
  end;
cobegin A(i,j):= M(i,j,1) □ 1≤i,j≤n    coend;
```

Example 3

G is the n vertex undirected graph with edges given by the $n \times n$ boolean matrix A. The following program L checks the connectivity of G in time $O(\log_2^2 n)$:

```
L:
cobegin A(i,i):=1 □ 1≤i≤n  coend
for i:=1 step 1 until log₂n do A:= A × A;
comment A is the transitive closure of the incidence
        matrix for G
s:=1;
while s≤log₂n do
  begin cobegin A(i,j):=A(i,2j-1)∧A(i,2j) □
                1≤i≤n∧1≤j≤n/2^s        coend ;
  end;
s:=1; while s≤log₂n do
        begin cobegin A(i,1):=A(2i-1,1)∧A(2i,1) □
                      1≤i≤n/2^s      coend; s:=s+1 end;
if A(1,1) = 1 then accept else reject;
```

VI. FINAL REMARKS

In the section III we stated only rather evident logical properties of synchronous parallel programs. It seems that this kind of problems should be studied more carefully, particularly as a construction of the axiomatization and the system of inference rules for formulas of the form K_α, where K is a parallel program.

The proof of the complexity theorem based among others on the fact, that the value of an arbitrary variable after i steps of computation depends only at most on c^i other variables. There could be more active processors, but the history of computation for every variable is restricted to the exponential amount of memory. An interesting question arises, what would happen if functions had an unlimited number of arguments and were computable in one step.

The parallel language PP can be a useful tool for programming so far intractable problems, for instance members of P-space not known to be in P-time. The parallel algorithms for these problems will run in polynomial time.

REFERENCES

[1] Banachowski, L. et al: An introduction to algorithmic logic, in Mazurkiewich, A., Pawlak, Z. /Eds/, *Mathematical Foundations of Computer Science*, PWN, Warszawa, 1977.

[2] Chandra, A., Stockmeyer, L.: Alternation, *Proceedings of the 17-th Annual Symposium on Foundations of Computer Science*, Oct. 1976, pp. 98-108.

[3] Dańko, W.: Programs with arrays, *Fundamenta Informaticae*, N.3, 1978, pp. 379-398.

[4] Flynn, M.: Very High-Speed Computing Systems, *Proc. IEEE*, Vol. 54, Dec. 1976, pp. 1901-1909.

[5] Goldschlager, L.: Synchronous Parallel Computations, Technical Report No. 114, University of Toronto, 1977.

[6] Hartmanis, J., Simon, J.: On the Power of Multiplication in Random Access Machines, *Proceedings of the 15 th Annual Symposium on Switching and Automata Theory*, 1974, pp. 113-123.

[7] Hartmanis, J., Simon, J.: Structure of Feasible Computations, *Advances in Computer Science*, Vol. 14, Academic Press, N.Y., 1975, pp. 1-43.

[8] Kung, R.: The new method for finding solution of equations, in Traub, Wozniakowski (Eds), *Complexity and Effectiveness*, Prentice Hall, 1976.

[9] Pratt, V., Stockmeyer, L.: A Characterization of the Power of Vector Machines, *Journal of Computer and System Science*, Vol. 12, (1976), No 2,

[10] Radziszowski, S.: Programmability and P=NP conjecture, Karpinski, M. (ed), *Foundamentals of Computation Theory*, Lecture Notes in Computer Science, Vol. 56, Springer-Verlag, Berlin, 1977, pp. 494-498.

[11] Stapp, L.: The Proof of Correctness of Jacobi Parallel Program, CC PAS Report, No. 323, 1978.

S. Radziszowski
Uniwersytet Warszawski,
Instytut Informatyki
00-950 Warszawa PKiN,
POLAND

COLLOQUIA MATHEMATICA SOCIETATIS JÁNOS BOLYAI
26. Mathematical Logic in Computer Science
Salgótarján (HUNGARY), 1978

A CERTAIN TYPE OF DEPENDENCY TREE TRANSFORMATIONS

A. Říha

We present here a definition of automata accepting and transforming trees instead of strings, and some of their fundamental properties. Automata proposed here are modified J. Thatscher's tree automata. From his work [1] only the main ideas were used, their further development is independent. The extension of the domain of definition of automata from strings to trees gives a formalism better suited to handle concepts arising in some parts of mathematical theory of formal languages and mathematical linguistics (see also [4]). A series of works were presented where the properties of tree automata were studied and the results applied to problems of semantics, transformations, syntax-directed translation and decidability. This work uses a different type of trees, another method of managing unbounded branching of trees and characterizes the power of the defined automata in a different way.

A dependency tree is a special type of directed rooted tree with labelled nodes (where a tree is a finite connected acyclic graph).

Definition

A dependency tree T over alphabet A (a DA-tree) is either the empty tree (denoted Λ) or an ordered sixtuple $(U,H,v,<,f,A)$, where (U,H,v) is a directed rooted tree with root v, $<$ is a linear order on U fulfilling the projectivity condition (H^+ is the transitive closure of H):

$\forall u,w,x \in U(((u,w)\in H \vee (w,u)\in H)\ \&\ u<x<w \Rightarrow (x,w)\in H^+ \vee (x,u)\in H^+)$,

A is a finite alphabet and f:U A is a function. The set of all dependency trees over alphabet A is denoted by A .

In the diagram of a dependency tree, e.g.

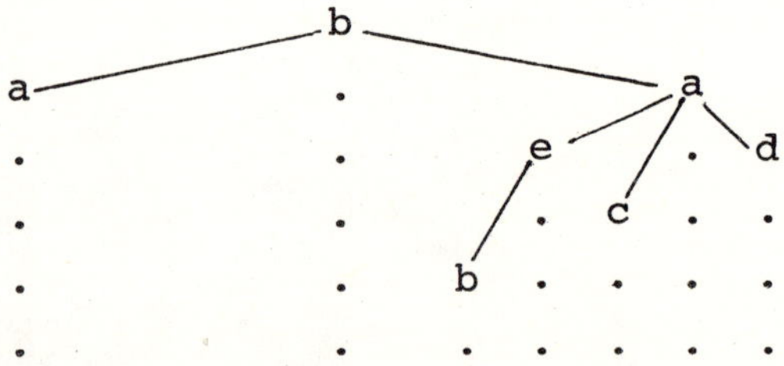

the linear order is represented by the distribution of nodes from left to right and the projectivity condition has the following meaning: the vertical drawn from any node downwards does not intersect with any edge.

So far, dependency trees were used to describe the structure of natural language sentences and, sometimes, of arithmetical expressions. Other applications are possible (in the area of information systems and programming languages).

In studying the relationship between tree automata and ordinary automata on strings, several types of strings corresponding in some way to trees are suitable - projections of trees, linear representations of trees etc.

A projection pr(T) of a dependency tree T is the

string of the labels of nodes written according to the linear order.

Definition

Let $T = (U,H,v,<,f,A)$ be a DA-tree, $U=\{u_{i_1} < \dots < u_{i_n}\}$. A projection $pr(T)$ of the tree T is $pr(T)=a_{i_1} \dots a_{i_n}$, where $a_{i_j}=f(u_{i_j})$ for $j=1,2,\dots,n$. A projection of the empty tree is the empty string Λ.

Notation

For dependency tree T and a node y of T, Ty denotes the dependency subtree of T at y. By $T = (Ty_1 \dots Ty_m a Tz_1 \dots Tz_n)$ we shall denote that $T = (U,H,v,<,f,A)$,

$$f(v)=a, \{y_1 < \dots < y_m\}=\{u \mid u \in U, (u,v) \in H, u<v\} \qquad (1)$$

$$\{z_1 < \dots < z_n\}=\{u \mid u \in U, (u,v) \in H, u>v\} \qquad (2)$$

Let the set (1) (resp. (2)) be empty, we shall take $m=1$ and $Ty_1 = \Lambda$ (resp. $n=1$ and $Tz_1 = \Lambda$).

Obviously, more than one tree can correspond to the same projection. It is, therefore, convenient to use linear representations of trees, that correspond to trees unambigously. There are several possible forms of such linear representation. Here, we shall use the following:

Definition

Linear representation $lr(T)$ of a DA-tree T is

$$lr(T)=\begin{cases} \Lambda & \text{for } T=\Lambda \\ (lr(Ty_1)\dots lr(Ty_m)a\ lr(Tz_1)\dots lr(Tz_n)) & \text{for } T=(Ty_1 \dots aTz_1 \dots Tz_n). \end{cases}$$

Example

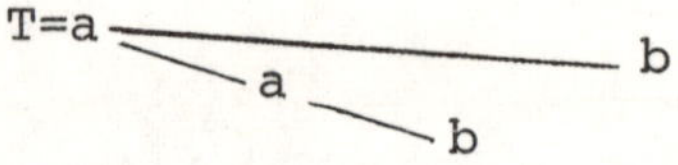

$pr(T)=aabb$, $lr(T)=(a(a(b))(b))$.

A tree automaton can be e.g. a generalization of a finite automaton and can traverse a tree either from the root to the leaves (a top-down automaton) or from the leaves to the root (a bottom-up a.). Here a bottom-up automaton will be considered.

Definition

A deterministic bottom-up automaton M accepting DA-trees is a system (Q,q_o,A,δ,F), where Q is a finite set of states, $q_o \in Q$ is an initial state, A is a finite alphabet, $F \subset Q$ is a set of final states, $\delta: Q\times A\times Q \to Q \cup Q\times Q \to Q$ is a transition function, $\delta(q_o,q)=q$ for $q \in Q$.

Definition

Let $sum: Q^* \to Q$, $state: A^{\#} \to Q$ be defined as follows: $sum(\Lambda)=q_o$, $sum(wq)=\delta(sum(w),q)$ for $w\in Q^*$, $q\in Q$, state $state(\Lambda)=q_o$, $state(T)=\delta(sum(state(Ty_1)\ldots state(Ty_m)),$ $a, sum(state(Tz_1)\ldots state(Tz_n)))$ for $T=(Ty_1\ldots Ty_m aTz_1\ldots$ $\ldots Tz_n)$. A DA-tree T is accepted by a bottom-up automaton M, iff $state(T)\in F$. The set of all DA-trees accepted by M is denoted by $L(M)$. □

For binary DA-trees (the trees of the form $T =$ $= (Ty_1 aTz_1)$, where $Ty_1, Tz_1 \neq \Lambda$ or $T=(a)$) it would be sufficient to define a bottom-up automaton with transition function of the form $\delta: Q\times A\times Q \to Q$ and with function state defined: $state(\Lambda)=q_o$, $state(T)=\delta(state(Ty_1),$ $a, state(Tz_1))$ for $T = (Ty_1 aTz_1)$. The function sum mentioned above is used in a computation of a bottom-up automaton to "transform" a general DA-tree to a binary DA-tree and then a step "up" is made just as in the case of binary trees.

Now we shall study the power of a bottom-up automaton. Two following theorems were proved. To make the presentation shorter we give only the main ideas of proofs. The detailed proofs were given in [2].

THEOREM 1.

For any bottom-up automaton M *there exists a context-free grammar* G, *such that* pr(L(M))=L(G).

Proof.

Let M be a bottom-up automaton, $M=(Q,q_o,A,\delta,F)$, where the function δ is defined by a finite number of "rules" of the following types:

1/ $\delta(q_o,q)=q$ 3/ $\delta(q_o,a,q_o)=q$ 5/ $\delta(q_o,a,q_1)=q$

2/ $\delta(q_1,q_2)=q$ 4/ $\delta(q_1,a,q_o)=q$ 6/ $\delta(q_1,a,q_2)=q$.

We construct a context-free grammar $G=(\{S\}\cup Q\cup\bar{Q},A,P,S)$, where $S\notin Q$, $\bar{Q}=\{\bar{q}\mid q\in Q\}$, $\{S\}\cup Q\cup\bar{Q}$ is a finite nonterminal alphabet, A is a finite terminal alphabet and P is a finite set of rewriting rules constructed as follows:

a/ for each occurences of a "rule" of type i/ in the definition of the transition function there is a rule of the type p_i/ in P, where

p_1/ $q\to\bar{q}$ p_3/ $\bar{q}\to a$ p_5/ $\bar{q}\to aq_1$

p_2/ $q\to q_1\bar{q}_2$ p_4/ $\bar{q}\to q_1a$ p_6/ $\bar{q}\to q_1aq_2$

b/ for each $q\in F$ there is a rule $S\to\bar{q}$ in P

c/ P does not contain any other rules.

Then pr(L(M))=L(G) holds (proof by induction). □

THEOREM 2.

For any context-free grammar G *there exists a deterministic bottom-up automaton* M, *such that* $L(G)=pr(L(M))$.

Proof.

First, to a given cf-grammar G we construct an equivalent cf-grammar $G'=(V_N,A,P,V_N')$, where $V_N' \subset V_N$ is a set of initial symbols, V_N is a finite nonterminal alphabet, A is a finite terminal alphabet, P is a finite set of rewriting rules in Greibach normal form and for each two rules $N_1 \to w_1$, $N_2 \to w_2$ holds $w_1=w_2 \Rightarrow N_1=N_2$.

Second, to the grammar G' we construct a bottom-up automaton $M=(Q,q_o,A,\delta,V_N')$, where $Q=\{N_1N_2\ldots N_k$ such that $\exists N \to aN_1N_2\ldots N_kN_{k+1}\ldots N_s \in P$ for some $s \geq k\} \cup \{q_o\}$, δ is defined as follows:

a/ for each $N \in Q$ $\delta(q_o,N)=N$

b/ if $N_1N_2\ldots N_k$, $N_1N_2\ldots N_kN_j \in Q$, then $\delta(N_1\ldots N_kN_j)=$ $=N_1\ldots N_kN_j$

c/ if $N \to a \in P$, then $\delta(q_o,a,q_o)=N$

d/ if $N \to aN_1\ldots N_s \in P$, then $\delta(q_o,a,N_1\ldots N_s)=N$

e/ δ does not contain any other "rule".

Then $pr(L(M))=L(G)$ (proof by induction). □

The states of the constructed automaton are all initial segments of strings of nonterminal symbols that occur on the righthand sides of rules of grammar G'. For each string $w \in L(G')$ trees corresponding to its derivations are in $L(M)$, e.g. for the derivation $S \Rightarrow aN_1\ldots N_n \Rightarrow^* aa_1N_1^1\ldots N_1^{n_1}\ldots a_nN_n^{n_n} \Rightarrow^* w$ the tree

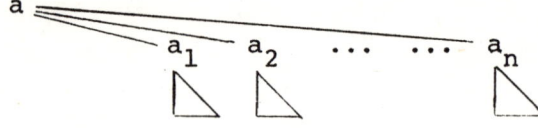

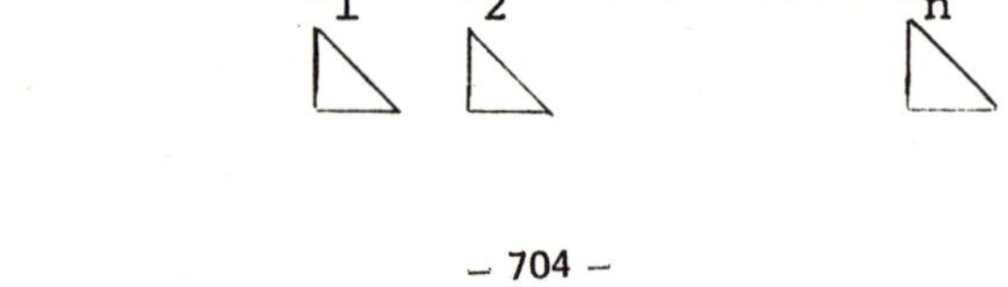

As for linear representations of dependency trees, theorem similar to Theorem 1 holds, but not that corresponding to Theorem 2,

THEOREM 1'

For any bottom-up automaton M *there exists a* cf-*grammar* G, *such that* $lr(L(M))=L(G)$.

Proof.

The proof of this theorem is quite similar to the proof of Theorem 1. The only difference is the form of rewrtiting rules of the cf-grammar:

$p_1'/\ q \to \bar{q}$ $\quad p_3'/\ \bar{q} \to (a)$ $\quad p_5'/\ \bar{q} \to (q_1 a)$

$p_2'/\ q \to q_1\bar{q}_2$ $\quad p_4'/\ \bar{q} \to (aq_1)$ $\quad p_6'/\ \bar{q} \to (q_1 a q_2)$. □

THEOREM 3.

There exists a context-free grammar G *that*

a/ *if* $w \in L(G)$, *then* w *is a linear representation of a* DA-*tree*

b/ *there exists no deterministic bottom-up automaton that would accept the set of* DA-*trees represented by the linear representations from* L(G) *(denoted by* $lr^{-1}(L(G))$ *)*.

Proof.

Let $G=(V_N,V_T,P,S)$, $V_T=\{a,b,c,(,)\}$, $V_N=\{S,B\}$, $P=\{S\to(c),\ S\to((a)B(b)),\ B\to c,\ B\to(a)B(b)\}$.

Then $L(G)=\{w \mid w=(u^n c v^n),\ n\geq 0,\ u=(a),\ v=(b)\}$.

The corresponding set of trees

$lr^{-1}(L(G))=$ 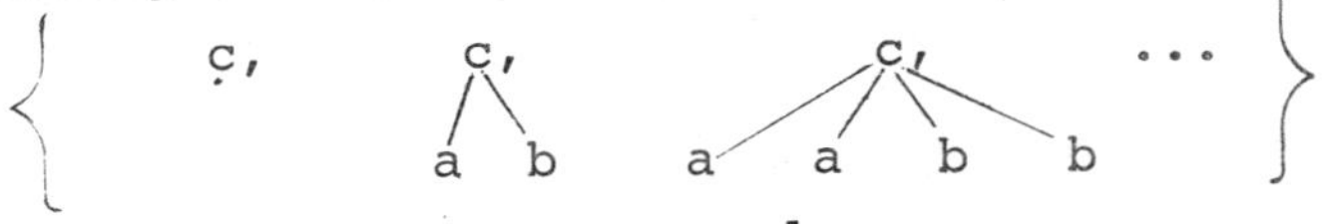

Let $M=(Q,q_o,\{a,b,c\},\delta,F)$, $L(M)=lr^{-1}(L(G))$, $\delta(q_o,a,q_o)=$ $=q_a \in Q$, $\delta(q_o,b,q_o)=q_b \in Q$. Then for all i, $i \geq 0$, $sum(q_a^i) \in Q$, $sum(q_b^i) \in Q$ and there exist $j,k \geq 0$,

$j \neq k$, that $sum(q_a^j)=sum(q_a^k)$. Therefore
$\delta(sum(q_a^k),c,sum(q_b^j)) = \delta(sum(q_a^j),c,sum(q_b^j)) \in F$
and the tree

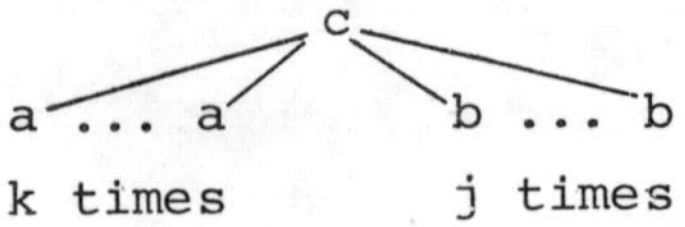

is in $L(M)$ - contradiction!

Thus the automaton cannot exist and Theorem 3 is proved. □

The sufficient condition for a set of DA-trees to be accepted by a bottom-up automaton is still an open problem, as well as the question whether an analogy to Theorem 2 holds for some special type of DA-trees (e.g. binary trees).

A bottom-up transducer can be defined that traverses DA-trees just in the same way as the bottom-up automaton does.

Definition

A deterministic bottom-up tree transducer is a system $M=(Q,q_o,A,\delta)$, where Q is a finite set of states, $q_o \in Q$ is an initial state, A is a finite alphabet and $\delta: Q\times A\times Q \to Q\times A \cup Q\times Q \to Q$ is a transition function. □

Definition

Let sum and state be functions defined in the same way as for a bottom-up automaton. Let $out: A^\# \to A^\#$ be defined as follows: $out(\Lambda)=\Lambda$,
$out(T)=(Ty''_1 \ldots Ty''_m a'' Tz''_1 \ldots Tz''_n)$ for $T=(Ty_1 \ldots Ty_m a Tz_1 \ldots$ $\ldots Tz_n)$, where $Ty''_i = out(Ty_i)$ for $i=1,\ldots,m$, $Tz''_j = out(Tz_j)$ for $j=1,\ldots,n$,
$a''= \delta_A(sum(state(Ty_1)\ldots state(Ty_m)),a,sum(state(Tz_1)\ldots$ $\ldots state(Tz_n)))$, where $\delta_A=\delta \circ \Pi_A$. We say that a bottom-

-up transducer M transduces a DA-tree T_1 into a DA-tree T_2, iff $out(T_1)=T_2$.

The deterministic bottom-up tree transducer performs a variant of finite state relabelling, it does not change the structure of a DA-tree. It is capable to perform every relabelling of a DA-tree, that could be realized by another variant of tree transducer, that traverses the tree in the following way

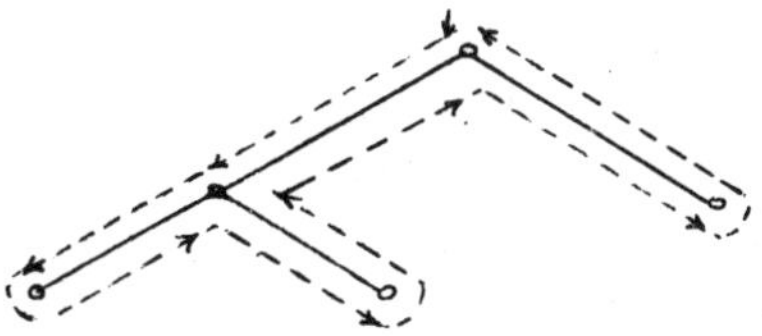

and can transform the label of a node only after the "up" step. That will be evident from the following definition and theorem.

Definition

Let $tab: A \times A \to A$ be a given function, let $\tilde{tab}: A^* \times A \to A$ be defined as follows: $\tilde{tab}(\Lambda, a)=a$, $\tilde{tab}(wb,a)=tab(b,\tilde{tab}(w,a))$, $w \in A^*, a,b \in A$, $V(T)$ denotes the label of the root of the tree T. Then a bottom-up transformation defined by the function tab is the following function:

$$transf(T)=\begin{cases} \Lambda & \text{for } T=\Lambda \\ (T'y_1 \ldots T'y_m a' T'z_1 \ldots T'z_n) & \text{for } T=(Ty_1 \ldots aTz_1 \ldots), \end{cases}$$

where $T'y_i=transf(Ty_i), i=1,\ldots,m$, $T'z_j=transf(Tz_j)$, $j=1,\ldots,n$, $a'=\tilde{tab}(V(T'z_1)\ldots V(T'z_n),\tilde{tab}(V(T'y_1)\ldots V(T'y_m),a))$.

THEOREM

For every bottom-up transformation there exists a deterministic bottom-up transducer that realizes it, i.e. $\forall T_1,T_2 \in A^{\#}(transf(T_1)=T_2 \iff out(T_1)=T_2)$.

Proof.

Let R be the relation on A^* defined as follows: $(w,w')\in R \iff \forall a\in A(\tilde{tab}(w,a)=\tilde{tab}(w',a))$. Obviously, R is an equivalence relation of a finite index. Let M be the set of equivalence classes of relation R, let $[w]$ denote an equivalence class of R containing $w\in A^*$. Let $\overline{tab}:M\times A \to A$ be defined: $\overline{tab}(m,a)=b \iff m=[w]$ & $\tilde{tab}(w,a)=b$, $m\in M$, $a,b\in A$, $w\in A^*$. Now let $N=(M,[\Lambda],A,\delta)$ be a bottom-up transducer, where the function $\delta:M\times A\times M \to M\times A \cup M\times M \to M$ is defined as follows:

a/ for every $a_1,a_2\in A$ and $m_1,m_2\in M$ such that $\exists b\in A(\overline{tab}(m_1,a_1)=b$ & $\overline{tab}(m_2,b)=a_2$ $\delta(m_1,a_1,m_2)= =([a_2],a_2))$

b/ for every $a\in A$ $\delta([\Lambda],a,[\Lambda])=([a],a)$

c/ for every $a\in A$, $m\in M$ $\delta(m,[a])=m'$, where $m'=[wa]$, $w\in m$

d/ δ does not contain any other "rule".

Then $\forall T_1,T_2\in A^\#$ $(transf(T_1)=T_2 \iff out(T_1)=T_2)$; (proof by induction). □

In this presentation only bottom-up automata, transducers and transformations were considered. A top-down one can be a generalization of a finite automaton, as well. For top-down automata similar theorems hold as for bottom-up ones (including a realization of a top-down transformation). Besides, there is a possibility of composition of transformations. Such problems are subjects to further research.

REFERENCES

[1] Thatcher,J.W., Tree Automata: An Informal Survey. In Aho,A.V. (ed.) *Currents in the Theory of Computing*, London 1973, 143-172.

[2] Řiha,A., Automata Accepting and Transforming Dependency Trees. Disertation, Charles University, 1977 (in Czech).

[3] Hopcroft,J.E., Ullman,J.D., *Formal Languages and Their Relation to Automata*, Addison-Wesley, 1969.

[4] Sgall,P., Hajičová,E., A "Functional" Generative Description (Background and Framework), PBML 14, Prague, 1970.

A.Řiha
Computing Center of Charles University
Prague
Czechoslovakia

COLLOQUIA MATHEMATICA SOCIETATIS JÁNOS BOLYAI
26. Mathematical Logic in Computer Science,
Salgótarján (HUNGARY), 1978.

SOME REMARKS ON CORRECTNESS PROVING FOR PARALLEL PROGRAMS

L. Stapp

1. *Formulation of a problem*

Let us recall the problem of solving the system of n iterative equations, i.e.

$$x_1 = \varphi_1 (x_1, x_2, \ldots, x_n)$$
$$\ldots\ldots\ldots\ldots\ldots\ldots\ldots\ldots$$
$$x_n = \varphi_n (x_1, x_2, \ldots, x_n)$$

We assume that our system may be solved in the finite number of iterations (the problem is exposed by Dijkstra in D78). There are many methods for solving such a system. In this paper we propose to use methods of parallel programming. Each component of the approximation of solution is computed concurrently with others by its own processor. After computing all components the stop condition is tested (i.e. if for every $i, x_i = \varphi_i(x_1, \ldots, x_n)$). If the stop condition holds, the solution is found, if not the next approximation is computed in the analogous way.

One may say that our solution is rather an uneffective one because computation of all components of solution in an asynchronous way seems to be a more interesting one. But the effectieness of both methods is an open question, and from the other point of view our parallel algorithm can be treated as a model of a simple operating system. The activity of this operating system is as follows:

1. all processes of the system work (i.e. all facilities are occupied)
2. processes, consecutively but in an arbitrary order, one after the other, stop their action and wait for
3. the last active process, which "awakes" (activates) the others
4. from the begining - all processes work, and so on.

It seems to be interesting to study such an operating system.

2. *Basic notions*

Our program will be written in LOGLAN 77 language. LOGLAN 77 is a new programming language, designed at the University of Warsaw. It is universal language, here we shall quote only some necessary definitions and properties needed to understand our program.

The most powerful operation in LOGLAN 77 is prefixing - as in SIMULA 67. This operation permits us to build more complicated constructions based on simple ones. The basic syntactic units are structured types called "class" and "procedure". The calling of the procedure P is of the form: "*call* P(...)". The generation of an object of a type T is of the form:

"*new* T(...)". It is an expression, and the result of the above operation (i.e. the body of a new object - copy of the type T/ may be assigned on the variable z, if only z was earlier declared to be a variable of the type T.

In LOGLAN 77 we have special tools for parallel computation. First there is a type called process; statements of every object of subtype of the "process" type (i.e. every object of the type prefixing the type "process") are executed sequentially, independently from the statements of other objects of program. There are two following procedures to service processes:

activate(x)

passivate.

If any process calls "activate (x)" the new process, say A, generated earlier on the variable x (i.e. instruction "x := *new* A(...)" had been executed) begins its action. The same effect one can gain using an instruction "*call* activate (*new* A(...))".

The process which calls "passivate" stops its action.

Every process works on its own data; to synchronize the work of processes there is another special tool -the "monitor". The monitor is a data structure (so called "shared" variables) and a set of procedures operating on it (so called "entry" procedures). The shared variables are common for all processes, but an access to these data is possible only by the use of the "entry" procedures. It is proved (see M77, M78, M78a) that in every moment of parallel program computation at most one "entry" procedure of the monitor can be executed.

In every "entry" procedure one can call two following standard procedures: delay(q) and continue(q), where q is a qeue idenfier (in LOGLAN 77 q is declared to be the variable of a standard type "head"). The above procedures have the following effects: the process which called delay(q) terminates the execution of "entry" procedure, enters the queue q and stops. The process which called continue (q) returns from the monitor (i.e. finished the execution of its "entry" procedure) and continues the action outside; simultaneously the first process from the queue q (if any) resumes its execution of this "entry" procedure call, in which this process had called deley(q).

We know (see M77, M78, M78a) that such defined monitor satisfies 3 basic properties:

1. mutual exclusion
2. system deadlock prevention
3. fairness of system scheduling

Both of the above types, process and monitor, are attributes of a standard type PARALLEL, therefore it is convenient to prefix any parallel program by this type.

Now we shall give some definitions and ideas of LOGLAN 77 semantics rather an intuitive way; for more precise description see L77.

1. By a snapshot of a parallel computation we shall mean the set of all objects existing at the given moment, some of their statements are under execution at the moment of the snapshot, some others are ready to be executed.
2. By a configuration we shall mean the snapshot of a parallel computation such that at least one statement is ready to be executed.

3. By a tree of all possible computation H of a given program P we shall mean the tree of configurations such that

   3.1. the root of H is a given initial configuration of the program P

   3.2. the configuration K′ is a son of the configuration K (K < K′) iff the following conditions hold:

      a/ configuration K′ arises from the configuration K by completion at least one instruction which has been under execution or ready to be executed in K,

      b/ instruction mentioned in (a) are not in conflict (i.e. they are not calling of "entry" procedure nor instruction of it),

      c/ instructions which are under execution or ready to be executed in K and are not mentioned in (a), are under execution in K′,

      d/ other objects do not change.

   (for more precise definition see L77, M78, M78a).

4. By a final configuration we shall mean a configuration in which there are no instructions under execution nor ready to be executed.

Now we introduce the definition of correctness.

DEFINITION 1. *Parallel program* P *is total correct with respect to given initial configuration* K_0 *and given condition* α *iff the two following conditions hold*

1. *every path of the tree of all possible computations is finite.*

2. *the condition* α *hold in an arbitrary final configuration.*

3. *The parallel program (some intuitions)*

The program is given in on the last two pages of this paper, here only some intuitions and ideas of it will be given. As it has been said before there are n subprocesses - objects of type declared as Jac. M-th object of Jac type computes the m-th component of the next approximation of the solution. To communicate between processes and to synchronize their work there is a subtype of the monitor type - declared as MONFORJAC.

An action of every process (i.e. a processor which is executing an object of JAC) is as follows: the process takes the "old" approximation of the solution (using the entry procedure RECEIPT), then it computes the value of its component and sends this value to the monitor MONFORJAC (using a parameter "effect" of an "entry" procedure DELIVERY). The action of DELIVERY is the most important point of our program and is as follows:

1. if the process called DELIVERY is not the last active one (i.e. the length of queue q is less than $n-1$) then this process is delayed in queue q.
2. if the process is the last active one (all other processes are in q) then the stop condition is tested. If the condition holds, a message is remained in the MONFORJAC monitor (the value of variable "boo" is changed) and the solution may be printed.

3. Next this process "awakes" the first one from queue q (*call* continue (q)), this one "awakes" the next one etc. If the solution has been found the message is sent to every process (using parameter "fin").

After returning from the monitor this message is tested, and either the process finished its action or carries its action once more: takes the value of the last computed approximation etc.

4. *Correctness of parallel program*

In this chapter we shall try to prove that the ideas of the previous chapter are realized by our program, and that it is sufficient for correctness of our program.

In the beginning some definitions we need: Let H be the tree of all possible computations of our program. For sake of simplicity and without loss of generality it is assumed that the initial configuration K_o (i.e. the root of the tree H) is a configuration describing a state of our program relization after generation of MONFORJAC and activation of all JAC processes. It is also assumed that some initial data are given.

Now we introduce the following notations:

1. $\overset{*}{<}$ is a transitive closure of the relation $<$; $K \overset{*}{<} M$ iff a configuration K is an ancestor of a configuration M iff a configuration M is a descendant of a configuration K.

2. $val_K(x)$ is a value of a variables x in a configuration K

3. $|q|_K$ is a length of the queue q in a a configuration K, i.e. a number of processes passivated in q.

4. for simplicity of notation $val_K(n)=val_{K_0}(n)=n$ for any configuration K.

DEFINITION 2. *A configuration* K *is a causing deadlock configuration iff there exists a process* A *such that for any descendant* M *of* K $(K \overset{*}{<} M)$, *the process* A *is passivated in the queue* q.

It is obvious that if K is a causing deadlock configuration, then for any final (if exists) configuration M, such that M is a descendant of K , process A has not finished its action. Therefore we must show that there is no one causing deadlock configuration in the tree H. To show it we will prove the following 4 lemmas.

LEMMA 1. *For any configuration* K, *such that none of the following instruction*

call delay (q)

call continue (q)

is under execution nor is ready to be executed, the number of processes in the queue q *is equal to the value of the variable* k , *i.e.*

$$|q|_K = val_K(k)$$

<u>Proof</u>: Proof goes by induction on the length of computation (see S78).

LEMMA 2. *For any configuration* K, *the value of the variable* K *in* K *is less than* n *and not less than* 0, *i.e.* $0 \le val_K(k) < n$

<u>Proof</u>: Proof goes by the induction on the length of computation.

DEFINITION 3. *By a tigger configuration we shall mean any configuration* K *of the tree* H *such that the following conditions hold in* K

1. *there exists one active process, other processes are passivated in the queue* q
2. *the instruction "call continue* (q)" *is ready to be executed.*

It is easy to see that if K is a trigger configuration then there is only one son of K in the tree H: a configuration K′ in which this active process has completed the execution of "continue".

LEMMA 3. *Let* K *be a trigger configuration, K -the root of the tree* H *and let* G *be a branch of the tree* H, *such that* $K_o \in G$ *and* $K \in G$. *Then there exists a sequence of configurations* $K_2, K_3, \ldots, K_n$ *such that the following conditions hold*

1. $K_j \in G$ $\qquad j = 2,\ldots,n$
2. $val_{K_j}(k) = n-j$ $\qquad j = 2,\ldots,n$
3. *for any configuration* M, $M \in G$, *if* K_j *is a descendant of* M *and is an ancestor of* K_{j+1} $(K_j \overset{*}{<} M \overset{*}{<} K_{j+1})$

then

$$val_M(k) = val_{K_j}(k)$$

<u>Proof</u>: The existance of $K_2, K_3, \ldots, K_n$ results immediately from the definition of the standard procedure "continue", lemma 2 and the decleration of DELIVERY.

The above lemma explains the meaning of the trigger configuration. Now we introduce the very important lemma,

LEMMA 4. *For any configuration* K *and for any path* G *of the tree* H, *such that the initial configuration* K_o *belongs to* G *and* K *belongs to* G, *if the value of the variable "boo" is true in* K, *then there exists a configuration* M *such that the following conditions hold:*

1. M *is a trigger configuration*

2. M *is a descendant of* K, $K \overset{*}{<} M$

3. $M \in G$

Proof of the lemma is here ommitted and can be found in S78.

From the above lemma results immediately

COROLLARY 1. *There is no causing deadlock configuration in the tree of all possible configurations of our program.*

The notion of the trigger configuration is useful also in proving correctness. It is seen from the following

LEMMA 5. *Let* K, K' *be two trigger configurations on the same path of the tree* H, $K \overset{*}{<} K'$, *such that*

(1) *the value of the variable "boo" is true in the configuration* K ($val_K(boo) = 1$)

(2) *for any configuration between* K *and* K' ($K \overset{*}{<} M \overset{*}{<} K'$), M *is not a trigger configuration*

Then, for any i, $1 \leq i \leq n$,

$$val_{K'}(Y[i]) = \varphi_i(val_K(Y[1]),\dots,val_K(Y[n]))$$

From the above lemma we obtain a

THEOREM *Our program computes a solution of the system of iterative equations using parallel iteration method.*

<u>Proof</u>: Let G be a path of the tree of all possible computations of our program, such that $K_0 \in G$ Let $K_1, K_2, \dots, K_j$, be a sequence of all trigger configurations belonging to G. From lemma 5 it is known that $val_{K_j}(Y)$ is the j-th approximation of the given system. However there exists such p, that the p-th approximation is just a solution /see the assumption in the first chapter/. Hence because $val_{K_p}(boo) = 1$, there exists such a configuration M, $M \in G$, that M is a final configuration. But $val_{K_{p+1}}(Y)$ is a solution of the given system and

$$val_M(Y) = val_{K_{p+1}}(Y).$$

Final remarks

The above method for parallel program correctness proving seems to be a simple one, and rather strongly depends on the form of program under investigation. But this method does not need any axilliary variable (Dijkstra in D78 needs n^2). Some methods and some tools used in this paper are useful for other programs, as e.g. in the asynchrounous solution of the system of iteration equations.

References

D78 Dijkstra, E.W., Finding the correctness proof of a concurrent program, in *Mathematical Foundation of Computer Science'78 Lecture Notes in Computer Science* no. 64, Springer Verlag, Berlin /1978/, 31-38.

L77 *LOGLAN 77 report - internal report of Institute of Mathematical Machines* /in Polish/, *July 1977.*

M77 MÜLDER,T., On properties of certain synchronizing tools for parallel computation, in FCT, *Lecture Notes in Computer Science* no.56. Springer Verlag, Berlin /1977/, 459-465.

M78 Mülder,T. Implementation and properties of certain tools for quasi-parallel and parallel computations, *ICSPAS report* no 356.

M78a Mülder,T. Implementation and properties of certain tools for parallel computation, to appear in *Fundamenta Informaticae.*

S78 Stapp,L., *Correctness of parallel program "Jacobi"*, ICSPAS report no 332.

APPENDIX

The program

```
PARALLEL begin
  variable n:integer;
  n:= read integer;
  begin
    type MONFORJAC: monitor class (n:integer);
         begin
           variable q:head, k:integer,boo,stop:boolean,
           Y: array [1:n] of real
           comment the shared variables;
           type RECEIPT: entry procedure (result W:array of real);
                         begin
                           W:=copy Y;
                         end RECEIPT;
           type DELIVERY: entry procedure (i:integer,effect:real,
                                           result fin:boolean);
                         begin
                           if stop then stop:=(Y[i]=effect);
                           Y[i]:=effect;
                           if k=/= n-1 then
                           begin
                             k:=k+1;
                             call delay (q);
                           end else
                           if ¬ stop then
```

```
                begin
                  boo:=false;
                  print (Y);
                end else stop:=true;
                fin:=boo;
                if k =/= 0 then
                begin
                  k:=k-1;
                  call continue (q);
                end
              end DELIVERY;
  comment statements initializing MONFORJAC follow;
  boo:=true;
  stop:=true;
  q:=new head;
  k=0;
end MONFORJAC;
type JAC:process class (n:integer,i:integer,Mon:MONFORJAC);
        begin
          variable active:boolean,s:real,
          Z: array [1:n] of real;
          active:=true;
          while active do
          begin
            call Mon.RECEIPT (Z);
            s:=FISUBI (Z);
            call Mon.DELIVERY (i,s,active);
          end
end JAC.
```

```
    variable Jacmon:MONFORJAC,I:integer;
    comment the main program follows;
    Jacmon:=new MONFORJAC (n);
    for i:=1 step 1 until n do
    call activate (new JAC (n,i,Jacmon));
  end
end of program;
```

L. Stapp
Institute of Mathematics
Warsaw Technical University
Poland

COLLOQUIA MATHEMATICA SOCIETATIS JÁNOS BOLYAI
26. Mathematica Logic in Computer Science
Salgôtarján (HUNGARY), 1978.

NORMAL FORM OF PROOF OF CERTAIN FORMULAS OF SITUATION CALCULUS

O. Štěpánková

Introduction

The situation calculus was deviced by Green [2] as a tool allowing to embed planning of goal oriented behaviour into the framework of predicate calculus. His technique is studied in [4,5] and it is compared there to another problem solving system STRIPS [1]. Mc Carthy and Hayes [3] search those important properties of problem solving systems which play a crucial role in description of a complex environment and in the formulation of its problems. This paper strongly influenced some refinements of situation calculus suggested in [5, 6].

There are only two different syntactical forms of the axioms used for the description of the static and dynamic laws of the environment in many cases. We show that there is a special normal form of proof for formulas specifying problems in such worlds. Moreover this normal form of proof allows an easy interpretation as a flow chart for robot's behaviour. These results imply a possibility of the use of situation calculus for the planning in the world with more than one acting entity

Definition

Let us call *simple situation language* a first order language with two sorts only - objects and situations, such that

a) there is just one situation constant - denoted s_o

b) object functions - denoted by $f, f_1, f_2, \dots$ - have object arguments only

c) situation functions - denoted by $\varphi, \psi, \dots, \varphi_1, \dots$ - have some object arguments and just one situation one

d) predicates - denoted by $P, P_1, P_2, \dots$ - have beside object arguments just one situation argument

Notation

Let object variables of a simple situation language be the elements of $\{x, y, z, \dots, x_1, \dots\}$, situation variables of $\{s, s_1, \dots\}$ and object constants of $\{c, c_1, \dots\}$, if not otherwise stated.

The object functions which are not dependent on situation arguments are intended to describe those functions which are supposed to be stable in time - "distance" between two towns, "colour of eyes" for example.

The situation functions stand for the actions of the acting entity which change the relations among the objects of the world, e.g. "move something somewhere" or "grab something".

Formulas referring to a single situation play a special role in simple situation languages - we call these formulas uniform. They will be defined by induction on the length of formulas.

Definition

Let s be a situation variable of a simple situation language L.

An atomic formula of L is *uniform with respect to* s, if its situation argument is s, i.e. it is of the form $P(a_1,\dots,a_k,s)$

(i) if A_1, A_2 are uniform with respect to s then $\neg A_1, A_1 \& A_2, \forall x A$, are also uniform with respect to s (wrt s)

(ii) there are no other formulae uniform wrt s

Let t be a situation term of L. A formula B is *uniform with respect to* t (wrt t) iff there is a formula B' uniform wrt s such that $B = B'|_t^s$ (the result of the substitution of t for all occurences of s in B').

The uniformity of a formula B wrt t is sometimes emphasized by writing $B\|[t]\|$ instead of B only.

Definition

A theory T with a simple situation language L is a theory of *situation calculus* iff it has axioms of two types only

a) $\forall s A$, where A is uniform wrt s - these axioms are called *core of* T

b) $\forall s\ ({}^1B\|[s]\| \to {}^2B\|[\varphi(\bar{x},s)]\|$, where $\varphi(\bar{x},s)$ is a situation function of L, ${}^1B$ and ${}^2B$ are uniform wrt s,

$\varphi(\bar{x},s)$ resp. and $\bar{x}$ is the sequence of all free object variables of the formula ${}^1B$ & ${}^2B$ - these axioms are called *transition axioms*, formulas of this particular form transition wrt $\varphi(\bar{x},s)$.

Denote $U(T)$ *the uniform restriction of* T - the theory with the same language as T but core axioms only. ⌑

The core aciom s are introduced to capture laws common for all descriptions of the world referring to a single situation - time instant, e.g. symmetry of the relation "near by" with respect to its object arguments

$$\forall s\ (NEAR(x,y,s) \rightarrow NEAR(y,x,s))$$

or existence of a minimal element in some well specified set of objects (whose elements can differ in various moments)

$$\forall s\ \exists x\ \forall y\ (ROOM(c,s)\ \&\ IN(x,c,s)\ \&\ BOX(x,s)\ \&$$
$$\&\ (BOX(y,s)\ \&\ IN(y,c,s) \rightarrow SMALLER(x,y,s)))$$

The transition axioms specify the conditions and the results of the actions which can be used to make changes in the world and they give the frame information about those relations, which are uneffected by them as well.

Suppose the enviroment and its possible changes are described by a situation theory T. We are interested in finding a sequence of possible actions, which would lead from an initial situation s specified by some formula ${}^1C|[s]|$ to a situation s in which the goal formula of a problem ${}^2C|[s]|$ holds. An obvious tactic to how to find this sequence is to try to prove the formula

$$ {}^1C|[s_0]| \rightarrow \exists s\ {}^2C|[s]| \qquad (1)$$

in the theory T.

Originally it was tacitly assumed that the theory T has a following property "if (1) is provable in T, then there exists a single variable free term t of the language of T such that

$$ {}^{1}C|[s_{o}]| \rightarrow ({}^{2}C|[s]|)|^{s}_{t} $$

is provable in T".

Obviously this case is very rare. The Hilbert-Ackerman theorem states that if T is open and the formula (1) has only existential quantifiers, then there exists a finite set τ of variable-free situation terms of T such that

$$ {}^{1}C|[s_{o}]| \rightarrow \bigvee_{t \in \tau} ({}^{2}C|[s]|)|^{s}_{t} \qquad (2) $$

is provable in T.

The need to handle the finite sets of terms motivated the definition of a branching plan [5] involving "if--then-else" constructs in contrast to the classically used linear plans. The same idea is reflected in the definition of a tree-proof of formulae of certain type [7].

What does it mean that a formula (2) has a tree proof, or that τ represents a branching plan for achieving ${}^{2}C$ form an initial situation ${}^{1}C$?

Suppose that the frame of your trip with picnic is roughly described by a theory T. Stepping out for a walk in woods you meet a dog Pluto - your friend, living in a house near a brook. Your primary goal is to get rid of Pluto (because he is not allowed to run about in woods freely). In other words, you look for a sequence τ of variable-free terms of T for which the following formula is provable in T

$$[\mathrm{IN}(bag,salami,s_o)\ \&\ \mathrm{MEET}(wood,pluto,s_o)\ \&\ \mathrm{DOG}(pluto,s_o)\ \&\ \&\ \mathrm{FRIEND}(pluto,s_o)] \to \bigvee_{t\in\tau} \mathrm{GETRIDOF}(pluto,t) \tag{3}$$

If you prove that the formula 3 has a tree-proof for

$$\tau = \{gooff(throw(stone,pluto,s_o)),\ gooff(close(home(pluto)walk(home(pluto),s_o))),\ leavetwater(give(salami,pluto,walk(brook,walk(home(pl.),s_o))))\}$$

then you can be sure that the tree

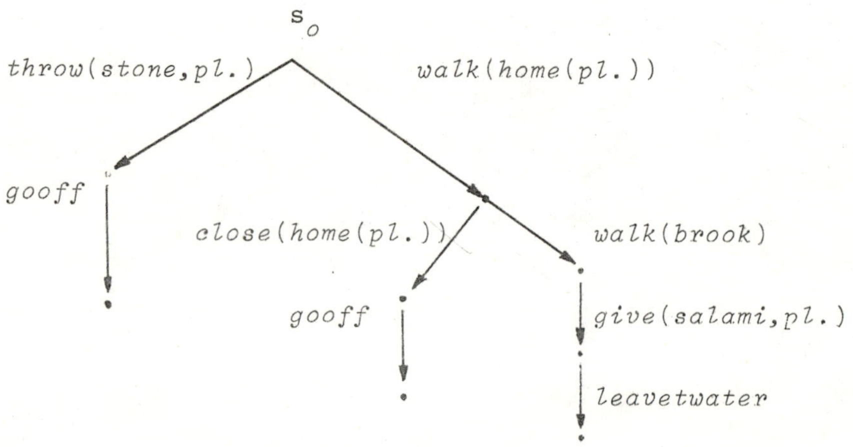

constructed from all situation subterms of terms in τ fully describes a plan which leads you to the primary goal considering present conditions and limitations.

Surely, the properties of the original situation allow to aply one of the actions "throw a stone on Pluto" or "walk Pluto to his home". If the conditions of the first one are met in the very case - you are non-sentimental enough to be able to throw a stone on your pet - you can do it and go off. The goal is reached. Else the conditions of the other action "walk Pluto to his home" are met. Proceeding this way you achieve a situation named "walk(home(pluto),s_o)". Now again the real world has to be consulted - either the door of Pluto's home is open (and you can close Pluto there) or you have to give him salami to keep him on one place for a while, walk to the brook and leave through water. All the situations on the leaves of the tree reach the goal.

⌑

Let τ be a finite set of variable-free situation terms of a simple situation language, denote

$$cl(\tau) = \{t : t \text{ is a situation subterm of some } t' \in \tau\}$$

Definition

Let K be a theory of situation calculus with a simple situation language. Let τ be a finite set of variable-free situation terms of the language of K ; X , $\{Y_t\}_{t\in\tau}$ let be non-open formulas uniform with respect to s_o and $t \in \tau$ resp.

$$X|[s_o]| \rightarrow \bigvee_{t\in\tau} Y_t|[t]| \tag{4}$$

has a *tree proof* in K iff
either

(i) card(cl τ)) = 1 (i.e. $\tau = \{s_o\}$) and

$$U(K) \vdash X|[s_o]| \rightarrow Y|[s_o]|$$

or

(ii) $\mathrm{card}(\mathrm{cl}(\tau))>1$ and for some $t_o = \varphi(\bar{a},t_1)$ maximal in $\mathrm{cl}(\tau)$ there exist finite sets $\mathcal{B}_1,\ldots,\mathcal{B}_k$ $(k \geq 1)$ of transition axioms transitive wrt $\varphi(\bar{a},t_1)$ such that

$$(\forall j \leq k)(\mathcal{U}(K) \vdash \bigwedge_{B\in\mathcal{B}_j} {}^2B \rightarrow Y_{t_o}|[t_o]|$$

and

$$X|[s_o]| \rightarrow \bigvee_{t\in(\tau\cdot\{t_o\})\cup\{t_1\}} Z_t|[t]|,$$

where

$$Z = \begin{cases} Y_t & \text{if } t \notin \{t_o,t_1\} \\ \bigvee_{i\leq k} \bigwedge_{B\in\mathcal{B}_i} {}^1B & \text{if } t = t_1 \ \&\ t_1 \notin \tau \\ \bigvee_{i\leq k} \bigwedge_{B\in\mathcal{B}_i} {}^1B \vee Y_{t_i} & \text{if } t = t_1 \ \&\ t_1 \in \tau \end{cases}$$

has a tree proof, too.

The following theorem [7] demonstrates the soundness of this definition. The formulas of the form (4) have a normal form of proof in situation calculus with a simple situation language.

Theorem

Let K be an open theory of situation calculus with a simple situation language such that it holds for any variable-free situation term $\varphi(\bar{a},t_1)$ of K and any finite set $\mathcal{B}$ of variable-free instances of transition axioms transition wrt $\varphi(\bar{a},t_1)$ "if $\mathcal{U}(K)[\bigwedge_{B\in\mathcal{B}} {}^1B]$ is consistent then $\mathcal{U}(K)[\bigwedge_{B\in\mathcal{B}} {}^2B]$ is consistent".

Let τ, X, $\{Y_t\}_{t\in\tau}$ be as in the preceding definition and let the prenex normal form of the formula

$$X\,|[s_o]|\; \rightarrow \bigvee_{t\in\tau} (Y_t\,|[t]| \qquad (5)$$

include only existentially quantified variables.

The formula (5) is provable in K iff either this formula has a tree proof or if there is a subset $\tau' \subseteq \tau$ and a situation term $t_o \in \tau'$ such that

$$\bigvee_{t\in\tau'} \mathrm{Rep}^{t_o}_{s_o} (Y_t\,|[t]|)$$

has a tree proof. (Rep $^{t_o}_{s_o}$ A is an operation replacing all occurences of t_o in A by s_o). ¤

Though this theorem itself is not suprising, its application led to some interesting results in the problem solving area. It implies, for example, that there exists to any STRIPS-like problem solving system a theory of situation calculus with the same problem solving power (5).

Theories with situation dependent object functions

Much more interesting are the theories with a simple situation language enriched by some object functions dependent on one situation argument beside the object ones. Such languages will be in the rest of the paper called *situation languages*.

The situation dependent object functions give information about some changing data - present position of a certain object, blood presure of a person, etc. Their use in the language causes new types of problems, having a feature time. All variable free terms of this language can not be easily interpreted in the real world, which can be studied only sequentially - as the situations arise, not seeing all possible situations of the past and future at once.

The situation terms of the situation language, which can be interpreted as plans for behaviour, were specified in [4] using the following definitions.

Definitions

The only **-subterm* of a situation constant s_o is s_o. A closed situation term t_1 is *-subterm of a closed situation term $\varphi(\bar{a},t_2)$ iff either $t_1 = \varphi(\bar{a},t_2)$ itself or t_1 is *-subterm of t_2. ⌑

Let t be a closed situation term. Any object constant is *regular with respect to t* (wrt t).

Let $a_1,\ldots,a_k$ be closed object and t_1 closed situation terms.

Let f,g be functions with the value object and with k-object arguments only, k-object and one situation argument, resp.

Let φ be a function with the value situation depending on k-object and one situation argument:

$f(a_1,\ldots,a_k)$ is regular wrt t iff $(\forall i \leq k)$ (a_i is reg. wrt t);

$g(a_1,\ldots,a_k,t_1)$ is regular wrt t iff

$(\forall i \leq k)$(a_i is reg wrt t_1) and t_1 is *-subterm of t;

$\varphi(a_1,\ldots,a_k,t_1)$ is regular wrt t iff

$\varphi(a_1,\ldots,a_k,t_1)$ is *-subterm of t and t_1 is reg. wrt t_1 and $(\forall i \leq k)$(a_i is reg. wrt t_1)). ⌑

A b term b is *regular* if there is a situation term the term b is regular with respect to. ⌑

The variable-free regular situation terms are just those ones, which can be interpreted as plans in the world, which the acting entity changes and learns to know using its actions.

Again we are interested in the proof of the formulas

of the form

$$X|[s_o]| \rightarrow \exists s\ (Y|[s]|) \qquad (6)$$

We try to find the conditions on the theory with the situation language, which would cause that if a formula (6) is provable in this theory then

a) there exists a set τ of regular situation variable-free terms such that

$$X|[s_o]| \rightarrow \bigvee_{t\in\tau} (Y|[s]|)|_t^s \qquad (7)$$

is provable

b) moreover (7) has a tree proof in this theory.

This means, τ can be interpreted as a plan for achieving Y from the initial situation specified by X.

Definitions

Let P be a predicate symbol of a situation language, let P have n object and one situation argument.

A closed atomic formula is *s_o-flat* iff it is of the form $P(a_1,\dots,a_n,s_o)$, where each a_i is regular wrt s_o and P is of the type specified above.

Let a_o be an object constant.
If the replacement of each object variable in an atomic formula $P(b_1,\dots,b_n,s_o)$ by a_o results in an s_o-flat formula, then $P(b_1,\dots,b_n,s_o)$ is s_o-flat formula, too.
If A,B are s_o-flat formulas and x is an object variable, then $\neg A$, $A\&B$ and $\forall xA$ are s_o-flat formulas. ¤

Let t be a situation term. A formula B is *t-flat* iff there is an s_o-flat formula B_1 such that B can be obtained from B_1, when every occurence of s_o in B_1 is replaced by t. ⌑

Let T be an open theory of situation language. Any axiom of T let be of one of the following forms:

a) $\forall sA$, where A is s-flat

b) $\forall s\ ({}^1A|[s]| \rightarrow {}^2A|[\varphi(\bar{x},s)]|)$, where ${}^1A$, ${}^2A$ are s-, $\varphi(\bar{x},s)$-flat formulas, resp., with no other free variables then $\bar{x}$

c) $({}^1B|[s]| \rightarrow {}^2B|[\varphi(\bar{x},s)]|)|^{\bar{y},s}_{\bar{b},t}$, where ${}^1B$, ${}^2B$ are s-, $\varphi(\bar{x},s)$-flat formulas with no free variables but $s,\bar{x},\bar{y}$ and t is closed situation term, $\bar{b}$ are closed object terms regular wrt t.

Let $\mathcal{U}(T)$ be the restriction of T with the same language but axioms of the first type (a)) only.

The theory T is *transition theory* iff for any non--open situation term $\varphi(\bar{d},t)$ of t and any finite set φ of variable-free instances of the axioms of T of the type

$$C = {}^1C|[t]| \rightarrow {}^2C|[\varphi(\bar{d},t)]|$$

holds

"if $\mathcal{U}(T)[\bigwedge_{C\in\varphi} {}^1C]$ is consistent then $\mathcal{U}(T)[\bigwedge_{C\in\varphi} {}^2C]$ is consistent" ⌑

The transition theories of situation calculus form an important class of theories with situation dependent object functions which preserve the property of normal form of proof - tree proof - of certain formulas.

Theorem

Let K be an open consistent transition theory of

situation calculus. Let X, Y be open s_O-, s-flat formulas of K, resp.

$$K \vdash \forall X|[s_0]| \rightarrow \exists s \quad \exists Y|[s]| \qquad (8)$$

iff there is a finite set τ of regular variable-free situation terms of K such that

$$\forall X|[s_0]| \rightarrow \bigvee_{t \in \tau} (\exists Y|[s]|)|_t^s$$

has a tree proof.
($\forall A, \exists A$ denote general, existential resp., closure over object variables of a formula A). ⌑

The proof of this theorem as well as some connected results can be found in [7].

The last theorem includes also an assertion that a proof of a formula of the type (8) certainly leads to a set of nonopen situation terms, which can be used as a basis for derivation of a plan how to achieve Y from X. Consequently, situation calculus with situation-dependent object functions can be considered as a tool not only for description of the world, but also as a tool for planning. The transition theories are rich enough to represent some actions and problems in a changing environment with more than a single acting entity. An example of such a transition theory is presented in [6]. On the other hand there is an example of an axiomatics for a card game given in situation language in [6], which forms a nontransition theory, whose proofs lead to inappliable solutions.

References

[1] R. Fikes, N.J. Nilsson: "STRIPS: A new approach to the application of theorem proving to problem solving"
Proceedings of the 2nd IJCAI, Imperial Coll., London 71

[2] C.C. Green: Theorem proving by resolution as a basic for question answering system
Machine Intelligence 4 (1969), 183-205

[3] J. Mc Carthy, P.J. Hayes: Some philosophical problems from the standpoint of artificial intelligence
Machine Intelligence 4 (1969), 463-505

[4] O. Štěpánková, I.M. Havel: Incidental and state dependent phenomena in robot problem solving
Proceedings of AISB Conf.,
University of Edinburgh, 1976, 266-278

[5] O. Štěpánková, I.M. Havel: A logical theory of robot problem solving
Artificial Intelligence 7, 1976, 129-161

[6] O. Štěpánková: Planning in uncertain environment
Proceedings of AISB Conference, 1978, Hamburg

[7] O. Štěpánková: Robot planning and the situation calculus (in czech)
Tech. report of the Inst. for Comp. Tech. 1978)

OLGA ŠTĚPÁNKOVÁ

Institute of computation techniques of ČVUT
Horská 3, Praha 2, Czechoslovakia

COLLOQUIA MATHEMATICA SOCIETATIS JÁNOS BOLYAI
26. Mathematical Logic in Computer Sicnece
Salgótarján (HUNGARY), 1978.

COMPLETENESS THEOREM FOR LOGIC OF EFFECTIVE DEFINITIONS

J. Tiuryn

0. Introduction

The aim of this paper is to present a logic to deal with properties of algorithmic processes in abstract structures. As we were not interested in the way in which a given process determines an object in a structure but rather in possibility of doing this - we adopt here for the notion of algorithmic process a slight generalization of the notion of effective definitional scheme due to H Friedman (cf. Friedman [1971]). Thus we deal here with effective definitions, and we call a logic proposed in that paper Logic of Effective Definitions (in abbr. LED).

The principles behind a development of LED are the following.

0.1. This logic should be able to express such basic propertiex of programs or definitions as: equivalence, partial correctness, total correctness.

0.2. The mathematical semantics for this logic should be easily definable in an arbitrary abstract structure, so that, hopefully, metamathematical properties will become easily understandable.

0.3. As we deal with processes that may give no result, i.e. they may not terminate, we have decided to base our logic on a three-valued propositional calculus to be able to distinguish in a natural way between three possible cases for truth-value of a formula in a given interpretation: true (t), false (f), undefined ($\uparrow$). We find as an appropriate candidate for such logic - the three-valued constructive logic with strong negation (cf. Rasiowa [1974]).

0.4. The choice of effective definitional schemes as terms for our logic is due to the following reasons.
0.4.1. They are very easy to handle with - this is important in establishing metamathematical properties.
0.4.2. It turns out that they are effectively equivalent to recursively enumerable infinite trees viewed as program schemes (for the latter notion see Kfoury [1972]). On the other hand many various program schemes can be "unfolded" into such trees.
0.4.3. Effective definitional schemes can be treated as program schemes of universal power (over total interpretations). A comprehensive discussion of this topic the reader may find in Shepherdson [1973].

The reader interested in other logical approaches to algorithmic processes is referred to Rasiowa [1977], Constable [1977], Engeler [1978], Grabowski and Kreczmar [1978], Parikh [1978], where also other references can be found.

The paper is divided into five sections. Section 1 is preliminary and it fixes some basic notations and definitions. In Section 2 we present three-valued constructive logic with strong negation.

Section 3 is devoted to describe the syntax and semantics of LED. We describe here schemes for defining functions and call them functional effective definitional schemes (in abbr. feds), and schemes for defining relations calling them relational effective definitional schemes (in abbr. reds). Feds's will stand for terms in our logic, while reds's we will treat as "atomic formulas". Then we define formulas analogously as in first order logic. Among all formulas of LED we distinguish those in which intuitionistic negation does not occur - we call them Lukasiewicz formulas.

After defining semantics we show that some basic properties of program schemes can be expressed in LED by open formulas.

We also introduce two notions of equivalnce for formulas: weaker (<=>) - two formulas are <=>-equivalent iff they determine the same classes of models, and stronger (≡) - two formulas are equivalent iff they have the same truth-values in all interpretations. In any logic based on two-valued propositional calculi both these notions coincide. It is not the case in LED. At the end of Section 3 we show that both these notions can be expressed syntactically in LED by open formulas.

In Section 4 we present some normal form results for LED. In particular we show that every open Lukasiewicz formula is strongly equivalent to some reds, and it is weakly equivalent to some formula of the form $S \doteq S$, where S is a feds. The meaning of the last formula is "the scheme S computes total function". We also show here that every open formula is weakly equivalent to a recursively enumerable set of formulas of the form $S \doteq S$, for some feds S.

Section 5 contains the main result of this paper. We introduce a notion of Consistency Property - analogous to that in model theory of $L_{\omega_1\omega}$ (cf. Keisler [1971]). Then we state Model Existence Theorem for LED, and as a corollary we derive Completeness Theorem for LED, for previously described system of axioms and rules of inference. Among schemes of inference rules there is one with infinite r.e. set of premises. This is very similar to the ω-rule.

Due to space limitations we present here our results without proofs. The extended complete version of this paper will appear soon (Tiuryn [1978]).

1. *Preliminary definitions and notations*

Throughout the paper we are dealing exclusively with countable languages, i.e. with those having countably many finitary predicate and function symbols, and with countably many constant symbols. Let L be such a language. It will be fixed throughout the paper if not specified otherwise. Also the set $X = \{x_0, x_1, \ldots,\}$, disjoint with L, is fixed throughout the paper. We treat X as the set of individual variables.

The notations we are going to use here, in principle, will agree with those introduced in Chang and Keisler [1973]. We recall only some of them.

The equality symbol to be used in our language will be denoted by $\doteq$, to differ it from the ordinary metalinguistic equality that will be denoted by =.

L-structures will be denoted by scribe characteres: $\mathcal{A}, \mathcal{B}, \mathcal{C}, \ldots$. Then carriers of those structures are: A, B, C, ..., respectively.

If $\mathcal{A}$ is an L-structure, α is a first order formula in language L), and $v \in A^X$ is a valuation, then "$(A,v) \models \alpha$" means "v satisfies α in A"; "$A \models \alpha$" means "for every $v \in A^X$, $(A,v) \models \alpha$"; "$\models \alpha$" means "for every A, $A \models \alpha$", i.e., α is a tautology. Writing $\alpha(x_{i_o},\ldots,x_{i_n})$ we indicate that all variables that occur in α are among $\{x_{i_o},\ldots,x_{i_n}\}$. The same convention we adopt for terms. If t is a term then $\alpha(x/t)$ denotes the formula obtained from α by replacing every occurence of x in α by term t. If A is a structure and t is a term, then t_A is a meaning of t in A. It can be presented as function $t_A : A^X \to A$. We will use similar notations for terms and formulas of LED.

If f, g: A → B are partial functions, then "$f \sqsubseteq g$" means "g extends f". Every n-ary partial relation r in A can be presented as a function $r : A^n \to \{t, f, \uparrow\}$, where "$r(a) = \uparrow$" means "r is not defined in $a \in A^n$".

2. *Three-valued constructive logic with strong negation*

Let us start with the following example. Suppose we are given two formulas α, β and two algorithms for computing truth values of α and β respectively in a fixed structure A and valuation $v \in A^X$. Suppose moreover that we have no guarantee that these algorithms will terminate by giving an answer. Our task is to compute the truth value of disjunction $\alpha \vee \beta$. The most natural procedure to do that is to start parallelly with the algorithms computing truth values of α and β. It we get an answer, say $(A,v) \models \alpha$, then we may stop computation with result $(A,v) \models \alpha \vee \beta$. However, if we get answer $(A,v) \models \neg\alpha$ then we cannot stop the computation of truth value of β because of obvious reasons.

This leads to the following truth-table for disjunction, where $\uparrow$ stands for "the process does not terminate".

$\vee$	f	$\uparrow$	t
f	f	$\uparrow$	t
$\uparrow$	$\uparrow$	$\uparrow$	t
t	t	t	t

One can draw truth-tables similarly for negation and conjunction.

x	$\sim$ x
f	t
$\uparrow$	$\uparrow$
t	f

$\wedge$	f	$\uparrow$	t
f	f	f	f
$\uparrow$	f	$\uparrow$	$\uparrow$
t	f	$\uparrow$	t

The above-draw figures coincide with truth-tables of three-valued logic of partial recursive functions developed by S.C. Kleene (cf. Kleene [1959], p. 332).

Observe that if we define on the set $C_o = \{f, \uparrow, t\}$ an order relation: $f \leq \uparrow \leq t$, then $\vee$ and $\wedge$ defined above become join and meet respectively, in the poset $\langle C_o, \leq \rangle$.

It turns out that the algebra $\langle C_o, \uparrow, t, \sim, \vee, \wedge \rangle$ is not functionally complete. For example, the operation $\neg$ defined below is not a polynomial in this algebra.

x	$\neg$x
f	t
$\uparrow$	t
t	f

This means that certain properties could not be expressed in logic based on this algebra. Fortunately one can prove (using for example functional completeness of three-element Post algebra) that the algebra $\langle C_0, t, \sim, \neg, , \wedge\rangle$ is functionally complete.

Propositional calculus based on this algebra is known as three-valued constructive logic with strong negation. Connective ~ is called *strong negation*, while ¬ is called *intuitionistic negation*. A comprehensive exposition of this logic the reader may find in Rasiowa [1974].

In Rasiowa [1974] the reader may also find a finite complete axiomatization of that predicate calculus.

3. *Logic of Effective Definitions (LED) - syntax and semantics*

3.1. First we define the functional effective definitional schemes. They are going to play a role analogous to that of ordinary terms in first order logic.

Denote by *OF* the set of all open formulas in first order language *L*. Let *T* denote the set of all terms. Fix some arithmetization of *OF* and *T* (cf. for example, Shoenfield [1967], Section 6.6).

By a *functional effective definitional scheme (feds)* we mean a (total) recursive function $S : \omega \to OF \times T$ such that the set of all distinct symbols from $L \cup X$ occuring in expressions in $S(\omega)$ is finite.

In a similar way we define *relational effective definitional schemes (reds)* - as (total) recursive functions $R : \omega \to OF \times OF$ such that the set of all distinct symbols from $L \cup X$ occuring in expressions in $R(\omega)$ is finite.

Formulas of the *logic* of *effective definitions* (LED) are defined inductively. This is the least set *F* of expressions satisfying conditions:

3.1.1. Every reds and every atomic formula of first order logic belongs to F;

3.1.2. If S, Q are feds's and t is a term then $S \doteq Q$, $S \doteq t$, $t \doteq S \in F$;

3.1.3. If $\alpha, \beta \in F$, *then* $\alpha \vee \beta$, $\sim \alpha, \neg \alpha, \alpha \wedge \beta \in F$;

3.1.4. If $x_n \in X, \alpha \in F$, *then* $(\exists x_n)\alpha$, $(\forall x_n)\alpha \in F$.

Formulas obtained by (3.1.1.) and (3.1.2.) only, are called *atomic formulas* of LED, while formulas obtained by (3.1.1) - (3.1.3.) are called *open formulas* of LED. Finally formulas in which $\neg$-negation does not occur we call *Lukasiewicz formulas*.

In the sequel we will use abbreviations:

$\alpha \rightarrow \beta$, for $\sim\alpha \vee \beta$;

$\alpha \Rightarrow \beta$, for $\neg\alpha \vee \beta$;

$\alpha \leftrightarrow \beta$, for $(\alpha \rightarrow \beta) \wedge (\beta \rightarrow \alpha)$;

$\alpha \Leftrightarrow \beta$, for $(\alpha \Rightarrow \beta) \wedge (\beta \Rightarrow \alpha)$.

For a feds S and $n \in \omega$, $S(n)_0$, $S(n)_1$ denote the first and second component of S(n), respectively. The same applies to reds.

3.2. Our next step is to define semantics for LED. Let A be an L-structure and S a feds. The meaning of S in A, denoted by S_A. will be a partial map from A^X into A defined as follows:

$$S_A(v) = S(n)_{1A}(v), \text{ if } (A,v) \models S(n)_0 \wedge \bigwedge_{i<n} \neg S(i)_0$$
$$= \text{undefined otherwise.}$$

Similarly the meaning of a reds R in A, denoted by R_A, is a partial relation in A^X defined as follows:

$$R_A(v) = t, \text{ if } (A,v) \models R(n)_0 \wedge R(n)_1 \wedge \bigwedge_{i<n} \neg R(i)_0;$$
$$= f, \text{ if } (A,v) \models R(n)_0 \wedge R(n)_1 \wedge \bigwedge_{i<n} \neg R(i)_0;$$
$$= \uparrow, \text{ otherwise.}$$

The meaning of atomic first order formulas is defined as usually. If α is an atomic formula of first order logic and $v \in A^X$, then $\alpha_A(v) = t$ if $(A,v) \models \alpha$, and $= f$, otherwise.

If S,T are either terms or feds's then

$$
\begin{aligned}
(S \doteq T)_A(v) &= t, \text{ if both } S_A(v),\ T_A(v) \text{ are defined and } S_A(v) = T_A(v);\\
&= f, \text{ if both } S_A(v),\ T_A(v) \text{ are defined and } S_A(v) \neq T_A(v);\\
&= \uparrow, \text{ otherwise.}
\end{aligned}
$$

If α, β are formulas of LED and $x_n \in X$, then

$$
\begin{aligned}
(\alpha \vee \beta)_A(v) &= \alpha_A(v) \vee \beta_A(v);\\
(\sim\alpha)_A(v) &= \sim(\alpha_A(v));\\
(\neg\alpha)_A(v) &= \neg(\alpha_A(v));\\
(\alpha \wedge \beta)_A(v) &= \alpha_A(v) \wedge \beta_A(v);\\
((\exists x_n)\alpha)_A(v) &= \sup\{\alpha_A(v(x_n/a)) : a \in A\};\\
((\forall x_n)\alpha)_A(v) &= \inf\{\alpha_A(v(x_n/a)) : a \in A\},
\end{aligned}
$$

where $v(x_n/a) \in A^X$ and is defined as follows

$$
\begin{aligned}
v(x_n/a)(x_i) &= v(x_i), && \text{if } i \neq n\\
&= a, && \text{if } i = n.
\end{aligned}
$$

This completes the semantics of LED. For a formula α of LED we write $(A,v) \models \alpha$ for $\alpha_A(v) = t$. With a given formula α of LED and structure A one may associate a set of "truth valuations" $\alpha^A = \{v \in A^X : (A,v) \models \alpha\}$. Thus α^A can be viewed as a total function $\alpha^A : A^X \to \{t,f\}$ defined $\alpha^A(v) = t$ if $(A,v) \models \alpha$, and $= f$ otherwise. It is easily seen that $\alpha^A = (\neg\neg\alpha)_A$. Observe also that an application of strong negation $\sim$ to a formula α semantically corresponds to "switching truth values of α without changing domain of definition", while an application of intuitionistic negation $\neg$ to α corresponds to "taking set-theoretical complement of α^{A}".

It is a routine matter to check that in a standard model of Peano arithmetic functions definable by feds's coincide with partially recursive functions, sets definable by Lukasiewicz open formulas coincide with r.e. sets, sets definable by open LED formulas are Δ_2^0 sets, and thus sets definable by arbitrary formulas of LED coincide with arithmetical sets.

Let S,Q be feds's. Denote by $S \sqsubseteq Q$ the formula $S \doteq S \Rightarrow (Q \doteq Q \wedge S \doteq Q)$. Denote by $S \sim Q$ the formula $(S \sqsubseteq Q) \wedge (Q \sqsubseteq S)$.

3.3. Proposition

For any L-structure A and feds's S,Q.

3.3.1. $A \models S \doteq S$ *iff* S_A *is a total function.*
3.3.2. $A \models S \sqsubseteq Q$ *iff* $S_A \sqsubseteq Q_A$, *i.e. if* Q_A *extends* S_A.
3.3.3. $A \models S \sim Q$ *iff* $S_A = Q_A$, *i.e. if S and Q are equivalent in* A.

Using Proposition 3.3 it is very easy to express other properties of program schemes as partial correctness, by means of open formulas of LED.

In a similar way for arbitrary formulas α,β of LED we denote by $\alpha \sqsubseteq \beta$ the formula

$$(\alpha \leftrightarrow \alpha) \Rightarrow ((\beta \leftrightarrow \beta) \wedge (\alpha \leftrightarrow \beta)).$$

Denote by $\alpha \equiv \beta$ the formula $(\alpha \sqsubseteq \beta) \wedge (\beta \sqsubseteq \alpha)$.

3.4. Proposition

For any L*-structure* A *and formulas* α, β *of LED.*

3.4.1. $A \models \alpha <-> \alpha$ *iff* α_A *is a total predicate in* A, *i.e. if for every* $v \in A^X$, $\alpha_A(v) \in \{t, f\}$.
3.4.2. $A \models \alpha => \beta$ *iff* $\alpha^A \subseteq \beta^A$.
3.4.3. $A \models \alpha <=> \beta$ *iff* $\alpha^A = \beta^A$.
3.4.4. $A \models \alpha \sqsubseteq \beta$ *iff* $\alpha_A \sqsubseteq \beta_A$.
3.4.5. $A \models \alpha \equiv \beta$ *iff* $\alpha_A = \beta_A$.

Therefore, by (3.4.3), $\models \alpha <=> \beta$ says that formulas α and β determine the same classes of models. By (3.4.5) stronger equivalence $\models \alpha \equiv \beta$ means that α and β are semantically equivalent.

The next result, due to C.E. Gordon, establishes a connection between absolute prime computability in the sense of Moschovakis (cf. Moschovakis [1969]) and definiability in LED.

3.5. Theorem

(Gordon [1971]) *For every finite language* L *and* L*- structure* A*:*

3.5.1. Functions definable by feds's coincide with functions absolutely prime computable in the sense of Moschovakis;
3.5.2. relations definable by reds's coincide with relations semiabsolutely prime computable in the sense of Moschovakis.

4. Some normal form theorems

4.1. Theorem

For every Lukasiewicz open formula α of LED there exists a reds R and a feds S with

4.1.1. $\models \alpha \equiv R$;
4.1.2. $\models \alpha \Leftrightarrow S \doteq S$.

4.2. Theorem

For every open formula α of LED there exists a recursively enumerable set of feds's $\{S_n, n \in \omega\}$ with the property that for every L-structure A, $A \models \alpha$ *iff* $A \models \{S_n \doteq S_n, n \in \omega\}$.

4.3. Theorem

For every open formula α of LED there exist reds's R_i, Q_i, P_i for $i < n$, and feds's S_i, T_i for $i < m$ such that

4.3.1. $\models \alpha \equiv \bigwedge_{i<n} (\neg R_i \Rightarrow (Q_i \Rightarrow P_i))$;
4.3.2. $\models \alpha \Leftrightarrow \bigwedge_{i<m} (S_i \doteq S_i \Rightarrow T_i \doteq T_i)$.

One can prove also that the prenex normal form theorem for formulas of LED holds.

5. *Model existence theorem and complete axiomatization for LED*

5.1. Compactness Theorem and Upward Löwenheim - Skolem Theorem fail in LED as the following example shows. Let S be a unary function symbol and O, c let be constant symbols. Define a feds S by $S(n) = (S^n(O) \doteq x_0, x_0)$ for $n \in \omega$, where $S^O(O)$ is O, and $S^{n+1}(O)$ is $S(S^n(O))$ for $n \in \omega$. Let Σ be the following set of sentences:

$$(\forall x_0)\ S \doteq S, \quad \neg S^n(O) \doteq c, \quad n \in \omega.$$

Then it is easily seen that every finite subset of Σ has a model, while Σ itselft has no model. It is also easily seen that the first sentence in the above list has a countable model but has no model of power greater then ω.

Because of a similar situation in the logic $L_{\omega_1\omega}$ we adopted one of the main tools in that logic for LED. This is Model Existence Theorem based on the notion of Consistency property (cf. Keisler [1979]).

First we introduce some notations. Let C be a countable set of new constant symbols and let L_C be the language obtained from L by adding each $c \in C$. Denote by LED(L), LED(L_C) the logic of effective definitions corresponding to L and L_C respectively.

First step is to define operations on formulas consisting of moving negations inside.

Let L be a formula of LED(L_C). We define $\alpha\sim$ and $\alpha\neg$ inductively:

5.1.1. If α is atomic then $\alpha\neg$ is $\neg\alpha$. To define $\alpha\sim$ consider two cases:

I. If α is a reds R then R$\sim$ is a reds Q which is strongly equivalent to $\sim$R. The construction of this reds is provided by Theorem 4.1, (4.1.1.). In fact the code number of Q can be effectively found from the code number of R.

II. In other cases we define $\alpha\sim$ to be $\sim\alpha$.

5.1.2. $(\sim\alpha)\neg$ is $\neg(\alpha\sim)$, and $(\sim\alpha)\sim$ is α.

5.1.3. $(\neg\alpha)$ and $(\neg\alpha)\sim$ is α.

5.1.4. $(\alpha \wedge \beta)\neg$ is $\neg\alpha \vee \neg\beta$, and
$(\alpha \wedge \beta)\sim$ is $\sim\alpha \vee \sim\beta$

5.1.5. $(\alpha \vee \beta)\neg$ is $\neg\alpha \wedge \neg\beta$, and $(\alpha \vee \beta)\sim$ is $\sim\alpha \wedge \sim\beta$.

5.1.6. $((\exists x_n)\alpha)\neg$ is $(\forall x_n)\neg\alpha$, and
$((\exists x_n)\alpha)\sim$ is $(\forall x_n)\sim\alpha$.

5.1.7. $((\forall x_n)\alpha)\neg$ is $(\exists x_n)\neg\alpha$, and
$((\forall x_n)\alpha)\sim$ is $\cdot(\exists x_n)\sim\alpha$.

Remark:

Observe that for every formula α of LED(L_C), $\models \neg\alpha \Leftrightarrow \alpha\neg$, and $\models \sim\alpha \Leftrightarrow \alpha\sim$.

As the next step we define $\doteq$-*reduction* for formulas, i.e. we transform each formula α into an equivalent formula $\alpha^{=}$ which has no subformulas of the form S $\doteq$ T, where S,T is either feds or term.

Let α be a formula of LED(L_C). Define $\alpha^{=}$ inductively:

5.1.8. If α is of the form S $\doteq$ T, where S,T is either feds or term, then $(S \doteq T)^{=}$ is a reds R which is strongly equivalent to S $\doteq$ T. The construction of the reds R is provided by Theorem 4.1, (4.1.1), and the code number of R can be effectively found from the code number of S $\doteq$ T.

5.1.9. If α is atomic and is not of the form in (5.1.8) then $\alpha^{=}$ is α.

5.1.10. $(\ )^{=}$ operation uniquely extends to all formulas by the property that it preserves: $\sim$, $\urcorner$, $\vee$, $\wedge$, $(\exists x_n)$, $(\forall x_n)$.

5.2. Now we can define the main notion of this section. Let P be a set of countable sets of sentences of LED(L_C). P is said to be *consistency property* iff for each $p \in P$ all the following hold.

(C 1) (Consistency rule) Either $\alpha \notin p$, or $\sim\alpha \notin p$ and $\urcorner\alpha \notin p$.

(C 2) (reds-rules) If R is reds and $R \in p$, the for some $n \in \omega$,
$p \cup \{\bigwedge_{i<n} \urcorner R(i)_0, R(n)_0, R(n)_1\} \in P$.
If $\urcorner R \in p$, then either $p \cup \{R\sim\} \in P$, or
$p \cup \{\urcorner R(n)_0\} \in P$ for every $n \in \omega$.

(C 3) (negation rules) If $\urcorner\alpha \in p$ then $p \cup \{\alpha\urcorner\} \in P$;
If $\sim\alpha \in p$ then $p \cup \{\alpha\sim\} \in P$.

(C 4) ($\doteq$-rule) If $\alpha \in p$ then $p \cup \{\alpha^{=}\} \in P$.

(C 5) ($\wedge$-rule) If $\alpha \wedge \beta \in p$ then $p \cup \{\alpha\} \in P$ and $p \cup \{\beta\} \in p$.

(C 6) ($\vee$-rule If $\alpha \vee \beta \in p$ then either $p \cup \{\alpha\} \in P$ or $p \cup \{\beta\} \in P$.

(C 7) (quantifier rules) If $((\forall x_n)\alpha) \in p$ then for all $c \in C$, $p \cup \{\alpha(x_n/c)\} \in P$.

If $((\exists x_n)\alpha) \in p$ then for some $c \in C$, $p \cup \{\alpha(x_n/c)\} \in P$.

(C 8) (equality rules) Let t be a basic terms and $c, d \in C$.
If $(c \doteq d) \in p$ then $p \cup \{d \doteq c\} \in P$.
If $(c \doteq t)$, $\alpha(x_n/t) \in p$, then $p \cup \{\alpha(x_n/c)\} \in P$.
For some $e \in C$, $p \cup \{e \doteq t\} \in P$.

Importance of the notion of consistency property lieas in the following result.

5.3. Theorem

(Model existence theorem) If P is a consistency property and $p \in P$, then p has a model.
Proof of this theorem is similar to a proof of analogous result for $L_{\omega_1\omega}$ (cf. Keisler [1971]). Using the notion of consistency property one constructs a model for p from the set of constants C. As a corollary one may derive the following result.

5.4. Theorem

(Downward Löwenheim-Skolem theorem) If Σ is a countable set of sentences in LED(L) and Σ has a model, then it has a finite or countable model.

Another corollary will be completeness theorem for LED.

5.5. *Axioms for LED*

AXIOM 1. All axioms for constructive logic with strong negation + the axiom $\alpha \vee \neg\alpha$.

AXIOM 2. $\sim\alpha \Leftrightarrow \alpha\sim$

AXIOM 3. $\neg\alpha \Leftrightarrow \alpha\neg$

AXIOM 4. $\alpha \Leftrightarrow \alpha^{=}$

AXIOM 5. $(\forall x_n)\alpha \Rightarrow \alpha(x_n/t)$, where t is a term and variables in t do not occur bound in α.

AXIOM 6. $(\alpha \wedge t \doteq x_n) \Rightarrow \alpha(x_n/t)$, where t is a term and variables in t do not occur bound in α.

AXIOM 7. $x_0 \doteq x_1 \Rightarrow {>}x_1 \doteq x_0$.

AXIOM 8. $x_0 \doteq x_0$.

5.6. *Inference rules*

1. From α, $\alpha \Rightarrow \beta$, infer β.
2. From $\alpha \Rightarrow \beta(x_n, \ldots)$, infer $\alpha \Rightarrow (\forall x_n)\beta$, where x_n does not occur free in α.
3. If R is a reds, then from

   $$\alpha \Rightarrow ((R(n)_0 \wedge R(n)_1) \Rightarrow \bigvee_{i<n} R(i)_0)$$

   for all $n \in \omega$, infer $\alpha \Rightarrow \neg R$.
4. If R is a reds, then from $\alpha \Rightarrow (\neg\sim R \wedge R(n)_0)$ for some $n \in \omega$, infer $\alpha \Rightarrow R$.

The set of *theorems* of LED(L) is the least set of formulas of LED(L) which contains all the axioms and is closed under inference rules. We write $\vdash_{LED(L)}\alpha$ for "α is a theorem of LED(L)".

5.7. *Theorem*

(Completeness theorem for LED(L)). For any sentence

of $LED(L)$, $\vdash_{LED(L)}\alpha$ *iff* $\models \alpha$.

Proof: The reader can easily check that all axioms are valid and all inference rules preserve validity. Therefore every theorem is valid.

For the proof in converse direction let P be the set of all finite sets of sentences in $LED(L_C)$ $\{\alpha_i, i < n\}$ such that not $\vdash_{LED(L_C)} \neg \bigwedge_{i<n} \alpha_i$. Using axioms and inference rules one proves that P is a consistency property. Now if α is a sentence of $LED(L)$ and not $\vdash_{LED(L)}\alpha$ then not $\vdash_{LED(L_C)} \neg\neg\alpha$, and it means that $\{\neg\alpha\} \in P$. By Model Existence Theorem tehere exists A with $A \models \neg\alpha$, i.e. not $A \models \alpha$.

References

C.C. Chang, H.J. Keisler: *Model theory* North-Holland Publishing Company [1973]

R.L. Constable: On the theory of programming logics Proc. 9th Annual ACM Symp. *Theory of Computing*, Boulder, Co. [1977]

E. Engeler: Generalized Galois theory and its applications to complexity *Berichte des Instituts für Informatik*, No. 24, ETH Zürich [1978]

H. Friedman: Algorithmic procedures, generalized Turing algorithms, and elementary recursion theory *Logic Colloquium' 69*, R.O. Gandy and C.M.E. Yates (eds), North-Holland Publ. Comp. [1971]

C.E. Gordon: Finitistically computable functions and realizations on an abstract structure *The Journal of Symbolic Logic* 36 [1971]

M. Grabowski, A. Kreczmar: Dynamic theories of real and complex numbers
Proceedings of MFCS'78, *Lecture Notes in Comp. Sci.* No. 64, J. Winkowski (ed.), Springer Verlag [1978]

H.J. Keisler: *Model theory for infinitary logic*
North-Holland Publishing Company [1971]

D.J. Kfoury: Comparing algebraic structures up to algorithmic equivalence
Automata, Languages and Programming, M. Nivat (ed.) North-Holland Publishing Company [1972]

S.C. Kleene: *Introduction to metamathematics*
North-Holland Publishing Company [1959]

Y.N. Moschovakis: Abstract first order computability I, II.
Transactions of the AMS 138 [1969]

R. Parikh: The completeness of propositional dynamic logic
Proceedings of MFCS'78, *Lecture Notes in Computer Science* No. 64, J. Winkowski (ed.), Springer Verlag [1978]

H. Rasiowa: *An algebraic approach to non-classical logics*
North-Holland Publishing Company [1974]

H. Rasiowa: Algorithmic Logic
ICS PAS Report, No. 281, Warsaw [1977]

J.C. Shepherdson: Computation over abstract structures: serial and parallel procedures and Friedman's effective definitional schemes
Logic Colloquium'73, J.C. Shepherdson/J. Rose (eds.), North-Holland Publishing Company [1973]

J.R. Shoenfield: *Mathematical logic*
Addison-Wesley Publishing Company [1967]

J. Tiuryn: Logic of effective definitions - completeness

theorem, and normal form results
Schriften zur Informatik und Angewandten Mathematik, RWTH Aachen (to appear) [1978]

J.Tiuryn

RWTH Aachen, Lehrstuhl für Informatik II
and Warsaw University, Institute of Mathematics